Eco-Design of Buildings and Infrastructure

Sustainable Cities Research Series
Book Series Editor: Bruno Peuportier

ISSN: 2472-2502 (Print)
ISSN: 2472-2510 (Online)

Centre for Energy Efficiency of Systems, MINES Paristech, Paris, France

Volume 4

Eco-Design of Buildings and Infrastructure

Developments in the Period 2016–2020

Editors

Bruno Peuportier
MINES ParisTech, Paris, France

Fabien Leurent
École des Ponts ParisTech, Marne-la-Vallee, France

Jean Roger-Estrade
AgroParisTech, Paris, France

CRC Press
Taylor & Francis Group
Boca Raton London New York

CRC Press is an imprint of the
Taylor & Francis Group, an **informa** business

Photo Credits (cover): © Ferrier Marchetti Studio / Photo Luc Boegly/
Métropole Rouen Normandie

Originally published in French
'Éco-conception des ensembles bâtis et des infrastructures – Tome 2'
edited by Bruno Peuportier, Fabien Leurent and Jean Roger-Estrade
© 2018 Presses de Mines, Paris

CRC Press/Balkema is an imprint of the Taylor & Francis Group, an informa business

© 2021 Taylor & Francis Group, London, UK

Typeset by codeMantra
Printed and bound in Times New Roman

Library of Congress Cataloging-in-Publication Data
[Applied for]

Published by: CRC Press/Balkema
                Schipholweg 107C, 2316 XC Leiden, The Netherlands
                e-mail: Pub.NL@taylorandfrancis.com
                www.routledge.com – www.taylorandfrancis.com

ISBN: 978-0-367-55770-6 (hbk)
ISBN: 978-0-367-55771-3 (pbk)
ISBN: 978-1-003-09507-1 (ebk)

DOI: 10.1201/9781003095071
https://doi.org/10.1201/9781003095071

Contents

Preface: Understanding the Future of the Urban World   viii
Introduction   x
Contributors   xvii

PART I
New Developments in the Methods for Eco-design   I

1   **Consequential Life Cycle Assessment Applied to Buildings**   3
    CHARLOTTE ROUX

2   **Study of the Quantification of Uncertainties in Building Life
    Cycle Assessment**   29
    MARIE-LISE PANNIER, PATRICK SCHALBART, AND BRUNO PEUPORTIER

3   **Life Cycle Assessment of Transportation Systems**   61
    ANNE DE BORTOLI, ADELAÏDE FERAILLE, AND FABIEN LEURENT

4   **Spatial Refinement to Better Evaluate Mobility and Its
    Environmental Impacts**   97
    NATALIA KOTELNIKOVA-WEILER, FABIEN LEURENT, AND ALEXIS POULHÈS

5   **Ecological Compensation in France: Towards a Territorial System?** 114
    JULIE LATUNE, HAROLD LEVREL, PAULINE DELFORGE, AND
    NATHALIE FRASCARIA-LACOSTE

6   **Biodi(V)strict®: A Tool for Incorporating Biodiversity within Development** 134
    ANGEVINE MASSON AND NATHALIE FRASCARIA-LACOSTE

PART 2
Incorporating the Human Factor within Eco-design   145

7   **Social Practices and Ways of Living**   147
    CHRISTOPHE BESLAY

8   **Modelling the Energy Consumption of Buildings**   161
    JEAN-PIERRE LÉVY AND FATEH BELAÏD

9   **Reconstruction of Building Occupation by Machine Learning**   181
    ERIC VORGER AND MAXIME ROBILLART

10 **What Future for Nature in the City?**   194
    ISABELLE RICHARD

11 **Housing Demand: Residential Paths and Inequalities in Comfort and
    Access to Property**   215
    VINCENT LASSERRE-BIGORRY, FABIEN LEURENT,
    AND NICOLAS COULOMBEL

PART 3
Simulation at the Service of Eco-design   249

12 **DREAM: An Urban Equilibrium Model**   251
    FABIEN LEURENT, NICOLAS COULOMBEL, AND ALEXIS POULHÈS

13 **Structural Design of a Hierarchical Urban Transport Network**   271
    FABIEN LEURENT, SHENG LI, AND HUGO BADIA

14 **Application of the *ParkCap* Model to Urban Parking Planning**   286
    HOUDA BOUJNAH AND FABIEN LEURENT

15 **Modelling of Microclimates**   318
    HELGE SIMON AND MICHAEL BRUSE

16 **Development of a Methodology to Guaranteed Energy Performance**   335
    SIMON LIGIER, PATRICK SCHALBART, AND BRUNO PEUPORTIER

17   **The Contribution of Prospective Energy Systems to the Life Cycle Assessment of Buildings**   351
     EDI ASSOUMOU AND JÉRÔME GUTIERREZ

PART 4
Towards a Renewal of Techniques and Systems   371

18   **Real-Time Energy Management Strategies for Buildings and Blocks**   373
     MAXIME ROBILLART AND MARIE FRAPIN

19   **Urban Farming: From Discovery to Knowledge in the EEBI Chair**   404
     CHRISTINE AUBRY, ANNE CÉCILE DANIEL, BAPTISTE GRARD,
     AGNÈS LELIÈVRE, NATHALIE FRASCARIA-LACOSTE, AND CLAIRE CHENU

20   **Collecting, Classifying and Visualising Big Data for Biodiversity and Mobility**   420
     MADJID MAIDI, HELA MAROUANE, SALMA REBAI, AND SÉBASTIEN HERRY

21   **Smart Mobility: A Landscape Under Development**   449
     FABIEN LEURENT, OLIVIER HAXAIRE, AND GAËLE LESTEVEN

     Conclusions and Perspectives   497

Preface: Understanding the Future of the Urban World

For more than a decade, together or separately, citizens, associations and elected representatives of local authorities have been questioning the future of cities and, more particularly, quality of life in the city. Many debates have taken place on eco-neighbourhoods, smart and sustainable cities, soft mobility, smart grids, etc.

For more than a decade, we have been steadfast in the belief that our clients and partners need to feed these concepts – which are a bit vague for some and perfectly marketed for others – with concrete accomplishments and solid guarantees in terms of environmental performance.

VINCI's trades cover the entire value chain, the life cycle of transport infrastructure and public facilities: design, financing, construction, operation and maintenance. We have therefore naturally adopted life cycle assessment (LCA) and eco-design – notions already widely used in the industrial world – to propose tools, methods and benchmarks that are useful for seeing our gamble through, namely to integrate the environment in building and infrastructure design to enable all stakeholders to make the right decisions at each stage of the life cycle.

We made this gamble ten years ago by creating this chair[1] with three schools – MINESParisTech, École des Ponts ParisTech and AgroParisTech. During the first five-year cycle of this scientific chair, researchers and entrepreneurs from the group worked together within their own fields of construction, mobility and biodiversity to produce response elements. Armed with these initial scientific successes and fully aware of the scale of the questions opened up by this approach, we chose to continue the experiment with a new five-year cycle, asking the researchers to continue their work and encouraging them to spread it as widely as possible.

This second period opened up new fields of study, such as urban agriculture, neighbourhood life cycle analyses, improved maintenance of transportation networks and user influence in the energy performance of buildings.

Regular conferences, discussion days, annual seminars and numerous workshops opened up debates, bringing together ideas from both the scientific world and business and community representatives. The results of the work were circulated widely, giving them an international reach.

1 www.chaire-eco-conception.org

Under the co-direction of Bruno Peuportier, Fabien Leurent and Jean Roger-Estrade, this collective publication details the work carried out during this second series from 2016 to 2020. Faced with the exciting challenges of urban demography, climate change and the preservation of the environment, new uses of digital techniques allow us to better understand the futures of the urban world in all their complexity.

Happy reading.

Xavier Huillard
CEO of VINCI

Introduction

1. Eco-Design, a Compelling Necessity

Throughout the twentieth century, and even more so in the twenty-first century, the presence and influence of the human species on planet Earth grew very strong. The environmental consequences are becoming increasingly known, including the anthropogenic contribution to the greenhouse effect and the impacts on health, biodiversity and resources. The awareness of this human imprint on the world calls for curative as well as preventive actions. Eco-design is part of this perspective.

Eco-design is a design approach that integrates an environmentally friendly approach with social and economic objectives. Its field of origin is industrial ecology, the aim of which is to design controlled and optimised production processes and products that use resources frugally and save on waste. One of the fundamental principles is to consider all the stages of a product's life: manufacturing and distribution, use, maintenance, end-of-life and recycling. Each stage of life has different consequences on the environment, and this should be anticipated from the design phase in order to make the product intrinsically sustainable.

The progressive spreading of eco-design is particularly high in areas of application relating to construction, that is the economic sector that includes building and public works. Assurances must be given of the environmental performance of the residential, production, service or leisure functions in the buildings that house them and in the associated infrastructures. These include the technical networks for the energy flows, water and information, the infrastructure and modes of transportation that serve the area.

One characteristic of the construction and public works sector is the long duration of the usage phase, and so these technical objects are designed to provide for the territory in the long term. The preferred functional unit in performance analysis is a unit of use for a certain period of time, which can be correlated with a year so as to calibrate the performance measurement. For example, the energy performance of a building is indicated by "labels" that reflect the kWh of energy consumed per square metre per year. For a user, this usage performance is added to the purchase or rental price in order to estimate the budget needed for tenancy of the place. For builders, developers or concessionaires, it is now necessary to consider environmental performance when conceiving a project, not only in terms of the construction phase but also in terms of the phases of use and deconstruction. This means they must consider the entire life cycle of the property, which can last for a century or more.

Old buildings that were designed and built in the "prehistory" of eco-design have poor energy performance, which makes the construction phase largely a minority aspect within the broader context of use lasting 40 years, let's say. The respective contributions of the two phases will be much more balanced for buildings designed with high-energy performance, resulting in a total budget that is one or two orders of magnitude lower than that of an old building with the same area. Eco-design can then be usefully applied to renovation projects and not only to new construction.

2. The Eco-Design Chair and Its Scientific Objectives

The Chair of Eco-design of buildings and infrastructures (Chaire d'Éco-conception des ensembles bâtis et des infrastructures – EEBI) has been contributing to the development of knowledge for designing places for human life and activities for ten years now. Its primary scientific purpose spans buildings, outdoor spaces and infrastructures in a broader perspective that integrates the conditions of the places, not only at the micro-local scale of a building considered in isolation, but also at the local scale of a district, and even the wider scale of an urban area or territory. It is a big topic, then, one in which the notion of environmental performance meets the challenges of urban and territorial coherence.

For example, urban forms influence the performance of buildings in terms of heat loss linked to density, air renewal, wind deflection and solar sources due to mask effects. On the other hand, buildings are connected to energy grids at different scales.

The modes of transport used to access buildings depend on the district, including factors such as the quality of public transport servicing, walking and cycling facilities, and facilitation of car sharing and electromobility.

The conciliation between nature and the highly artificial environment of the city is also conceived at several scales. It concerns various assets in various forms, from revegetation of the ground, facades or roofs of buildings to the permeability of the land and the building configuration properties in terms of evapotranspiration (which are fundamental in heat island phenomena), passing through the relationship between population and nature by means of social gardens, green pathways, and green and blue belts, as well as the site's contribution to biodiversity.

In order to get to know and integrate this great diversity of issues, three "French-style" engineering schools came together through the Eco-Design Chair, each bringing their "research specialties": building and infrastructure energy and LCA for MINES ParisTech, mobility systems and territorial planning for École des Ponts ParisTech, and biodiversity, nature in the city and urban agriculture for Agro ParisTech.

Within the framework of the Chair, these specialties have been in dialogue with one another for ten years now, thanks to researchers interacting with the engineers and managers of the Vinci group, acting as designers and constructors of buildings and infrastructures, promoters of real estate projects, and especially eco-districts and concessionaires of transport infrastructures. The knowledge and questions that have arisen have been shared within a very active scientific and technical community.

3. The Achievements of the Chair in the 2008–2013 Period

This second collective publication reports the developments of the Chair's second five-year period between 2013 and 2018. Before sketching the outlines, let's go back a few years and recall the contributions of the first five-year phase, which took place between 2008 and 2013.

When the Chair first originated, each of the three research teams upholding it was already well established in their respective scientific fields. They clearly share the analytical approach of modern engineering: the constitution and the use of a technical object produces impacts of an economic as well as an environmental and social nature. These impacts raise issues of performance that must be addressed through the object's design. This performance is measured by objective and, if possible, quantitative evaluation methods. The mechanisms of the impacts, their causalities, are modelled by "physical laws" in the broad sense, including not only the physics of inanimate objects but also the dynamics of living and the economic and social behaviour of the individuals. Finally, physical laws are digitally simulated to test different variants, exploring various possible forms for the technical object in question and thus improving the design with regard to issues of performance.

Thus, the shared technical culture of engineering provided the starting point for the Chair. However, the three initial positions were also broad ranging in terms of their evaluation methods. The industrial orientation of the École des Mines has led the Centre Energétique et Procédés (Energy and Processes Centre) to invest in building energy modelling since the 1980s, and in LCA since the 1990s.

At École des Ponts, the development and subsequent fitting out of the local area are long-standing concerns. The evaluation of large investment projects is subject to socio-economic methods that are progressively enriched in order to integrate the environment in several respects: the effects of the interruption and fragmentation of habitats in the construction phase, and consumption of energy, emissions of pollutants and greenhouse gases in the use phase.

At AgroParisTech, the concept of life cycle is omnipresent among the fields of ecology, agronomy and life sciences and the environment, but management is primarily planned on the scale of agricultural territories. Biodiversity has been subject to assessment indicators that are still very reductive in the face of the complexity of their subject matter, as are specialised impact studies for certain living species or certain places. The presence and multiple roles of nature in the city is a field of investigation that only emerged in 2008.

As such, in 2008, the three schools represented highly diverse fields of investigation and cultures of evaluation. This diversity was gradually organised into a complementary whole. The approaches were opened to each other through recognition of their respective complexities, by means of a constructive and stimulating debate to study the opportunity for transposition, intersection and hybridisation. The very nature of subjects dominates, of course, and continues to impose itself on researchers and engineers alike.

From 2008 to 2013, the École des Mines (Centre d'Efficacité énergétique des Systèmes – Centre for Energy Efficiency of Systems) developed a method for modelling the occupancy of buildings through their users and their visitors, as a key factor of energy consumption and environmental footprints in the usage phase. The dynamic aspects of LCA began to be studied, in order to take the temporal variation of the impacts into account in a more precise way. Optimisation techniques were implemented to develop energy management strategies that take into account the interactions between buildings and energy grids, in collaboration with the Centre Automatique et Systèmes (Automatic and Systems Centre). The issue of model reliability has already been explored.

AgroParisTech is interested in life in the city but also in controlling the ecological impacts of infrastructure. The first research was launched into the indicators of the state of biodiversity within a neighbourhood. The ecosystem services provided by revegetation in cities were also the subject of preliminary work. Finally, work also addressed the issue of assessing the impact of an infrastructure on the functioning of the ecosystems affected.

École des Ponts ParisTech invested in LCA by conducting life cycle inventories for important construction materials, including glass, steel, concrete and wood, taking into account the production processes, sites and distribution routes of these heavy materials to the construction site (Navier laboratory). At the other end of the spectrum, Laboratoire Ville Mobilité Transport modelled the interactions between the transport system and urbanisation by constituting an original model for the formation of real estate prices based on local real estate capacities, modes of transport, the population, and the figures concerning employment, places and income. Between these two extremes, sensitive mobility models were developed for modes of transport in an urbanised context: for public transport, for car traffic and associated parking, giving priority to system effects that impose traffic conditions (capacity, congestion) and quality of service (time spent, discomfort) that determine users' choices (in terms of travel mode, route and schedule).

These varied developments were gathered together in a shared case study for a district of Cité Descartes, in the eastern part of the urban area of Paris. For this neighbourhood, urbanisation of which is still ongoing, the Chair studied the design of a set of buildings for residential or tertiary uses, accessibility by different means of transport, as well as local traffic conditions and car parking, and the impermeability and local profile of the biodiversity, linked in particular to the nearby presence of the Bois de Grâce.

4. Achievements in the 2013–2018 Period

The scientific and technical capital thus amassed in the first phase was developed and amplified by the Chair in its second five-year phase of activity, from 2013 to 2018, in four closely complementary areas: methods for eco-design, consideration of the human factor, simulation models and the study of technical solutions.

In terms of eco-design methods, the Chair refined the LCA of buildings in two complementary theses: firstly, the dynamic modelling of the energy mix and its consequences for the evaluation of prospective scenarios of energy transition (Charlotte Roux), then a sensitivity analysis method and quantification of uncertainties in LCA (Marie-Lise Pannier). An LCA was also developed for transport systems, integrating not only the infrastructure but also the vehicles that use it and their mutual interactions (Anne de Bortoli). In addition, the LCA for transport systems was incorporated into a holistic assessment that also addresses social and economic impacts, in particular the creation of values for actors and the circulation of value streams in the inter-sectoral production system. With regard to impacts on biodiversity, the "Avoid,

Reduce, Compensate" strategy required a complex evaluation process, for which a study was launched. One thesis focused on the organisation of the actors involved in the management of the "Avoid, Reduce, Compensate" sequence of infrastructure projects (Julie Lombard-Latune). Finally, the Biodivstrict method for diagnosing the state of biodiversity in a district was the subject of an industrial valuation by the creation of Urbalia, a joint venture between AgroParisTech and Vinci Construction France.

The environmental performance of urban systems is very much linked to the "human factor", that is users' lifestyles and behaviours. Collaborations were then carried out with teams in the human sciences and three sociological studies were conducted: a review of knowledge relating to people's lifestyles and everyday uses in relation to environmental issues (Christophe Beslay), a reflection on the intersection between sociology and engineering sciences for modelling building energy consumptions (Jean-Pierre Lévy), and a survey of the population's perception of the stakes and courses of action aimed at preserving nature in the city (Isabelle Richard). Engineering sciences, in particular connected sensors and machine learning techniques, were also mobilised to identify certain parameters relating to the occupation of buildings. The building occupancy model, developed by Eric Vorger in the first sequence and enriched by the incorporation of data from dynamic sensors, was also the subject of industrial development with the creation of the Koclicko start-up.

The *a priori* evaluation of impacts implements models to simulate the use of equipment. At the overall urban area level, the automobile and public transport networks were the subject of a general supply, demand and demand-side use model, in which optimisation of an objective function representing the collective interest makes it possible to calculate the size of the structural composition (distance of lines, number of stations, size of fleets) and to determine an ideal usage tariff (Sheng Li). The ParkCap simulation model of parking and car traffic was applied to the evaluation of prospective scenarios for managing mobility in Cité Descartes (thesis by Houda Boujnah). A dynamic parking model was conceived to theorise the supply–demand balance of this service, according to the spatial distribution of the lots of spaces and the distribution of the demand in space and time. In order to study multimodal mobility in a neighbourhood, the model underwent a significant spatial refinement, detailing the locations at the urban block level. With regard to the local environmental impacts of the means of transport on the environment, an attributive method was developed in order to evaluate a proportionate share for each individual in movement based on the journeys made and their speed conditions, the vehicles used according to their motorisation and how they are filled, and to assign quantities of the impacts, respectively, generated or suffered for each location. Finally, the housing system was the subject of an in-depth econometric analysis. A demographic model made it possible to characterise the households' progression on the housing ladder. The critical step of first accessing the property was modelled econometrically (thesis by Vincent Lasserre-Bigorry). An outside team (TU Mainz) finely modelled the actual or perceived outdoor temperatures in the Cité Descartes neighbourhood in space and time, using a day in July as a reference situation (heatwave of 2003) and in prospective scenarios for 2030, pairing a planning programme with the expected effects of global warming. Building energy simulation was enriched to take into account certain technical innovations (thermodynamic systems, for example). A process that guarantees energy performance based on energy simulation and the propagation of uncertainties was developed (Simon Ligier).

The simulation was finally mobilised to carry out prospective studies on the long-term evolution of electrical systems. The long lifespan of buildings and infrastructure requires LCAs to be made over a reference period of 50–100 years. On the other hand, 70% of electricity is consumed in buildings. The production mix of this electricity and its evolution over time therefore have a strong influence on the environmental impact of the projects.

Beyond the previously mentioned potential to use sensors to better regulate systems, energy management strategies were developed as technical solutions for buildings and blocks of buildings (Maxime Robillard and Marie Frapin). The Chair focused on two other fields of innovation: Urban Agriculture and Intermediate Modes of Mobility. The technical forms of urban agriculture were the subject of an international review and a typology, as well as systematic experiments to design micro-ecosystems with a high productive output (thesis by Baptiste Grard). Intermediate modes of mobility, taxis and private car, carpooling, car sharing and bike sharing are developing halfway between private vehicles and public transport, and the Chair began to explore their areas of relevance and the potential impact on the environment and on the economy, both in the metropolitan context and as intercity connections (Paris-Bordeaux). Two other studies were launched to understand the digital transformation taking place: one to develop a mobile application to collect on the indices related to the Chair's subjects in the field, and the other to characterise the forms of digital transformation in the realm of transport, from the availability of computer technology and data collection, to the main functions in the use of the service, passing through the processing of information by artificial intelligence algorithms.

5. Purpose and Content of the Publication

The body of this book brings together most of the research contributions made within the framework of the Chair between 2013 and 2018. The work does not aim to report all the contributions exhaustively, but rather to present the essentials, in a form accessible to engineers and managers as well as researchers and students.

The topic is organised into four parts from eco-design methods to technical innovations, including consideration of the human factor and the use of digital simulation.

The first part deals with methods for eco-design. It has first place in this book, reflecting the position held by the evaluation and decision-making assistance for supporting choices in the design stage. It first presents the methodological developments of LCA: consequential LCA applied to buildings (Chapter 1), sensitivity analysis and quantification of LCA for uncertainties in buildings and districts (Chapter 2), and the holistic evaluation of mobility systems (Chapter 3). We then present the attributive analysis of the local environmental impacts of means of transport according to the vehicles, individual trips and places (Chapter 4), the problem of compensation as an environmental management tool in a system of social actors (Chapter 5) and the development of the Biodivstrict tool for the environmental diagnosis of neighbourhoods (Chapter 6).

The second part deals with the human factor, which is to say behaviours and conditions of use, as theorised in evaluation models. Chapter 7 shows the socio-technical nature of energy performance and proposes several modelling principles. The modelling of energy consumption in buildings, at the crossroads of sociology and engineering sciences, is presented in a review of recent advancements (Chapter 8). The occupation of the buildings is modelled by "machine learning" using measurements taken from sensors (Chapter 9). The sensitivity to nature and the courses of action envisaged by the citizens are then discussed (Chapter 10), as well as the relationship to housing on the part of households moving through the property ladder throughout their life cycle (Chapter 11).

The third part shows how digital simulation contributes to eco-design. A model of the housing system represents supply and demand interacting with one another, as the state of equilibrium determines both property prices and the consumption of each household in housing (Chapter 12). The simplified modelling of a multimodal mobility system in an agglomeration allows an optimal plan to be determined (Chapter 13). The demand for parking also depends on the demand for activity in the places served, and is distributed in space according to locally established capacities and management methods (Chapter 14). The detailed representation of buildings, vegetation and roads, and the fine physical modelling of energy flows, humidity and air movements enable microclimates to be simulated within a neighbourhood (Chapter 15). Energy simulation and the propagation of uncertainties contribute to progress towards the guarantee of performance (Chapter 16). The contribution of prospective energy systems to the building LCA is shown in the case of the electrical system (Chapter 17).

The fourth part presents innovative technical solutions: energy management strategies for buildings and urban blocks (Chapter 18), technical forms of urban agriculture (Chapter 19), participatory collection of indices in the field by means of a mobile application on a smart phone (Chapter 20) and digital transformation within transportation and new management potential (Chapter 21).

Finally, some conclusions lead to the proposal of paths for future works.

Contributors

Edi Assoumou, Mines ParisTech, CMA – PSL, Paris, France

Christine Aubry, AgroParisTech, Paris, France

Hugo Badia, KTH Royal Institute of Technology, Stockholm, Sweden

Fateh Belaïd, École des Ponts ParisTech, Marne-la-Vallée, France

Christophe Beslay, BESCB

Anne de Bortoli, École des Ponts ParisTech, Marne-la-Vallée, France

Houda Boujnah, École des Ponts ParisTech, Marne-la-Vallée, France

Michael Bruse, University of Mainz, Mainz, Germany

Claire Chenu, AgroParisTech, Paris, France

Nicolas Coulombel, École des Ponts ParisTech, Marne-la-Vallée, France

Anne Cécile Daniel, AgroParisTech, Paris, France

Pauline Delforge, EEBI Chair

Adelaïde Feraille, École des Ponts ParisTech, Marne-la-Vallée, France

Marie Frapin, Mines ParisTech, CES – PSL, Paris, France

Nathalie Frascaria-Lacoste, AgroParisTech, Paris, France

Baptiste Grard, AgroParisTech, Paris, France

Jérôme Gutierrez, Mines ParisTech, CMA – PSL, Paris, France

Olivier Haxaire, École des Ponts ParisTech, Marne-la-Vallée, France

Sébastien Herry, ESME Sudria, Ivry-sur-Seine, France

Natalia Kotelnikova-Weiler, École des Ponts ParisTech, Marne-la-Vallée, France

Vincent Lasserre-Bigorry, École des Ponts ParisTech, Marne-la-Vallée, France

Julie Latune, EEBI Chair

Agnès Lelièvre, AgroParisTech, Paris, France

Gaële Lesteven, École des Ponts ParisTech, Marne-la-Vallée, France

Fabien Leurent, École des Ponts ParisTech, Marne-la-Vallée, France

Harold Levrel, AgroParisTech, Paris, France

Jean-Pierre Lévy, École des Ponts ParisTech, Marne-la-Vallée, France

Sheng Li, École des Ponts ParisTech, Marne-la-Vallée, France

Simon Ligier, Mines ParisTech, CES – PSL, Paris, France

Madjid Maidi, ESME Sudria, Ivry-sur-Seine, France

Hela Marouane, ESME Sudria, Ivry-sur-Seine, France

Angevine Masson, AgroParisTech, Paris, France

Marie-Lise Pannier, Mines ParisTech, CES – PSL, Paris, France

Bruno Peuportier, Mines ParisTech, CES – PSL, Paris, France

Alexis Poulhès, École des Ponts ParisTech, Marne-la-Vallée, France

Isabelle Richard, ENVIRONNONS

Salma Rebai, ESME Sudria, Ivry-sur-Seine, France

Maxime Robillart, Kocliko, Toulouse, France

Charlotte Roux, Mines ParisTech, CES – PSL, Paris, France

Patrick Schalbart, Mines ParisTech, CES – PSL, Paris, France

Helge Simon, University of Mainz, Mainz, Germany

Eric Vorger, Kocliko, Toulouse, France

Part I

New Developments in the Methods for Eco-design

Chapter 1

Consequential Life Cycle Assessment Applied to Buildings

Charlotte Roux

Mines ParisTech, CES - PSL

1.1 Introduction

Today, new environmental assessment methods and approaches are being developed in connection with the central concept of the life cycle. They make it possible to improve the accuracy of the inventory by taking into account dynamic phenomena (DLCA, *dynamic life cycle assessment*) and economic sectors that are usually neglected (EIOLCA, *environmental input–output-based LCA*), and to extend the method to the economic field (LCC, *life cycle* cost) or the social sphere (SLCA, *social life cycle assessment*), etc. These new developments aim to fill the current gaps and limitations of LCAs (Finnveden et al. 2009, Guinée et al. 2010, Reap et al. 2008).

In parallel with these developments, LCA studies are generally classified into two large families in the scientific literature: the "attributional" approach (ALCA) and the "consequential" approach (CLCA), which serve to meet different objectives (Halvgaard et al. 2012). The attributional approach seeks to allocate part of a responsibility to a given system, while the consequential approach seeks to model the environmental consequences of a decision (Finnveden et al. 2009). When using life cycle assessment in the design phase, the project is by definition not yet realised. The addition of a new district, or the rehabilitation action on an existing district, is therefore a decision and impacts the existing one, which corresponds to a consequential approach. This framework can be opposed to the normative framework of building certification, whereby the building evaluated is already completed.

For Suh and Yang (2014), there are no well-defined boundaries between ALCA and CLCA, but rather, LCA studies introduce complementary approaches (data or economic models, scenarios, land-use change, system dynamics, etc.) to improve the realism of the environmental analysis. They also stress that no "complete" CLCA exists since it is impossible to model all the consequences of a decision exhaustively.

It is therefore important to determine whether the objective of the study imposes a "consequential" point of view. Then, in practice, an "attributional" model can be seen as a first approximation of a consequential question. For example, according to some researchers, it is possible to rely on an attributional "base" to carry out a consequential study (Yang 2016).

The application of the "consequential" philosophy is a paradigm shift that displaces the focus of the study: it is not "the building/district" that is evaluated, but rather "the decision to build or renovate the building/district". The implementation of this proposed new approach for the eco-design of buildings is based on the identification

Table 1.1 Main Features of the Consequential Project Approach Compared to the Classic Attributional Approach

Modelling Hypotheses	Attributional	Consequential Project
Manufacture of materials/ processes	Average technology(ies)	Marginal technology(ies)
Biogenic carbon	Neutral balance sheet	Differentiated balance sheet
Use of recycled materials – integrated benefit	Inventory method 100%	Consideration of market constraints: between 0% and 100%[a]
Recycling/recovery at the end-of-life of materials – integrated benefit	Inventory method 0%	Consideration of market constraints: between 0% and 100%[b]
Energy export/ co-production	Co-product method: allocation of production infrastructure	Avoided burden method: substitution of marginal technologies
Database modelling system for the background	Cut-off	Allocation at the point of substitution (APOS)**

[a] 50-50 by default.
[b] This would currently require more transparency in database inventories.

of marginal technologies and processes, and the expansion of the system, which integrates the substitution effects and the impacts avoided (see Table 1.1).

The "consequential project" approach proposed here justifies neglecting the market effects (use of economic equilibrium models, rebound effects, elasticities, experience curve) generally associated with consequential studies, given the limited size of the project compared to the size of the national economic system. The analysis of market effects is more suited to the study of an economic sector than to the study of a project.

The introduction of new mechanisms (economic, techno-economic) by the consequential approach requires more information on the system studied and its environment, and information that can be obtained in part thanks to a dynamic LCA.[1]

In particular, the issue of electricity and energy mix is complex and requires an additional modelling effort. On the one hand, more than two-thirds of electricity in France is consumed in buildings (SOeS 2016). On the other hand, electricity consumption can be one of the main contributors in the environmental assessment of buildings. It is then necessary to finely model this component of the system studied. The long lifespan of buildings also requires the introduction of a prospective approach. Finally, a reflection on the neighbourhood level must integrate the issues of transport and waste management.

The consequential project approach was applied to the development project of a new district, Cité Descartes.

The following sections of this chapter take up these different hypotheses, which were developed characterising the consequential project approach: electrical system modelling, transport and waste problems. Their impact on design aid is analysed through the example of Cité Descartes. The interest of a prospective approach is discussed in Chapter 12.

---

1 For a definition and a comprehensive conceptual framework of dynamic LCA, see Collinge et al. (2013).

1.2 Modelling the Electrical System

1.2.1 Model Overview

The study of the existing models of electrical system simulation has shown the need for the development of a new model, which is specifically created for the study of buildings and neighbourhoods. In particular, one of the main obstacles to the use of the existing models was the impossibility of ensuring consistency between the data repository used for the study of the electrical mix and that used for the study of buildings (weather, reference consumption, etc.).

The model presented here represents the current French situation (e.g. French meteorological data, installed capacities, plant characteristics) but can potentially be calibrated for other national contexts. It is an explicit model that allows a production mix to be associated at each hour of the year, according to a demand that depends significantly on the meteorological conditions. The model generates mixes that are consistent with the data used in building energy efficiency, independent of climatic or economic hazards inherent in real years (e.g. cold wave, heat wave, strikes, other hazards).

Three large submodels are considered (Figure 1.1):

- Electricity demand, the sum of national consumption, exports and consumption of pumping stations;
- Incidental productions, independent of electricity demand, but dependent on meteorological factors (e.g. wind) or unknown local economic variables (e.g. household waste incinerator);
- Modular production, which covers the residual demand once the incidental productions are subtracted.

Incidental productions, independent of the level of demand, are first entered in the model and subtracted from the national consumption. The demand to be met is then adjusted with border exchanges (addition of exports, subtraction of imports) and increased by consumption due to pumped energy transfer stations (PETS).

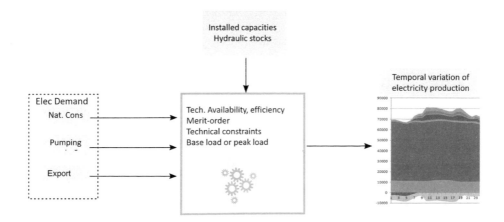

Figure 1.1 Principle of simplified modelling of the electricity generation system.

Modular production units must then meet the residual demand according to their cost and technological constraints.

The model was calibrated on the actual years 2012 and 2013, then validated on 2014.[2] In a second stage, as well as when real weather data are used to construct typical meteorological data like TRYs (test reference years) (Lund 1985), a model is constructed for determining electricity production and its hourly variations in a "type" frame. For example, a representative demand for typical data used in dynamic thermal simulation (DTS) is used, as well as average historical availability of production units, average hydraulic productibility, etc.

This allows the model to be used for assisting building and neighbourhood design: consistency is ensured between the input data used (meteorological data in particular) and the average model of the incidents for actual years (availability of water resources for hydraulic production, decrease in availability due to a disruption at a nuclear power station, cold wave or heat wave …).

The final model gives an electrical production for each hour representative of the current functioning of the system with respect to a given request. It is possible to study the sensitivity of the mix with respect to the integrated parameters, such as the installed capacities or the hydraulic generation[3] potential to simulate hazards (dry or rainy year, for example).

The model comes close to other models existing in the literature, such as the EnergyPLAN model developed by Professor Lund (Lund 2007) or the dispatch model developed by Raichur, Callaway and Skerlos (2015). However, unlike the two models cited, the model presented here explicitly models the link between climate data, electricity consumption and electricity production (particularly photovoltaic and wind energy). In addition, it is adapted to the French context, in which nuclear power plays a central role. In this study, certain production constraints could not be explicitly integrated, like reserve constraints (unused available capacity mobilised in the event of an incident), in contrast to others (Raichur, Callaway, and Skerlos 2015). The model is centred on the electrical system, unlike the EnergyPLAN model, which allows the modelling of an entire energy system. A full description of the model is available in Roux, Schalbart and Peuportier (2016).

This model serves as a basis for the development of dynamic and marginal evaluation methods for the impacts of electricity. These methods improve the accuracy of a life cycle assessment, regardless of the objective (certification or eco-design) and the associated methodological approach (attributive or consequential project).

1.2.2 Determination of Marginal Electricity Production

There are several existing methods for quantifying marginal effects and determining the "marginal" power generation of a group of technologies, depending on whether consideration is made of one or more marginal technologies, the short or long term,

---

2 Years for which hourly technology-by-technology production data are available thanks to the recent RTE Eco2mix transparency scheme http://www.rte-france.com/en/sustainable-development/eco2mix.

3 According to the definition by RTE, "producible energy is that which would be derived from natural inputs if the factories were continuously exploited under optimum conditions. It is estimated from average input flows over a long period" (RTE 2012).

and the dynamics of the production systems or only the costs (Mathiesen, Münster, and Fruergaard 2009). Two methods are proposed here.

The first, called "GHG-P", is based on the recommendations of the GHG Protocol[4] for grid-connected projects (GHG Protocol 2007). The second is based on the use of the electrical system operating model presented previously, called "derived marginal". According to the classification given by Mathiesen, Münster, and Fruergaard (2009), both methods correspond to the determination of a complex mix of marginal technologies, based on the modelling of production at hourly intervals.

The first approach (GHG-P) consists of classifying production technologies according to their merit order[5] and then extracting the production at the top of this merit order (the last production units started up because they are more expensive). The marginal production threshold recommended by the GHG Protocol is 10%.

The marginal "derivative" approach implies a rather different use of the electricity production model presented above. The model for simulating electricity production is used twice: once to meet "typical" French consumption (sum of national consumption, exports and pumping) and a second time to meet the sum of this same typical consumption and net consumption of the project studied. The difference between the two results constitutes the marginal production resulting from the project's net consumption.

Only nuclear technologies, modular hydraulics, centralised coal and gas, and peaks can be mobilised to satisfy the marginal demand, and the other productions are considered incidental. Table 1.2 presents the marginal mixes obtained for the two methods as well as a comparison with the average annual reference mix for the model.

Table 1.2 Average Marginal Mix over the Year Considering a Threshold of 5% or 10% of Marginal Production (Relative to the Total Production) for the GHG-P Method and a Coal Price Higher or Lower than Gas. For the Derived Method, Marginal Mixes Are Evaluated Corresponding to Constant Demand or Demand Equivalent to 0.1% of Total Demand

Annual Average of Marginal Mixes	Peak (%)	Coal (%)	CCG (%)	Nuclear (%)	Other (%)
GHG-P method: 10% coal > CCG	1	18	12	70	0
GHG-P method: 5% coal > CCG	1	35	13	51	0
GHG-P method: 10% coal < CCG	1	18	12	70	0
GHG-P method: 5% coal < CCG	1	24	24	51	0
GHG-P method: 2% coal < CCG	3	15	48	35	0
Derived method, constant demand	1.0	37.1	24.8	37.1	0
Derived method, demand 0.1% of total demand	1.3	39.9	26.6	32.3	0
Annual mix average of reference	0.1	2.0	1.4	75.9	20.7[a]

[a] Hydraulic (12.7), wind (3.3), PV (1.2), decentralised thermal power (3.6).

4 Greenhouse Gas Protocol: a set of methodological guides developed by the World Resource Institute (WRI) and the World Business Council on Sustainable Development (WBCSD) for reporting greenhouse gas emissions from projects, companies, organisations and activities. http://www.ghgprotocol.org/.
5 By-product technologies (renewable and non-renewable energy decentralised production, run-of-river hydroelectricity), followed by hydroelectricity, nuclear, gas, coal, advanced technologies (net imports, combustion turbine, fuel oil).

The marginal approaches provide a more realistic evaluation of the consequences of additional electricity consumption imposed on the production system compared to the average methods. With regard to the modelling of avoided impacts, the use of marginal methods does not introduce inconsistencies relating to the type of production avoided, as is the case for average methods (e.g. photovoltaics avoiding photovoltaics). This modelling technique, also called "system expansion", is also recommended in the context of consequential studies to avoid allocation issues (Ekvall and Weidema 2004). However, this method is more sensitive to the uncertainties of the model and to the assumptions made, notably the share of coal in the flexible fossil production.

The marginal demand constituted by the electricity consumption of a building or the GHG-P method is applicable to a reference form electricity production that would come from another modelling method, for example prospective modelling. It is easier to implement and faster than the "derived marginal" method. Speed can be important in the case of sensitivity and uncertainty analysis or when there is a large number of simulations, like those greater than 1,000.

1.2.3 Determination of Environmental Impacts

The evaluation of the impacts generated by the consumption of electricity or avoided by the local production of electricity must take all the stages in the life cycle of the electricity production into account. This includes the extraction of the raw materials necessary for operating the power stations, and also the construction of power stations and the transmission (high-voltage) and distribution (low- and medium-voltage) power grids. The impacts generated per kWh delivered on the low-voltage grid and produced by i technology are represented by the following equation:

$$i_i = C_f \times (\text{ICV}_1 \times (1 + l_i) + \text{ICV}_{\text{net}}(i), \tag{1.1}$$

where C_f represents the matrix of the calculated impact characterisation factors, the life cycle inventory of i technology electricity generation, the grid loss rate and the grid infrastructure inventory, which is variable depending on the technology considered.[6] The flow inventory is based on generic data from the ecoinvent V3.2 database (Frischknecht and Rebitzer 2005, Frischknecht et al. 2007, ecoinvent 2010), following the recommendations of Itten, Frischknecht and Stucki (2012) concerning the consideration of grid infrastructures, for example, and the information on on-line losses published by RTE[7] and ERDF.[8] Regarding the production of electricity, many processes are available in the ecoinvent database, including 26 specific to France (Treyer and Bauer 2014). Table 1.3 presents the results of the environmental impact

---

6 Infrastructures are allocated to technologies according to energy production (energy allocation) and the level of connection between the technologies (e.g. thermal power plants are connected to the transmission grid, and they need the transmission grid and the distribution grid to reach the final consumer, unlike PV systems, which are supposed to be connected to the distribution grid and therefore only use that one).

7 RTE – Customer portal: http://clients.rte-france.com/lang/fr/visiteurs/vie/vie_perte_RPT.jsp RTE's HV network corresponds to the ecoinvent HV+MT network.

8 ERDF: http://www.erdf.fr/Compensation_des_pertes, accessed in July 2015.

Table 1.3 Attributional Dynamic Approach and Derived Marginal Approach

Per kWh	Unit	Average Approach	Marginal Approach	Difference[a] (%)
Acidification	kg SO$_2$ eq	4.49E−04	3.17E−03	150
Biodiversity	PDF.m².an	3.39E−02	4.31E−02	24
Primary energy	MJ	1.29E +01	1.37E+01	6
Radioactive waste.	dm^3	6.02E−05	3.13E−05	−63
Greenhouse effect	kg CO$_2$ eq	7.89E−02	6.05E−01	154
Depletion of resources	kg Sb eq	5.91E−04	4.99E−03	158
Eutrophication	kg PO$_4^{3-}$	1.56E−04	4.68E−04	100
Human health	DALYs	2.26E−07	5.80E−07	88
Smell	mm^3	9.42E+02	1.68E+03	56
Waste	t	5.14E−02	2.06E−01	120
Photochemical ozone	kg C$_2$H$_4$	2.13E−05	1.44E−04	148
Water	dm^3	5.98E+00	4.47E+00	−29

[a] Difference (marginal − average) divided by the mean of the two mixes.

assessment per kWh of electricity consumed, with a medium (majority nuclear) or marginal (majority gas and coal) approach.

The differences between average and marginal methods are much greater than those observed between the average annual and hourly methods (Roux, Schalbart, and Peuportier 2016).

The impacts of the marginal mix are significantly higher than the average mix for 9 out of 12 indicators. The indicators that increase the most are those related to the production of electricity from fossil fuels (mainly gas and coal): acidification, the greenhouse effect, depletion of resources and photochemical ozone. For these indicators, the difference between the two approaches is around one order of magnitude. On the other hand, the primary energy indicator varies little because the efficiency of the gas, coal and nuclear power plants is fairly close. The absence of renewable technologies and the presence of gas and coal thermal technologies cause the indicator to increase, but this hike is offset by a smaller contribution from nuclear power. Since nuclear energy is less present in the marginal mix than in the average mix, the indicators of nuclear waste generation and water consumption are lower for the marginal mix.

1.2.4 Application to Cité Descartes

Two variants of Cité Descartes building concepts are compared to the reference project (BBC, compliant with RT2012):

- a "BBC+PV" variant with energy performance identical to the reference variant, but also including a renewable electricity generation system (photovoltaic system);
- a "PAS + PV" variant with heating requirements of less than 15 kWh.m^{-2}.an-1 per building and very good airtightness in addition to the photovoltaic system.

The monthly balance of useful energy requirements per station and end-use per vector is given in Figure 1.2. Dual flow ventilation is installed in the PAS+PV variant, which

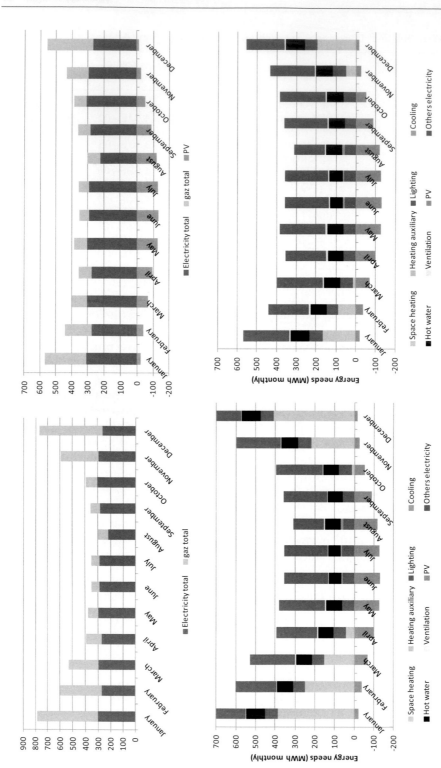

Figure 1.2 Monthly assessment of needs (above) and final energy consumption (below) on all zones, variant BBC/BBC+PV (left) and PAS+PV (right).

leads to a higher ventilation consumption than for the BBC variant. The seasonality of uses is much more marked for the BBC variant than for the PAS+PV variant. In addition, the maximal production period of the photovoltaic system coincides with the air conditioning needs of the offices.

Figure 1.3 presents the comparison of the three variants, namely, BBC, BBC+PV and PAS+PV, according to the project consequential approach (CLCA-P). The PAS+PV variant has less impact than the BBC+PV variant for these indicators, except for damage to human health, where both variants are equivalent.

The use of the consequential approach underlines the interest of setting up on-site renewable production in a context where the electricity mix is not 100% composed of renewable electricity: the two variants with photovoltaic energy (BBC+PV and PAS+PV) can be considered as less impacting overall than the reference variant.

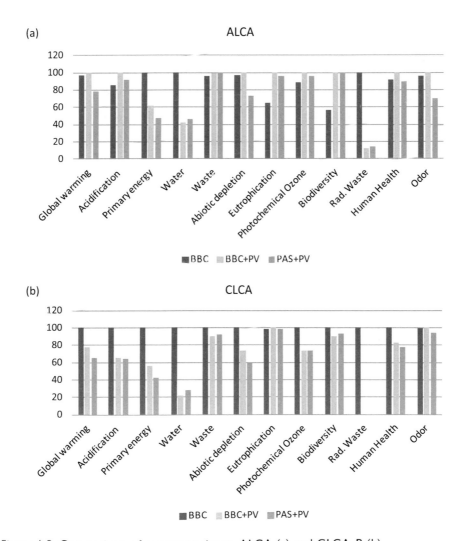

Figure 1.3 Comparison of energy variants, ALCA (a) and CLCA-P (b).

For most indicators, the energy produced by the solar system and the production avoided consequently supplants the impacts of its manufacture (and renewal).

By keeping an attributional approach and therefore, in particular, by using an average mix for the evaluation of the impacts of electricity, the classification of the variants is especially different. The passive variant (PAS + PV) is more favourable than the BBC variant (BBC + PV). However, the ranking of the two BBC variants depends strongly on the indicator considered.

Photovoltaic production makes it possible to reduce electricity consumption, but it has very little impact on many indicators (except the indicators related to nuclear generation: radioactive waste, water used, primary energy). For many indicators (e.g. abiotic resources, acidification, damage to health), the impacts of the manufacture of the photovoltaic system are assessed as more significant than the impacts avoided by the production of renewable energy.

The use of an attributional approach in design may favour the BBC variant over the BBC+PV variant, while the consequential analysis, which takes into account the existing production constraints, shows that on-site electricity generation has a certain environmental relevance, related to the constraints currently weighing on the electrical system.

1.3 From Buildings to Neighbourhoods: Transport

1.3.1 Everyday Mobility

When conducting LCA building studies for the purpose of aiding design, the environmental consequences of occupants' mobility are rarely integrated into the study. However, these decisions can strongly influence transport needs when multiple development sites are possible, for example. At the neighbourhood scale, road layouts, parking spaces and the construction of soft-mobility routes contribute to design decisions and influence the occupants' daily mobility (Beirão and Sarsfield Cabral 2007, Gärling and Schuitema 2007).

1.3.2 Purpose and Method of Study

This section proposes three modelling approaches depending on the purpose of the study, in order to integrate the daily mobility in the LCA of buildings, districts or territories. These approaches are presented in Figure 1.4. Depending on the objective of the study and the data available, the environmental impacts associated with the neighbourhood design will be different, and it is not always possible to obtain all data necessary to carry out a consequential project study.

1.3.2.1 Mobility Diagnosis: The Attributional Approach (ALCA)

The mobility diagnosis corresponds to the creation of a current, past or projected assessment of all the environmental impacts associated with the daily movements of the users of an urban area, a neighbourhood or a building. All the infrastructures used by these users must be taken into account. For example, a diagnosis of this type makes it possible to analyse the contribution of mobility in all environmental impacts

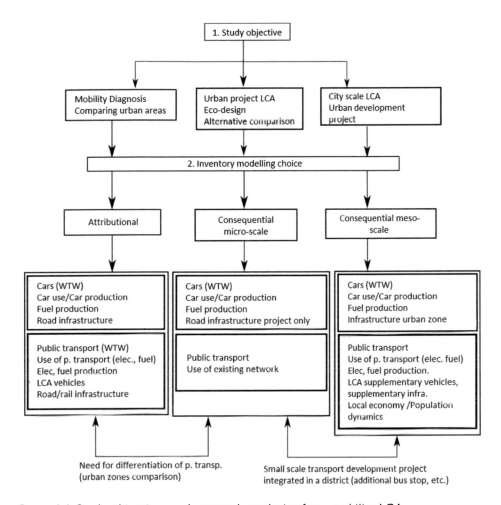

Figure 1.4 **Study objectives and system boundaries for a mobility LCA.**

in relation to other themes such as food, housing and waste treatment. It can also reveal significant average deviations for some urban areas, for example, if the bus fill rate is very low, which has a high impact per passenger.km.

1.3.2.2 Eco-Design of Neighbourhoods Consequential Project Approach (CLCA-P)

The approach consists of determining the consequences of the addition or rehabilitation of a neighbourhood on the urban environment in which it will take place. The design decisions that can be made concern the roads, the spatial organisation of the neighbourhood (zoning, site plan) and buildings (construction principles, energy performance). These developments can influence the mobility of occupants, and especially the distribution of public transport and the use of private vehicles among users.

The fact that transport is "avoided or created" by the transfer of households from their previous location to the new district is not taken into account. This "avoided or created transport" will be identical for all design variants and is not linked to district-level decision elements, as long as the implementation of the programme is not put into question (the project will be carried out, and the objective is to choose between different design variants of the same project).

With regard to infrastructure integration, as with electricity consumption, several modelling assumptions can be made:

- The use of existing infrastructure can be considered "free" in the sense of the LCA (no associated impact). If an occupant of the new district gets on the bus, the variation of the impacts of the bus with or without this occupant may be considered negligible.[9]
- The use of existing infrastructures implies their maintenance and renewal. As in ALCA, a part of manufacturing, maintenance and end-of-life infrastructure can be associated with each use of these infrastructures.

In cases where several project development sites are possible, the performance of the public transport associated with each zone is a differentiating factor between the sites. The usage and infrastructures associated with public transport must be integrated.

In cases where the settlement site is fixed, the first option, excluding the existing public transport system boundaries, corresponds to the actual consequences of the project's location in the area. On the other hand, the roads and road infrastructures created as part of the neighbourhood design are integrated into the environmental study. The increase in the number of private vehicles and the potential associated congestion can be integrated, depending on the data available.

1.3.2.3 Urban Eco-Design: Consequential Territory Approach (CLCA-T)

Land-use planning projects have a sphere of action that goes beyond the district scale. The environmental assessment of these complex projects can be done at the local level (impact study) or through a life cycle assessment methodology. Again, the examination of an urban development project corresponds to a study of a consequential nature. However, the meso-scale and not the micro (district)-scale may need to take into account additional consequences, which go beyond the developments proposed here:

- Influence on the property market: for example, a study integrated these effects in the context of the remediation of polluted soils (Lesage, Deschênes, and Samson 2007);
- Influence on population dynamics (relocation in accordance with the job pool, gentrification, etc.);
- Marginal modification of the local/regional transport system (new station, bus/ metro station, etc.).

---

9 An accurate assessment would include an increase in bus fuel consumption based on the weight of the passenger(s).

1.3.3 Determination of the Environmental Impacts of Daily Mobility in the CLCA-P Approach

Using the outputs of a transport modelling tool can take effects like congestion in urban areas into account. The contextualisation of transport data, according to the approach proposed in Figure 1.5, has a significant influence on the results, particularly with regard to the bus fleet used in the Paris region compared to the generic data contained in the Swiss ecoinvent database (Roux, Schalbart, and Peuportier 2016). In particular, the model made it possible to evaluate distances travelled in terms of mode of transport and direct emissions (CO, NO_x, PM).[10] It is therefore useful to correct the generic inventories taken from the LCA databases in order to take these local specificities into account.

The precise knowledge of the trip zones also makes it possible to refine the calculation of certain impacts, like the effects of the local emission of fine particles on health (Peuportier et al. 2017). The health impact (EcoIndicator 99 (Goedkoop and Spriensma 2001)) of a typical vehicle travelling in Ile de France is presented in Figure 1.6, for the case of the average urban density taken into account in Humbert et al. (2011). The variability of this impact as a function of the population density of urban areas in Paris and Marne la Vallée is represented by an uncertainty bar. The impact of transport is all the more important as the area is dense. Taking urban characteristics into account significantly modifies the results (difference greater than 10% for the indicator calculated without the participation of climate change (CC)).

Transport models generally consider an initial state, which notably makes it possible to calibrate the model or to carry out a mobility diagnosis of the urban area studied. In the case of the evaluation of an urban development project, including a transport infrastructure part, the impacted area is planned in a future state in which the urban development project has been carried out. It is not common to have this future state without the planned urban development. However, in the case of a consequential study, the analysis concerns the variation between the situation with and without the plan, especially if the project involves both the built grounds and the urban transport network. The variation between the initial state and the future state often has a broader scope than the project (change in vehicle fleet characteristics, demographic evolution). However, the environmental analysis is not able to legitimise the planned project on its own. The most important thing is to have different design variants, which is not always the case.

A change of scale, from a district to a city or part of a city, is a research axis in its own right. The complexity of this change of scale goes far beyond the building-to-district change, and additional analysis tools may be needed, like ACV input–output hybrid modelling.

1.3.4 Application to Cité Descartes

For daily transport, only the impacts of private vehicles are included in the consequential project approach. The increase in the impact of public transport generated by the users of the neighbourhood is assumed to be negligible, for example the impact of additional passengers in the RER. A high-speed train service is considered negligible.

---

10 Taking the fleet of vehicles in Ile-de-France and congestion phenomena into account.

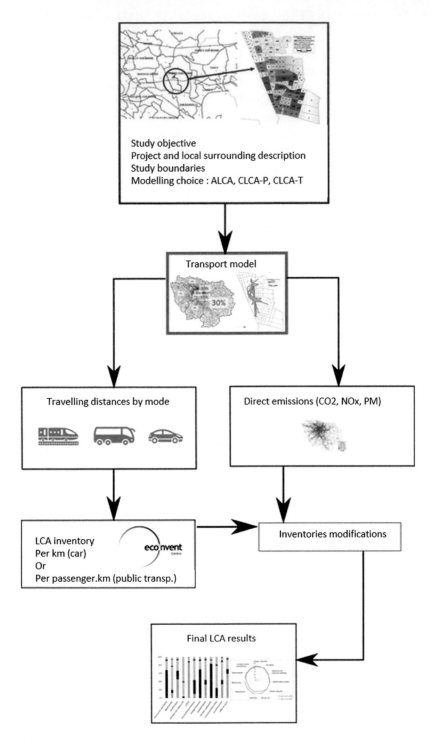

Study objective
Project and local surrounding description
Study boundaries
Modelling choice : ALCA, CLCA-P, CLCA-T

Transport model

30%

Travelling distances by mode

Direct emissions (CO2, NOx, PM)

LCA inventory
Per km (car)
Or
Per passenger.km (public transp.)

ecoinvent

Inventories modifications

Final LCA results

Figure 1.5 Summary of the procedure for determining transport impacts.

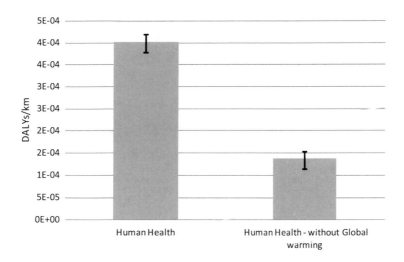

Figure 1.6 Variation in the human health indicator with and without the contribution of climate change (CC) by spatialising the impact linked to fine particles extracted from Peuportier et al. (2017).

This assumption is valid if the decisions do not concern specific adjustments to public transport (e.g. the addition of a bus stop). The model used to produce the results of the distance travelled by users according to mode (Figure 1.7) is the MODUS v2.2 model[11] developed by DREIF[12] and LVMT

The programme currently anticipated for Cité Descartes has a low functional diversity per block: three of the blocks mainly comprise offices, while the other four are residential buildings. With the aim of allowing heat exchange between offices and dwellings, a programmatic variant was studied by considering an identical functional diversity for each of the blocks, while preserving the total surface envisaged for each type of use (housing and offices, the businesses being fewer in number and located on the ground floor of the building in both cases). The change of programmatic variant has consequences for the modal choices. A decrease in the flow of soft transport methods (walking, cycling) is observed, as well as a slight decline in public transport flows and a slight increase in the number of private vehicle flows.

The final increase in the annual distance travelled by car is estimated at 7% more than the initial project, which leads to a corresponding deterioration in the environmental assessment of transport in a consequential approach. Offices generate a greater need for mobility than housing. Placing these areas further from public transport has therefore resulted in a net increase in the distance travelled by car.

The zoning of uses also has an impact on energy consumption. A DTS of the mixed project was carried out, placing dwellings in the upper floors of office buildings and

11 http://www.driea.ile-de-france.developpement-durable.gouv.fr/IMG/pdf/Documentation_ MODUSv2-1-2_cle25e7bd.pdf.
12 Regional Administration of Equipment Ile-de-France.

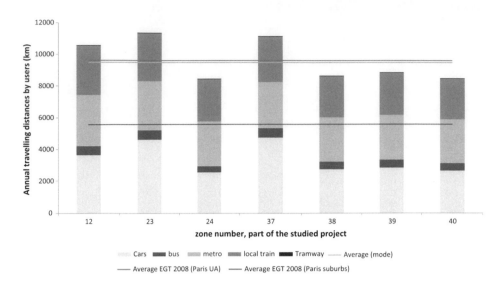

Figure 1.7 Distance travelled by neighbourhood user (attributes) by mode. The average of the 2008 global transport surveys (enquêtes globales transport, EGT) for the urban area of Paris and the Paris suburbs correspond, respectively, to 38 and 22 km per person per weekday working day (see the review by the CGDD (French General Commission for Sustainable Development)): The mobility of the French, Panorama issue on the national transport and travel survey 2008, p. 12 (www. developpement-durable.gouv.fr/IMG/pdf/Rev3.pdf) and 253 working weekdays per year. The subzones correspond to blocks of buildings with a different usage profile (more offices in zone 12/23/37, more housing in 24/38/39/40).

integrating office areas in the first floors of residential buildings. The results show a stability in the district's overall heating needs, which decreased by 1.6% thanks to better exposure of the dwellings (higher floors, south-facing buildings). Mixed zoning also has a significant influence on air conditioning needs, which decreased by 8.8% due to lower solar supplies to offices located on lower floors, which are less exposed. Energy sharing between homes and offices has not been studied in this simulation, but it could constitute an additional trigger.

The variation in the environmental assessment of the project resulting from the zoning change is presented in Figure 1.8, taking the changes in transportation and energy performance into account. Without factoring in the energy pooling, the increase of the transport takes precedence over the improvement of the buildings' environmental performance.

This exercise shows that the zoning of the site plan influences the district users' mobility. When developing the master plan, the effects of zoning on the energy performance of buildings (masking effects, exposure of buildings according to their internal provisions, etc.) and the effects on transport must be studied jointly in order to optimise the overall environmental performance of the project.

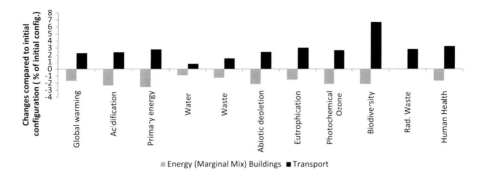

Figure 1.8 Effect of mixed zoning on the environmental performance of the project.

1.4 From Building to District: Household Waste

1.4.1 Household Waste

The development of a new district generates an influx of occupants (inhabitants, businesses) that produce waste. This will have to be "treated" by the community according to the possibilities available: are there recycling plants or an incinerator nearby? Where will the final waste be disposed of? Will all the waste be "treated" in the same way? What collection effort is/will be provided by the community?

At the level of a building project, the management and treatment of household waste is often excluded from the system boundaries because it is not influenced by the design of the budget or energy systems.[13] At the district level, there are triggers that can influence the environmental performance of the project: organic recovery (collective composting facilities, setting up shared gardens, etc.), deposit (glass), separate collection of packaging, heating system, including a household waste incinerator (UIOM), etc.

The diversity of options and the materials involved in the treatment of household waste impose a systemic view of the district and an analysis of the constraints weighing on waste incinerators (Figure 1.9).

1.4.2 System Boundaries, "Consequential Project" Approach

The consequential project approach aims to evaluate all the environmental consequences of a decision. The household waste treatment system can have an influence on design decisions. For example: should the new district be connected to the district heating system by recovering waste from incineration? What is the effect of setting up a collective compost bin at the foot of a building in terms of the environmental impact of waste treatment?

---

13 On the other hand, household waste management can be integrated in a behavioural study for an awareness-raising policy. It depends on the objectives of the study.

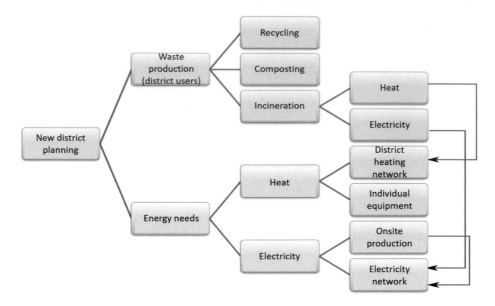

Figure 1.9 Systemic view of the district – waste *vs.* energy needs.

The influence of district design on the amount of waste generated by its occupants is considered negligible. The behaviour of the district's occupants in terms of consumption serves as input data for the design problem. The variability of behaviours can be treated by sensitivity and uncertainty analyses, but all the design variants must be analysed by considering the same behaviour (or a distribution of behaviours) and thus a fixed quantity of waste leaving the buildings for each type. Even if it is conceivable to treat some of the wastes directly in the neighbourhood (e.g. organic waste recovered as compost), this does not influence the quantity of waste at the start (e.g. the amount of organic waste before it is sent to compost).

The initial environmental load of discarded consumer products is therefore excluded from the system boundaries studied. It is not an environmental consequence of the realisation of the project, but rather preliminary information.

Finally, the various elements to consider for the treatment of a type of waste leaving a building for a "consequential project" district LCA are as follows:

- Waste flow characteristics: quantity, composition, calorific value;
- Possible treatment paths: recycling, energy recovery, material recovery, landfilling;
- The availability and economic competitiveness of the different treatment paths: are there any constrained paths (e.g. maximum capacity reached, insufficient outlets)? What are the marginal modes of treatment?

The first two elements are needed in the LCA whatever the methodological approach, but the latter is very specific to studies of a consequential nature.

Table 1.4 Characterisation of Waste Generated by Cité Descartes Annually

	Total Waste Quantity (t/year)	% Incinerated with Recovery	% Recycled	% Composted or Deposited Glass	% Landfill	PCI (MJ/kg)
Glass	63	53.6	39.8	0	6.6	0.1
Paper/cardboard	471	19.5	69.7	0	11.6	17.9
Plastics	97	53.6	29.2	0	6.7	34.1
Organic	318	67.1	3.3	28.7	0.9	4.3
Metals	21	76.5	20.1	0	3.4	0
Textiles	55	100	0	0	0	19.8
Others	55	100	0	0	0	10

1.4.3 Application to Cité Descartes

The estimate used for this study of the type and quantity of waste generated by Cité Descartes annually is presented in Table 1.4.[14] In this mixed-use district, with a high proportion of offices and businesses, paper/cardboard accounts for more than half of the tonnages collected (52%). It is assumed that energy generation avoids the use of natural gas (primary energy indicator and negative abiotic resource depletion).

The "waste" contribution analysis by type of waste is presented in Figure 1.10. The categories of plastics and paper/cardboard dominate all the indicators evaluated.

The inert waste indicator provides information on the potential for improving this overview: a large part of this indicator consists of glass waste, which can be recycled.

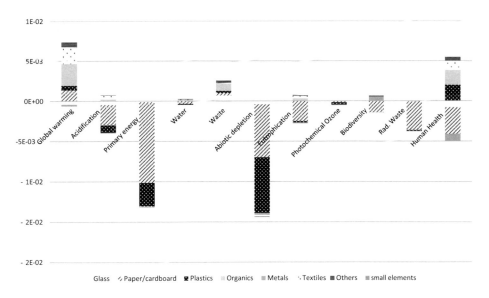

Figure 1.10 Waste treatment contribution analysis, by waste type, standardised impacts.

14 Data partly taken from Grimaud (2015), especially with regard to the quantity of waste.

The assessment of glass in relation to the other indicators shows the environmental effects of this type of recycling (global warming potential, for example). Organic waste can be recovered (methanation or composting, for example).

Plastics recycling mainly concerns polystyrene from offices and businesses. Despite the high calorific value of this type of waste, the assessment of incineration is very mixed: combustion-related impacts are not always offset by the impacts avoided by energy recovery. It should be noted that the human toxicity indicator would be lower in a context where the use of fuel oil is avoided. On the other hand, the recycling assessment is very favourable.

It is important to emphasise the scope of study chosen here, which excludes the environmental burden from the production of consumer goods that have become waste. This perimeter is suitable for the study of design variants, but it is absolutely not suitable for the study of management scenarios, including awareness campaigns to reduce the amount of waste generated, for example (e.g. food waste). As shown in Figure 1.11 in relation to the case of plastics, in this case, the perimeter proposed here would lead to a paradox, or an increase in the amount of waste would result in a reduction of the impact.

1.4.3.1 Potential Case of a Heat System Near Cité Descartes

At the time of designing the district, it may be necessary to study the following questions, through the lens of the consequential project approach:

- **Q1**: Is it better to connect the new buildings to the district heating network or to install collective (gas) boilers in each building?

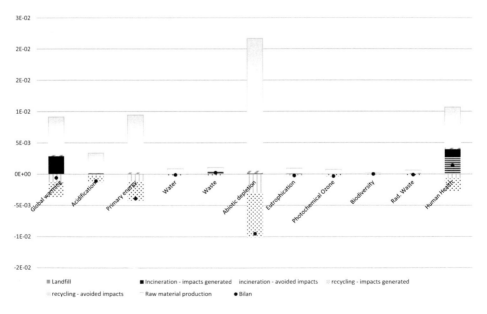

Figure 1.11 Impact of plastics processing, and contribution analysis according to treatment types.

- **Q2**: What is the impact of setting up collective neighbourhood composting facilities for a shared garden, considering that 50% of the putrescible waste would be composted?

A bi-energy system comprising a gas boiler and a household waste incinerator is considered here as an example.

Q1: District heating or collective boiler?

The district heating network or the collective boilers will make it possible to satisfy the energy needs of heating and Domestic Hot Water (DHW) The impacts of the waste incineration on the district activity will be effective whether the buildings are connected to the heating network or not, as they are not a consequence of the connection to the network. In this specific case, these impacts can therefore be excluded from the boundaries of the comparative study. The quantity of waste generated and its calorific value can, nevertheless, be the subject of a sensitivity study, representing the significant variability of the behaviours on this subject (quantities of waste, sorting, etc.).

The incinerator is considered to have available capacity (it can incinerate additional waste generated by new buildings). Two scenarios are possible:

- **scenario 1**: Gas is the marginal heat generation technology for the whole year;
- **scenario 2**: The incinerator produces surplus (non-recovered) heat during the summer, and the use of this heat by the new buildings installed gives rise to no environmental consequences in this case. Gas is the marginal technology during the heating period only (winter).

Depending on the saturation level of the incinerator and the marginal heat production technology, the connection to the heat system can be more or less favourable than the installation of collective boilers. It is therefore important to consider marginal heat generation technologies in order to evaluate the environmental consequences of connection to the heat system.

This approach can also influence the evaluation of the installation of thermal solar panels. This installation is less relevant in the event that some of the incidental heat is lost in the summer.

This assessment process may be linked to the implementation of a planning strategy over time: solar panels may not be a priority today (if there is surplus heat in the summer) but they will be when there is no more incidental heat to recover (the new buildings are connected, evolution of the heat system, decrease in the amount of incinerated waste, etc.). In this case, access to the roof and additional space for the solar storage tank can be integrated from the design stage and thus anticipate the future evolution of the building and its surroundings.

Q2: Impact of district composting.

Starting from this same study context, the question of the impact of setting up a collective compost system in the district can also be studied. The composted waste was previously incinerated. The installation of compost bins reduces the amount of putrescible waste incinerated by half. The heat incinerator will produce less, and this decrease in production will be offset by gas if the production is lower than the demand for heat. Compost avoids the use of artificial fertilisers. According to data from the ecoinvent database, 1 kg of compost is equivalent to 7 g of nitrogen fertiliser, 4 g of phosphorus and 6 g of potassium. The chain of consequences associated with the placement of a compost bin is shown in Figure 1.12. The environmental impacts of each final consequence must be determined.

The environmental assessment of the introduction of collective composting in the district is presented in Figure 1.13 for the two options considered above:

- *scenario 1*: Gas is still the marginal technology in the heat system (any decrease in production equals an equivalent increase in gas consumption);
- *scenario 2*: The incinerator is in a surplus situation in summer; reducing the amount of waste incinerated at that time does not generate an increase in gas consumption by the network. On the other hand, the reduction in the amount of waste incinerated during marginal gas periods (in winter) leads to an increased consumption of gas in the district heat system.

Again, taking into account the constraints on the district heat system is likely to change the final assessment. While both options favour the establishment of a

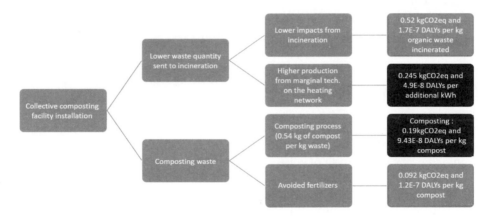

Figure 1.12 Consequences of setting up collective composting.

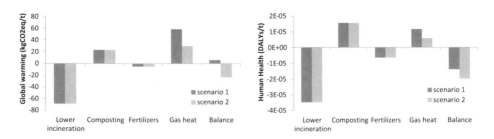

Figure 1.13 Assessment of the impact variation for one tonne of composted household waste for scenario 1 (marginal gas year-round on the heating network) and scenario 2 (marginal gas only in winter), compared to a reference without composting (incineration).

collective composting system for the indicator of damage to human health, the results are more contrasted for the case of the greenhouse effect.

1.5 Conclusion and Perspectives

To compare different variants of a project with an eco-design objective, it is necessary to evaluate the environmental consequences of each variant. A consequential LCA approach is thus more appropriate than an "attributional" study. This change in perspective relative to the attributional LCA used so far has significant implications for the development of life cycle inventories and ultimately for the results of the study. The implementation of a consequential approach requires the existing methods to be adjusted and a homogeneous study framework to be established. This work highlights the importance of defining a study area consistent with the objectives. Depending on the spatial scale (building, neighbourhood, territory), the actors involved, the scale of the project and the associated driving forces, the scope of study as well as the modelling assumptions must be adjusted. The same perimeter can hardly be used for different purposes, as has been illustrated in particular in the case of transport and household waste.

On the other hand, the application of the consequential approach to buildings involves taking the peculiarities of this sector into account: interactions with urban infrastructures (especially in the case of districts), the importance of energy consumption in the overall performance and the long life of the systems studied. The term "project" illustrates the small scale studied in relation to the national or world economy. The approach considers the existing and known production constraints (the recycling market, for example) but neglects market effects (price effect, rebound effects) which can only be studied in the context of a meso- or macro-economic study. Focussing on buildings involves reinforcing the knowledge of interactions between buildings and other sectors (electricity generation, transport, waste).

The dynamic simulation model of the electrical system developed allows for a time-based evaluation of the impacts of electricity consumption. This assessment can be based on the determination of an average mix or a marginal mix, and can thus improve the accuracy of attributional and consequential LCA studies.

Methodologies specific to the case of the transportation for inhabitants and waste treatment at the neighbourhood level have been developed. Coupling with a local mobility simulation tool makes it possible to take traffic conditions into account, and therefore provides more accurate estimates of the distances travelled by mode, as well as the direct emissions of CO_2, NO_x and fine particles. This work was carried out thanks to a collaboration with the ENPC City Mobility Transport laboratory.

The importance of interdisciplinary collaboration for studying districts emerges on numerous occasions, and the working bases have been laid (transport, heat system, electrical grid, prospective modelling – see Chapter 12). The integration of transport tools could be more developed, especially with regard to the spatialisation of local emissions. This subject is an important research avenue for improving the accuracy of flow inventories. The main difficulty lies in the amount of data that must be processed to manage the large number of characterisation factors.

Life cycle assessment is gradually expanding to increasingly complex subjects and systems: from mass-produced manufactured products in the 1970s to buildings in the

1990s, and districts in the 2000s. One of the next steps is to move on to the scale of the city and/or territory. Initial work on this topic is available (Larsen and Hertwich 2010, Loiseau et al. 2012). A consequential analysis of urban development projects will have to integrate the local economy, probably through regional hybrid models. IO or "input–output" tools are used to analyse different household scales across the country, passing through the city (Munksgaard et al. 2005). Their hybridisation with life cycle assessment allows for the evaluation of technological changes. There are examples at the urban scale concerning the treatment of wastewater (Lin 2009). This promising methodological approach should be explored for urban or territorial development projects, integrating both buildings and public spaces (the construction sector), energy, transport, waste treatment and water, addressing global and local impacts in terms of human health and biodiversity.

Bibliography

Beirão, G., and J. A. Sarsfield Cabral. 2007. "Understanding Attitudes Towards Public Transport and Private Car: A Qualitative Study." *Transport Policy* 14 (6): 478–89. https://doi.org/10.1016/j.tranpol.2007.04.009.

Collinge, W. O., A. E. Landis, A. K. Jones, L. A. Schaefer, and M. M. Bilec. 2013. "Dynamic Life Cycle Assessment: Framework and Application to an Institutional Building." *The International Journal of Life Cycle Assessment* 18 (3): 538–52. https://doi.org/10.1007/s11367-012-0528-2.

Ecoinvent. 2010. *International Reference Life Cycle Data System Handbook*. Ecoinvent Data v2.2. Swiss Centre for Life Cycle Inventories. Www.Ecoinvent.ChILCD.

Ekvall, T., and B. Weidema. 2004. "System Boundaries and Input Data in Consequential Life Cycle Inventory Analysis." *The International Journal of Life Cycle Assessment* 9 (3): 161–71. https://doi.org/10.1007/BF02994190.

Finnveden, G., M. Z. Hauschild, T. Ekvall, J. Guinée, R. Heijungs, S. Hellweg, A. Koehler, D. Pennington, and S. Suh. 2009. "Recent Developments in Life Cycle Assessment." *Journal of Environmental Management* 91 (1): 1–21. https://doi.org/10.1016/j.jenvman.2009.06.018.

Frischknecht, R., N. Jungbluth, H. J. Althaus, G. Doka, T. Heck, S. Hellweg, R. Hischier, T. Nemecek, G. Rebitzer, and M. Spielmann. 2007. "Overview and Methodology." Ecoinvent Report, no. 1. http://www.ecoinvent.org/fileadmin/documents/en/01_ OverviewAndMethodology.pdf.

Frischknecht, R., and G. Rebitzer. 2005. "The Ecoinvent Database System: A Comprehensive Web-Based LCA Database." *Journal of Cleaner Production, Life Cycle Assessment Life Cycle Assessment* 13 (13–14): 1337–43. https://doi. org/10.1016/j.jclepro.2005.05.002.

Gärling, T., and G. Schuitema. 2007. "Travel Demand Management Targeting Reduced Private Car Use: Effectiveness, Public Acceptability and Political Feasibility." *Journal of Social Issues* 63 (1): 139–53. https://doi.org/10.1111/ j.1540-4560.2007.00500.x.

GHG Protocol. 2007. "Guidelines for Quantifying GHG Reductions from Grid-Connected Electricity Projects." http://ghgprotocol.org/files/ghgp/electricity_final.pdf.

Goedkoop, M., and R. Spriensma. 2001. "The Eco-Indicator99: A Damage Oriented Method for Life Cycle Impact Assessment: Methodology Report." http://irs.ub.rug.nl/dbi/4581696db734f.

Grimaud, J. 2015. "ACV Eau et Déchets - Cité Descartes." Internal report. Champs-sur- Marne: Efficacity.

Guinée, J. B., R. Heijungs, G. Huppes, A. Zamagni, P. Masoni, R. Buonamici, T. Ekvall, and T. Rydberg. 2010. "Life Cycle Assessment: Past, Present, and Future†." *Environmental Science & Technology*. 45 (1): 90–6. https://doi.org/10.1021/es101316v.

Halvgaard, R., N. K. Poulsen, H. Madsen, and J. B. Jorgensen. 2012. "Economic Model Predictive Control for Building Climate Control in a Smart Grid." In *Innovative Smart Grid Technologies (ISGT)*, 2012 IEEE PES, 1–6. https://doi.org/10.1109/ ISGT.2012.6175631.

Humbert, S., J. D. Marshall, S. Shaked, J. V. Spadaro, Y. Nishioka, P. Preiss, T. E. McKone, A. Horvath, and O. Jolliet. 2011. "Intake Fraction for Particulate Matter: Recommendations for Life Cycle Impact Assessment." *Environmental Science & Technology* 45 (11): 4808–16. https:// doi. org/10.1021/es103563z.

Itten, R., R. Frischknecht, and M. Stucki. 2012. *Life Cycle Inventories of Electricity Mixes and Grid*. ESU-Services Ltd., Uster. http://www.esu-services.ch/fileadmin/download/publicLCI/itten-2012-electricity-mix.pdf.

Larsen, H. N., and E. G. Hertwich. 2010. "Implementing Carbon-Footprint-Based Calculation Tools in Municipal Greenhouse Gas Inventories." *Journal of Industrial Ecology* 14 (6): 965–77. https://doi.org/10.1111/j.1530–9290.2010.00295.x.

Lesage, P., L. Deschênes, and R. Samson. 2007. "Evaluating Holistic Environmental Consequences of Brownfield Management Options Using Consequential Life Cycle Assessment for Different Perspectives." *Environmental Management* 40 (2): 323–37. https://doi.org/10.1007/s00267-005-0328-6.

Lin, C. 2009. "Hybrid Input–Output Analysis of Wastewater Treatment and Environmental Impacts: A Case Study for the Tokyo Metropolis." *Ecological Economics, Methodological Advancements in the Footprint Analysis* 68 (7): 2096–105. https://doi.org/10.1016/j.ecolecon.2009.02.002.

Loiseau, E., G. Junqua, P. Roux, and V. Bellon-Maurel. 2012. "Environmental Assessment of a Territory: An Overview of Existing Tools and Methods." *Journal of Environmental Management* 112 (December): 213–25. https://doi.org/10.1016/j.jenvman.2012.07.024.

Lund, H. 1985. *Short Reference Years and Test Reference Years for EEC Countries: Final Report*. Thermal Insulation Laboratory, Technical University of Denmark. Lyngby, Denmark.

Lund, H. 2007. *EnergyPLAN-Advanced Energy Systems Analysis Computer Model-Documentation Version 7.0-Http://Www. EnergyPLAN. Eu*. Aalborg University, Aalborg.

Mathiesen, B. V., M. Münster, and T. Fruergaard. 2009. "Uncertainties Related to the Identification of the Marginal Energy Technology in Consequential Life Cycle Assessments." *Journal of Cleaner Production* 17 (15): 1331–38. https://doi.org/10.1016/j.jclepro.2009.04.009.

Munksgaard, J., M. Wier, M. Lenzen, and C. Dey. 2005. "Using Input-Output Analysis to Measure the Environmental Pressure of Consumption at Different Spatial Levels." *Journal of Industrial Ecology* 9 (1–2): 169–85. https:// doi.org/10.1162/1088198054084699.

Peuportier, B., C. Roux, M.-L. Pannier and N. Kotelnikova. 2017. "Using LCA to Assess Urban Projects, a Case Study." *Journée Thématique "La spatialisation en ACV"*. Réseau EcoSD.

Raichur, V., D. S. Callaway, and S. J. Skerlos. 2015. "Estimating Emissions from Electricity Generation Using Electricity Dispatch Models: The Importance of System Operating Constraints." *Journal of Industrial Ecology*, April. https://doi.org/10.1111/jiec.12276.

Reap, J., F. Roman, S. Duncan, and B. Bras. 2008. "A Survey of Unresolved Problems in Life Cycle Assessment." *The International Journal of Life Cycle Assessment* 13 (5): 374–88. https:// doi.org/10.1007/s11367-008-0009-9.

Roux, C., N. Kotelnikova-Weiler, P. Schalbart, F. Leurent, and B. Peuportier. 2016. "LCA of Urban Development Projects: Coupling LCA and Transport Simulation Tools." https://hal.archives-ouvertes.fr/hal–01463775.

Roux, C., P. Schalbart, and B. Peuportier. 2016. "Development of an Electricity System Model Allowing Dynamic and Marginal Approaches in LCA—Tested in the French Context of Space Heating in Buildings." *The International Journal of Life Cycle Assessment*, December. https://doi.org/10.1007/s11367-016-1229-z.

RTE. 2012. "Bilan Prévisionnel 2012 de l'équilibre Offre-Demande: La Sécurité de l'alimentation Électrique Assurée Jusqu'en 2015." http://www.rte-france. com/fr/actualites-dossiers/a-la-une/bilan-previsionnel-2012-de-l-equilibre-offre-demande-la-securite-de-l-alimentation-electrique-assuree-jusqu-en-2015-1.

SOeS. 2016. "Bilan Énergétique de La France Pour 2015." http://www.statistiques. developpement-durable.gouv.fr/fileadmin/documents/Produits_editoriaux/Publications/Datalab/2016/datalab-bilan-energetique-de-la-france-pour-2015-novembre2016.pdf.

Suh, S., and Y. Yang. 2014. "On the Uncanny Capabilities of Consequential LCA." *The International Journal of Life Cycle Assessment* 19 (6): 1179–84. https://doi.org/10.1007/s11367-014-0739-9.

Treyer, K., and C. Bauer. 2014. "Life Cycle Inventories of Electricity Generation and Power Supply in Version 3 of the Ecoinvent Database—Part II: Electricity Markets." *The International Journal of Life Cycle Assessment*, January, 1–14. https://doi.org/10.1007/s11367-013-0694-x.

Yang, Y. 2016. "Two Sides of the Same Coin: Consequential Life Cycle Assessment Based on the Attributional Framework." *Journal of Cleaner Production* 127 (July): 274–81. https://doi.org/10.1016/j.jclepro.2016.03.089.

Chapter 2

Study of the Quantification of Uncertainties in Building Life Cycle Assessment

Marie-Lise Pannier, Patrick Schalbart, and Bruno Peuportier

MINES PARISTECH, CES – PSL

2.1 Introduction

Buildings have been identified as an important lever of action for achieving national and international objectives for the environment. Hence, there is a need to act on this sector, which is at the heart of our way of life. To limit the environmental impacts of the building sector, it is necessary to apply an eco-design approach to buildings, whether they are new or are being renovated. This approach must make it possible to take into account all the potential impacts occurring during the life cycle of the buildings, in order to avoid shifting the emissions of pollutants from one environmental problem to another, from one place to another or even from one stage of the life cycle to another. Life cycle assessment (LCA) is particularly well suited for doing this. In addition to evaluating the environmental performance of a product, LCA offers the possibility of comparing the environmental impacts of products with the same function over their entire life cycle, and thus helps decision-makers to choose the design alternatives that are the most environmentally friendly.

In order to guide choices towards the most sustainable built alternatives, reliable and robust tools are needed. However, environmental modelling of buildings using LCA requires a large number of hypotheses to be made, related to the building and the possible evolutions that may take place during its long life cycle, and related to the environmental data used for the evaluation. Uncertainties affect all of these elements (Huijbregts, 1998; Björklund, 2002), potentially changing results and decisions based on LCA. This may call the reliability of the method into question, as the uncertainties are considered to be significant (Heijungs and Huijbregts, 2004). Nevertheless, they should not be seen as a weakness. Indeed, taking the uncertainties into account makes it possible to improve the knowledge of the impacts associated with a product, to identify the key points on which further research should be carried out (Heijungs, 1996), to improve the understanding of the model's behaviour (Lacirignola et al., 2017) and to assist the decision-making (Andrianandraina, 2014; Muller, 2015). To disregard uncertainties is to dispense with information that is rich and useful for interpreting LCA results.

The origins and effects of uncertainties on the results should be studied in order to increase the confidence of the user in LCA. Uncertainty quantification methods can be applied to better understand these uncertainties. They rely on performing a large number of simulations and on a statistical treatment of the results, to determine the uncertainty level of the outcome or knowledge of the factors having the most influence on the results. Thus, taking uncertainties into account leads to a change of practices by moving from a deterministic approach (carrying out a simulation of the life cycle to get an impact value for each indicator) to a statistical approach (implementation of numerous simulations to get a distribution of the impact values).

The LCA community has recognised the need to study the effects of uncertainties on the results. However, uncertainties and the ways to take them into account have not often been addressed in building LCA. The uncertainty quantification methods to be applied in order to improve the quality of decision-making must be adapted to the context of LCA of buildings. It is essential that they can be applied to the complete life cycle of buildings, including the use phase – usually the major life cycle contributor – and the temporal dynamics associated with it. These methods must make it possible to take into account the numerous sources of the uncertainties affecting the results, and also the multi-criteria nature of the LCA. In order to help in choosing the most sustainable alternative, these methods need to be adapted so that they can characterise the uncertainties in a context of alternative comparison. Finally, given the limited time that is typically dedicated to LCA in the design process, these methods must give accurate results in terms of uncertainty quantification but they must also have a low computation time.

This chapter presents a contribution to the implementation of a methodology that makes it possible to take uncertainties into account in the context of the building LCA. The objective is to progress towards more reliable building LCA tool, (i) by studying the main sources and effects of uncertainties on results, and (ii) by taking advantage of uncertainties to provide additional information that can help in choosing the most sustainable alternatives. For more details on the methods and assumptions, the reader can refer to the thesis by Pannier (2017), on which the work presented in this chapter is based.

2.2 Tools for Taking Uncertainties into Account

2.2.1 Vocabulary and Notations Related to Uncertainties

Physical phenomena are transcribed into numerical models using uncertain K factors, $x = (x_1 \ldots x_k)$ equations of varying complexity, f, and providing one or more outputs, Y, such that $f(x) = Y = (y_1, \ldots, y_G)$, with G, the number of outputs. In building LCA, the numerical model corresponds to the modelling of one or more buildings carried out thanks to specific simulation tools.

Uncertain factors constitute information needed for numerical simulation. They group input variables and model parameters. The input variables are the boundary conditions of the system being studied, like the solicitations affecting the building (heating power, meteorological variables, occupancy scenarios, etc.). The parameters are divided into design parameters and internal model parameters. The design parameters require a choice (explicit or implicit) of modelling tools (geometry, materials, etc.) made by the user. The internal parameters of the model are linked to modelling assumptions made by the modelling tool designer (correlations, databases, …) and, by default, are not accessible to the user of these tools. The uncertain factors take different forms: they can be scalar, vectorial or matrix; stochastic or static; continuous, discrete or categorical (discrete and unordered).

The outputs are obtained by numerical simulation. In the case of LCA, they correspond to the environmental impacts, that is problems or damage caused to the environment during the life cycle of the building. The outputs that are exploited in a study are called "quantities of interest".

Several sources of uncertainty affect the quantities of interest. They reflect the lack of knowledge of a system or the stochastic variations of the system (Björklund, 2002). These two notions refer to the concepts of uncertainty and variability, which are defined as follows:

*   Uncertainty reflects a lack of knowledge about the exact value of the model's parameters. By improving our knowledge of the system being studied, it is possible to reduce the uncertainty and to know the output(s) of the model more accurately (Figure 2.1a and c), that is without bias, and more precisely (Figure 2.1a and b), that is with a low dispersion of values.
*   Variability relates to the stochastic variation or natural heterogeneity of the model's input variables. It is not about reducing it but rather of improving the knowledge of it by using more complete samples to determine the dispersion of values of the quantity.

In the following, we will use the term "uncertain factors" to refer to all the elements resulting in uncertainties in the outputs of a model.

2.2.2 Uncertainty Quantification Method

Uncertainties can be quantified using statistical tools: uncertainty analysis (UA) and sensitivity analysis (SA). These complementary analyses make it possible to answer various questions concerning the sources and consequences of the uncertainties within a model.

The objective of UA is to quantify the level of uncertainty in the outputs of the model caused by the model's uncertain factors; the uncertainties are then propagated through the model to know the range of the possible output variations. In applying this UA approach, decision-making will be based on more precise knowledge of the environmental impacts of the product under consideration.

SA corresponds to "the study on how the uncertainties in the output of a model [...] can be attributed to different sources of uncertainty in the input factors of the model" (Saltelli et al., 2004). The objective is to prioritise the uncertain factors according

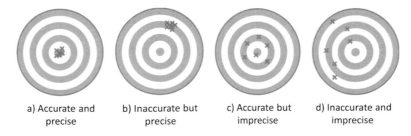

a) Accurate and precise   b) Inaccurate but precise   c) Accurate but imprecise   d) Inaccurate and imprecise

Figure 2.1 Accuracy and precision.

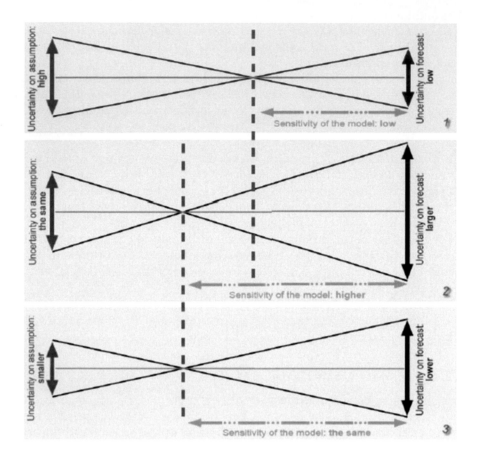

Figure 2.2 Relationship between sensitivity and uncertainty (Chouquet, 2007).

to their importance (Faivre et al., 2013). This makes it possible to identify the key points on which further research should be carried out (Heijungs, 1996), to improve the understanding of the model's behaviour (Lacirignola et al., 2017) or to help the decision-making (Andrianandraina, 2014).

The link between uncertainty and sensitivity has been illustrated by Chouquet (2007) and is presented in Figure 2.2. For a given uncertainty on a factor, the uncertainty on the output is all the more important as the model is sensitive to this factor (cases 1 and 2). Moreover, for a fixed model sensitivity (regarding one factor), the greater the uncertainty on the factor, the higher the output uncertainty (cases 2 and 3).

2.2.2.1 Uncertainty Analysis (UA)

2.2.2.1.1 UA METHODS

In UA, factor uncertainties are propagated through the model to determine the range of possible output values. Several UA methods can be used, and they are listed in Table 2.1 according to the approach considered. The so-called internal approaches

Table 2.1 Uncertainty Analysis Methods

Approach	Methods
Internal	Analytical uncertainty propagation, Taylor series expansion Interval analysis: interval propagation Possibility theory: fuzzy sets
External	Methods based on sampling: random (Monte Carlo), Stratified (Latin hypercube sampling), Importance sampling (low-discrepancy sequences) Reliability theory: FORM and SORM

are intrusive. They require the computer code to be modified in order to propagate the uncertainties. However, once implemented, the results can be obtained in one simulation. External approaches do not require a change in the computer code. They are based on:

1. The characterisation of the uncertainty level on each factor using probability distributions that are constructed from surveys, literature reviews or expert judgement.
2. The sampling within these distributions, that is drawing a large number of probable value sets for the uncertain factors (all of the sets constituting a design of experiment or DoE).
3. The calculation of the model outputs for each set of uncertain factor values.
4. The determination of the uncertainty in the output(s) (determining the probability distributions of the outputs).

External approaches are applied when the operation of the model is not accessible or if it is not possible to modify it because of its complexity. The first step in the definition of the probability distribution is very important because it will affect the sampling and thus the uncertainties obtained at the output. However, it can be tricky to define distributions for uncertain factors. The sampling step must allow for good coverage of the factor variation space. The available sampling methods can be distinguished by the way they explore the space and by their speed of convergence.

2.2.2.1.2 APPLICATION TO BUILDING LCA

Analytical uncertainty propagations have been used in building LCA to determine the share of uncertainties associated with building materials (Hoxha et al., 2017). However, the method only applies to linear models and does not take energy consumption into account, although it is influenced by the choice of materials. In most cases, uncertainty propagation methods based on random sampling have been applied to quantify the uncertainty of the model output (Verbeeck and Hens, 2007; Wang and Shen, 2013; Pomponi et al., 2017).

LCA is often used in a comparative context. However, in the presence of uncertainties, the comparison of alternatives is complicated. Indeed, the uncertainties could potentially change the ranking of alternatives. The propagation of uncertainty then has real meaning in LCA in order to study the possible alternative ranking changes, and to guide the decision-maker towards the most sustainable alternatives in the presence of uncertainties. Comparative studies assessing two alternatives have been conducted in building LCAs (Huijbregts et al., 2003; Hoxha et al., 2014; Heeren et al., 2015).

2.2.2.2 Sensitivity Analysis (SA)

2.2.2.2.1 SA METHODS

SAs are the statistical methods that simplify or allow for a better understanding of numerical models by prioritising uncertain factors according to their influence, and in some cases to their interactions with other factors. The calculation of sensitivity indices (SIs) for each factor and each quantity of interest of the model allows the factors to be ranked according to their influence. The higher the value of the SI, the more influential the factor. Many SA methods exist, and the choice of one over another is based not only on the objectives of the study (Saltelli et al., 2004) but also on the complexity and regularity (linearity, continuity) of the model (Iooss, 2011).

The steps for carrying out an SA are the same, regardless of the methods. They consist of:

1. Characterising the uncertainty of the uncertain factors (definition of probability distributions or ranges of variation of uncertain factors).
2. Generating uncertain factor value sets by random draw or by using DoE.
3. Calculating model outputs for each set of uncertain factor values.
4. Calculating SIs to find out the influence of the factors.

For most SA methods, the number of model simulations to be carried out directly depends on the number of uncertain factors taken into account. In addition, some methods require a greater number of simulations to be carried out in order to compute the SI calculations. Therefore, computing time can increase considerably in SA.

A review of the SA method was proposed by Iooss (2011) and then detailed in a book published by the MEXICO network – Méthodes pour l'EXploration Informatique de modèles COmplexes (Methods for the Numerical Exploration of Complex Models, Faivre et al., 2013). The existing SA methods are summarised in Table 2.2.

In local SAs, small perturbations are applied around the factor reference values. The computation of the partial derivatives near the reference point enables the model sensitivity to be determined when there is a small variation in the factor value. SIs are calculated for each uncertain factor. While easy to implement, local SAs do not take into account the effects of variations on a large area. Moreover, these methods require model linearity assumptions (Iooss, 2011) and do not provide insights into the interactions between uncertain factors (Tian, 2013). Screening or global SA methods have therefore been developed to address these issues.

The purpose of screening is to explore the behaviour of a model quickly through discretisation of the variation range of uncertain factors into a number of finite points called "levels". Screening is applied for models with many uncertain factors: from a few dozen to a few hundred (Faivre et al., 2013). Often, screening methods are used

Table 2.2 Sensitivity Analysis Methods

Type of SA	Approach	Method
Qualitative SA	Graphical	Scatterplot, Cobweb plot
Local SA	Local	Partial derivatives
Screening	High-dimensional screening	Supersaturated designs
		Screening by group, sequential bifurcations
	Design of experiment (DoE)	Full factorial design
		Fractional factorial design: OAT[a], simplex designs, Plackett and Burman designs, centred composite DoE, Box and Behnken, hybrid DoE.
	Elementary effects	Morris method
Global SA	Regression	Linear regression: Pearson, SRC, PCC
		Rank-based regression: Spearman, SRRC, PRCC
	Statistical tests	Based on averages, medians, variances, regionalised SA
	Based on functional decomposition	Calculation of Sobol indices
		Periodic sampling: FAST, E-FAST, RBD-FAST, EASI, COSI
Deep model exploration	Smoothing methods	Smoothing splines, local polynomials (LPs).
	Metamodel	Polynomial chaos, kriging.
Other	Reliability theory	FORM and SORM

[a] OAT: one-factor-at-a-time.

prior to a more precise SA in order to eliminate the least influential factors and thus reduce the number of simulations and calculation times for these future analyses.

Finally, global SAs allow us to explore the entire variation space of uncertain factors (Saltelli et al., 2008). For this type of analysis, it is therefore necessary to know the probability distributions in addition to the variation range of the factors. Global SAs are designed to measure the importance of uncertain factors more precisely, and to do this, it is necessary to have a sufficiently large output sampling. Within global SAs, variance-based methods can be used to determine the influence of factors by means of SIs, which can represent the proportion of the variance attributable to a factor (first-order index), or the proportion of variance attributable to a factor and the effects of its interaction with other factors (total indices). These latter methods are recognised and often applied because of their accuracy. Moreover, they are valid whatever the behaviour of the model.

2.2.2.2.2 APPLICATION TO BUILDING LCA

In building LCA, the influence of factors has mainly been studied using regression methods or local SAs (Huijbregts et al., 2003; Verbeeck and Hens, 2007; Peuportier et al., 2013; Suh et al., 2014; Heeren et al., 2015). Many of these studies identify the lifespan of building components and occupant behaviour as two influential uncertain factors. Other factors are also influential in dynamic energy simulation (ventilation, material thickness, heating system).

The LCA community has taken a strong interest in SA by providing guides for good practice (Cucurachi et al., 2016; Lacirignola et al., 2017) and studies of the performance of methods (Groen et al., 2016).

2.3 Presentation of the Case Studies

The uncertainty quantification studies detailed below are based on two of the passive houses of the INCAS platform, located on the INES site at Le Bourget-du-Lac. The first one is a shuttered concrete house called "IBB" (concrete envelope and external insulation with extruded polystyrene), and the second one is a wooden-frame house called "IOB" (wood wool for insulation). Their geometries are almost identical, but only their constructive systems differ (Figure 2.3).

2.3.1 Modelling Tool

The Pléiades software suite was used for the environmental modelling of the building. The Pléiades ACV building LCA tool (Polster, 1995; Popovici, 2005) is coupled with the DBES tool (Dynamic Building Energy Simulation) Pleiades+COMFIE (Peuportier and Blanc-Sommereux, 1990). This allows the energy consumption taking place during the use phase of the building to be determined more precisely.

Methods for quantifying uncertainties require many energy and environmental model simulations. The use of a multiple simulation platform, undertaken by the statistical tool R, facilitates the automated management of these calculations, and the statistical analysis of UA and SA results.

2.3.2 Modelling Hypotheses

The objective of this LCA is to study the design of a detached house by comparing several constructive alternatives, in order to determine the most sustainable alternative. A consequential-oriented approach is preferred for design purposes, as it makes it possible to evaluate the effects of design choices on the system and its context. The methodological framework of the "consequential project" approach presented in Chapter 1 was therefore followed.

The functional unit is a detached house of 90 m² studied over 80 years. This house is inhabited and the occupation corresponds to the average occupancy in France for a house with four main rooms. The comparison is based on the equivalent thermal resistances for the insulation material in order to study the constructive systems, all things being equal elsewhere.

In this study, the system boundaries correspond to the envelope, and the networks and systems necessary for the building use. We consider the complete life cycle of the building, including the building materials, their manufacture and transportation to

Figure 2.3 Houses of the INCAS platform (IBB (a), double wall constructive system (IDM) (b), IOB (c) – not presented in this chapter).

the site, their potential replacement during the life of the building and their disposal. Energy and water consumption, as well as the equipment necessary for the building use (electricity grid and water network transmission), are taken into account in the building-use phase. Furthermore, the occupants' transport and the household waste produced during the use phase are not included. Indeed, the goal is not to compare different sites for the house, nor to study the consequences of the occupation, but to study the effects of the design choices of the building.

The inventory of construction materials is based on the precise descriptions provided by several project deliverables ANR Maison Passive, ANR HABISOL-SIMINTHEC and ANR Fiabilité. A 5% material waste rate is taken into account to represent the losses during construction.

In the use phase, data on energy consumption are simulated using Pleiades+-COMFIE. Meteorological data from the city of Mâcon are used. The occupancy (presence of residents, temperature setpoint and internal gains) corresponds to the average occupancy in France for a house with four main rooms. The double flow-controlled mechanical ventilation provides a fresh air flow of 110 m^3/h, and the 0.9 efficiency heat exchanger is bypassed in the summertime. The heat balance is based on a multi-zone model where all the rooms are in different thermal zones. The result of this modelling shows the annual heating loads to be around 15.5 kWh/m^2 for IBB and 16.1 kWh for IOB. Electricity is used for air heating and Domestic Hot Water (DHW) production. The impacts of electricity generation are characterised using a dynamic approach: the consumption during one hour of the year is multiplied by the impacts of the electricity generation mix during the same hour. For water consumption during the usage phase, a volume of 137 l/day, of which 34 l is DHW, is considered per occupant (Vorger, 2014) and the efficiency of the water network is set at 74%.

In terms of renovation, joinery (doors and windows) are changed every 30 years, flooring, coverings and paints every ten years and technical equipment every 20 years. Depending on their nature, end-of-life materials are landfilled, incinerated or recovered (energy or material recovery). In the latter case, they avoid the use of new materials or energy, and the avoided or generated impacts of the energy recovery are recorded in the environmental report, in accordance with the framework of the consequential project approach.

Once the materials and processes used during the building lifetime are inventoried, environmental data from the ecoinvent database are associated with them. This makes it possible to calculate the environmental indicators.

The modelling assumptions and choices identified above lead to uncertainties. Their effects on the results can be studied by applying the methodology to take into account uncertainties in building LCA, which is presented in the following paragraph.

2.4 Identification of Influential Uncertain Factors

SA methods are used in order to determine the most influential factors in building LCA. Knowledge of the factors' influence makes it possible to determine on which ones to act in order to reduce the uncertainty of the results or to improve the knowledge of their variability. This also allows LCA practitioners to limit the data collection time as they can focus their efforts on these influential factors during the inventory phase.

Several SA methods are available. Each has different computational cost/accuracy trade-offs and sometimes only apply to certain categories of models (depending on the linearity, monotony and the presence of interactions). Within the framework of building LCAs, it is necessary to consider the dynamic aspects related to the thermal behaviour of the building in order to increase the accuracy of the studies. However, this results in potentially long computation times and potentially uses nonlinear models, in which several factors can interact. As the time available for carrying out a study is quite short in the design step of a project, accurate methods that give results quickly are required.

In order to identify the SA methods that have a good computational cost/accuracy compromise in the context of LCA of buildings, six methods were applied to the case of the IBB house, considering a set of 22 uncertain factors. One of the SA methods serves as a reference and allows the accuracy of the other methods to be studied according to different criteria.

2.4.1 Sensitivity Analysis Methods Studied

Six of the SA methods presented in Table 2.2 were applied. They were chosen as they represent a wide range of SA methods (three screening methods, two global SAs and one smoothing method) and because they can be used in the presence of discrete factors. The functioning of the selected methods is presented in Table 2.3.

The number of simulations to be carried out depends on the DoE associated with each method, but it is also adjusted to obtain stable SIs. Thus, the confidence intervals on the SIs were determined using a *bootstrap*[1] method on the number of repetitions for PB and Morris, and on the sample size (number of sets of values in the sample) for global SAs. This makes it possible to determine the settings of the methods offering results with the greatest precision. Table 2.3 also summarises the settings used in the comparative study: number of levels, number of repetitions, type of sampling and sample size.

The Sobol index calculation method was defined as the reference method for studying the performance of SA methods. Indeed, SA methods based on variance were preferred for quantifying the importance of uncertain factors. In addition, Sobol indices give the most accurate results among the variance-based methods (Saltelli and Bolado, 1998).

2.4.2 Criteria for Comparing Methods

In order to compare the accuracy of the methods, the relative influence *RI* of a factor for a given output is calculated. RI is defined as the ratio of the SI of one factor to the sum of the SIs of all factors:

$$RI_i = \frac{IS_i}{\sum_i IS_i}. \tag{2.1}$$

---

1 The bootstrap corresponds to the creation of a new sample by sampling with replacement in an existing sample. It allows a new set of values to be obtained, which are used to calculate sensitivity indices without the need to perform new simulations of the model.

Table 2.3 SA Methods Applied. In the Graphics, One Point Corresponds to One Simulation

Type	Method	Drawing	Characteristics	Sensitivity Index of Factor i
Screening	MMSA DoE (SA minimum - maximum)	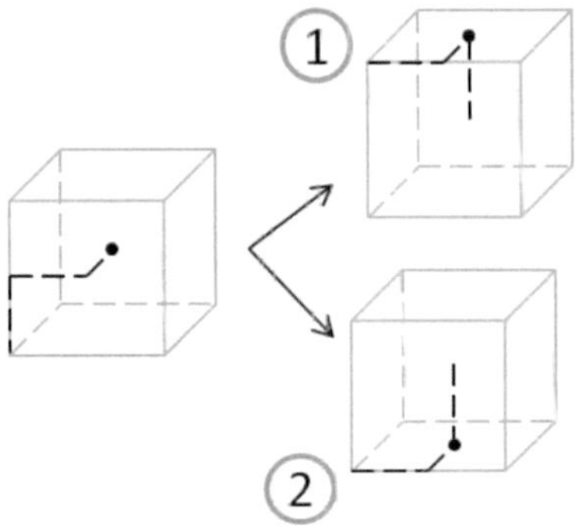	OAT DoE: change of the value of only one factor at a time. Calculation of the effect, on each quantity of interest of the model, of jumping from one boundary (minimal value) of the variation space to the other (maximal value), while keeping all other factors at their reference values. Limited exploration of the variation range of the factors.	$S_i = f(1) - f(2)$
	Plackett and Burman (PB) DoE (1946)	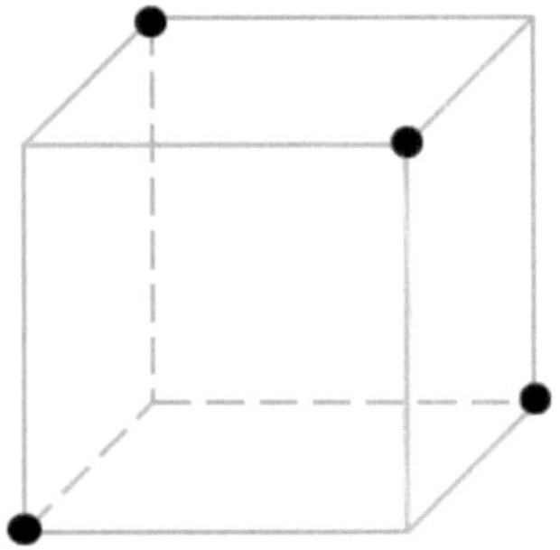	Specific DoE. Change the value of all factors at each simulation. Repetition of PB DoE (changing the order of factors) to avoid aliasing the effect of one factor with the interaction effect of two other factors.	Si obtained by analysis of variance • 20 repetitions of PB DoE

(Continued)

Table 2.3 (Continued) SA Methods Applied. In the Graphics, One Point Corresponds to One Simulation

Type	Method	Drawing	Characteristics	Sensitivity Index of Factor i
	Morris (1991) method	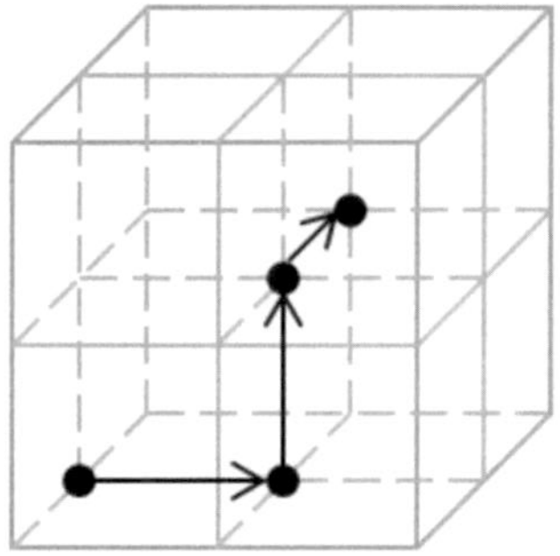	Specific DoE: repetition of OAT DoE. At each OAT, starting from an initial simulation (random value of all factors), only the value of a factor is modified at each simulation. Calculation of EE: effect on the output of the change in the value of a factor. Mean μ_i^* of $|EE|$ of factor i on trajectory = importance of factor standard deviation σ_i of the EE = nonlinear effects or interactions	$d_i^* = \sqrt{\left(u_i^{*2} + \sigma_i^2\right)}$ Combined nonlinear and interaction effects • 6 levels • 50 repetitions of OAT DoE
Global SA	Linear regression SRC	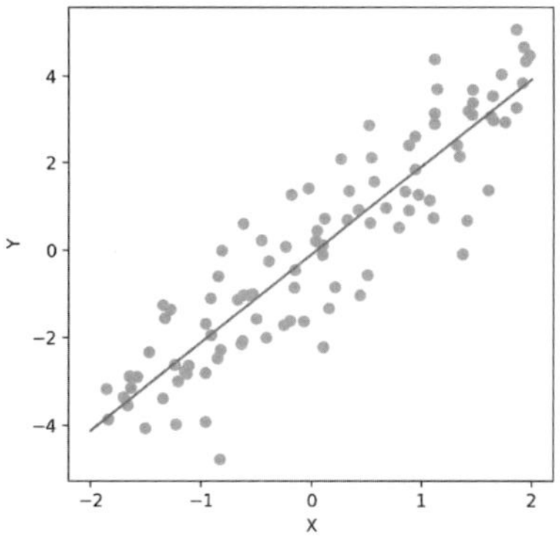	Not specific DoE. Linear regression based on uncertainty propagation. Approximation of a linear model: $$Y = \beta_o + \sum_{i-1}^{K} \beta_i X_i + \varepsilon$$	• Random sampling of size 2,000

(Continued)

Table 2.3 (Continued) SA Methods Applied. In the Graphics, One Point Corresponds to One Simulation

Type	Method	Drawing	Characteristics	Sensitivity Index of Factor i
	Calculation of Sobol indices (1993)	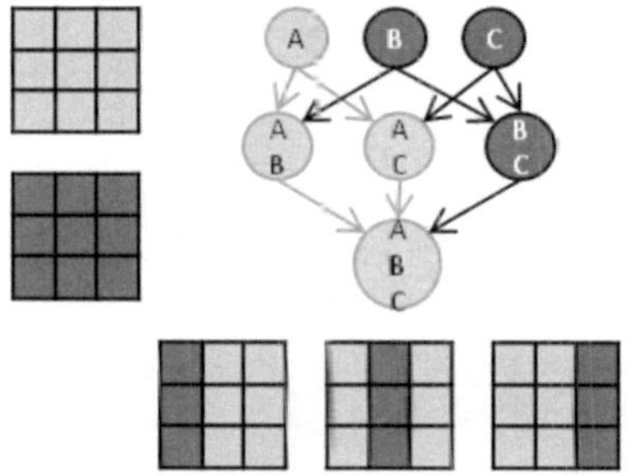	Specific DoE: two independent samples 1 and 2 are created by drawing in the distributions of K factors. K other samples are then constructed. They are identical to 1 apart from the values of the ith factor, which comes from 2. The cross-study of the results of simulations of the K + 2 samples leads to the first-order indices (study of the output variation when the value of all the factors except the ith change) or total indices (study of the variation when the value of the ith factor changes, the value of the others being unchanged).	TSi: total indices of Sobol-Jansen (1999) • Sampling by Latin hypercube of size 5,000
Deep model exploration	Smoothing based on an LP technique (Veiga et al., 2008)	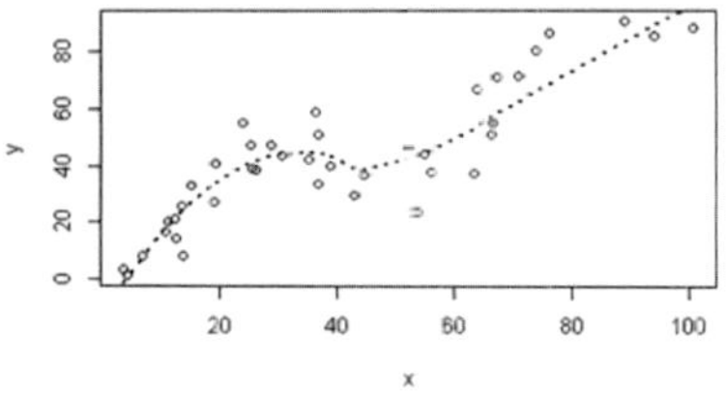	Not specific DoE LP regression based on uncertainty propagation (first sample). Use of the regression model (instead of the physical model) to simulate the sets of factor values of a second sample and to estimate the first and total Sobol indices.	$\widehat{s}_i^{(1)}$ or $\widehat{s}_i^{(2)}$: first-order indices • Sampling with Sobol sequences of size 1,000

Table 2.4 Factors and Distributions Considered

Uncertain Factors			Reference Value	Probability Distribution	Unit
Parameter uncertainty	Envelope	Insulation material thickness (in walls)	20 cm	N (μ =20; σ = 0.25; min = 19; max = 21)	cm
		Concrete thickness (in walls)	15 cm	N (μ =15; σ = 0.25; min = 14; max = 16)	cm
		Windows U-value	1.27–1.45 W/ m²/K depending on the window	Nr (μ =0; σ = 2.5; min = −30; max = 30)	%
		Windows solar factor	0.6 for double and 0.45 for triple glazing	Nr (μ =0; σ = 2.5; min = −30; max = 30)	%
		Thermal bridges	Values from the French thermal regulation	Nr (μ =0; σ = 25; min = −75; max = 75)	%
		Construction material waste	5%	N (μ =5; σ =3; min = 0; max = 15)	%
	Lifetime[a]	Window lifetime	35 years	N (μ =35; σ =12; min = 5; max = 100)	years
		Building finishes (covering) lifetime	22 years	N (μ 22; σ =7; min = 5; max = 200)	years
	Site	Albedo	0.35	Ur (a = −15; b = 15)	%
		Construction material transportation	75 km	N (μ =75; σ =30; min = 0; max = 200)	km
		Waste material transportation	25 km	N (μ =25; σ =10; min = 0; max = 100)	Km
		Water network efficiency	74%	N (μ =74; σ =13; min = 37; max = 100)	%
Model uncertainty		Electricity mix	-	C [yearly; hourly]	-
Uncertainty due to choice		Building lifetime	80 years	N (μ =80; σ =20; min = 40; max = 500)	years

(Continued)

Table 2.4 (Continued) Factors and Distributions Considered

Uncertain Factors			Reference Value	Probability Distribution	Unit
Temporal variability	Climate	Outdoor temperature	Hourly climate file of Mâcon from the French thermal regulation (DJ18 = 2022)	Nr (μ=0; σ = 0.25; min = −3; max = 3) Deviation to the outdoor temperature, constant at all time steps	°C
		Global horizontal radiation		Nr (μ =0; σ =5; min = −20; max = 20)	%
	Scenario	Indoor temperature setpoint	Hourly typical scenarios for a dwelling with four main rooms	Nr (μ =0; σ = 0.25; min = −3; max = 3) Deviation to the temperature setpoint, constant at all time steps	°C
		Number of occupants		Nr (μ =0; σ =5; min = −60; max = 60)	%
		Internal gains (appliances)		Nr (μ =0; σ =5; min = −75; max = 75)	%
		Ventilation rate	110 m³/h	Nr (μ =0; σ =5; min = −30; max = 30)	%
	Choice	Time horizon (IPCC)	-	C [20 years; 500 years]	-
Variability between source and object		Concrete type	-	C [normal; exacting]	-

N = normal distribution, U = uniform distribution and C = categorical variable, r = relative variation, min and max = upper and lower boundaries of the distribution.
[a] Extrapolation of data from the INIES database (accessed in September 2015).

The ranking of factors according to their importance was also studied. We also propose determining the most relevant group of influential factors, that is the one for which further research must be carried out prior to complementary LCA studies in order to better define their values or distributions. To this end, the factors are ordered according to their RI and a group of factors that explain a certain part of RI are identified. In this way, it is possible to observe whether a method selects too many factors compared to the reference one or whether, on the contrary, it fails in the selection of influential factors.

The calculation time required for the SIs is also a criterion to take into account. This time corresponds to the model simulation time, depending on the number of uncertain factors for the methods requiring a particular experimental design, but also on the calculation time required for the SA methods for computing the indices, which directly depends on the number of uncertain factors.

2.4.3 Uncertain Factors Taken into Account

The study of the influence of factors was conducted on a set of 22 uncertain factors listed in Table 2.4. The classification of the factors adopted refers to the usual LCA classification: distinction between parameter uncertainty, model uncertainty and uncertainty due to choice, and between spatial variability, temporal variability and variability between sources and objects[2] (i.e. variability in the performance and environmental characteristics of equivalent products) (Huijbregts, 1998; Björklund, 2002).

The factors relate to the aspects of the building energy or building LCA. In most cases, they describe uncertainties related to assumptions about the building or its context, but also uncertainties about the methodology, the life cycle inventory or the life cycle impact assessment. The uncertainty on these factors was characterised based on data from the literature. Generally, a low uncertainty was considered for the factors reflecting energy aspects because of the good knowledge of the building's characteristics.

In the table, the uncertainties are presented as relative variations (indicated by the index "r") or as absolute variation around a reference value. The probability distributions are:

- Truncated normal distributions N, defined by expected, standard deviation and the minimum and maximum values that can be taken by the factors.
- Uniform distributions U, defined by the boundary of the distribution.
- Discrete distributions C, defined by the values that can be taken by the categorical variables.

For three of the factors – the building lifespan, water network efficiency and time horizon for global warming potential (GWP)[3] – the probability distributions selected are presented in Figure 2.4. A truncated normal distribution is used for the first two factors.

---

2 This corresponds to the variability that appears between the materials used in a building and the materials whose inventory is used to calculate the environmental impacts, for example. For example, the composition of concrete used in the building may be different from that taken into account in the concrete environmental data considered in the LCA.

3 Since the greenhouse effect depends on the rate at which the gases degrade in the atmosphere, the chosen time horizon varies with the results.

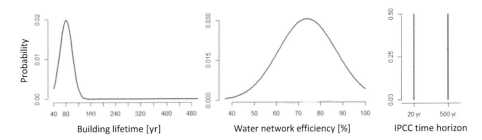

Figure 2.4 Probability distributions for the building lifetime, water network efficiency and time horizon for GWP.

Table 2.5 Number of Simulations and Calculation Times

	MMSA	PB	Morris	SRC^2	LP	Sobol (Ref.)
Number of simulations	42	480	1,150	2,000	1,000	120,000
Computation time	4 minutes	30 minutes	1:20	2:20	1:50	180 hours

Truncation avoids extreme or unrealistic values being drawn during sampling. For the third factor, the two values that can be taken by the categorical variable are presented with the same occurrence probability.

2.4.4 Results

The simulations were conducted on a powerful desktop computer with 6 cores and 16 GB of memory. The number of simulations to be performed (which depends on the method and its setting) and the computation times (simulation time and time for evaluating the indices) are summarised in Table 2.5.

If computation times alone were taken into account, the choice would lead to the MMSA method. In order to compare the precision of the methods with the reference method (Sobol), the RIs of the factors for the six methods are presented in Figure 2.5 for two environmental indicators: climate change and cumulative energy demand (primary energy). In this figure, each bar represents the results of an SA method. The RI of each factor is presented using a different pattern in each bar.

For the climate change indicator, the most influential factors are the building lifetime, the GWP time horizon and the electricity mix. For primary energy, the building lifetime and some factors influencing the energy balance (thermal bridges, occupation) appear to be influential. Regarding all the environmental indicators studied,[4] the variability of the model is mainly explained by the electricity mix and the building lifetime.

The results show that the uncertain factors identified by the reference method (Sobol) are also identified by the other methods. However, their RI may vary. Overall, global and smoothing methods give very similar results in terms of the RI and ranking

---

4 The set of indicators studied is that implemented in the Pléiades LCA eco-design tool.

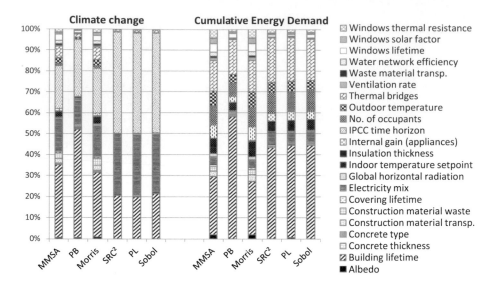

Figure 2.5 Comparison of the RI of uncertain factors for the six SA methods.

of factors. Differences in the RI given to each factor may result in changes in the rank-ing for the screening methods, however. For example, this is the case for the climate change indicator, for which the time horizon is the most influential factor for global methods, whereas this factor is ranked after the building lifetime for the screening methods. In addition, the study of the group of factors to be used to reach a certain part of the RI (e.g. 80%–95%) shows that screening methods tend to select a larger number of factors. Collecting additional data to better describe these particularly in-fluential factors will then require more time for LCA practitioners.

The comparison of the first order and total indices for the Sobol method, as well as the study of the validity of the model's linearity assumption for the SRC method, showed that the model presents few interactions and is fairly close to the linearity. This can be explained by the low variation ranges considered for some of the factors, inducing nonlinearities. This would no longer be the case if the sensitivity study was taken to optimise the design (large variation ranges).

The differences observed between the results of the six SA methods can be explained by the nature of the methods used. Interactions and nonlinearities are not quantified with all methods. The way of quantifying the uncertainties differs. The number of levels is not always the same in the screening methods. Finally, the factor distribu-tions change according to the type of methods because of the implicit assumption of uniform distributions in the screening methods. The observed differences are summa-rised in Table 2.6.

In order to improve the performance of the screening methods, adaptations were carried out on two of them: PB and Morris. For PB, the number of levels was increased from 2 to 6. Repetitions of the DoE were carried out to overcome the aliasing effects and for a good coverage of all the factor levels. For Morris, adaptation consists of calculating a variance instead of an effect. For this, the square of the difference of the

Table 2.6 Study of Differences between SA Methods

	Interactions and Nonlinearities	Quantification of Uncertainty	Level	Underlying Distribution
MMSA	No	Effect	2	Uniform
PB	No	Variance	2	Uniform
Morris	Yes	Effect	6	Uniform
SRC²	No	Variance	∞	Normal
LP	No (because first-order indices are computed)	Variance	∞	Normal
Sobol (Ref.)	Yes	Variance	∞	Normal

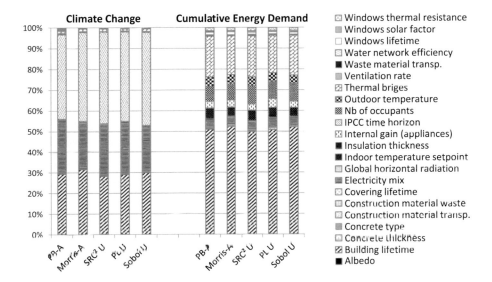

Figure 2.6 Comparison of the RI of uncertain factors for the adapted methods.

output values is calculated at each change in the value of a factor, instead of the elementary effect (EE). This metric is then used as an SI. The adaptations tend to increase the computation times of these methods, which exceed 1:25 and 2:40, respectively, for the PB adaptation (noted as PB A) and the Morris adaptation (noted as Morris A). The results in terms of RI are shown in Figure 2.6.

The adaptations of the Morris and PB methods strongly affected the results and make it possible to obtain RIs close to those given by the Sobol method but with a reduced computation time. Moreover, for the four methods studied, the ranking of the factors and the group of factors to be used to explain a certain proportion of the RI are always very close to those of the reference method.

The remaining differences observed between the screening methods and the other three methods can be explained by the difference in factor distributions. This has been established by performing additional simulations in which uniform distributions are taken into account in global SA and smoothing methods. An additional adaptation of the two screening methods is envisaged to allow for the management of non-uniform distributions.

In this case study, it is thus possible to obtain results of the same order of precision as with the reference method, with greatly reduced computation times.

2.4.5 Assistance with the Choice of Methods

Although the standard versions of the six SA methods studied give different RIs to the factors and rank them differently, they all highlight the most influential factors. Indeed, the factors identified by the reference method are almost always indicated by the other methods. Thus, if the aim of a study is to determine which factors have the most influence on the outputs, the quick, less precise methods (MMSA, PB) may be sufficient. Since the quickest methods tend to select a larger number of influential factors, attention should be paid to the amount of time spent to collect additional LCA data to improve the characterisation of the distributions of these uncertain factors.

If it is necessary to have a precise knowledge of the influence of the uncertain factors, the use of the Sobol method is preferred, despite the high calculation time required. However, the SRC^2, smoothing or adapted methods allow rather precise results to be obtained regarding the set of influential factors to reach a certain part of the variance and regarding the ranking of the most influential factors, within a reduced computation time. Another option may be to limit the computational cost of the Sobol method. This involves employing methods with a lower cost upstream of the Sobol method, in which only the most influential factors are included.

Many of the methods used are valid for linear models and do not allow the effects of interactions to be quantified. They should therefore be used with caution in the case of a nonlinear model or a model with strong interaction effects. In our case, the LCA model of the buildings is not linear. However, we considered small variation ranges on the factor of the energy model that induces nonlinearities. Thus, the model comes close to a linear model, and the interactions effects are low; however, this would doubtless no longer be valid for wider variation ranges, or if other factors are taken into account in the design optimisation phase (the interaction between thermal inertia and solar opening, for example). Given this approximation of the model's linearity, methods such as PB A and SRC^2 make it possible to evaluate the variances of the factors quite precisely. In practice, if a building is well known, the linearity assumption can be made and methods valid for linear models can be used. In addition, if the model is well known, and if the interactions between factors are perceived to be weak, methods that do not quantify the interactions can be used. In the opposite case, for less well-known buildings (e.g. those with large ranges of variation on several of their factors), the linearity assumption or that of weak interactions may no longer be valid. The performance of PB A and SRC^2 still need to be studied in this context, where Morris, Morris A or PL could be more suitable for higher-order indices.

Based on the recommendations for the choice of methods in the field of statistics (Iooss, 2011; Faivre et al., 2013) and on the results of the methods in our case study, we propose a decision tree, shown in Figure 2.7. Its purpose is to assist in the choice of SA methods in the context of building LCA. A method is chosen according to the objectives of the study, the precision aimed for and the time available to carry out the study. In addition, the calculation time associated with each method is indicated in

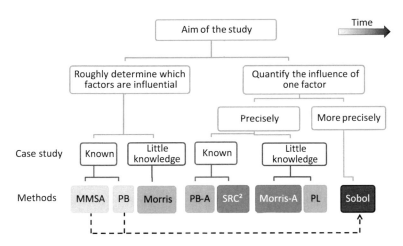

Figure 2.7 Selection of an SA method.

Figure 2.7 by a colour code. The least expensive methods in terms of calculation time are in light grey, and the more costly methods tend towards dark grey. We stress that the reliability of this tree still has to be tested with other case studies, in particular, in the presence of nonlinearities or more significant interactions.

In the early design phase, practitioners have many options for improving the performance of projects. SA studies can help determine which actions are the most effective. However, designers have little time for such studies. Thus, the SA methods that are the most expensive in terms of calculation time cannot be considered, despite their accuracy. On the other hand, in the early design stage, the characteristics of the building are not well known and wide ranges of variations need to be taken into account to define the uncertainties on the factors. In this case, it seems important to test the linearity assumption of the model before choosing the method to be used.

Lastly, we recall that only 22 uncertain factors were considered in this study. However, other sources of uncertainty affect the results in building LCA. This tends to increase calculation times, as the methods' computational cost (simulation time and calculation time of the indices) depends directly on the number of uncertain factors. Thus, in the presence of a large number of uncertain factors, it may be necessary to use methods with a lower computation time, at least as a first approach, even if they are less precise.

2.5 Comparison of Alternatives in the Presence of Uncertainties

Taking uncertainties into account provides an additional insight that can help choose a better alternative by indicating how much confidence the decision-maker can have in the results of the study. In this context, we propose a methodology to compare building alternatives in the presence of uncertainties based on the use of SA and UA. The approach is then illustrated through the case of the design of a detached house.

2.5.1 Methodology for the Comparison of Alternatives in the Presence of Uncertainties

2.5.1.1 UA in a Comparison Context

The approach often used to compare alternatives in LCA is to perform one uncertainty propagation for each alternative and present the results in the form of histograms with uncertainty bars, as shown in graphs a and b of Figure 2.8. The conclusion reached by this type of study is that the results could potentially change the conclusion in case (i), since the uncertainty bars overlap and the ranking of alternatives is retained despite the presence of uncertainty in configuration (ii), as the uncertainty bars do not overlap. However, this representation does not provide a satisfactory answer to the effect of the uncertainties on the results, as it does not take the dependences between alternatives into account. In order to solve this problem, Henriksson et al. (2015) recommend using the same DoE for the factors common to several alternatives to be compared. Thus, the alternatives are actually comparable as they are studied with all things being equal.

Moreover, studying UA results in the form of a metric reflecting the alternative comparison context (ratio or difference between alternatives) helps to understand the results. In order to determine which alternative has the lowest impacts given the uncertainties, the probability that one alternative is better than another may be studied, as in graph c of Figure 2.8. When the probability calculated for an output is close to 0 or 1, it is possible to conclude that one alternative is better than another (in the example, A is better in case 2 and B is better in case 1). If the probability is close to 0.5 (case 3), the choice of a better alternative is more difficult.

In addition, it is also important to know whether the impact values of the compared alternatives are close or not. This can be studied by representing the distribution of the impact difference as in graph d) of Figure 2.8. If the impact values are close

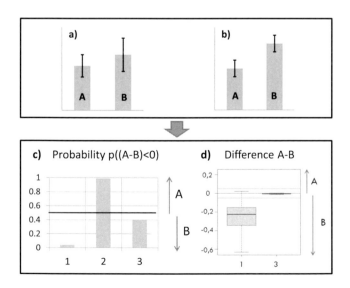

Figure 2.8 UA in a comparison context.

(A-B close to 0 as in case 3), then the choice of one alternative or another will have similar environmental consequences. In order to help choose the most sustainable alternative, Prado-Lopez et al. (2015) propose first and foremost to study the indicators for which the impact values of the alternatives under study are highly contrasted.

2.5.1.2 SA in a Comparison Context

In order to carry out an SA in an alternative comparison context, we propose following the same principles as in UA. The SAs are first conducted using the same DoE for all the factors shared by several alternatives. Then, instead of studying which factors have the most influence on the results for each alternative separately, the SA is carried out on the difference between the impact values of the alternative. This makes it possible to identify the factors influencing the comparison of alternatives and thus potentially calling the results into question.

2.5.1.3 Approach for the Quantification of Uncertainties in a Comparison Context

The five stages of the proposed approach for quantifying uncertainties in an alternative comparison context are summarised in Figure 2.9.

Firstly, the uncertain factors are identified. The uncertainties on these factors are characterised using variation range (minimum and maximum values) in order to simplify the collection of information regarding the uncertainties.

The way uncertainties are characterised can have a strong influence on the results. A step is therefore proposed that involves determining the most influential factors by performing an SA before propagating the uncertainties on each alternative. The same DoE is used for all the alternatives to be compared, and the SA is carried out on the difference between the impact values of the alternatives. Then, for the factors identified as being the most influential, the characterisation of the uncertainty is modified in order to describe the values that can be taken by these factors more realistically. The SA method to be applied can be chosen as proposed in Section 2.4.5, taking into account the number of uncertain factors and knowledge of the case study.

After having improved the distributions of uncertain factors, the uncertainties are propagated on all the alternatives to be compared using the same DoE. Based on the results of the UA, the performance of the alternatives is studied pairwise by calculating the probability that one alternative is better than another, as well as the difference between the impact values of the alternatives.

In addition, the robustness of the results can be studied in order to determine the effect of the choice of the factor distributions on the characterisation of the uncertainties and on the ranking of alternatives.

2.5.2 Application

The proposed methodology is applied to the comparison of two constructive alternatives of INCAS houses: the IBB shuttered concrete house and the IOB wooden-framed house. A set of 161 uncertain factors was taken into account to reflect uncertainties concerning the site, building envelope, systems, component lifetime, environmental data and methods for analysing the life cycle impacts. The influence of the variability

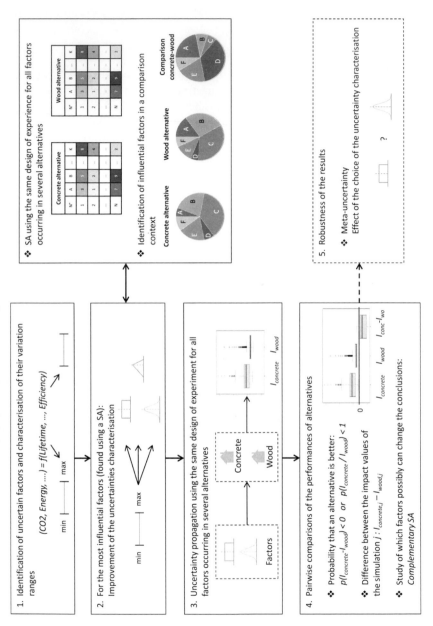

Figure 2.9 Approach for the comparison of alternatives in the presence of uncertainties.

of climate and occupation has also been studied. The full list of factors is provided in Pannier (2017).

The uncertainties were characterised based on data from the literature. In addition, complementary models were coupled with the software suite to account for uncertainties in the environmental data (Brightway LCA tool and uncertainties provided in ecoinvent), the variability in weather conditions (Ligier et al., 2017) and the variability of occupation (Vorger occupation scenario generator (2014)).

The adapted Morris method is used as an SA to determine for which factors the characterisation of uncertainty should be improved. 100 repetitions are carried out, and the factor variation space is discretised into six levels. For UA, 5,000 simulations are carried out following Sobol sequence sampling.

2.5.2.1 Results

The results of the SA, obtained after 22 hours of calculation, are presented in Figure 2.10 for four environmental indicators: climate change, cumulative energy demand, resource depletion and water consumption. To avoid overloading the graphs, only the names of the most influential factors are specified in the legend: the characterisation factors of the climate change indicator (denoted as CF CO_2 in the caption), the environmental data (probability distributions provided by the ecoinvent base), the type of polystyrene (R134a or CO_2 extrusion), the product lifetimes, etc.

The most influential factors on the separate alternatives are not the same as those identified in the comparison of the alternatives. For example, the variability of the occupancy has a large influence on the separate alternatives but only a small influence on the difference between the impacts of IBB and IOB. For some materials, the quantities involved in the two constructive alternatives differ. Factors involving these materials then appear to be influential. This is the case for the process for extruding polystyrene regarding climate change, for example. Other factors have important nonlinear effects

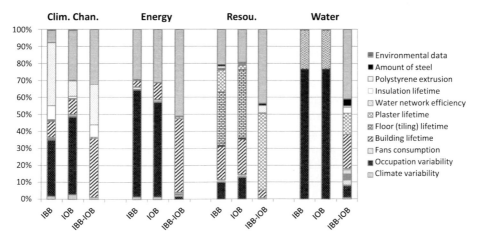

Figure 2.10 Results of the adapted Morris screening for IBB and IOB, and the difference between IBB and IOB.

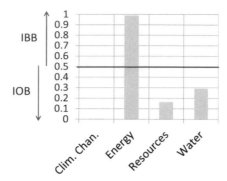

Figure 2.11 Probability that the impacts of IBB are lower than those of IOB.

or interactions. This concerns the lifespan of the building and uncertainties about environmental data in particular.

The 15 most influential factors on the difference between impact values have been identified for all environmental indicators. Additional information is collected on these factors in order to improve the characterisation of their uncertainty before the UA stage.

The results of the uncertainty propagation were obtained within 7 hours of simulations for each alternative. However, studies on the most adapted sample size for UA have shown that accurate results can be obtained with 1,000 simulations, that is in five times less computation time. In order to determine which is the best alternative, in Figure 2.11, we present the probability that the impacts of IBB are lower than those of IOB.

IBB is almost always the best alternative for the cumulative energy demand: the probability that the IBB alternative is better is close to 1. This is related to the heating needs, which are higher for IOB than for IBB, which benefits from greater thermal inertia. The probability is very close to 0 for climate change, so the IOB alternative is almost always preferred. This is explained by the higher impact of the concrete material on this indicator. Although the probability that the IOB alternative is better remains high for resources and water consumed, it becomes more difficult to conclude which is the better alternative.

The distribution of the relative differences between the alternatives is presented in Figure 2.12. For each indicator, these relative differences are calculated as the difference in the impact value of the two alternatives relative to the maximum of the impact values of the two alternatives. In the figure, the impact differences are positive when IBB has lower impacts, and negative when IOB is the best alternative. The line in the middle indicates zero difference, an equivalence between the alternatives, and the lines around it indicate the relative differences of −5% and 5%. For climate change, the relative difference between alternatives is almost always significant. This means that the impact values are quite different for all the simulations for this indicator, and that the choice of one alternative or another will have different effects on the environment. In addition, the sign of the difference shows that IOB is the best alternative on this indicator. On the contrary, the relative difference for the water consumed is

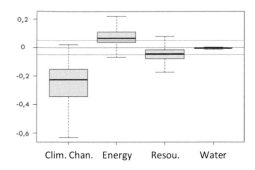

Figure 2.12 Distribution of the relative differences between the IBB and IOB alternatives.

always very small, which means that the choice of one alternative or another will have similar environmental consequences. Indeed, the major contributor to the water consumed indicator corresponds to the water required by occupants in the usage phase. The quantities of water involved for this item are equivalent in both cases. For the other two indicators, the relative difference is small. For these indicators, the choice of an alternative will have less effect on the results but the trends show that IBB is preferred for energy and IOB for resources. In order to practically reduce the impacts of the project, it is better to use the indicators that have a large difference between the impact values of the alternatives when choosing the best alternative.

K nowing the best alternatives for the indicators where the choice of alternative has a significant impact on the results, an alternative can be chosen using multi-criteria decision analysis methods, for example. These methods aim to find the best compromise among several alternatives, based on the preferences of the decision-makers. It would also be useful to normalise indicators using inhabitant equivalents, which leads to the prioritisation of indicators contributing significantly to the building studied in relation to the average impact per capita. For example, if a building corresponds to the energy consumption of 100 inhabitants but only five inhabitants with regard to eutrophication, priority will be given to the least energy-consuming alternative, even if the impact of this last alternative is greater in terms of eutrophication (this should be tailored according to the local context and the actors' priorities).

2.5.2.2 Robustness of the Results

The previously obtained results directly depend on the assumptions made, particularly with regard to the uncertainty characterisation on the most influential factors. The validity of these results has been studied by modifying elements in the uncertainty characterisation. This is illustrated here by showing the effect of choosing the expected value for the building lifetime: from 80 to 100 years. This factor was chosen because it has an influence on the comparison of alternatives for several indicators, as shown in Figure 2.10. The probability that IBB is the best alternative is presented in Figure 2.13 and the relative differences between the alternatives are presented in Figure 2.14.

For the indicators for which it was possible to conclude regarding the choice of a better alternative, modifying the uncertainty characterisation on the building lifetime

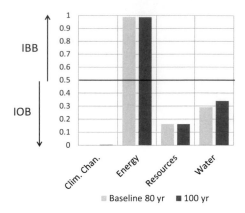

Figure 2.13 Probability that IBB impacts are lower for two lifetimes.

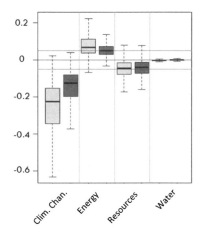

Figure 2.14 Distribution of the relative differences between the alternatives for two lifetimes.

was not likely to change the probability that one alternative is better than another in our case. On the other hand, modifying the expectancy for this factor causes the differences between the alternatives to evolve in the foreseen manner: the increase in the expected value on the lifespan tends to reduce the differences between the alternatives. This is related to the importance of the use phase, where the energy consumption and the environmental impacts of the two alternatives are close. This is due to the hypothesis of equivalent insulator thermal resistance, defined in the functional unit.

Further studies on the robustness of the results have shown that the definition of factor expectancy has a greater effect than a change in the form of the probability distribution or in the standard deviation value. Thus, to achieve a high level of robustness in the results, it appears to be important to know the expected values of the factors. Determining these values requires a good knowledge of all the possible values that the factors can take, especially for the most influential factors. This highlights the importance of the phase aimed at improving the characterisation of the uncertainties.

2.5.2.3 Impact of Uncertainties on the Alternative Comparison Results

For the indicators for which the choice of one alternative or another leads to very different impacts, the best alternatives are identified with and without taking the uncertainties into account, provided that the values taken for the factors in the deterministic model correspond to the expectancies considered for the same factors in the UA. Thus, when the user has confidence in the expected value, it is quite unlikely that the uncertainties will cause the results to be brought into question. Ranking can be modified in some cases as the probability that an alternative is better never reaches 100%. Taking uncertainties into account, nevertheless, provides two items of information that are useful for improving understanding of the results and the decision-making. On the one hand, this makes it possible to highlight the indicators to rely on for choosing the best alternative (large relative difference). On the other hand, the uncertainties give a certain level of confidence in terms of the choice made, as the probability that the classification of the alternatives has been modified is indicated.

Although this configuration has not been illustrated in this chapter, it is possible to compare more than two alternatives in the presence of uncertainty. The alternatives under study are then compared pairwise. Based on the results of these comparisons, it is possible to study which alternatives are best for each indicator although visualising the results is more complex. When there is a large number of alternatives to be compared, these results can become difficult to interpret. As proposed in Henriksson et al. (2015), it becomes necessary to group the alternatives according to their performance in order to make comparisons. For this purpose, statistical tests can be used to determine the groups of alternatives with different behaviours in terms of their environmental performance.

2.6 Conclusions and Perspectives

The necessity of taking uncertainties into account in LCA is recognised. However, in building LCA, these uncertainties, their quantification and their impact on the results remain rarely addressed.

In order to develop the practice of uncertainty quantification in building LCA, the performance of SA methods was compared in order to determine which of them could be used to obtain precise results, in terms of the factors' influence, within limited computation times. In our case, the results showed that the building lifetime and the methodological approach used to take the electricity production mix into account are two very influential factors. Assistance in choosing the appropriate SA methods for building LCA is proposed on the basis of the methods' applicability conditions (quantification of nonlinear effects, interactions, number of simulations required).

However, the robustness of the decision tree introduced would need to be validated by carrying out additional studies on more complex cases. The results of the SA method(s) selected *via* this tree can be useful both from a research perspective (to identify research areas that require further work to reduce uncertainties) and from an operational design perspective (to identify the most influential design factors that require special attention).

A methodology for taking uncertainties into account for building LCA has also been proposed. It is adapted to the specific context of alternative comparison. This methodology is based on the integration of SA and UA, to make a decision at a given confidence level by focusing on the environmental aspects that differ significantly between the alternatives. Thus, information is drawn from uncertainties to improve the quality of decision-making. The application of the methodology to a case study has shown that the results can be affected by the characterisation of uncertainties and the choice of the most probable value in particular. For the most influential factors, the step of improving the characterisation of uncertainty is then important for a better representation of the values these factors can take, when possible.

In order to obtain the most realistic results, particular attention should be paid to characterising the uncertainties of the factors. In addition, the various sources of uncertainty affecting the results should be included in the study. Some types of uncertainty have not been studied in the work presented in this chapter. For example, it would be necessary to study the effects of uncertainties on the long-term scenarios. This involves the systematic development of prospective scenarios regarding the evolution of the building and its context, for different points in time. The development of such scenarios calls for a large number of assumptions, especially for distant time horizons, and also requires expertise in prospective modelling to avoid unrealistic results.

The proposed methodology could also be extended to study larger spatial scales (blocks, districts, cities or territories). At these scales, environmental modelling integrates complementary modelling tools to take aspects such as transport, networks or waste management into account. This results in an increase in the number of uncertain factors to be studied. These aspects tend to increase the calculation times significantly, and it is important to have reliable methods for taking uncertainties into account that do not require overly large computation times.

Finally, it is important to develop visualisation tools to aid the comparison of several alternatives in the presence of uncertainties, and thus best support the decision-makers in their choices.

Bibliography

Andrianandraina, A. (2014). Approche d'éco-conception basée sur la combinaison de l'Analyse de Cycle de vie et de l'Analyse de Sensibilité, PhD Thesis, Université Nantes Angers Le Mans, 413.

Björklund, A.E. (2002). Survey of approaches to improve reliability in LCA. *Int. J. Life Cycle Assess.* 7, 64–72 DOI: 10.1007/BF02978849.

Chouquet, J. (2007). Development of a method for building life cycle analysis at an early design phase - Implementation in a tool - Sensitivity and uncertainty of such a method in comparison to detailed LCA Software, PhD Thesis, University of Karlsruhe, 452.

Cucurachi, S., Borgonovo, E.M., Heijungs, R. (2016). A protocol for the global sensitivity analysis of impact assessment models in life cycle assessment. *Risk Anal.* 36, 357–377 DOI: 10.1111/ risa.12443.

Faivre, R., Iooss, B., Mahévas, S., Makowski, D., Monod, H. (2013). *Analyse de sensibilité et exploration de modèles - Application aux sciences de la nature et de l'environnement*, Ed. Quae, ISBN: 978-2-7592-1906-3, 324.

Groen, E.A., Bokkers, E.A.M., Heijungs, R., and Boer, I.J.M. de (2016). Methods for global sensitivity analysis in life cycle assessment. *Int. J. Life Cycle Assess.* 13 DOI: 10.1007/s11367-016-1217-3.

Heeren, N., Mutel, C.L., Steubing, B., Ostermeyer, Y., Wallbaum, H., Hellweg, S. (2015). Environmental impact of buildings—What matters? *Environ. Sci. Technol.* 49, 9832 9841 DOI: 10.1021/acs.est.5b01735.

Heijungs, R. (1996). Identification of key issues for further investigation in improving the reliability of life-cycle assessments. *J. Clean. Prod.* 4, 159–166 DOI: 10.1016/S0959-6526(96)00042-X.

Heijungs, R., and Huijbregts, M.A.J. (2004, June). *A Review of Approaches to Treat Uncertainty in LCA*. University of Ösnabrück, Osnabrück, 332–339.

Henriksson, P.J.G., Heijungs, R., Dao, H.M., Phan, L.T., de Snoo, G.R., and Guinée, J.B. (2015). Product carbon footprints and their uncertainties in comparative decision contexts. *PLoS One* 10, 1–11 DOI: 10.1371/journal.pone.0121221.

Hoxha, E., Habert, G., Chevalier, J., Bazzana, M., and Le Roy, R. (2014). Method to analyse the contribution of material's sensitivity in buildings' environmental impact. *J. Clean. Prod.* 66, 54–64 DOI: 10.1016/j.jclepro.2013.10.056.

Hoxha, E., Habert, G., Lasvaux, S., Chevalier, J., and Le Roy, R. (2017). Influence of construction material uncertainties on residential building LCA reliability. *J. Clean. Prod.* 144, 33–47 DOI: 10.1016/j.jclepro.2016.12.068.

Huijbregts, M.A.J. (1998). Application of uncertainty and variability in LCA - Part I: A general framework for the analysis of uncertainty and variability in life cycle assessment. *Int. J. Life Cycle Assess.* 3, 273–280 DOI: 10.1007/BF02979835.

Huijbregts, M.A.J., Gilijamse, W., Ragas, A., and Reijnders, L. (2003). Evaluating uncertainty in environmental life-cycle assessment. A case study comparing two insulation options for a Dutch one-family dwelling. *Environ. Sci. Technol.* 37, 2600 2608 DOI: 10.1021/es02971.

Iooss, B. (2011). Revue sur l'analyse de sensibilité globale de modèles numériques. *J. Société Fr. Stat.* 152, 3–25.

Jansen, M.J.W. (1999). Analysis of variance designs for model output. *Comput. Phys. Commun.* 117, 35–43 DOI: 10.1016/S0010-4655(98)00154-4.

Lacirignola, M., Blanc, P., Girard, R., Pérez-López, P., et Blanc, I. (2017). LCA of emerging technologies: addressing high uncertainty on inputs' variability when performing global sensitivity analysis. *Sci. Total Environ.* 578, 268–280 DOI: 10.1016/j. scitotenv.2016.10.066.

Ligier, S., Robillart, M., Schalbart, P., and Peuportier, B. (2017). Energy performance contracting methodology based upon simulation and measurement. *Conf. IBPSA.* (San Francisco), 10.

Morris, M.D. (1991). Factorial sampling plans for preliminary computational experiments. *Technometrics.* 33, 161–174 DOI: 10.1080/00401706.1991.10484804.

Muller, S. (2015). Estimation de l'incertitude sur les flux d'inventaire du cycle de vie – modélisation et développement de facteurs empiriques pour l'approche Pedigree, PhD Thesis, Université de Montréal, 250.

Pannier, M.-L. (2017). Étude de la quantification des incertitudes en analyse de cycle de vie des bâtiments, PhD Thesis, MINES ParisTech PSL, 484 p.

Peuportier, B., and Blanc-Sommereux, I. (1990). Simulation tool with its expert interface for the thermal design of multizone buildings. *Int. J. Sol. Energy.* 8, 109–120 DOI: 10.1080/01425919008909714.

Peuportier, B., Thiers, S., and Guiavarch, A. (2013). Eco-design of buildings using thermal simulation and life cycle assessment. *J. Clean. Prod.* 39, 73–78 DOI: 10.1016/j.jclepro.2012.08.041.

Plackett, R.L., and Burman, J.P. (1946). The design of optimum multifactorial experiments. *Biometrika*. 33, 305–325 DOI: 10.2307/2332195.

Polster, B. (1995). Contribution à l'étude de l'impact environnemental des bâtiments par analyse du cycle de vie, PhD Thesis, École nationale supérieure des mines de Paris, 268 p.

Pomponi, F., D'Amico, B., and Moncaster, A.M. (2017). A method to facilitate uncertainty analysis in LCAs of buildings. *Energies*. 10, 524 DOI: 10.3390/en10040524.

Popovici, E. (2005). Contribution to the life cycle assessment of settlements, PhD Thesis, École Nationale Supérieure des Mines de Paris, 244.

Prado-Lopez, V., Wender, B.A., Seager, T.P., Laurin, L., Chester, M., and Arslan, E. (2015). Tradeoff evaluation improves comparative life cycle assessment: A photovoltaic case study. *J. Ind. Ecol*. 9 DOI: 10.1111/jiec.12292.

Saltelli, A., and Bolado, R. (1998). An alternative way to compute Fourier amplitude sensitivity test (FAST). *Comput. Stat. Data Anal*. 26, 445–460 DOI: 10.1016/S0167-9473(97)00043-1.

Saltelli, A., Tarantola, S., Campolongo, F., and Ratto, M. (2004). *Sensitivity Analysis in Practice: A Guide to Assessing Scientific Models*, Ed. Wiley, ISBN: 978-0-470-87095-2, 232.

Saltelli, A., Ratto, M., Andres, T., Campolongo, F., Cariboni, J., Gatelli, D., Saisana, M., and Tarantola, S. (2008). *Global Sensitivity Analysis: The Primer*, Ed. Wiley, ISBN: 0-470-72517-6, 304.

Sobol, I. (1993). Sensitivity estimates for nonlinear mathematical models. *Math. Model. Comput. Exp*. 1, 407–414 DOI: 1061-7590/93/04407-008$9.00.

Suh, S., Tomar, S., Leighton, M., and Kneifel, J. (2014). Environmental performance of green building code and certification systems. *Environ. Sci. Technol*. 48, 2551–2560 DOI: 10.1021/es4040792.

Tian, W. (2013). A review of sensitivity analysis methods in buildings energy analysis. *Renew. Sustain. Energy Rev*. 20, 411–419 DOI: 10.1016/j.rser.2012.12.014.

Veiga, S.D., Wahl, F., and Gamboa, F. (2008). Local polynomial estimation for sensitivity analysis for models with correlated inputs. *Technometrics*. 51, 452–463 DOI: 10.1198/TECH.2009.08124,.

Verbeeck, G., and Hens, H. (2007). Life cycle optimization of extremely low energy dwellings. *J. Build. Phys*. 31, 143–177 DOI: 10.1177/1744259107079880.

Vorger, E. (2014). Étude de l'influence du comportement des occupants sur la performance énergétique des bâtiments, PhD Thesis, École nationale supérieure des Mines de Paris, 474.

Wang, E., and Shen, Z. (2013). A hybrid data quality indicator and statistical method for improving uncertainty analysis in LCA of complex system – application to the whole-building embodied energy analysis. *J. Clean. Prod*. 43, 166–173 DOI: 10.1016/j. jclepro.2012.12.010.

Chapter 3

Life Cycle Assessment of Transportation Systems

Anne de Bortoli, Adelaïde Feraille, and Fabien Leurent

École des Ponts ParisTech

3.1 Introduction

Rail transportation flourished as early as the nineteenth century, and was then joined by automobile and aerial modes in the twentieth century, transforming the world into a global network well equipped with air, sea and rail links, and especially with roads and cars. To a large extent, road transportation has grown to become the main vector of mobility. In retrospect, the development of transportation over the course of a century and a half seems prodigious. It was integrated with the socioeconomic development of human societies in countries that were able to equip themselves. There is another side to the coin, however, which concerns both society and the environment.

3.1.1 Environmental Impacts of Transportation and Well-Being

Going for bread by car, ordering a product from China on the Internet, jumping on a bus for a weekend in Europe or traveling 10,000 km by plane to spend a week in Asia: these are ideas that would have left our ancestors stunned even a few decades ago. Indeed, although mobility is central to our lives today, for better or for worse, these practices have evolved quickly, as have their impacts on the environment. Emitters of greenhouse gases (GHGs) and atmospheric pollutants, consumers of space and non-renewable resources, producers of waste and harmful sounds – these are the "prices to pay" that make transportation services a challenge when it comes to the well-being of our societies. We should remember that they rank in the "top 3" of the world's household GHG emissions, after food and housing. Already responsible for 14% of GHG emissions in 2010 according to the IPCC (Edenhofer et al. 2014), and despite various environmental policies adopted around the world, the volume of transportation emissions seems to be growing, apart from occasional declines during the economic recessions in 2008 and 2013 (European Environment Agency 2017). Take the case of France (Figure 3.1): while the country's total GHG emissions decreased by 16% between 1990 and 2015, mainly through the energy transformation and manufacturing sectors, transportation[1] grew by 12% between 1990 and 2015, making it the largest contributor (29%), mainly due to road vehicles. While it is difficult to estimate the extent of these expected consequences of these rising emissions with certainty, it is nonetheless alarming, with cascading effects and likely complex feedback within the

---

1 Direct emissions, either related to the use stage only

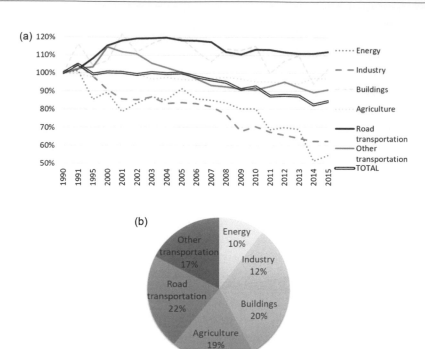

Figure 3.1 Evolution of total GHG emissions in France by sector relative to 1990–
2015 (a) and respective contributions by sector (b). (SECTEN, Citepa 2016
(https://www.citepa.org/images/III-1_Rapports_Inventaires/SECTEN/
SECTEN-Fichiers_ avril2017v2.zip).)

global system. All the continents are already affected by adverse changes attributed to
climate change: those of the physical systems (glaciers, snow, ice and/or permafrost,
rivers, lakes, floods and/or droughts – coastal erosion and/or sea levels), biological
systems (terrestrial ecosystems, forest fires, aquatic ecosystems) and human systems
(food production, lifestyles, health and/or economy) (Field et al. 2014).

Emissions of pollutants, meanwhile, are no less harmful, with air pollution ranked
as the third biggest cause of death in France after tobacco and alcohol (49,000 deaths
annually), not to mention the effect on the ecosystems, which, let us remember, we
depend on for our well-being. According to the French Court of Auditors (Cour des
Comptes 2015), 60% of the French inhabitants inhale poor-quality air, at a social cost
estimated between 20 and 30 billion Euro per year. Despite the strong media coverage of pollution peaks, the vast majority of deaths (85%–100%) would be thought
to be due to chronic pollution that does not exceed the regulatory thresholds. Pro-
longed exposure to harmful aerosols appears to be the most damaging mode of air
pollution for humans (Cour des Comptes 2015). For its part, the European Union has
reported that 400,000 deaths in its territory are supposedly caused by air pollution
each year. Transportation's responsibility within this air pollution (in terms of direct
emissions) is significant: in 2014 in France, it was estimated to account for 61% of
nitrogen oxide emissions (CITEPA 2016a) and 19% of fine particulate matter (PM2.5)

(CITEPA 2016b). On the other hand, there have been noteworthy improvements in the sector, for example, in terms of sulfur dioxide emissions (<3% of national emissions in 2014), heavy metals and other harmful emissions: between 1990 and 2013, the transportation sector recorded −97% lead emissions, −87% cadmium emissions, −85% mercury emissions, −79% nickel emissions, −69% volatile organic compound emissions non-methane, −62% arsenic emissions and −54% PAH emissions (carcinogens) (Cour des Comptes 2015). Nitrogen oxides, which are problematic for ecosystems (acidification, eutrophication), climate change (formation of tropospheric ozone) and human health (deep penetration of the pulmonary tree), also underwent a sharp decline (−68% between 1990 and 2014), notably thanks to the European emission standards for road vehicles, as did PM2.5 (−47%), PM10 (−40%), carbon monoxide (−95%) and NMVOCs (−96%) (Citepa data).

Apart from their emissions, some quantitative elements relating to other types of the environmental impacts of transportation can be put forward, although the data available are fragmentary. In terms of energy, the sector accounts for almost a third of global and French consumption. However, it is extremely dependent on oil, which accounts for 98% of its consumption.[2] The rail sector is the only one that has massively electrified (more than two-thirds of activities), while road-based electromobility is still struggling to penetrate the market. As the main mobility infrastructures consume space, roads occupy about 1.2% of the territory (Berger 2006). Of the 330 million tons of aggregates consumed in France each year, public works consume 80% (and roads 50%) versus 20% for the building sector. Although aggregates are not considered to be scarce resources, local shortages are already appearing,[3] as in the Paris Basin, Aquitaine and Picardy, thus increasing the transportation distances and, finally, the impacts of the sector. Of the mineral resources considered "critical" and consumed by transportation, in order of decreasing criticality, we highlight the rare earth (strong criticality) present in batteries and electric motors; aluminum, bauxite, chromium, iron, manganese, nickel and zinc (average criticality according to the European Union) found in vehicle bodies, electronics, tires and batteries; and finally copper (vehicles, signage).

To conclude, we note that the building and public works sector accounts for two-thirds of waste production in France, including 94% of inert waste, 5% of ordinary industrial waste and 1% of hazardous waste (CGDD 2015). The transportation sector has significantly improved its practices, yet it still accounts for almost 20% of hazardous waste emissions in construction materials, used solvents and oils, scrap vehicles, batteries and accumulators (CGDD 2015). Reduction of waste production was also one of the objectives of the voluntary agreement signed by the public works sector following the Grenelle de l'Environnement debate. For roads, this led to the development of French tools allowing the environmental comparison of variants (de Bortoli 2015). Finally, in France, a study co-piloted by ADEME and the audit and consultancy firm EY (EY 2016) estimated the social cost of transportation noise – impacts on sleep, daytime discomfort, increased cardiovascular risks, decrease in property value,

---

2 techniques-ingenieur.fr/actualite/articles/chiffre-cle-98-des-energies-de-transport-sont-issues-du-petrole-5909/
3 usinenouvelle.com/article/demain-les-granulats-vaudront-de-l-or.N139431

productivity losses and learning disabilities – at 20 billion euros per year, more than a third of the total cost of noise. According to this report, 25 million people in France are significantly affected by transportation noise, including 9 million who are exposed to levels that are critical for their health.

To face this situation, how is the environment taken into account within the decision-making process in the transportation sector in France?

3.1.2 Decision Support Tools for Transportation

Given the environmental improvement of certain aspects, and transfers of impacts on others, the evidence shows that the public authorities cannot avoid an environmental assessment in the key sector embodied by transportation. This assessment spans several levels: in diagnostics, to carry out an environmental "inventory" of transportation services; in forward planning, to explore several possible futures for freight and mobility and to guide development; in comparative design, leading towards eco-innovation, where social needs meet respect for global environmental limits.

Currently, two main methods of ex ante evaluation shed light on the public utility and environmental impacts of major transportation projects in France: the socioeconomic assessment and the environmental impact study. Nevertheless, the environment remains only partially considered through these methods, as do the impacts considered and the perimeter evaluated. In the socioeconomic evaluation, which aims to quantify the utility of the project, the environmental assessment focuses on certain effects linked to the exploitation stage only (emissions of GHGs, air pollutants and noise, in particular), while the environmental impact study concentrates mainly on local effects, such as the loss and fragmentation of habitats related to new infrastructure.

However, while the emission impact of transportation (GHG and pollutants) is very important on the French and global assessment, for example, the figures previously presented for the sector are those resulting from the so-called UNFCCC classification, relating to the Kyoto protocol. Again, only emissions from the operating stage of transportation vehicles (direct emissions) are accounted for. Emissions from the rest of the transportation service production life cycle are therefore omitted, such as the life cycles of the infrastructures and vehicles in nonuse stages (production, maintenance, end-of-life), as well as ancillary services (insurance, design, etc.). These emissions are mainly accounted for in the manufacturing sector (chemicals, metallurgy, mineral products, electronics). The truncated environmental assessment made on the transportation projects in this way makes it difficult to get out of a "silo" optimization logic, that is, subsystem by subsystem. However, in the theorem bearing his name, the American mathematician Bellman proved that the sum of individually optimized subsystems does not equate to the optimum of the corresponding overall system (Bellman 1952). In other words, better environmental performance could be achieved through a tool for comprehending the impacts of a complex system as a whole. This tool already exists and could be used to support the decision-making process concerning transportation.

Indeed, life cycle assessment (LCA) is a standardized method for quantifying the potential impacts of a product, project or system throughout its life cycle, and it makes it possible to produce[4] an environmental balance sheet of the system, also called an "environmental assessment". This environmental balance sheet then allows for a multi-criteria environmental comparison of several system variants. These two steps are the key to the eco-design approach. The methodological framework for LCA is set by the ISO 14040 and ISO 14044 standards and was designed to carry out the environmental assessment of industrial products during the 1990s (Jolliet et al. 2005). The use of LCA as one of the responses to the desire to rationalize environmental decision-making in transportation, whether public or private, has required and still requires methodological adaptations, which an ever-growing international community have been contributing to for 20 years now.

3.1.3 Our Objectives: To Create an LCA Model for Decision Support

Given their practical importance and complexity, environmental impacts constitute a major *knowledge objective* for transportation systems. This knowledge must necessarily be *integrative* in order to distinguish the components and subsystems within a larger system and to consider the different stages in the life cycle of each component, as well as the different natures of the impacts.

Due to the complexity of their parts, each actual system has distinctive features: depending on the mode or modes of transportation involved, depending on the particular social forms, from demand uses to organizations of the actors concerned, depending on the local conditions in the territory concerned and, more broadly, depending on its physical and human geography (which populations are exposed in which places) (climate, relief), and its biogeography (ecosystems in places). The peculiarities of the actual systems determine the possibility and mode of knowing their environmental performance. This knowledge will not take the form of an immediate datum, but rather a construction, a work of composition developed on the basis of theoretical resources and elementary models to describe, explain and simulate the elements.

The result is a shift in the issue: the objective of practical knowledge gives way to a goal that is both theoretical (the science of the environment for transportation systems) and methodological, involving modeling an actual system and its impacts on the environment.

In practice, an evaluation method of this kind will be applied for management purposes, to inform planning or operating decisions with regard to the system in question. This is why the knowledge objective is completed by a pragmatic objective: to integrate the evaluation method within a management perspective, and in particular to study the desirability of such a project or a transportation policy targeting the system in order to transform it.

We will address these two objectives of methodological development and planning application. We will construct an environmental assessment model for a particular transportation system chosen as a case study, and apply it to a planning management perspective. Through this experience, we will have investigated the feasibility of the two objectives, and we will have explored and marked out a path for achieving them.

---

4 Usinenouvelle.com/article/demain-les-granulats-vaudront-de-l-or.N139431

3.1.4 Outline of the Chapter

The rest of this chapter is organized in five parts. First, we will present our literature review on the LCA of transportation systems in the retrospective form of a chronicle of scientific contributions, and also point out the main centers of knowledge development at the international level (Section 3.2).

We will then study the case of the Martinique BRT (bus rapid transit) line, constituting a model for its LCA and numerically calculating a set of quantitative indicators to assess the major environmental impacts: GHG emissions, primary energy consumption, nonrenewable resource depletion, solid or radioactive waste disposal, acidification and eutrophication of natural environments, and different types of ecotoxicity (Section 3.3).

We will then address the second objective: to use the evaluation method in a managerial perspective, to shed light on decision-making within a planning problem. We developed project alternatives to fulfill the desired transportation function and evaluated them one by one, to compare them against the different environmental indicators in a multi-criteria manner. Our comparison highlights the prominent role of filling passenger vehicles: among buses, automobiles and even trams/streetcars, the occupancy rate is the key factor in a mode's environmental performance, beyond any ideological consideration (Section 3.4).

Our experience was based on the well-established methodological resources: the EcoInvent database for processes, on the one hand, and the OpenLCA software to implement the model and perform calculations, on the other hand. These are materials and tools for crafting LCA models of transportation systems, well ahead of any hypothetical standardization and industrialization. To promote this kind of craftsmanship, the Ecodesign Chair is engaged in establishing specific training possibilities for engineering students at École des Ponts ParisTech (Section 3.5).

In conclusion, we will draw on scientific and technical lessons for the environmental assessment of transportation systems, and we will consider possible extensions, especially in view of incorporating LCA in the procedures to appraise transportation projects and policies (Section 3.6).

3.2 LCA and Transportation Systems: An Overview

3.2.1 Three Decades of Development

Here, we propose a simplified chronology of LCA for transportation systems, from the beginnings in the late 1990s to the present day. Figure 3.2 traces three stages of development spanning roughly a decade each, which we will discuss in turn.

3.2.1.1 1990s: Initiation – LCA of Roads and Automobiles

LCA appeared in industrial ecology in the 1970s after being foreshadowed by the Material Requirements Planning and Manufacturing Resource Planning methods, which emerged in the 1960s to better manage resources in a productive business environment. We date the first use of the method for transportation subsystems to the 1990s, in conjunction with the standardization of LCA. The first case studies (and up until

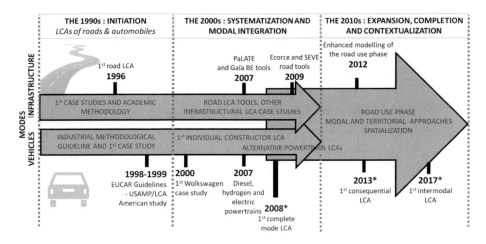

Figure 3.2 Simplified chronology of LCA for transportation systems: major development stages and milestones (*: modal milestones).

now, most of the LCA work produced in the transportation field) focused on roads, or more precisely on the roadway. Concentrating first on the comparison between pavements with hydrocarbon binders and hydraulic binders (Häkkinen and Mäkelä 1996; Horvath and Hendrickson 1998; Roudebush 1999; Berthiaume and Bouchard 1999), the question of the development and quality of environmental inventory data quickly became paramount (Stripple 2001). It was also at this time that a guide to the application of LCA to the automotive sector was published by the European Council for Automotive R&D (EUCAR) (Ridge 1998). Pioneering studies were carried out, one by the United States Automotive Materials Partnership (USAMP/LCA 1998) and the other by the Kobayashi et al. (1998).

3.2.1.2 2000s: Systematization and Modal Integration

The 2000s saw the diversification of academic studies for road cases. Most were restricted to the construction stage, without including the essential use stages (de Bortoli 2014; Araújo, Oliveira and Silva 2014) and end-of-life (EoL) stage of the road (Nisbet et al. 2001; Park et al. 2003; Treloar, Love and Crawford 2004; Zapata and Gambatese 2005; Athena Institute 2006; Chan 2007; Muga et al. 2009; Huang, Bird and Heidrich 2009; White et al. 2010; Abdo 2011). On the other hand, LCAs of motor vehicles were produced by manufacturers (Schweimer and Levin 2000; Volkswagen AG 2008a, 2008b). It was also during this decade that the first "institutional" LCA tools for roads appeared: PaLATE at the University of Berkeley[5] and GAiA.BE[6] in the French road construction company Eurovia, followed in France by the tools SEVE (USIRF) and

---

5 http://faculty.ce.berkeley.edu/horvath/palate.html
6 http://www.batiactu.com/edito/gaia-un-logiciel-environnemental-chantiers-routiers-6203.php

ECORCE (IFSTTAR) (Ventura et al. 2012) to name a few (de Bortoli 2015). Studies on other types of infrastructure also appeared, as well as those for complete modes of transportation (Chester 2008), particularly in relation to rail (Rozycki, Koeser and Schwarz 2003; Spielmann and Scholz 2005; Spielmann, Bauer and Dones 2007; Horvath 2006; Svensson 2006; Chester 2008).

3.2.1.3 2010s: Proliferation, Completion and Contextualization

Finally, the 2010s were very fertile in terms of LCA case studies applied to transportation systems. AzariJafari et al. (2016) made an inventory of around 125 LCAs on the road mode exclusively, which were completed between 2010 and 2015. However, the authors of systematic reviews report a lack of consensus in the methods (Inyim et al. 2016). On the other hand, although we can see a strengthening of the transparency related to life cycle inventories, which provide crucial data for the quality of evaluations, they are still not automatically disseminated. There was strong interest in modeling the usage stage of the road (Wang et al. 2012; Akbarian et al. 2012; Greene et al. 2013; Louhghalam, Akbarian, and Ulm 2014; Wang, Harvey, and et Kendall 2014; Bryce, Katicha et al. 2014; Bryce, Santos, et al. 2014; Xu, Gregory and Kirchain 2015; Reyna et al. 2015; Santos 2015; Lin, Chien and Chiu 2016; Chong and Wang 2017; Trupia et al. 2017). Various studies have looked at comparing the traction modes of vehicles (Ally and Pryor 2007; Kliucininkas, Matulevicius and Martuzevicius 2012; Sundvor 2013; Cooney, Hawkins and Marriott 2013; Xu et al. 2015; Ercan and Tatari 2015) with a particular interest in road electromobility (Warburg et al. 2013; Messagie, Macharis and Van Mierlo 2013; Hawkins et al. 2013; Helmers, Dietzet, and Hartard 2015; Cox et al. 2018), with the usage stage of a mode often having the biggest environmental impacts on the life cycle. Developments in LCA for modes of transportation are continuing (Dave 2010; Chester, Horvathet, Madanat 2010; Chester and Horvath 2010; Chester et al. 2012; Miyoshi and Givoni 2013; Le Féon 2014; Yue et al. 2015; Banar and Özdemir 2015; Bauer et al. 2015; Jones, Moura and Domingos 2016; de Bortoli, Féraille and Leurent 2017), with a shift towards intermodal travel (Hoehne and Chester 2017), a focus on the contextualization of inventories (Jullien et al. 2012) and going beyond simple environmental diagnosis, thus favoring consequential approaches: under the effect of modal postponements, what are the environmental benefits relating to the implementation of a new transportation offer (Chester et al. 2013; Saxe, Miller and Guthrie 2017)? What is the potential environmental impact of electromobility across the United States (Onat et al. 2017)?

3.2.2 Scientific Centers Addressing Decision-Making Problems

While there are currently numerous contributions to LCAs of transportation systems from diverse origins, we can identify some academic centers that are particularly active in the approach according to modes of transportation: a "LCA transportation[7]" center in the United States which is led by Dr. Chester's team at Arizona State University, a spin-off from UC Berkeley (original work by A. Horvath); a European center that is particularly renowned for its applications to transportation, the Norwegian

---

7 http://transportationlca.org/

University of Science and Technology (NTNU), founder of industrial ecology, and the Luxembourg Institute of Science and Technology (LIST); a more widespread Chinese community is gradually being formed, which regularly produces case studies geared towards decision-making for transportation.

Various environmental studies of transportation policies using LCA can be identified, and these are often based on a collaboration between an academic laboratory and the various "producers" of transportation services (public authorities, project owners or managers) based on the need for quality input data. In the United States, LCA has notably been used to study the impact of land-use planning (urban forms) on the environment (Chester et al. 2013), the interest of Transit-Oriented Development (TOD) for the Los Angeles (Nahlik and Chester 2014) and Phoenix regions (Kimball et al. 2013), transit and high-speed transit policies in California (Matute and Chester 2015), high-speed rail policies (Michalek et al. 2011; Chester and Ryerson 2014) and the organization of American freight (Nahlik et al. 2016), as well as the comparison of air transportation and high-speed rail and the parking policy in Los Angeles County (Chester et al. 2015). Some studies highlight points of concern to be clarified in the LCA of transportation decision-making: the importance of dynamic LCAs (taking time into account) (Chester and Cano 2016), the importance of taking into account uncertainties about the results to be dealt with by developing tools adapted to each system, for example, that of pavements (Gregory et al. 2016).

3.3 Case Study: The Environmental Balance Sheet of a Bus Rapid Transit Service

The founding principle of collective modes of transportation is the massification of flows of passengers by means of high-capacity vehicles. This results in economies of scale in the service production, and the production costs can be shared among the customers. The same applies to the environmental plan: the quantities of impacts are correlated with the number of passengers.

It is still necessary to fill enough vehicles to achieve these high economic and environmental returns. However, conventional bus lines are generally less efficient in terms of travel time than private cars. This is because of the stops for passengers to board and alight, the travel distances between the travelers' points of origin and destination and the stops, waiting times at stops, distances and connection times between network lines, not counting the transaction times needed to acquire information and buy a ticket. Railways win in terms of travel time by running on their own tracks, which can offset the additional costs over time, especially if the time intervals are high and reduce waiting times in the station. BRT lines share the principle of a faster route through dedicated sections and junction priority on the roads. This last provision limits the frequency of service, however, in order to regulate the flow capacity for the routes covered by the line.

We report the LCA environmental balance sheet of a BRT line here, demonstrating an LCA of a transportation system integrating infrastructures and vehicles. After presenting the transportation solution (Section 3.3.1), we will indicate the principles and specify the evaluation method (Section 3.3.2). We will then present the evaluation data and hypotheses Section 3.3.3), followed by the results of the simulation Section 3.3.4). We will conclude with a summary accompanied by a discussion Section 3.3.5).

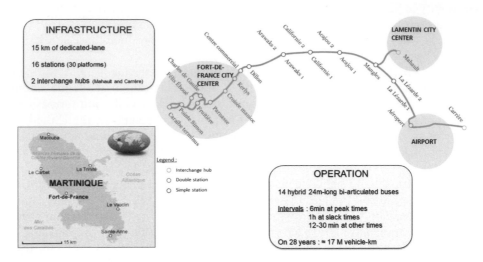

Figure 3.3 Diagram of the line and specific conditions.

3.3.1 Presentation of the Transportation Project

Our case study concerns a dedicated-lane public transportation line located on the French Caribbean island of Martinique. This forked line has a total length of about 15 km and connects the center of the city of Fort-de-France (capital of the island) to the city of Lamentin ("Mahault" interchange) and the airport ("Carrère" interchange) (see the diagram in Figure 3.3). It has 16 stations, 2 interchange hubs, and 1 maintenance center (outside the perimeter), and is traversed by a fleet of 14 hybrid double-articulated buses measuring 24 m.

The project was carried out within the framework of a public–private partnership, and the Caraïbus company has been the concession holder of the line for a period of 20 years. The opening of the line was scheduled for the end of 2015. Each interchange hub has bus stations, bus terminus locations, local collective taxis (a popular form of public transportation), as well as 110 parking spaces in the Mahault hub and 450 in the Carrère hub, a place for the driver and a commercial agency. The Mahault hub also has spaces for school buses. On weekdays, the service runs from 5.30 AM–10 PM. In terms of the frequency of services, as of 2017 there is a bus scheduled every 6 minutes during peak hours (6.30 AM–8.30 AM and 4 PM–6 PM) in the core area, and every hour after 8 PM; intervals between peak hours are intermediate, with buses running every 12 or 30 minutes.

3.3.2 Principles and Method of Evaluation

We carried out a "process-based LCA" for a BRT line: a study based on industrial data whose scope of evaluation is wider than that of previous studies (see Figure 3.4). Our original inputs came from applying a process-based LCA to a real complex system, and also from creating associated macro-processes: earthworks, pavement, etc. Relying on the EcoInvent V2.2 database, we used the OpenLCA 1.4 open source software.

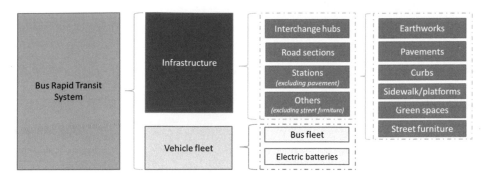

Figure 3.4 Description of the system studied: modeling by subsystem (column 3) and by construction batch (column 4).

OpenLCA is a modular LCA software released in 2007 by GreenDeltaTC that allows researchers to use EcoInvent environmental data and also to adapt and modify them as needed. The EcoInvent database is the most used in Europe. It contains international industrial life cycle inventory data from many production sectors: energy, resource extraction, manufacturing, chemicals, agriculture, waste management and transportation.

3.3.2.1 Framework and Objectives of the Study

Objective: Our study aims to carry out a process-based LCA on the Martinique BRT line over a 28-year period, which corresponds to the structural design of the infrastructure, in order to quantify the environmental impacts and allocate them by subsystem and by life cycle stage. The function studied is passenger transportation on the line during the 28 years of the concession, with a 5% level of infrastructure risk at the expiry of the concession. The lifespan of a road infrastructure is difficult to establish, and we choose a duration of observation during which the maintenance operations to be carried out have been planned with precision.

System boundaries: The studied system is delimited (notion of perimeter) and described in Figure 3.4 and Table 3.1. We envisage two sequences to analyze the infrastructure subsystem: one according to the "topological" functions of the node or arc (the nodes ensure different functions: intermodality at the interchange hubs, access stations to the bus service, passing through the intersections and crossroads with other flows , and the arcs for passing through the space in the running section), and the other according to construction batches as distinguished in the operational assembly of the project (Figure 3.4).

As shown in Table 3.1, the scope of the evaluation is both broad and homogeneous across the technical subsystems. The following elements were overlooked due to a lack of data: changing bulbs, cleaning lamp posts and maintaining street furniture; incorporation of bituminous mixes in new asphalt, managing and maintaining green spaces, as well as maintaining curbs (possible changes caused by being struck by buses, cleaning with pressurized water jets).

Table 3.1 Life Cycle Stages Considered

Subsystem	Construction	Use	Maintenance	End-of-Life	Transport
Earthworks	X	N/A	X *	X *(100)	Included
Pavement sections	X	N/A	X	X *(>28)	Included
Curbs	X	N/A	N/A	X *(>30)	Included
Sidewalks	X	N/A	N/A	X *(30)	Included
Green spaces	X	N/A		N/A	Included
Street furniture	X	X	X *	X *(>28)	Included
Bus	X	X	N/A	X (14)	Europe to Martinique
Batteries	X	N/A	N/A	X (10)	Europe to Martinique

X * (α): inclusion not necessary over 28 years because lifespan "α" considered higher than 28 years.
N/A: not applicable.

Table 3.2 Study Indicators and Characterization Methods

Indicator	Unit	Characterization Method
Consumption of primary energy	MJ eq	CED
Depletion of nonrenewable resources	kg eq antimony	CML 2001
Solid waste	kg	EDIP
Radioactive waste	kg	EDIP
Climate change (100 year horizon)	kg CO_2 eq	CML 2001
Atmospheric acidification – generic	kg SO_2 eq	CML 2001
Stratospheric ozone depletion – constant	kg CFC-11 eq	CML 2001
Photochemical oxidation	kg ethylene eq	CML 2001
Eutrophication – generic	kg PO_43- eq	CML 2001
Freshwater aquatic ecotoxicity (100-year horizon)	kg 1.4-DCB eq	CML 2001
Marine aquatic ecotoxicity (100-year horizon)	kg 1.4-DCB eq	CML 2001
Terrestrial ecotoxicity (100-year horizon)	kg 1.4-DCB eq	CML 2001
Human toxicity (100-year horizon)	kg 1.4-DCB eq	CML 2001

3.3.2.2 Impact Categories and Characterization Methods

Environmental impacts are assessed through the impact categories and characterization methods presented in Table 3.2.

With regard to ecotoxicity, indicators were chosen that cover the entire biosphere, including all natural compartments (freshwater, salt water, soils) and all categories of living species (humans and other species). The eutrophication and atmospheric acidification indicators were calculated neglecting the specific geographical context. The consumption of energy resources is calculated according to the cumulative energy demand (CED) method. Waste and radioactive waste is calculated using the Environmental Design of Industrial Products (EDIP) method. The indicators for climate change, natural resource depletion, atmospheric acidification, stratospheric ozone depletion, photochemical ozone formation, eutrophication, ecotoxicities and human toxicity were calculated using the equivalence method. The principle is to convert the flows of substances that may contribute to these impacts into a reference substance flow specific to each impact category. This conversion method was developed by the Institute of Environmental Sciences (CML) of Leiden University (the Netherlands).

3.3.3 Data and Assumptions

Our case study is predominantly based on industrial data (Figure 3.5) and life cycle inventories from the EcoInvent database, version V2.2. The quantitative estimates are made on the basis of the Caraïbus partnership dossier, the bill of quantities and cost estimates provided by the road builder Eurovia and documents from the bus manufacturer VanHool. The construction equipment was used 7 hours a day, and their diesel consumption comes from the professional "building machine" bases of the eco-comparators SEVE and Gaïa (supplied by Eurovia), which take idle working time into account. The net calorific value of the diesel taken is equal to 38.08 MJ/L. The depreciation of the machines (machinery and small equipment) and equipment was not taken into account. The rates were provided by Eurovia. The electrical mix chosen as a first approximation is that of Poland as it is very close to that of Martinique, with a thermal source of about 95%.

3.3.3.1 Transportation of Materials

Unless otherwise stated, we have excluded the transportation of materials upstream of suppliers. Supplier-site distances were either provided by our industry partners or calculated from the supplier list using Google Maps. For the running section and the stations, we took the barycenter of the total route as a point of origin for calculating distances from suppliers. For supplies to the site, we considered that 16- to 32-ton trucks meeting EURO4 standard were used, except for the green spaces (see details). We made the assumption that the site managers each drive 30 km/day in vehicles on site. Finally, we considered the transportation related to the disposal of the site materials according to different storage locations.

3.3.3.2 Bus Fleet

The fleet consists of 14 hybrid double-articulated buses measuring 24 m with a 14-year lifespan. The fleet was renewed once during the period studied, and we considered this second batch of buses to have the same performance (consumption, emissions) as

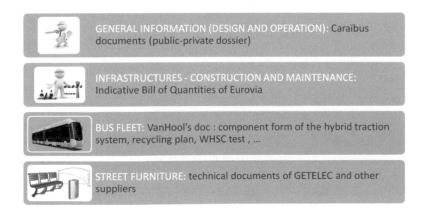

Figure 3.5 Sources of key industrial data.

they meet the most stringent emissions standards currently in force (Euro6). Due to an overly significant percentage of unidentified materials in the VanHool data in relation to the mass cutoff threshold tolerated by the NF P 01-010 standard to carry out an LCA in France (98%), we used the EcoInvent life cycle inventories on the construction, maintenance and EoL of buses. We identified the materials and transformation processes relating to the engine; then, we carried out a first ICV calculation in proportion to the total masses of the VanHool bus and the EcoInvent bus by subtracting the mass of their respective internal combustion engines because it was not proportional to the mass/size of the bus. We then completed this process with that of the electric battery of the hybrid system (*"electric motor, electric vehicle, at plant, RER"*). Regarding the battery, the information provided by the manufacturer was reduced to the mass, which is why we calculated the mass ratio between the electric cells and the rest of the battery based on the expert opinion of the car manufacturer Opel, which is corroborated by the data from the study by Li et al. (2013), and the battery materials outside the electric cells (Li, Li and Yuan 2013). We considered that the batteries have a lifespan of ten years, which equates to three sets of batteries per bus but two EoL during our observation period of 28 years.

Regarding the bus use stage, we considered two successive stages: a first implementation stage from 2015 to 2017 with 56 journeys per direction and per day, then a second stage of 26 years from 2018, with 69 journeys on the core area of the line per direction per day. During this period, the annual increase in passenger traffic is estimated at 0.42%. The consumption measured on the SORT cycle per bus was 56 L/100 km. We adapted EcoInvent's *"operation, regular bus, CH"* process to the consumption and emissions (for NO_x, CO_2 and CO) of the VanHool bus, measured by the WHSC "World Harmonized Stationary Cycle" procedure dedicated to heavy-duty engines.

3.3.3.3 Road Sections

Of the total 15 km of the line, 12.5 km had been previously constructed. Eurovia constructed the remaining 2.5 km, as well as all the stations. Nevertheless, in this study, we considered that the 15 km was built and maintained according to technical choices made by Eurovia. The material volumes of the concrete elements were calculated based on the standardized product sheets. The sizing and maintenance assumptions were provided by Eurovia. The systematically neglected elements were the labor of specialized workers and the transfer of equipment, the various elements not quantified in the bill of quantities and cost estimates, as well as the small construction equipment that can be reused.

3.3.3.4 Street Furniture

This item encompasses the production of the equipment and its transportation, as well as the energy consumed by the lamp posts and the access barriers. When no specific data were provided, we chose standard elements in supplier catalogues, produced a material balance and selected the most suitable EcoInvent processes, including the manufacturing processes as far as possible. For lighting, the line has 1 lamp post every 250 m per direction in the running section, electrical cabinets, electrical wiring and station lighting. For the stations, and for each dock, this lot has a bus shelter (12 × 1.40 m),

4 benches (70 kg of steel), a video surveillance mast (and its 881 kg foundation block made of concrete), 2 garbage bins (20 kg of stainless steel each), a technical cabinet, lighting and, if the configuration allows it, guard rails.

Each interchange hub has 10 bicycle hoops (10 kg of steel per item), barriers demarcating heights (10 kg of steel) and lifting barriers (70 kg of steel), 10 arches to protect trees and lamp posts (10 kg of steel each) and 25 kg steel safety rails.

The material balance of the lamp posts was made on the basis of the catalogue from the supplier Alunox, that of the LED lights (stations and poles) and that of the Onyx bulb lights on the catalogue Eclatec (running sections) on Cegelec data, and the balance of the electrical cabinets was based on the Environmental Product Declaration of the Schneider Electric Prisma Plus cabinet. The material balance of the bus shelters is extrapolated from the Techni-contact catalogue. The energy consumed for the installation of the lamp posts and railings was given by GETELEC: 50 liters of diesel for 10 masts and 50 L for 100 linear meters of railings. With regard to the use stage, each light has an operating time of 4,500 hour/year, the power of the LEDs being 50 W and the ONYX lamps 250 W. The consumption of the automatic barriers totals 15,000 kWh at a power of 0.18 kW.

Finally, it is assumed that all elements are manufactured in Clamart, France, except those made from concrete. They are transported by road to the port of Le Havre, then by boat to Martinique.

Note that the street furniture for green spaces is included in the following item.

3.3.3.5 Green Spaces

In this category, account was taken of the use of topsoil, plantations and their street furniture, separators, marking paints, etc. The concrete rails were provided by Satrap, including their bases, separators and extra foundations, and were transported over 5 km. The rails were provided by Getelec and transported over 3 km. The signage equipment and materials were provided by Serr and transported over 24 km. The signage and information signs are aluminum and weigh 1.46 kg. The plantation rubble was transported over 20 km and the topsoil over 25 km. A planted tree weighs 200 kg on average, and the lawn was brought in rolls (6 kg/m²). In the green spaces, the bins weigh 25 kg of steel, the bicycle racks 16 kg and the lamp posts 12 kg.

3.3.4 Results

3.3.4.1 Overall Results by Subsystem

Figure 3.6 presents the contributions of the environmental impacts according to seven subsystems: (i) running sections, (ii) stations, (iii) multimodal hubs, (iv) nonfuel and battery buses, (v) bus fuel, (vi) bus batteries and (vii) others, which includes roundabouts and emergency exits. Three of these subsystems contribute significantly to the majority of impact categories: the running section, the buses and the fuel they consume. The bus's electric batteries are responsible for a notable share for three impact categories: 16% of the solid waste (this being linked to the current difficulty of recycling batteries), as well as 21% and 18% of the aquatic and marine ecotoxicity indicators, respectively.

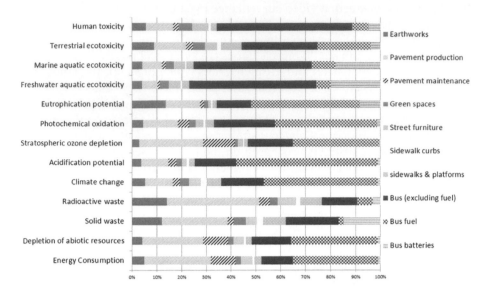

Figure 3.6 Contribution of each subsystem to the different impact categories.

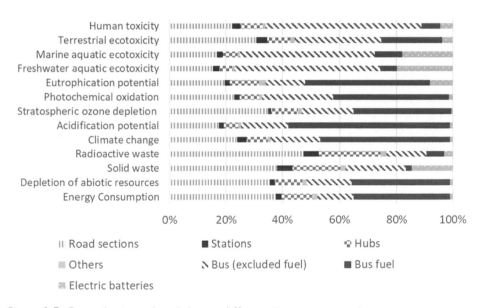

Figure 3.7 Contribution of each lot to different impact categories.

3.3.4.2 Results by Lot

Figure 3.7 highlights the respective contributions of street furniture, green spaces, and earthworks, pavement production and maintenance, curbs, sidewalks and embankments. It shows that the lots for "green spaces" and "street furniture" have low impacts that never exceed 8% each on any of the indicators. In terms of infrastructure, production of the roadway generates the greatest impact on the indicators – to the order of 30%, both in total primary energy consumption and in the destruction of the ozone layer, as well as in the depletion of nonrenewable resources and the production of solid and radioactive waste. Roadway maintenance contributes more than 10% to energy consumption and resource depletion.

3.3.4.3 Focus on Rolling Stock

Figure 3.8 presents the contributions of each stage of the bus's life cycle, isolating those concerning the batteries. It shows that the production of buses, their energy consumption during the use stage and the production of batteries are the three items that account for most of the environmental impacts. Maintenance of the bus fleet has a smaller but not negligible share, amounting to roughly 10% of the impacts according to our model.

Production of the buses has a particular impact on human toxicity, accounting for 70% of the total, as well as aquatic (56%) and marine ecotoxicity (52%).

Fuel consumption makes up 82% of the primary energy consumption and 86% of the depletion of natural resources. It is also the leading contributor to indicators of eutrophication (65%), formation of photochemical ozone (77%), ozone depletion (76%), acidification (88%) and climate change (88%).

The contribution of batteries to photochemical ozone, stratospheric ozone depletion, acidification and climate change indicators is very low (less than 5%). However, they generate a significant proportion of the toxicity and ecotoxicity indicators of the bus system (between 10% and 30%), as well as 53% of solid waste.

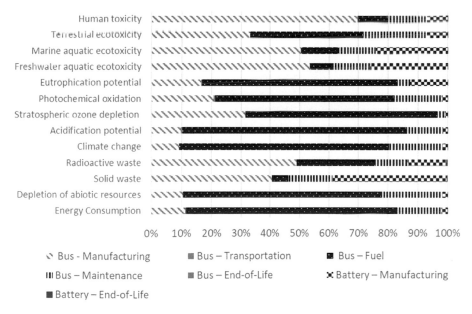

Figure 3.8 Contribution of each stage of the life cycle of the bus and its electric batteries to the different impact categories.

3.3.4.4 Focus on Street Furniture

Figure 3.9 identifies the contribution of each stage of the life cycle of street furniture to the environmental impacts. It highlights the dominance of the usage stage, that is, lighting consumption, on all indicators apart from human toxicity and the production of solid and radioactive waste.

3.3.5 Uncertainties and Sensitivity Tests

We have assessed the composition of the environmental impacts of the service rendered by the BRT. The results of the model are sensitive to various exogenous parameters: the lifetimes specified for the subsystems, the fuel consumption of the buses, lighting during operation. Any uncertainty regarding any of these exogenous parameters is reflected on the results of the evaluation *via* the model. To measure the uncertainties affecting the results, we conducted some simple analyses of local sensitivity as a first approach, parameter by parameter. This method provides information on the robustness of the model although it remains superficial regarding the overall uncertainty affecting an LCA model (Saltelli and Annoni 2010).

Let us first question the bus fuel consumption hypothesis, which is an important factor in the results. By moving from 56 L/100 km to 42 L/100 km (consumption of similar rolling stock in the city of Metz, France), environmental impacts are reduced by about 10% for two-thirds of the indicators. On the other hand, extending the bus lifespan use would have little impact on the overall balance sheet. Similarly, the use of batteries for only five years instead of 10 would only have a significant impact on solid waste generation and aquatic and marine ecotoxicity (Table 3.3).

Next, the lighting subsystem: our model allocates the impacts of a lamp post every 250 m to the BRT (in fact, only the 105 lamp posts erected in the new 2.5 km of the

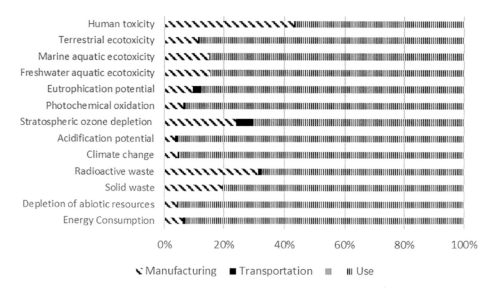

Figure 3.9 Contribution of each stage of the life cycle of urban furniture to the different impact categories.

Table 3.3 Some Sensitivity Analyses: Variation of p.km Impacts Compared to Base Case

Scenario	Base	Lamp Post/50 m	Bus Used 18 Years	Batteries Used five Years	Conso 421
Total primary energy consumption	1	+16%	−0.39%	+1.1%	−9.0%
Depletion of natural resources	1	+15%	−0.16%	+1.1%	−9.8%
Solid waste	1	+13%	−0.23%	+16.5%	−0.6%
Radioactive waste	1	+19%	−2.37%	+2.7%	−1.3%
Climate change	1	+17%	−0.24%	+1.3%	−13%
Acidification	1	+6%	−0.10%	+1.3%	−16%
Destruction of the ozone layer	1	+7%	−0.12%	+0.6%	−9.3%
Photochemical ozone formation	1	+10%	−0.14%	+1.7%	−12%
Eutrophication	1	+3%	−0.24%	+8.2%	−11%
Freshwater aquatic ecotoxicity	1	+13%	−0.56%	+22.2%	−1.6%
Marine ecotoxicity	1	+12%	−0.53%	+20.0%	−2.6%
Terrestrial ecotoxicity	1	+15%	−0.29%	+4.7%	−6.3%
Human toxicity	1	+20%	−0.25%	+5.9%	−2.1%

running section), in addition to the lighting at the stations. In fact, the dedicated lane is attached to the existing mixed traffic lanes that also require lighting, at a rate of about one lamp post every 25 m. If we allocated a lamp post to this transportation system every 50 m, its impacts would be much greater, by +3% to +20% depending on the indicator studied.

3.3.6 Summary: Results of an Eco-Profile

We have drawn up an environmental balance sheet of the BRT mode, including linear and nodal infrastructures, rolling stock and their use. This helps to identify "environmental black spots", that is, life cycle stages or elements of the system that are most harmful to the environment. The results show that the most critical parts are the running sections, the buses and the fuel consumed by these buses. Regarding infrastructure, pavement construction contributes most to the impacts measured by the indicators – about a quarter of total primary energy consumption, depletion of non-renewable resources and solid waste, among others. With regard to street furniture, the most significant part of the impact is due to the use stage and therefore to the consumption of lighting.

With regard to pollutant emissions associated with fuel consumption by vehicles, although they are the main contributors to climate change and acidification indicators, their share does not exceed 50% and 60%, respectively. However, the socioeconomic calculation mandatory in France only considers pollutant emissions in the use stage of the environmental assessment of a transportation project. In our case, it means this method neglects half of the environmental impact.

The shares of infrastructure and rolling stock are even higher given that the transportation mode has an exclusive right of way. Indeed, the infrastructure is much less used than a roadway shared by several types of vehicles. This leads to a very different distribution of environmental impacts between infrastructure and vehicles compared to a conventional road profile (de Bortoli 2014).

Thus, our study provides elements for considering the "eco-design" of a BRT mode of transportation. Given the importance of pavement construction, it would be interesting to understand which materials and processes are most harmful to the environment, in order to strive to optimize them. In this respect, the use of recycled asphalt mixes or lowering the temperature of asphalt mixes would provide two different well-known solutions, as long as it does not reduce the lifespan of the roadway in a way that counterbalances the initial environmental benefits. On the other hand, the energy consumption related to lighting plays a significant role. It is therefore possible to envisage optimization of this subsystem by playing on the technology of the bulbs used, the functioning of the lamp posts or joint optimization of the lighting and reflective properties of the road surface.

3.4 From the Eco-Profile Sheet to the Intra- and Inter-Modal Comparison

The eco-profile makes it possible to "put things in perspective", to attribute the respective contributions of the system's components to the various environmental impacts. It also identifies the environmental black spots of the transportation project and reveals targets for eco-design. At this stage, it is up to the eco-designer to imagine variants to improve the environmental performance of the project, or in other words to produce better eco-design. The evaluation method will make it possible to calculate the environmental performances of each variant quantitatively. It will constitute the test bench of the different variants, and its results will guide the progressive design and improvement of solutions throughout the project.

As a demonstration, we will continue to study the Caribbean BRT line in this section by designing and evaluating several variants: first, a variant of rolling stock, thus remaining within the same mode of transportation; then, after having defined a functional unit for the transportation service in multimodal context, we will consider variants belonging to other modes of transportation, that is, cars and trams/streetcars.

3.4.1 Variations of Rolling Stock

In the case studied thus far, the planned fleet of buses for the Caribbean line includes 14 double-articulated buses of 24 m in length, which are qualified as hybrids because they recover their braking energy. We could consider replacing them with standard 13 m thermal buses. A fleet of 28 standard buses would be required to maintain the same hourly passenger capacity.

Figure 3.10 shows the comparison of the two options, excluding the impact in terms of infrastructure, which does not change. The maximum impact value is shown in the figure before the indicator name. The complete life cycle of the rolling stock is different, in terms of the manufacturing (body, traction system), operating (different energy consumption) and maintenance, as well as EoL (especially for batteries). The two variants are not classified in the same order according to all the impact categories considered. While hybrid buses prove to be more beneficial than a fleet of shorter thermal buses that is twice the size in number – with increases ranging from 10% to 40% in terms of contribution to climate change (GHG emissions), primary energy consumption, acidification, eutrophication, formation of photochemical ozone, destruction of

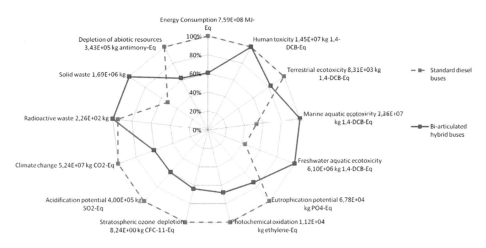

Figure 3.10 Environmental comparison of the life cycle of 14 double-articulated hybrid buses and 28 standard diesel buses.

the ozone layer, depletion of natural resources and terrestrial ecotoxicity – the classic rolling stock flect is more virtuous for the indicators concerning the production of solid waste (−45%), radioactive waste (−3%), freshwater aquatic ecotoxicity (−60%) and marine ecotoxicity (−43%).

Of course, when taking the infrastructure into account, the environmental differences are relatively weaker, as it adds some impacts that are significant for the entire service,

3.4.2 *Functional Equivalence of Transportation in a Multimodal Context*

A mode of transportation is a particular way to provide a service: a transportation service. To improve the performance of a transportation system at the environmental level (and, more broadly, according to the various aspects of sustainable development), it is necessary to use a mode of transportation within its domain of relevance, considering all the modes of transportation as options for planning the mobility system (Figure 3.11).

In such an expanded multimodal framework, the comparison of options must be based on a common standard measure: the functional unit. The function of a transportation service is to move a person or a good, allowing them to pass through a space, measured by a distance. This is why the unit of account for the "passenger transportation" function is therefore the product of a "passenger" unit by a unit of distance traveled: passenger. kilometer, also known as p.km.

To assign a quantity of impacts to a quantity of p.km, it is still necessary to know the numbers of passengers and the distances traveled along the line, in order to multiply them and to obtain the quantity of p.km. This quantity can be predicted by a traffic study.

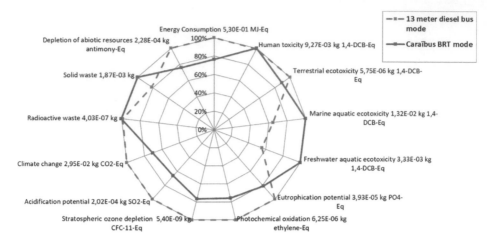

Figure 3.11 Environmental impacts of the standardized line per p km over a 28-year observation period, according to the average occupancy rate: intermodal comparison.

Before considering other modal options, we first normalized the impacts of the Caraïbus line per passenger-kilometer, according to three assumptions of the number of passengers using the line: optimistic, moderate and pessimistic. This makes it possible to frame the environmental impacts on the line per p.km over 28 years (Figure 3.12: strictly speaking, the number of passengers transported by bus, and thus the mass, varies according to the occupancy rate, but we did not take this into account in the evaluation of the bus consumption).

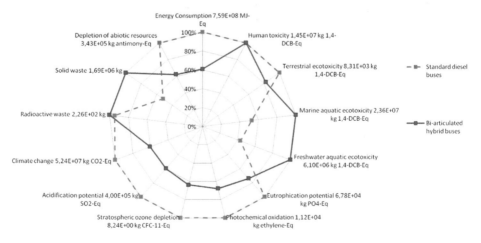

Figure 3.12 Eco-comparison of several modes of transportation: BRT according to two types of rolling stock (occupancy rate: 50%) and two automobile modes with contrasting occupancy rates.

3.4.3 Comparison with Private or Multi-Occupant Cars

We selected both variants of the bus fleet under the median assumption of a 50% passenger load, to compare them to two automotive-based transportation solutions. The first automobile option is a private car mode (PC), called "standard PC mode", modeled according to EcoInvent processes and assuming an occupancy rate of 1.2 people present in the car on average. The second automobile mode is more shared: multi-occupied cars with three people on average.

We modeled these two automobile modes by contextualizing them in relation to the route followed by the BRT line. Figure 3.13 shows that bus services are both much more efficient than standard PC for all indicators considered. The multi-occupancy car is quite competitive against buses that are only half full.

N.B. The results are approximate since it would be necessary to specify the velocity and acceleration profiles, which are important in the operational stage of vehicles for the kind of use.

3.4.4 Multimodal Comparison: BRT, Buses, Cars, Trolleybus, Trams

Let us complete our exploration by putting the BRT mode according to the Caraïbus specification in comparison with private cars, a standard bus mode, a trolleybus mode and a tram mode. For these last four modes, we simply imported their impacts from the EcoInvent database for a European (car) or Swiss context (other modes), without adapting them to the Martinican context.

The environmental performance of these different modes is evaluated in terms of energy consumption and also the impact on global warming in Figure 3.14. Assuming an occupancy rate of 50%, BRT remains more efficient than the standard PC by a factor of 5, the latter consuming 3.81 MJ eq and emitting 248 g CO_2 eq per p km. Compared to the classic bus mode, the gains of the BRT are very significant: 54% and 65% for the two indicators, respectively. Compared to the trolleybus in a Swiss context, BRT

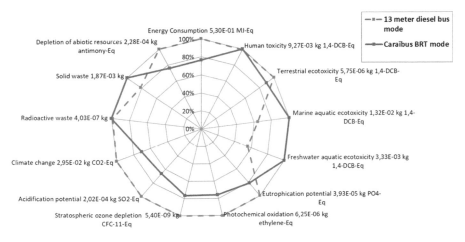

Figure 3.13 Energy and "carbon" comparison of competing modes of BRT. (European average inventories. (EcoInvent version V2.2).)

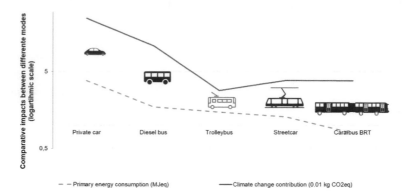

Figure 3.14 Energy and "carbon" comparison of competing modes of BRT. (European or Swiss average inventories. (EcoInvent version V3, CML and CED methods).)

is twice as energy efficient but contributes more to global warming (+35%). Lastly, in relation to trams, the 50% filled BRT is a better solution according to the two impact criteria: by 38% for the energy aspect, but only by a little in terms of the climate change aspect.

3.4.5 Summary and Discussion

Let's first address the comparison in terms of LCA's aptitude for assessing the environmental performance of transportation systems, and second the general need to contextualize the evaluation and consider a range of indicators.

The indicators do not unanimously decide between the fleet of long hybrid buses and the fleet of twice as many short diesel buses, far from it. Between the Caraïbus BRT and automobiles, everything depends on the respective occupancy rates. Multi-occupant cars (three people) seem to be a competitive challenger against a half-filled BRT, better than a classic mode of bus, a trolley bus or a tram in the Swiss version. Admittedly, cars driven alone come last in all respects.

Strictly speaking, more details should also be given on how the traffic works, on the demand side as well as on the supply side. On the demand side, the size of the bus and that of the fleet condition the frequency of service, and therefore the waiting time and attractiveness of the mode for the demand, thus *ultimately* affecting the level of usage and the filling rate. On the supply side, a larger fleet of vehicles exerts stronger impacts on the general operation of traffic: this is evident at intersections where these vehicles have priority. Indirect effects of congestion follow from this. To evaluate things more finely, the impact assessment would ideally be coupled with the simulation of supply and demand.

3.5 What Training for Eco-Design of Transportation?

Eco-design of transportation systems is a promising topic that deserves to be mastered by professionals trained in the discipline. To promote the development of LCAs for transportation systems, and eco-design engineering for these systems and even for territories more broadly, ParisTech Engineering School has created specific training courses in its engineering and other master's programs.

3.5.1 Eco-Design and Training in Urban Engineering

First developed in industrial ecology in the 1970s, eco-design is an approach to designing a product or service that is oriented towards a more resource-friendly use over the entire life cycle, by aspiring to frugality, recycling, and, more generally, minimizing negative impacts on the environment, such as consumption of materials and energy, emissions of polluting substances, damage to the natural environment and health of populations, as well as harm to biodiversity. It is spreading progressively in all sectors of economic activity thanks to the general awareness of the issues of sustainable development and through the broad mobilization for energy transition and the fight against climate change. Training students in the concepts and methods of eco-design in order to prepare them to carry out their profession at the service of companies and society is a commitment of the ParisTech Engineering School (ENPC) in particular, because of its traditional contribution to the technical planning of territories and its state-run administration. Mastering the complexity of the urban environment, or at least recognizing its hyper-complexity and respecting it in the design methods taught, constitutes a multidisciplinary educational challenge combining urban engineering, construction, transportation systems, urban services and environmental engineering.

3.5.2 Eco-Design Courses Offered at ParisTech Engineering School

At ENPC, eco-design concepts and more applied techniques of LCA are taught at several points of the students' progress, and across the school's three main training programs: the engineering course, other master's degrees and the specialized master's degrees. Awareness can be first raised during the first-year "scientific projects" of the engineering program, where the school's researchers propose a choice of different subjects to study. Thus, in 2015, three themes relating to eco-design more broadly were addressed by about 15 students: the study of the operating costs of road vehicles according to the investment in road maintenance, the life cycle cost analysis of pavements in partnership with IFSTTAR and CEREMA, and the LCA of French roads.

An "opening week" – a themed training week that takes place at the beginning of the semester for students in the second and third year of the engineering program – is given by the Ville, Environnement, Transportation department (VET, City, Environment and Transportation) as an "Introduction to Life Cycle Assessment". Through both theoretical and practical training, it aims to introduce students to the methodologies to quantitatively assess the environmental impacts of various activities (products, services, constructions), but remains primarily oriented towards LCA of buildings since it is based on the practical use of the dedicated EQUER tool. The knowledge is applied through practical work and a workshop on producing an ecological balance sheet for a detached house.

After this week, the VET students can take specialized training courses in eco-design, which we will present below. Since the start of the 2017 academic year, students from the "Civil Engineering and Construction" training department in the engineering or master's program have also been offered an introductory course on the LCA of building systems.

A well-established course is taught to various types of students in the "Transport et Développement Durable" (TraDD, Transportation and Sustainable Development) master's program. Its objectives are the general understanding of LCA, including its principles, methodology and applications, knowledge of the environmental profiles of the main means of transportation, learning simplified calculations and taking critical distance.

Lastly, the specialized master's degree in Action Publique pour le Développement Durable (PAPDD, Public Policies and Actions for Sustainable Development), which is mainly attended by engineering students from the Corps des Ponts, des Eaux et des Forêts (Corps of Bridges, Waters and Forests) and engineering students from the "Services Techniques de la Ville de Paris" (Technical Services of the City of Paris), also provides an opportunity to educate public engineers on the eco-design of transportation. In 2015–2016, the first group of students investigated the environmental performance of the revegetation of construction projects, like transportation projects, with a particular focus on quantifying it in environmental balance sheets through LCA. In 2016–2017, a second group of students focused on environmental performance criteria in public procurement and the introduction of LCA as a decision support tool in the road sector. The interactions of the stakeholders were analyzed through a refined understanding of the legal, technical and administrative frameworks, and a supplier market survey.

3.5.3 The Specialized Stream in "Eco-Design for Sustainable Cities"

The challenge of integrating eco-design into urban engineering training was raised in 2014 with the creation of a specialized stream: the "Eco-design for Sustainable Cities" advanced stream proposed in the last year of the engineering program in the VET department (Kotelnikova-Weiler et al. 2016). Supported by the ParisTech-VINCI industrial research chair, the training is based on five specialized courses. Prominent among these is a course in LCA of transportation systems, in a project-oriented form that offers each promotion of students a real case study in which they can apply LCA to a concrete transportation project. After having passed on the methodological bases of LCA to the students, the objective of the module is to get them to build an LCA model of one or more transportation modes for the practical case being considered, in order to finally evaluate the environmental performances of one or several variants. The models use the EcoInvent database, a set of "process" models consisting of inventories of physical elementary substances entering and exiting a unitary system throughout its life cycle, which can be industrial product or a service in a territorial context. Transportation system models are constructed through a detailed physical analysis of the mode of transportation in its context, including the nature and quantities of the substances used and emitted. In this chapter, we have given a detailed presentation of the case study proposed for the first promotion of students (2014–2015) and then reworked by the teachers of the module. This module can also serve as an opportunity to open the teaching up to a wider audience through lectures. The first lecture, which researchers and students on campus were able to attend, was organized at ParisTech engineering school in October 2014 to present eco-design at Eurovia and the use of eco-comparators and LCA in the road sector. Likewise, the ParisTech-Vinci Chair in eco-design regularly holds conferences for a professional audience, such as quarterly technical evenings and annual autumn schools. Our specialized students participate in these events in order to benefit from the views of the industrial figures present, to establish their professional networks or to present their own work.

Other studies have been carried out since 2014: the evaluation of the TZen3 BRT project in the northeast of Paris, in partnership with Ile-de-France Mobilités (the public authority for transportation across the Paris region), the eco-profile of the construction of the SEA high-speed train line between Tours and Bordeaux in partnership with the concession holder LISEA, and the comparative LCA of alternative modes of transportation – plane, conventional or low-cost train, bus, individual car use and carpooling – on the Paris–Bordeaux corridor in collaboration with CEREMA.

These four different studies have been highlighted in many French and international scientific conferences: the 2015 World Road Congress in Seoul, the SETAC specialized conference on LCA case studies in Montpellier in 2010, the sustainable building systems (SBE) conference in Hamburg in 2016, or the Conference on Life Cycle Management (LCM) in Luxembourg in 2017. They also provided an opportunity for some of our students to publish their first research articles (e.g., de Bortoli et al. 2015).

3.6 Conclusion

3.6.1 Summary and Key Messages

We constructed an LCA model for a passenger transportation line. We evaluated the environmental impacts both of the version of the project that was actually implemented in practice and of design variants. We were able to compare these variants in light of environmental issues.

Through this journey, we experienced the practicability of LCA for a transportation system. This experience has been corroborated by other research teams in several countries. Over the past 20 years, LCA of transportation systems has become a relatively lively scientific field supported by an international academic community. The various applications carried out have favored subsystems related to the road mode (infrastructure and/or vehicles) for the most part, but not exclusively.

Based on our practical experience of the study presented here, as well as others carried out with our students and the academic literature still being developed, we have drawn several lessons. Here, they are in the form of key messages:

- In comparison with the traditional socioeconomic assessment of projects and policies, LCA offers a real revolution in terms of assessing environmental impacts.
- The modeling of a project, service, policy or transportation system through LCA must be contextualized, that is, adapted to the territory in which the transportation in question is implemented. This contextualization, called "spatialization" in LCA, leads to a better understanding of the interaction between the project in this territory and the local and global environment, but it has a certain degree of methodological and technical complexity.

- LCA modeling crosses and goes beyond spatial scales. It addresses environmental impacts in their entirety, from small to large impact sources – from the bike racks to motorway sections.
- Thanks to attributional analysis, LCA modeling makes it possible to trace the sources of environmental impacts, their causal processes and their levers. It provides an opportunity to identify action targets and to simulate the magnitude of their environmental effects, by playing on design variables taken as sliders.

Faced with a growing social demand for a public policy that reconciles the three pillars of sustainable development, it seems necessary to develop adapted decision support tools in response and to also reinforce the dissemination of knowledge, especially in terms of evaluation – and environmental assessment in particular – among the actors involved in the decision-making chain. Technically, LCA is now a sufficiently mature method of quantifying environmental impacts that is appropriate for the transportation sector and could be applied in a systematic way. As illustrated in Figure 3.15, it can be used as an environmental decision support tool at any point in the process of producing transportation services, from multiscale planning by public authorities to operation, passing through all the major stages of the transportation project – planning, design and construction. There is still no integrated operational tool to guide investments towards sustainable mobility through global optimization of transportation systems instead of the classic silo approach, which is intrinsic to the organization of the stakeholder system that separates vehicles on one side and infrastructures on the other.

As demonstrated by a conclusive test conducted at ParisTech engineering school on the TZen3 BRT project in northeast of Paris (de Bortoli 2016), LCA could nevertheless be used in the socioeconomic calculation that is mandatory for projects financed by the state. The aim would be to widen its scope of environmental evaluation, which is currently restricted to the use stage, as well as the range of environmental impacts taken into account, since this restricted perimeter can lead to biased decision-making.

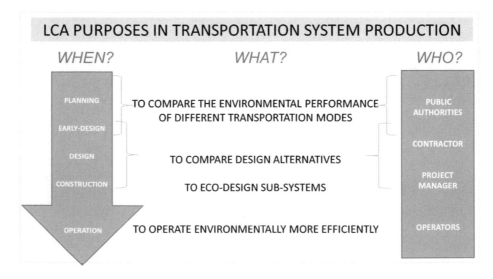

Figure 3.15 Use of LCA in the transportation system production process.

3.6.2 The Need for Further In-Depth Scientific and Technical Studies

Although it is the most complete method for environmental assessment available to-day, LCA can still be improved. It offers a panel of impact categories and indicators chosen by the analyst, omitting other such local impacts on the landscape or noise; its implementation involves uncertainties; its results depend on the choice of the system boundaries and assumptions, and are variable according to the model's degree of specificity (via the spatialization of life cycle inventories, for example). We could also mention the feasibility of LCA within the constrained times of transportation project, and the significantly heightened problems of contextualization and therefore of the quality of environmental inventories during ex ante studies.

In the absence of an LCA tool specifically tailored to transportation projects – which should be based on case studies such as those presented in this chapter – the need for expert evaluators combined with quality data now constitutes an obstacle to the systematic analysis of this type of project. Establishing detailed data is crucial and requires the contribution and cooperation of multiple actors: designers, builders, subcontractors, etc. (Figure 3.14).

An evaluation method called "Environmental Input-Output" (Kitzes 2013) would reduce both the time required for environmental comparisons and the potential complexity of data centralization, while maintaining a broad scope of evaluation. Nevertheless, it remains rather adapted to rough average assessments in the territory covered by the input–output matrices that link the environmental impacts to the activity sectors, and thus to the perimeter of the territorial planning rather than that of the local transportation project. An LCA implementation track for the latter scale would involve the production of more spatialized inventories related to transportation projects and their inclusion into a multimodal transportation LCA tool, ideally with a user-friendly and intelligently parameterized interface, which could be modified by the user to better address the case being studied. This would accelerate the completion of LCA studies on behalf of both the project owners and prime contractors.

As shown by our sensitivity tests, it also seems important to estimate the uncertainties and the variability of the calculated impacts, in order to exclusively retain the results and computation that are sufficiently robust. In particular, the forward-looking dimension of LCA seems to be a crucial asset for rationalizing a technical appraisal on objects with a long life span, the characteristics and therefore the environmental impacts of which are likely to evolve over time. However, this should be accompanied by a consideration of the uncertainties.

3.6.3 Towards Operationalization and Institutionalization

While research continues to refine the methods for applying LCA to transportation systems, other actors, both public and private, are already taking advantage of LCA to develop operational tools in order to increase the robustness of the results produced. In France, two eco-comparators, ECORCE (Ventura et al. 2012) and SEVE (IDRRIM 2013), were set up under the impetus of the "Grenelle de l'Environnement" (Grenelle Roundtable) in 2009 (de Bortoli 2015). These initiatives deserve to be pursued, in synergy with the development of the training offered in LCA and the spreading of knowledge in LCA among wider audiences.

Indeed, the relevance and power of LCA for environmental assessment deserve the full attention of the social actors who carry out system transformation projects. This is especially true in the case of transportation systems given their major economic and social challenges and the heavy environmental burden they constitute.

The public authorities hold an eminent position among the actors concerned, by virtue of their responsibility to represent the community and to arbitrate and manage the social issues, and to prepare the future for generations to come. National authorities have been able to develop the socioeconomic assessment of transportation projects and policies, to provide a regulatory framework and to gradually add an environmental component to it. Some partially calculated environmental impacts have already been incorporated into the regulatory assessment method, which aims to discern public utility. However, there is still a need to incorporate other categories of significant environmental impacts, and to consider the full life cycle of projects.

Operationalization seems feasible at the scientific and technical level, and this has already been outlined by the various contributions made. A detailed cartography still needs to be established to detect and fill gaps, to constitute a directory of process models and to allow for diversification within the territorial context: this is a whole development program, in short.

It will also be necessary to make the issues associated with the impacts perceptible and intelligible for the actors concerned. This is an educational challenge that is fundamental for the institutionalization of LCA in environmental assessment, which is necessary in order for it to be articulated to the socioeconomic evaluation of projects as practiced by the public authorities. It would be counterproductive to make decisions based on misunderstood and incorrectly interpreted indicators. The need for training is clear.

Finally, one operational rather than research-based issue remains, namely, that expertise needs to be established regarding the trade-offs between the multiple environmental stakes and the economic and social stakes. The values fixed to monetize the equivalent ton of CO_2 are emblematic of such arbitrations. It is by balancing the different issues of the real cases, and by expressing the collective preferences in favor of this or that option on a case-by-case basis, that we can determine appropriate values intended to simplify the quest to optimize the eco-design of transportation systems.

Bibliography

Abdo, J. 2011. "Analyse Du Cycle de Vie de Structures Routières." T89. Collection Technique CimBéton - BÉTON ET DÉVELOPPEMENT DURABLE. CIMBETON. http://www.infociments.fr/publications/route/collection-technique-cimbeton/ct-t89.

Akbarian, M., S. S. Moeini-Ardakani, F.-J. Ulm, and M. Nazzal. 2012. "Mechanistic approach to pavement-vehicle interaction and its impact on life-cycle assessment." *Transportation Research Record: Journal of the Transportation Research Board* 2306 (1): 171–79. https://doi.org/10.3141/2306-20.

Ally, J., and T. Pryor. 2007. "Life-cycle assessment of diesel, natural gas and hydrogen fuel cell bus transportation systems." *Journal of Power Sources* 170 (2): 401–11. https://doi.org/10.1016/j.jpowsour.2007.04.036.

Araújo, J. P. C., J. R. M. Oliveira, and H. M. R. D. Silva. 2014. "The importance of the use phase on the LCA of environmentally friendly solutions for asphalt road pavements."

Transportation Research Part D: Transport and Environment 32 (October): 97–110. https://doi.org/10.1016/j.trd.2014.07.006.

Athena Institute. 2006. *A Life Cycle Perspective on Concrete and Asphalt Roadways: Embodied Primary Energy and Global Warming Potential.* Ottawa: Cement Association of Canada.

AzariJafari, H., A. Yahia, and M. B. Amor. 2016. "Life cycle assessment of pavements: Reviewing research challenges and opportunities." *Journal of Cleaner Production* 112 (January): 2187–97. https://doi.org/10.1016/j.jclepro.2015.09.080.

Banar, M., and A. Özdemir. 2015. "An evaluation of railway passenger transport in Turkey using life cycle assessment and life cycle cost methods." *Transportation Research Part D: Transport and Environment* 41 (December): 88–105. https://doi.org/10.1016/j.trd.2015.09.017.

Bauer, C., J. Hofer, H.-J. Althaus, A. D. Duce, and A. Simons. 2015. "The environmental performance of current and future passenger vehicles: Life cycle assessment based on a novel scenario analysis framework." *Applied Energy* 157 (November): 871–83. https://doi.org/10.1016/j.apenergy.2015.01.019.

Bellman, R. 1952. "On the theory of dynamic programming." *Proceedings of the National Academy of Sciences* 38 (8): 716–19. https://doi.org/10.1073/pnas.38.8.716.

Berger, A. 2006. "Les Impacts Du Réseau Routier Sur l'environnement." Monthly Themed Letter of the Institut Français de l'environnement, no. 114 (October). http://www.statistiques.developpement-durable.gouv.fr/fileadmin/documents/Produits_editoriaux/ Publications/ Le_Point_Sur/2006/de114.pdf.

Berthiaume, R., and C. Bouchard. 1999. "Exergy analysis of the environmental impact of paving material manufacture." *Transactions of the Canadian Society for Mechanical Engineering* 23: 187–96.

Bryce, J., S. Katicha, G. Flintsch, N. Sivaneswaran, and J. Santos. 2014. "Probabilistic life-cycle assessment as network-level evaluation tool for use and maintenance phases of pavements." *Transportation Research Record: Journal of the Transportation Research Board* 2455 (December): 44–53. https://doi.org/10.3141/2455-06.

Bryce, J., J. Santos, G. Flintsch, S. Katicha, K. McGhee, and A. Ferreira. 2014. "Analysis of rolling resistance models to analyse vehicle fuel consumption as a function of pavement properties." In *Asphalt Pavements*, by Y. Kim, 263–73, Boca Raton, FL: CRC Press. https://doi.org/10.1201/b17219-39.

CGDD. 2015. "Bilan 2012 de La Production de Déchets En France." 615. Chiffres & Statistiques. CGDD.

Chan, A. W.-C. 2007. *Economic and Environmental Evaluations of Life Cycle Cost Analysis Practice: A Case Study of Michigan DOT Pavement Projects. Master of Science Thesis Dissertation in the Natural Resource and Environment*, Ann Arbor, MI: University of Michigan.

Chester, M. 2008. *Life-Cycle Environmental Inventory of Passenger Transportation in the United States.* Dissertations. Berkeley, CA: Institute of Transportation Studies, UC Berkeley.

Chester, M., W. Eisenstein, S. Pincetl, Z. Elizabeth, J. Matute, and P. Bunje. 2012. *Environmental Life-Cycle Assessment of Los Angeles Metro's Orange Bus Rapid Transit and Gold Light Rail Transit Lines.* (Arizona State University Report No. SSEBE- CESEM-2012-WPS-003) Tempe, AZ: Arizona State University. https://repository.asu.edu/attachments/94226/content/ chester-ASU-SSEBE-CESEM-2012-WPS-003.pdf.

Chester, M., A. Fraser, J. Matute, C. Flower, and R. Pendyala. 2015. "Parking infrastructure: A constraint on or opportunity for urban redevelopment? A study of Los Angeles county parking supply and growth." *Journal of the American Planning Association* 81 (4): 268–86. https://doi.org/10.1080/01944363.2015.1092879.

Chester, M., and A. Horvath. 2010. "Life-cycle assessment of high-speed rail: The case of California." *Environmental Research Letters* 5 (1): 014003. https://doi.org/10.1088/1748-9326/5/1/014003.

Chester, M., S. Pincetl, Z. Elizabeth, W. Eisenstein, and J. Matute. 2013. "Infrastructure and automobile shifts: Positioning transit to reduce life-cycle environmental impacts for urban sustainability goals." *Environmental Research Letters* 8 (1): 015041. https://doi.org/10.1088/1748-9326/8/1/015041.

Chester, M. V., and A. Cano. 2016. "Time-based life-cycle assessment for environmental policymaking: Greenhouse gas reduction goals and public transit." *Transportation Research Part D: Transport and Environment* 43 (March): 49–58. https://doi.org/10.1016/j.trd.2015.12.003.

Chester, M. V., A. Horvath, and S. Madanat. 2010. "Comparison of life-cycle energy and emissions footprints of passenger transportation in metropolitan regions." *Atmospheric Environment* 44 (8): 1071–79. https://doi.org/10.1016/j.atmosenv.2009.12.012.

Chester, M. V., M. J. Nahlik, A. M. Fraser, M. A. Kimball, and V. M. Garikapati. 2013. "Integrating life-cycle environmental and economic assessment with transportation and land use planning." *Environmental Science & Technology* 47 (21): 12020–28. https://doi.org/10.1021/es402985g.

Chester, M. V., and M. S. Ryerson. 2014. "Grand challenges for high-speed rail environmental assessment in the United States." *Transportation Research Part A: Policy and Practice* 61 (March): 15–26. https://doi.org/10.1016/j.tra.2013.12.007.

Chong, D., and Y. Wang. 2017. "Impacts of flexible pavement design and management decisions on life cycle energy consumption and carbon footprint." *The International Journal of Life Cycle Assessment* 22 (6): 952–71. https://doi.org/10.1007/s11367-016-1202-x.

CITEPA. 2016a. "Oxydes d'azote - NOx." CITEPAd. https://www.citepa.org/fr/air-et-climat/polluants/aep-item/oxydes-d-azote.

CITEPA. 2016b. "Poussières En Suspension." CITEPAa. https://www.citepa.org/fr/air-et-climat/polluants/poussieres-en-suspension.

Cooney, G., T. R. Hawkins, and J. Marriott. 2013, April. "Life cycle assessment of diesel and electric public transportation buses: LCA of diesel and electric buses." *Journal of Industrial Ecology* https://doi. org/10.1111/jiec.12024.

Court of Auditors. 2015. "Les Politiques Publiques de Lutte Contre La Pollution de l'air." Court of Auditors. https://www.ccomptes.fr/Publications/Publications/Les-politiques- publiques-de-lutte-contre-la-pollution-de-l-air.

Cox, B. L., C. L. Mutel, C. Bauer, A. M. Beltran, and D. v. Vuuren. 2018, March. "The uncertain environmental footprint of current and future battery electric vehicles." *Environmental Science & Technology.* https://doi.org/10.1021/ acs.est.8b00261.

Dave, S. 2010. *Life Cycle Assessment of Transportation Options for Commuters.* Cambridge, MA: Massachusetts Institute of Technology.

de Bortoli, A. 2014. "Eco-Concevoir l'entretien Routier: Contexte National et Identification Des Enjeux Énergétiques Du Système Routier." *Revue Générale Des Routes et de l'Aménagement*, no. 920 (June).

de Bortoli, A. 2015. "Assessing environmental impacts of road projects: The recent development of specialized eco-comparators in France." In *Proceedings of the XXVth World Road Congress.* Seoul: World Road Association.

de Bortoli, A. 2016. "Consequential environmental life cycle assessment and socio-economic analysis - hybridization test on a Parisian project of bus rapid transit." https://doi.org/10.13140/rg.2.2.33786.70084.

de Bortoli, A., A. Féraille, and F. Leurent. 2017. "Life cycle assessment to support decision-making in transportation planning: A case of French bus rapid Transit." In *Proceedings of the Transportation Research Board 2017.* Washington, DC. https://trid.trb.org/view.aspx?id=1439665.

de Bortoli, A., A. Feraille, G. Thing-Leo, A. Mokssit, Y. Qi, Z. Li, S. Luo, and F. Leurent. 2015, November 2–6. "Performance environnementale des modes de transport urbain sur cycle de vie: quels choix techniques pour le bus à haut niveau de service [Environmental performance of urban transportation modes on life cycle: What technical choices for the Bus Rapid Transit?]." In *Proceedings of the XXVth World Road Congress*. Seoul: World Road Association. https://hal.archives-ouvertes.fr/ hal–01261862.

Edenhofer, O., R. Pichs-Madruga, Y. Sokona, K. Seyboth, P. Matschoss, S. Kadner, T. Zwickel, P. Eickemeier, G. Hansen, S. Schloemer, C. von Stechow 2014. "Climate change 2014: Mitigation of climate change." *Fifth assessment report. United Nations report – IPCC - Working group III contribution to AR5*. http://www.ipcc.ch/publications_and_data/publications_and_data_reports.shtml.

Ercan, T., and O. Tatari. 2015. "A hybrid life cycle assessment of public transportation buses with alternative fuel options." *The International Journal of Life Cycle Assessment* 20 (9): 1213–31. https://doi.org/10.1007/s11367-015-0927-2.

European Environment Agency. 2017. "Environmental indicator report 2017 in support to the monitoring of the seventh environment action programme." EEA report No 21/2017. European Environment Agency. https://www.eea.europa.eu/publications/environmental-indicator-report-2017/at_download/file.

Field, C. B., V. R. Barros, D. J. Dokken, K. J. Mach, M. D. Mastrandrea, T. E. Bilir, M. Chatterjee, et al., eds. 2014. *Climate Change 2014: Impacts, Adaptation, and Vulnerability; Summaries, Frequently Asked Questions, and Cross-Chapter Boxes; A Working Group II Contribution to the Fifth Assessment Report of the Intergovernmental Panel on Climate Change*. Geneva, Switzerland: Intergovernmental Panel on Climate Change.

Greene, S., M. Akbarian, F.-J. Ulm, and J. Gregory. 2013. "Pavement roughness and fuel consumption." https://cshub.mit.edu/sites/default/files/ documents/PVIRoughness_v15.pdf.

Gregory, J. R., A. Noshadravan, E. A. Olivetti, and R. E. Kirchain. 2016. "A methodology for robust comparative life cycle assessments incorporating uncertainty." *Environmental Science & Technology* 50 (12): 6397–6405. https://doi. org/10.1021/acs.est.5b04969.

Häkkinen, T., and K. Mäkelä. 1996. "Environmental adaption of concrete: Environmental impact of concrete and asphalt pavements." *VTT Tiedotteita-Meddelanden-Research Notes 1752. Valtion teknillinen tutkimuskeskus (VTT): Technical Research Centre of Finland*, ESPOO.

Hawkins, T. R., B. Singh, G. Majeau-Bettez, and A. H. Strømman. 2013. "Comparative environmental life cycle assessment of conventional and electric vehicles: LCA of conventional and electric vehicles." *Journal of Industrial Ecology* 17 (1): 53–64. https://doi. org/10.1111/j.1530-9290.2012.00532.x.

Helmers, E., J. Dietz, and S. Hartard. 2015, July. "Electric car life cycle assessment based on real-world mileage and the electric conversion scenario." *The International Journal of Life Cycle Assessment*. https://doi.org/10.1007/s11367-015-0934-3.

Hoehne, C. G., and M. V. Chester. 2017. "Greenhouse gas and air quality effects of auto first-last mile use with transit." *Transportation Research Part D: Transport and Environment* 53 (June): 306–20. https://doi.org/10.1016/j.trd.2017.04.030.

Horvath, A. 2006. "Environmental assessment of freight transportation in the U.S. (11 Pp)." *The International Journal of Life Cycle Assessment* 11 (4): 229–39. https://doi.org/10.1065/lca2006.02.244.

Horvath, A., and C. Hendrickson. 1998. "Comparison of environmental implications of asphalt and steel-reinforced concrete pavements." *Transportation Research Record* 1626 (1): 105–13. https://doi.org/10.3141/1626-13.

Huang, Y., R. Bird, and O. Heidrich. 2009. "Development of a life cycle assessment tool for construction and maintenance of asphalt pavements." *Journal of Cleaner Production* 17 (2): 283–96. https://doi.org/10.1016/j.jclepro.2008.06.005.

IDRRIM. 2013. "SEVE, Système d'Evaluation de Variantes Environnementales V 2." No. 160. Technical advice. http://www.idrrim.com/link/dl?site=fr&objectId=1936.

Inyim, P., J. Pereyra, M. Bienvenu, and A. Mostafavi. 2016. "Environmental assessment of pavement infrastructure: A systematic review." *Journal of Environmental Management* 176 (July): 128–38. https://doi.org/10.1016/j.jenvman.2016.03.042.

Jolliet, O., Saadé, M. and Crettaz, P. 2005. *Analyse Du Cycle de Vie - Comprendre et Réaliser Un Écobilan. Gérer l'environnement.* Presses polytechniques et universitaires romandes.

Jones, H., F. Moura, and T. Domingos. 2016, August. "Life cycle assessment of high-speed rail: A case study in Portugal." *The International Journal of Life Cycle Assessment.* https://doi.org/10.1007/s11367-016-1177-7.

Jullien, A., C. Proust, T. Martaud, E. Rayssac, and C. Ropert. 2012. "Variability in the environmental impacts of aggregate production." *Resources, Conservation and Recycling* 62 (May): 1–13. https://doi.org/10.1016/j.resconrec.2012.02.002.

Kimball, M., M Chester, C. Gino, and J. Reyna. 2013. "Assessing the potential for reducing life-cycle environmental impacts through transit-oriented development infill along existing light rail in Phoenix." *Journal of Planning Education and Research* 33 (4): 395–410. https://doi.org/10.1177/0739456X13507485.

Kitzes, J. 2013. "An introduction to environmentally-extended input-output analysis." *Resources* 2 (4): 489–503. https://doi.org/10.3390/resources2040489.

Kliucininkas, L., J. Matulevicius, and D. Martuzevicius. 2012. "The life cycle assessment of alternative fuel chains for urban buses and trolleybuses." *Journal of Environmental Management* 99 (May): 98–103. https://doi.org/10.1016/j.jenvman.2012.01.012.

Kotelnikova-Weiler, N., A. de Bortoli, A. Feraille, and F. Leurent. 2016. "Integrating Urban Ecodesign in French Engineering Curricula: An Example at École Des Ponts ParisTech." In *Proceedings of the Sustainable Built Environment Conference 2016*, Hamburg, Germany. https://hal.archives-ouvertes.fr/hal-01586997/document.

Le Féon, S. 2014. "Evaluation Environnementale Des Besoins de Mobilité Des Grandes Aires Urbaines En France - Approche Par Analyse de Cycle de Vie." Ecole Nationale Supérieure des Mines de Saint-Etienne. http://tel.archives-ouvertes.fr/docs/00/98/01/87/PDF/ Le_FA_on-Samuel-diff.pdf.

Li, B., J. Li, and C. Yuan. 2013. "Life cycle assessment of lithium ion batteries with silicon nanowire anode for electric vehicles." In *Proceedings of the International Symposium on Sustainable Systems and Technologies*. Figshare. http://dx.doi.org/10.6084/m9.figshare.805147.

Lin, T.-H., Y.-S. Chien, and W.-M. Chiu. 2016. "Rubber tire life cycle assessment and the effect of reducing carbon footprint by replacing carbon black with graphene." *International Journal of Green Energy* 14 (November): 97–104. https://doi.org/10.1080/1 5435075.2016.1253575.

Louhghalam, A., M. Akbarian, and F.-J. Ulm. 2014. *Pavement Infrastructures Footprint: The Impact of Pavement Properties on Vehicle Fuel Consumption.* Cambridge, MA: Massachusetts Institute of Technology. https://cshub.mit.edu/sites/default/files/documents/louhghalam-Euro-C-2014.pdf.

Matute, J. M., and M. V. Chester. 2015. "Cost-effectiveness of reductions in greenhouse gas emissions from high-speed rail and urban transportation projects in California." *Transportation Research Part D: Transport and Environment* 40 (October): 104–13. https://doi.org/10.1016/j.trd.2015.08.008.

Messagie, M., C. Macharis, and J. Van Mierlo. 2013. Key Outcomes from Life Cycle Assessment of Vehicles, a State of the Art Literature Review. In *2013 World Electric Vehicle Symposium and Exhibition (EVS27)*, 1–9. Barcelona: IEEE. https://doi.org/10.1109/EVS.2013.6915045 Barcelona.

Michalek, J. J., M. Chester, P. Jaramillo, C. Samaras, C.-S. N. Shiau, and L. B. Lave. 2011. "Valuation of plug-in vehicle life-cycle air emissions and oil displacement benefits." *Proceedings of the National Academy of Sciences* 108 (40): 16554–58. https://doi.org/10.1073/pnas.1104473108.

Miyoshi, C., and M. Givoni. 2013. "The environmental case for the high-speed train in the UK: Examining the London–Manchester route." *International Journal of Sustainable Transportation* 8 (2): 107–26. https://doi.org/10.1080/15568318.2011.645124.

Muga, H. E., A. Mukherjee, J. R. Mihelcic, and M. J. Kueber. 2009. "An integrated assessment of continuously reinforced and jointed plane concrete pavements." Edited by Charles Egbu. *Journal of Engineering, Design and Technology* 7 (1): 81–98. https://doi.org/10.1108/17260530910947277.

Nahlik, M. J., and M. V. Chester. 2014. "Transit-oriented smart growth can reduce life- cycle environmental impacts and household costs in Los Angeles." *Transport Policy* 35 (September): 21–30. https://doi.org/10.1016/j.tranpol.2014.05.004.

Nahlik, M. J., A. T. Kaehr, M. V. Chester, A. Horvath, and M. N. Taptich. 2016. "Goods movement life cycle assessment for greenhouse gas reduction goals: Goods movement LCA for GHG reduction goals." *Journal of Industrial Ecology* 20 (2): 317–28. https://doi.org/10.1111/jiec.12277.

Nisbet, M. A., M. L. Marceau, M. G. Van Geem, and I. L. Gajda. 2001. *Environmental Life Cycle Inventory of Portland Cement Concrete and Asphalt Concrete Pavements.* 2489. PCA R&D Serial. Skokie, IL: Portland Cement Association.

Onat, N. C., M. Noori, M. Kucukvar, Y. Zhao, O. Tatari, and M. Chester. 2017. "Exploring the suitability of electric vehicles in the United States." *Energy* 121 (February): 631–42. https://doi.org/10.1016/j.energy.2017.01.035.

Park, K., Y. Hwang, S. Seo, and H. Seo. 2003. "Quantitative assessment of environmental impacts on life cycle of highways." *Journal of Construction Engineering and Management* 129 (1): 25–31. https://doi.org/10.1061/(ASCE)0733-9364(2003)129:1(25).

Reyna, J. L., M. V. Chester, S. Ahn, and A. M. Fraser. 2015. "Improving the accuracy of vehicle emissions profiles for urban transportation greenhouse gas and air pollution inventories." *Environmental Science & Technology* 49 (1): 369–76. https://doi.org/10.1021/es5023575.

Ridge, L. 1998. "EUCAR – automotive LCA guidelines - phase 2." https://doi.org/10.4271/982185.

Roudebush. 1999. *Environmental Value Engineering Assessment of Concrete and Asphalt Pavement.* 2088a. PCA R&D Serial. Skokie, IL: Portland Cement Association.

Rozycki, C. v., H. Koeser, and H. Schwarz. 2003. "Ecology profile of the german high-speed rail passenger transport system, ICE." *The International Journal of Life Cycle Assessment* 8 (2): 83–91. https://doi.org/10.1007/BF02978431.

Santos, J. 2015. *A Comprehensive Life Cycle Approach for Managing Pavement Systems.* Coimbra: Universidade de Coimbra. https://estudogeral.sib.uc.pt/bitstream/10316/30093/1/A%20Comprehensive%20Life%20Cycle%20Approach%20   for%20Managing%20Pavement%20Systems.pdf.

Saltelli, A., and P. Annoni. 2010. "How to avoid a perfunctory sensitivity analysis." *Environmental Modelling & Software* 25 (12): 1508–17. https://doi.org/10.1016/j.envsoft.2010.04.012

Saxe, S., E. Miller, and P. Guthrie. 2017. "The net greenhouse gas impact of the Sheppard subway line." *Transportation Research Part D: Transport and Environment* 51 (March): 261–75. https://doi.org/10.1016/j.trd.2017.01.007.

Schweimer, G., and M. Levin. 2000. *Life Cycle Inventory for the Golf A4.* Versmold: Volkswagen AG. http://www.wz.uw.edu.pl/pracownicyFiles/id10927-volkswagen-life- cycle-inventory.pdf.

Kobayashi, O., H. Teulon, P. Osset, and Y. Morita. 1998. *Life Cycle Analysis of a Complex Product, Application of ISO 14040 to a Complete Car.* Graz. https://doi.org/10.4271/982187.

Spielmann, M., C. Bauer, and R. Dones. 2007. *Transport Services: Ecoinvent Report No. 14.* 14. EcoInvent Report. Dübendorf: Swiss Centre for Life Cycle Inventories.

Spielmann, M., and R. Scholz. 2005. "Life cycle inventories of transport services: Background data for freight transport (10 Pp)." *The International Journal of Life Cycle Assessment* 10 (1): 85–94. https://doi.org/10.1065/lca2004.10.181.10.

Stripple, H. 2001. "Life cycle assessment of road. A pilot study for inventory analysis." 2nd revised Edition. Report from the IVL Swedish Environmental Research Institute. http:// www.ivl. se/download/18.343dc99d14e8bb0f58b734e/1445515385608/B1210E.pdf.

Sundvor, C. F. 2013. *Life Cycle Assessment of Road Vehicles for Private and Public Transportation*. Master Research Thesis. Trondheim: Norwegian University of Science and Technology - Department of Energy and Process Engineering.

Svensson, N. 2006. *Life-Cycle Considerations for Environmental Management of the Swedish Railway Infrastructure*. Thesis manuscript No. 1064, Linköping: Linköping Studies in Science and Technology, Linköping University.

Treloar, G. J., E. D. Love Peter, and R. H. Crawford. 2004. "Hybrid life-cycle inventory for road construction and use." *Journal of Construction Engineering and Management* 130 (1): 43–49. https://doi.org/10.1061/(ASCE)0733-9364(2004)130:1(43).

Trupia, L., T. Parry, L. C. Neves, and D. L. Presti. 2017. "Rolling resistance contribution to a road pavement life cycle carbon footprint analysis." *The International Journal of Life Cycle Assessment* 22 (6): 972–85. https://doi.org/10.1007/s11367-016-1203-9.

USAMP/LCA. 1998. *Life Cycle Inventory of a Generic U.S. Family Sedan - Overview of Results USCAR AMP Project*. Society of Automotive Engineers, Inc, 339.

Ventura, A., M. Dauvergne, P. Tamagny, A. Jullien, A. Feeser, C. Vincent, S. Goyer, L. Beaudelot, H. Odeon, and L. Odie. 2012. "L'outil Logiciel ECORCE: Cadre Méthodologique et Contexte Scientifique." http:// hal.archives-ouvertes.fr/hal-00908345/.

Volkswagen A. G. 2008a. *The Golf – Environmental Commendation – Background Report*. Wolfsburg: Volkswagen AG.

Volkswagen A. G. 2008b. *The Passat– Environmental Commendation – Background Report*. Wolfsburg: Volkswagen AG. file:///C:/Users/anne.de-bortoli/Downloads/Volkswagen_ Passat_livsl_psananlyse.pdf.

Wang, T., J. Harvey, and A. Kendall. 2014. "Reducing greenhouse gas emissions through strategic management of highway pavement roughness." *Environmental Research Letters* 9 (3): 034007. https://doi.org/10.1088/1748–9326/9/3/034007.

Wang, T. I.-S. Lee, A. Kendall, J. Harvey, E.-B. Lee, and C. Kim. 2012. "Life cycle energy consumption and GHG emission from pavement rehabilitation with different rolling resistance." *Journal of Cleaner Production* 33: 86–96. https://doi. org/10.1016/j.jclepro.2012.05.001.

Warburg, N., A. Forell, L. Guillon, H. Teulon, and B. Canaguier. 2013. "Élaboration Selon Les Principes Des ACV Des Bilans Énergétiques, Des Émissions de Gaz à Effet de Serre et Des Autres Impacts Environnementaux Induits Par l'ensemble Des Filières de Véhicules Électriques et de Véhicules Thermiques, VP de Segment b (Citadine Polyvalente) et VUL à l'horizon 2012 et 2020." ADEME.

White, P., J. S. Golden, K. P. Biligiri, and K. Kaloush. 2010. "Modeling climate change impacts of pavement production and construction." *Resources, Conservation and Recycling* 54 (11): 776–82. https://doi.org/10.1016/j.resconrec.2009.12.007.

Xu, X., J. Gregory, and R. Kirchain. 2015. *Role of Use Phase and Pavement-Vehicle Interaction in Comparative Pavement Life-Cycle Assessment*. Washington DC.

Xu, Y., F. E. Gbologah, D.-Y. Lee, H. Liu, M. O. Rodgers, and R. Guensler. 2015. "Assessment of alternative fuel and powertrain transit bus options using real-world operations data: Life-cycle fuel and emissions modeling." *Applied Energy* 154 (September): 143–59. https://doi. org/10.1016/j.apenergy.2015.04.112.

Yue, Y., T. Wang, S. Liang, J. Yang, P. Hou, S. Qu, J. Zhou, X. Jia, H. Wang, and M. Xu. 2015. "Life cycle assessment of high speed rail in China." *Transportation Research Part D: Transport and Environment* 41 (December): 367–76. https:// doi.org/10.1016/j.trd.2015.10.005.

Zapata, P., and J. A. Gambatese. 2005. "Energy consumption of asphalt and reinforced concrete pavement materials and construction." *Journal of Infrastructure Systems* 11 (1).

Chapter 4

Spatial Refinement to Better Evaluate Mobility and Its Environmental Impacts

Natalia Kotelnikova-Weiler, Fabien Leurent, and Alexis Poulhès

École des Ponts ParisTech

4.1 Introduction

4.1.1 Background and Literature review

Much of the environmental impact of a district development project is due to mobility. This either takes place inside the district, as in internal "intra-zonal" mobility or transit traffic, or is induced by the district in relation to the rest of the urban area. Therefore, within the framework of the socio-economic and environmental analysis of different planning parties, it is important to estimate the specific environmental impacts correctly.

A literature review has been conducted on this subject, from which three main lines emerge. Firstly, in many countries, including France (see Quinet, 2013), environmental evaluation is an integral part of the cost–benefit analysis of transport projects and, more broadly, of any large infrastructure project. In order to study the effects of regulations and also urban policies on the overall scale of the region (see Mestayer, 2012; IRSTV, 2017), spatial objects and physical phenomena are described in space through a whole model chain, which associates transport demand models with environmental impact calculation models, notably including pollutant emissions. In particular, different modelling chains combining transport models at micro-, meso- and macro-scales, emission models and air quality models make it possible to evaluate the local extent of air pollution (see the review by Fallah Shorshanni et al., 2015).

Secondly, among the different combinations of transport models, emissions and air quality, recent developments have been moving towards multi-agent models (e.g. Hülsmann et al., 2014), and indeed, the disaggregated processing of individuals and situations allows for better modelling of the population's exposure to environmental conditions (see also Beckx et al., 2009). However, the increased spatial resolution goes hand in hand with an increase in the computational load.

Thirdly, an ambition has emerged to cover a wider range of environmental impacts in a global framework for evaluating local development projects: life cycle assessment (LCA). In principle, an LCA model integrates transfers between different spatial and temporal scales and ecosystems. Transport is then considered to be one of the factors contributing to district impacts (for a recent study, see Lotteau et al., 2015). The impacts of transport are generally treated in an aggregated way, ignoring local specificities.

However, certain transport-specific LCA models address the various phases of the vehicle life cycle, while taking some urban context indices into account (e.g. François et al., 2017). Other LCA models go further by considering a mode of transport in terms of infrastructure, vehicles and the operational functioning of traffic (see De Bortoli et al., 2016). Recent efforts have been aimed at combining the benefits of LCA and traffic evaluation based on a travel demand model (e.g. Mailhac et al., 2016).

4.1.2 Objectives

This work aims to keep the modelling at an intermediate level of complexity by focusing on spatial refinement of demand model that incorporates users' individual behaviours: the purposes of their trip, the choice of transport mode and route on a modal network. We intend to maintain coherence between the spatial trip distribution, modal choice and network assignment stages by refining the spatial description in each of these steps. Our key objective is to identify certain important environmental impacts, namely air pollutants and greenhouse gas (GHG) emissions, as well as energy consumption. In addition, spatial refinement also offers the opportunity to improve the assessment of several socio-economic indicators, in particular those of accessibility to employment or amenities.

4.1.3 Methodology

The method presented here makes it possible to quantify and qualify the mobility associated with a district by distinguishing the specific conditions and functions of the buildings that comprise it. It also makes it possible to quantify the environmental impacts of this mobility. It is based on common tools for consulting engineering, thereby facilitating replication and transferability.

Building on an existing model of the demand for mobility at the urban level, our main contribution consists of refining the spatial description of things within this model, in order to better describe the mobility of a district. We can thus identify and differentiate the mobility attributed to a given building in the district; quantify and characterise district mobility (in particular, the share of internal mobility); analyse the impact of local mobility on the ambient conditions within the district (pollutant emissions, local congestion); and attribute external impacts to the district's individual components. In order to achieve these goals, several modifications and adaptations have been made to the original city-level model. The idea of refining the representation of space in a zonal agglomeration model is certainly not new: see Manout et al. (2017) for a recent review.

Here, we plan to carry out this kind of spatial refinement at the scale of an individual district, to better represent the physical and economic functioning of the mobility affecting this district, and therefore to better assess the environmental and socio-economic impacts relating to a development project for this district.

We consider a case study using a pre-existing mobility model that covers the entire Ile-de-France region: the MODUS model by DRIEA, the hypotheses of which are described in Simonet (2012). It is a macroscopic, static, four-step model (trip generation, spatial trip distribution, modal choice and network traffic assignment): see Ortuzar and Willumsen (2011) for the theoretical position of such a model. The first three steps can be considered as demand modelling, while the fourth stage of user equilibrium traffic assignment pertains to traffic modelling. A fifth stage of environmental evaluation was added to complete this modelling framework. The methodology was applied to a district located in the eastern part of Greater Paris: the district of Cité Descartes (CD), which extends over 1 km² (see Figure 4.1b).

MODUS is a "zonal" model in which the entire region is divided into 1,300 traffic analysis zones (TAZs) to study the trip flows exchanged between them. One of the initial TAZs, CD, was further refined into nearly 100 sub-zones (see Figure 4.2b). In the following, we will refer to the initial aggregated TAZ, called the "CD TAZ", and the spatially refined district, called the "disaggregated CD district".

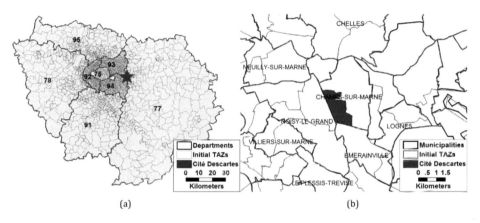

(a)   (b)

Figure 4.1 (a) The TAZs of the Ile-de-France region and the position of CD. (b) Location of the district in relation to the neighbouring municipalities.

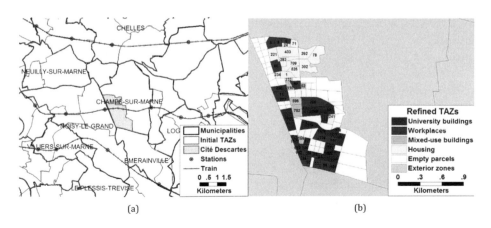

(a)   (b)

Figure 4.2 Refining the transport analysis zones in the CD district. (a) Initial TAZs in the vicinity of CD. (b) Disaggregated CD district. (Vanhille (2015).)

4.1.4 Structure of This Chapter

In the rest of this chapter, Section 4.2 presents the refinement of spatial representation. Then, the four mobility modelling steps and their adaptations are described, distinguishing between demand modelling (Section 4.3), on the one hand, and traffic modelling (Section 4.4), on the other. Section 4.5 then describes the additional step of the environmental evaluation and how it was specifically adapted to assign the impacts to the individual components of the neighbourhood. Thus, the spatial refinement renders the district mobility evaluation methodology sensitive to the layout and programming of the local planning project.

In conclusion, Section 4.6 summarises the contribution and indicates avenues for future research.

4.2 Spatial Modelling

4.2.1 Refinement of the Zone System

On average, the area of each sub-zone is roughly 1.5 ha (approx. 120 × 120 m), which is close to the scale of a building. The area of the largest subzone is 9 ha (approx. 300 × 300 m) and could be crossed on foot in less than 4 or 5 minutes. Each TAZ in the region, including the sub-zones, is characterised by four land-use variables: the working and non-working population, the number of jobs and the number of places available in educational institutions. The sub-areas of CD are almost exclusively mono-functional (see Figure 4.2b): housing, university buildings, offices, etc., while mixed-use occupations are described by an appropriate combination of the four land-use variables. The spatial refinement enabled a precise description of the local planning project's programme and spatial configuration.

4.2.2 Network Refinement

The relationship between sub-zones and transport networks was also specified by refining the road network and through the creation of specific connectors. In the initial city-level MODUS model, the road and transit networks are simplified – their accuracy matches that of the TAZs. In order to analyse and accurately represent local mobility within the CD district, the simplified model of the local road network was replaced by a detailed description, with the addition of access routes that correspond to the secondary level in the hierarchy of routes on the concrete network (see Figure 4.3). Although it is simplified in some parts of the agglomeration, we did not have to modify the public transport system as the public transport stations and stops in the district we were studying had already been exhaustively modelled and precisely located.

Once the sub-zones are created and the network refined, it is necessary to connect them. To do this, we defined a "centroid" in each zone, that is a special node that models the "trip injection point" of the area to aggregate its population and its uses virtually. Each centroid was located in the geographical centre of its sub-zone. We then connected the centroid to the road and public transport networks through links called "connectors". Thus, the specific access conditions of each sub-zone are finely described.

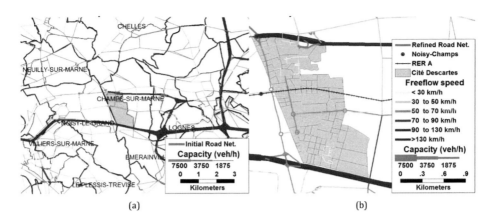

(a)   (b)

Figure 4.3 Refinement of the local road network. (a) Initial road network in the vicinity of CD. (b) Locally refined network in the CD district. (Vanhille (2015).)

4.3 Travel Demand Modelling

4.3.1 *Generation*

Trip generation modelling aims to quantify the incoming and outgoing flows for each zone for an average period over a set of days. Spatial disaggregation enables one to simulate the flows emitted and received by each building or group of buildings, attributing the share of the district's overall demand to each micro-zone (Figure 4.4). This mobility demand is also segmented according to the purpose of the trip (six purposes are distinguished in the MODUS model) and the individuals' characteristics (captive or not captive users of public transportation).

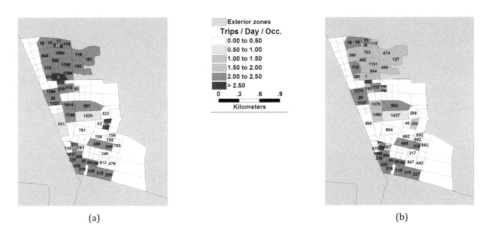

(a)   (b)

Figure 4.4 Daily flows entering or exiting the sub-zones of CD for all purposes. (a) Total attractions and attraction rates per occupant. (b) Total emissions and emission rates per occupant. (Vanhille (2015).)

MODUS deals with individual trips (as opposed to daily activity programmes or round-trip loops), combining a statistical and macroscopic principle for the physical description of things and a microeconomic principle for the simulation of individual mobility behaviours. Its trip generation sub-model is based on the land-use characteristics of each zone (the four variables mentioned above) and relies on trip production and attraction rates that were first calibrated at the regional level and then specified according to the characteristics of the demand segments (in particular, the rates of captive and non-captive users vary within the region). A certain imbalance was noted between incoming and outgoing trips during the period simulated, especially for the residential areas. This is due to regionally calibrated attraction and emission rates aiming at a global balance for mixed-use TAZs. The aggregated TAZ of CD and the disaggregated CD district clearly account for the same share of the overall land-use characteristics – total working and non-working population, jobs and places in educational institutions. Since the generation model follows a linear formulation, the total mobility demand is the same whether the district is aggregated or disaggregated into sub-zones.

As the micro-zones exhibit greater functional specialisation, there could be an additional opportunity to describe the sub-zones more precisely in terms of jobs and populations. However, the current calibration of the MODUS model is not precise enough for such specialisations. A new calibration would therefore be necessary to take into account the breakdown of job types and population segments (e.g. the mobility behaviour of the students of CD might differ significantly from that of a non-working person in Ile-de-France).

4.3.2 Spatial Distribution of Trips

The trip spatial distribution stage makes it possible to distribute the trips between the origin–destination (O-D) relations. Each O-D relation corresponds to an origin zone and a destination zone. The distribution model shares the flows of an origin between the different destinations: it simulates the users' choice of destination starting from their places of origin. In the MODUS model, the choice of destination is based on two aspects. The first of these is the usefulness of a destination zone for a given trip purpose, reflecting the spatial configuration of activities at the regional level. The second concerns the multimodal disutility of the destination area in relation to the zone of origin, depending on the regional multimodal network.

Spatial refinement within the district has a twofold effect on the spatial distribution modelling stage: on internal trips within the district, on the one hand, and other trips exchanged with the rest of the territory, on the other.

With regard to internal trips, spatial refinement leads to the individualisation of the micro-zones within the district, specifying multimodal access conditions between them. We can therefore explicitly model the short trips that take place within the district, as shown in Figure 4.5b. Thus, the internal movements are not only quantified but located in space, establishing the relations between the district's buildings.

This refined modelling also modifies the ratio between the district's external and internal flows. Although the overall size of the flows emitted and received by the areas is the same for both zonal systems, in the disaggregated configuration the total of the district's internal flows is lower than in the aggregated configuration: approximately

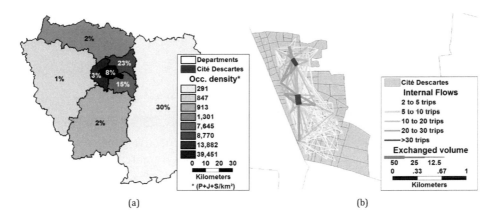

Figure 4.5 (a) Presentation of the flows exchanged between the district and the vicinity by key sectors. (b) "Desire paths" of internal flows. (Vanhille (2015).)

4,700 versus 8,400 daily movements. This observation can be explained by the fact that the initial MODUS model (at least its academic implementation in the TransCAD simulation software) does not take into account the intra-zonal impedance in the spatial distribution stage. This leads to an overestimation of the flow of internal trips in the district, to the detriment of the flows exchanged between the district and the other zones. The spatial subdivision of the district reintroduces the impedances in the model between the micro-zones, and therefore within the district.

Consequently, the disaggregated configuration will induce more exchange trips between the districts and the vicinity than the more aggregated configuration, to an amount equivalent to the reduction of the internal trips within the district. In the disaggregated simulation, the trips transferred from one category to another in this way are especially revealed in the vicinity of the district: the district's internal micro-zones are competing for the attribution of the flows emanating from the whole agglomeration. Strictly speaking, the distribution model should be recalibrated by introducing internal impedance for all the TAZs in the territory. Finally, as the size of the micro-zone becomes smaller and closer to that of an actual building, its own internal flow is destined to vanish.

It should be remembered that since MODUS relates to individual trips, it does not allow activity chains within a district to be taken into account. As a consequence, the explanation of the functional diversity of the district caused by micro-zoning will not produce any consequences in the transport demand model other than that within the district.

With regard to the impact of the spatial disaggregation on trips exchanged with the rest of the territory, two main aspects emerge. On the one hand, the spatial disaggregation leads to a specialisation of the micro-zones compared to the aggregate representation (e.g. certain micro-zones exclusively contain jobs and no population). This then results in a specialisation of the associated trip purposes and in repercussions on the choice of the destination (first-order effect, modifying the flows exchanged with an external zone up to 100%). On the other hand, disaggregation makes it possible to specify the network access conditions for each micro-zone compared to the aggregated configuration, accessibility to the rest of the territory and therefore destination choices (second-order effect, modifying the flows up to 10%).

4.3.3 Choice of Travel Mode

The so-called modal split or modal choice stage makes it possible to distribute the flow of users' trips between the available modes of transport by O-D relation and by behaviour segment. In the modal choice model, at the level of an individual trip situation, the modes are compared according to their respective generalised cost, which is formed by a combination of time and cost variables weighted by the specific coefficients. These coefficients were calibrated for each demand segment in Ile-de-France on the basis of the household mobility survey: the 2009 EGT (Enquête Globale de Transport, or Global Transport Survey).

The MODUS model addresses three modes of transport: active modes (the associated description actually corresponds to walking); private cars (the description of which includes travel time but not time spent looking for parking, and likewise traffic costs but not parking fees, and does not allow the possibility of intermodal combinations with the public transport); and public transport, which covers several sub-modes (train, suburban train, metro, tram and bus) integrated into a public transport network that is exclusively accessed by foot (since other modes of access are not modelled).

As mentioned in the generation step, finer treatment of the micro-zones leads to a more precise segmentation of demand, which in turn would require recalibration of the modal choice behaviour model. For example, it would be good to describe students aged 18 and over, who predominate in CD, as a specific demand segment, and to differentiate it from the rest of the segment of individuals aged 6–25, to better model mobility behaviours for study purposes. Additionally, consideration of a refined local network could also improve the description and modelling of public transport modes, both for the routes taken by buses and trams on the road, and for pedestrian access between centroids and stations. Furthermore, as in the distribution step, the sequences of activities in trip loops or activity programmes are neglected, which artificially simplifies the mode choice. In fact, the mode choices for trips carried out within the same activity chain are interdependent and are made by considering the dynamic conditions throughout the day. These neglected aspects leave room for improvement within future developments.

Despite the limitations mentioned above, spatial refinement addresses two major issues in district-scale transportation modelling. On the one hand, specification of the conditions for pedestrian access between the access point (the micro-zone centroid) and the transport networks is refined for each micro-zone. On the other hand, the distinction of the modal trip conditions for each micro-zone is given in relation to the complementary zone at the other end of the trip. This, in turn, leads to refined modal choices. These refined processes mark a significant progress compared to the initial aggregated model, which represents the district as a single homogeneous zone.

Figure 4.6a illustrates the model's sensitivity to the location within the district. It shows the modal split for daily commutes to work: in the northern part of CD, users favour modes of public transport, while the use of private cars is more significant in the south of the district. These modelling results are consistent with the configuration of the multimodal transport network: the southern part enjoys privileged access to the motorway network, while the northern part is further away from it but enjoys faster access to the railway station, which is served in a privileged manner by bus lines. The accessibility indicator evaluation, shown in Figure 4.6b for the home-work purpose and non-captive demand segment, is also improved and spatially refined, which results in greater sensitivity to the location of public transport stations in the district. The accessibility indicator follows the Hansen–Poulit formula:

$$A_z - \frac{1}{\theta} \ln\left(\Sigma_d O_d \exp(\theta U_{zd}) \right),$$

where A_z is the accessibility indicator, equal to the monetary value of access to jobs (or places of study in educational institutions) *via* transport networks, per trip and per day; Od is the number of opportunities in a destination zone; U_{zd} is the multimodal utility of trip making between the area and the destination area, according to the formula:

$$U_{zd} = \frac{1}{\mu} \ln\left(\Sigma_m \exp\left(\mu U_{zd}^{(m)} \right) \right).$$

4.4 Traffic Modelling

The model assigning traffic on a network compares the transport supply (infrastructure and service networks) and the mobility demand (trip flows) for a given period of the day (usually the morning peak period, MPP, or the evening peak period, EPP). Firstly, the daily demand is distributed over time and the demand volumes are derived by period, and for each peak period in particular, using specific average ratios per sub-region and per mode. Secondly, peak demand is assigned. The assignment model provides quality of service characteristics for each of the O-D relations (for MPP and FPP), by demand segment along the routes that the users "choose". This quality of service characterisation is used in the spatial distribution and modal choice steps, through feedback. The MODUS model assigns the flows of vehicles and passengers to links of the road network and the public transport network, respectively, whereas the active modes trips are not assigned to the road network (although their modal share has been quantified at the level of the modal choice model). The spatial refinement of the model influences the simulation quality for both car and public transport traffic.

4.4.1 Assignment of Private Car Traffic on the Road Network

MODUS makes it possible to specify the number of occupants per car for each O-D relation, based on average observations per sub-region. Thus, vehicle flows can be deduced from the traveller flows obtained in the modal choice step. Congestion is represented by the relationships between local flow rates and travel times. Moreover, as MODUS is a macroscopic model that deals with aggregated flows, average congestion conditions are determined for each road network link. Since MODUS is also a static model, hyper-congestion peaks and their consequences are thus smoothed out.

Once the actual travel times, subject to congestion phenomena, are determined for all network links, the trip flows are assigned to the shortest routes between the origin

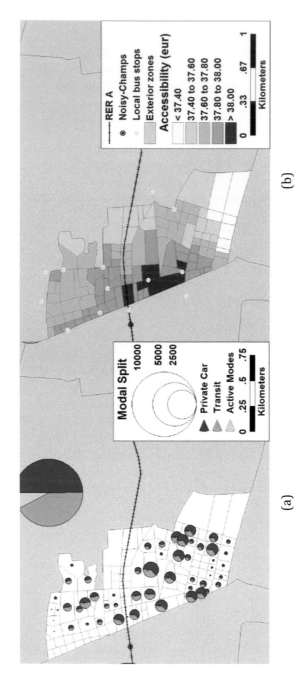

Figure 4.6 Multimodal access conditions at the micro-zone level in CD. (a) Modal split of daily commute to and from the district. (Vanhille (2015)). (b) Multimodal accessibility indicator for the "mode flexible" segment of the home-work purpose.

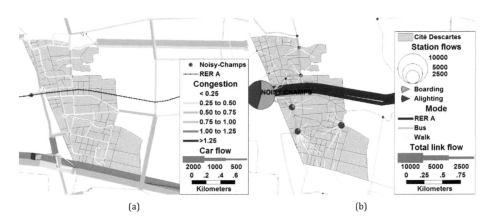

(a)   (b)

Figure 4.7 Assignment at the micro-zone level. (a) Assignment of cars on the local road network, PPE, in veh/h. (b) Assignment of passengers on the public transport network, PPM, in passengers/h. (Vanhille (2015).)

and destination zones. The criterion for determining the shortest route is a "generalized cost" indicator that summarizes the time and financial cost of the trip, weighted by specific coefficients. The assignment model puts supply and demand into balance: route times determine local flows, which themselves determine local times, which in turn determine route times. The calculation process is iterative.

The assignment model of private car traffic to the road network provides the vehicle flow, congestion level (w.r.t. link capacity) and average vehicle speed, at the basic link level.

The spatial refinement of the CD district makes it possible to identify the flows exchanged between the buildings within the district and to allocate them on a local network described more finely than the initial network. As a result, the secondary road links support their share of traffic, whereas only the main arteries were involved in the simplified description of the network.

Figure 4.7a represents the results of the road assignment for CD during the EPP, the most problematic period for private cars. The motorway flows have been visually diminished (they actually number approximately 3,500 vehicles/hour in each direction). The figure shows local traffic that contributes to ambient conditions, noise levels and pollution levels. Congestion can be observed on the motorway and its ramps in the southern part of the neighbourhood, while the main road in the northern part of the neighbourhood is underutilised. As expected, congestion levels in the neighbourhood itself are rather low, with the exception of the major North–South axis and the secondary East–West transit route.

4.4.2 Assignment of Traffic on the Public Transport Network

The transit trip flows are also assigned to the shortest routes, which are themselves calculated on a transport network assumed to be uncongested. Indeed, the MODUS model does not take the constraints on the capacity of public transport into account, which can induce longer travel times and discomfort, and increases the generalised cost in a twofold way. Interactions between road traffic and public transport operations are also neglected, like the impact of traffic congestion on buses that do not benefit from reserved lanes.

The assignment of traffic to public transport provides passenger flows by link and sub-mode, as well as boarding and alighting flows for each service and each station. Optimal routes are searched for between each origin and destination, which provide the travel time, including on-board, waiting, access and connection times. This description characterises the quality of the transport service.

Spatial refinement makes it possible to distinguish public transport use according to land use (local functions) and the location of the public transport stations in the district. Figure 4.7b shows the results for CD during the MPP, when residents leave their homes to go to work and workers arrive at their workplace. There is a noticeable contrast between the north and south of the district: the flows in the northern, residential part mainly involve boarding the network, whereas they are mainly alighting in the southern part, the part of the district where employment is concentrated.

4.5 Environmental Impact Evaluation

4.5.1 Hypothesis of the Calculation of Average Environmental Impacts

As MODUS disregards congestion in public transport, private cars and public transport were treated differently.

For private cars, overall impacts such as fuel consumption and CO_2 emissions were taken into account, as well as local emissions of PM10 and NO_X pollutants. Link-level emission factors were considered to be dependent on a number of factors: traffic conditions (represented by the average speed on the link, resulting from the assignment step) and fleet composition (based on the average regional composition of the fleet). The emission functions were based on the COPERT IV methodology, following the technical note from the French administration (Sétra et al., 2009).

The public transport modes were subdivided into electrical sub-modes and buses. The electrical sub-modes include the train, suburban train, metro and tram. Their traction electricity consumption was calculated. It represents approximately 75% of the operating energy. The associated CO_2 emissions, which take the French production mix into account, were also calculated (53 g/kWh). The average consumption was calculated at the level of each link, disregarding the dependence on the distance between the stations, the actual load and the speed of the vehicles.

For buses, the set of environmental impacts calculated was similar to that of private cars: fuel consumption, CO_2, NO_X and PM10 emissions. Average emission factors that disregard the dependence on speed and load could be found in the literature and adapted to the composition of the fleet provided by the RATP (the main public transport operator in Greater Paris) to obtain average emissions factors for buses operating in the region.

By multiplying the unit emission rates by the unit of distance and vehicle, the length of the link and the actual flow of vehicles (which also results from the assignment step), we obtain the total impacts produced on this link for each sub-mode.

4.5.2 Allocating Environmental Impacts to Individuals

The impacts per person are first calculated at the link level (network element) by dividing the total impacts produced by the vehicles of a sub-mode by the number of users of that sub-mode, that is by the traveller or passenger flows on this link (resulting from the allocation stage). These per-person link-level impacts are then aggregated along the paths that connect the origin and destination zones, providing an individual impact for each O-D pair. Finally, the individual impact of an O-D pair is multiplied by its effective demand flow, to obtain a total impact per O-D pair.

4.5.3 Attribution of Zone-Level Impacts

Several attribution schemes have been proposed and discussed in a related article (Kotelnikova-Weiler et al., 2017). Here, we will summarise the attribution scheme at the area level.

In an attempt to build an attribution scheme that avoids possible overlapping (and double counts) when evaluating several neighbourhoods in the same territory and guarantees transferability to any development project, whatever the programme, the attribution scheme proposed relies on two main principles:

- Any trip whose extremity (origin or destination) is located within a zone is associated with this zone without a priori assumption on what purposes should be associated with what types of building programmes;
- To avoid double counts, 50% of a trip's impacts are attributed to the origin zone and 50% are attributed to the destination zone. When the two zones coincide, 100% of the impacts of the movement are attributed to it.

Pragmatically, these principles result in the following formulas, which enable the environmental impacts of a zone to be calculated:

$$Q_z = \frac{1}{2}Q_z^+ + \frac{1}{2}Q_z^-, \text{ where}$$

$$Q_z^+ = \Sigma_{d \in D} Q_{zd} \text{ and } Q_z^- = \Sigma_{o \in O} Q_{oz}.$$

We read that the total impacts of a zone, z, Q_z, equal 50% of the total impacts of all the trips in which the zone z is the origin and 50% of the total impacts of all the trips in which zone z is the destination.

Based on this attribution scheme, several impacts were calculated for the micro-zones of the CD district: total energy consumption, total CO_2 emissions (combining private car, bus and electric mode contributions), total NO_X and PM10 emissions (combining bus and private car contributions). For illustration purposes, reflecting particular peak traffic conditions experienced by the neighbourhood users, Figure 4.8a shows the total morning peak hour NO_X emissions generated by each micro-zone. Larger NO_X emissions might indicate higher transport demand and/or poor access to public transportation. Buildings of educational institutions in the centre of the district thus have a high impact, whereas housing and office buildings on the North and South

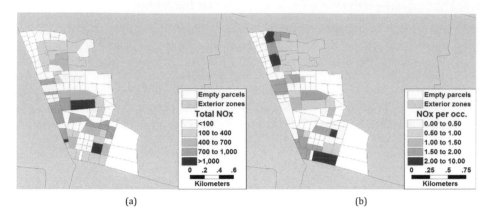

(a)   (b)

Figure 4.8 CD micro-zones as generators of NO_X emissions at the regional level combining private car and bus contributions, morning peak hour. (a) Total NO_X per zone (g). (b) NO_X per occupant (g). (Vanhille (2015).)

periphery of the district, respectively, have relatively low contributions. In order to better understand the structure of the impacts, another indicator may be useful: the ratio of total impacts to a total occupation indicator (population+jobs+places in educational institutions). This is presented in Figure 4.8b and tells a different story: indeed, the per occupant contribution, contingent to the settled urban function, is very low for educational buildings and high for residential and office buildings. This dual approach of the total impact and the impact per occupant helps to identify potential for improvement: either zones where incremental improvement might produce high reduction of impacts or zones requiring structural modifications of the transport offer. To get a more physical sense of the environmental impacts, daily emissions can also be computed: both peak period and off-peak period simulations need to be carried out, then corresponding hourly emissions can be combined in the following aggregation formula, to obtain the daily emissions:

$$Q_z^{Day} = aQ_z^{PPM} + a'Q_z^{PPE} + bQ_z^{OffPeak},$$

where, for private cars, $a=2h$, $a'=3h$ and $b=10h$, while for public transport, $a=1.5h$, $a'=2.5$ and $b=11h$ (values for Paris).

It should be remembered that the emission of pollutants is only one factor of local air quality among others that include background pollution, spatial and temporal propagation of pollutants, and their chemical transformation. However, our methodology aims to evaluate the contribution of a neighbourhood project to the regional emissions of atmospheric pollutants.

4.6 Discussion and Further Research

In this chapter, we have proposed a pragmatic methodology to evaluate some social and environmental impacts of a neighbourhood's mobility. Despite methodological limitations, including the necessity to improve the model's segmentation and calibration, it has allowed us to achieve our main goals:

- To provide a more precise description of the network topology;
- To simulate internal traffic on the local road network;
- To provide more precise localisation of local emissions within the neighbourhood;
- Thus, to pave the way to a more specific evaluation of local exposures inside the neighbourhood as both the emitting links and the receiving micro-zones are refined. However, a complementary model of air quality would be needed;
- To more specifically attribute the environmental impacts of mobility among the micro-zones that generate them, to evaluate a development project more finely.

These contributions make the evaluation method sensitive to both the function and location of the building, and to heterogeneous multimodal transport conditions within the district.

Our evaluation method is therefore applicable to a neighbourhood development project, as it is both sensitive to the programme and layout of the project, allowing for several design options to be compared. Transportation-specific design alternatives can also be modelled to predict their impacts: the creation or modification of a transport line, such as a shuttle service, layout and capacity dimensioning of local road network or hypotheses of the structure of the fleet according to its sizes and engines (including the energy mix between thermal and electrical engines).

The method is also sensitive to regional-level context, such as the spatial distribution of activities in the territory and the network configuration impacting regional accessibility. The simulation of a long-term project must therefore necessarily rely on sound hypotheses regarding regional planning, both in terms of the evolution of urbanisation and in terms of the development of transport networks. Additionally, as the environmental evaluation is also performed based on fleet composition and energy production mix, their future evolutions must also be hypothesised.

This methodology can be deployed in major urban agglomerations for which transport demand models are available and have been calibrated. The treatments presented here were implemented using the basic features of a commercially available transport-oriented GIS tool, TransCAD, making them replicable.

An original framework for assessing and comparing road and public transit impacts is provided, as well as an original proposal for the attribution methodology at the micro-zonal level. It could be further discussed and improved. Indeed, the 50–50 attribution scheme adopted here does not correctly reflect the hierarchy of activity purposes in the daily programme. For instance, let us consider a secondary activity carried out "on the way" to or from a main activity. It is often true that it is the main activity rather than secondary one that determines the distance to be travelled (the secondary activity would not motivate a long trip in itself), the mode to be used (taking into account the accessibility of the main activity's location) and sometimes the reason for making the additional secondary activity (since the person has already left the house, they can do both activities). Therefore, a major direction for further research consists of the implementation and adaptation of this evaluation methodology for tour-based or activity-based travel demand models.

Acknowledgements

This work has taken place at the junction of two research programmes: the ParisTech Chair of Eco-design of buildings and infrastructure, sponsored by the Vinci group, and Efficacity, the French R&D institute dedicated to urban energy transition. We would also like to thank the DRIEA (the state Department for Regional Planning) for providing us with the regional data and MODUS travel demand model.

The illustrations are mainly based on the master's thesis by Emilie Vanhille (2015), who worked at the LVMT under the joint direction of the authors.

Bibliography

Beckx, C., Int Panis, L., Uljee, I., Arentze, T., Janssens, D., Wets, G. (2009). Disaggregation of nation-wide dynamic population exposure estimates in The Netherlands: Application of activity-based transport models. *Atmospheric Environment, 43*, 5454–5462.

De Bortoli, A., Féraille, A., & Leurent, F. (2016). Environmental performance of urban transit modes: A life cycle assessment of the bus rapid transit. *Proceedings of the International Conference on Sustainable Built Environment*, Hamburg, March 7–11.

Fallah Shorshanni, M., André, M., Bonhomme, C., Seigneur, C. (2015). Modelling chain for the effect of road traffic on air and water quality: Techniques, current status and future prospects. *Environmental Modelling & Software, 64*, 102–123.

François, C., Gondran, N., Nicolas, J.-P., Parsons, D. (2017). Environmental assessment of urban mobility: Combining life cycle assessment with land-use and transport interaction modelling – Application to Lyon (France). *Ecological Indicators, 72*, 597–604.

Hülsmann, F., Gerike, R., Ketzel, M. (2014). Modelling traffic and air pollution in an integrated approach – the case of Munich. *Urban Climate, 10*, 732–744.

Institute for Research on Urban Sciences and Techniques, IRSTV (2017, January 26). Eval-PDU. Retrieved from http://irstv.fr/fr/recherche/contrats-de-recherche/acheves/42-eval-pdu.

Kotelnikova-Weil`er N., Leurent F., Poulhès A. (2017). Attribution methodologies for mobility impacts. *Transportation Research Procedia, 26/C*, 131–143.

Lotteau, M., Loubet, P., Pousse, M., Dufrasnes, E. (2015). Critical review of life cycle assessment (LCA) for the built environment at the neighborhood scale. *Bilding and Environment, 93*, 165–178.

Mailhac, A., Herfray, G., Schiopu, N., Kotelnikova-Weiler, N., Poulhès, A., Mainguy, S., Grimaud, J., Serre, J., Sibiude, G., Lebert, A., Peuportier, B., Valean, C. (2016). LCA applicability at district scale demonstrated throughout a case study: Shortcomings and perspectives for future improvements. *Proceedings of the Systainable Built Environment (SBE) Regional Conference*, Zurich, June 15–17.

Manout, O., Bonnel, P., Ouaras, H. (2017). L'impact de l'agrégation zonale sur la modélisation des transports: une approche statistique, *4èmes Rencontres interdisciplinaires doctorales de l'architecture et de l'aménagement durables*, Vaulx-en-Velin, 25 January 2017.

Mestayer, P. (Ed.) (2012). Eval-PDU. Évaluation des impacts environnementaux d'un PDU et de leurs conséquences socio-économiques: développements méthodologiques et tests sur le PDU de Nantes Métropole. *Final Scientific Report. ANR Villes Durables Program ANR-08-VILL-0005.*

Ortuzar, J.D., Willumsen, L.G. (2011). *Modelling Transport* (fourth ed.). Chichester: John Wiley, Sons Ltd.

Quinet, E. (Chair.) (2013). *Cost Benefit Assessment of Public Investments. Final Report. Summary and Recommendations.* Paris: CGSP (Policy Planning Commission).

Sétra, Cété de Lyon, Cété Normandie-Centre (2009, November). Emissions routières de polluants atmosphériques. Courbes et facteurs d'influence. (Sétra Information Note – Economy, Environment, Design series no. 92).

Simonet, T. (2012, August). Modèle regional multimodal de déplacements MODUS. *Hypothèses et Données. Spécifications* (Technical Note, Direction Régionale et Interdépartementale de l'Equipement et de l'Aménagement d'Ile-de-France).

Vanhille, E. (2015). Modélisation et évaluation environnementale de la mobilité. *Application à un projet d'aménagement de la Cité Descartes* Elargie (Master's thesis, École des Ponts Paris-Tech, France).

Chapter 5

Ecological Compensation in France

Towards a Territorial System?

Julie Latune

Université Paris-Saclay, AgroParisTech, ESE Lab

Harold Levrel

University Paris-Saclay, AgroParisTech, CIRED Lab

Pauline Delforge

Université Paris-Saclay, AgroParisTech, ESE Lab

Nathalie Frascaria-Lacoste

Université Paris-Saclay, AgroParisTech, ESE Lab

5.1 Introduction

In 1976, the law on the protection of nature was passed in parliament. Its purpose is to "protect natural spaces and landscapes, animal and plant species and natural balances" (Law No. 76–629 – art. 1). Articles 3 and 4 of the same law prohibit the destruction of protected species and set the conditions under which these lists of protected species are established. Article 2 introduces environmental impact studies (EIS) that "should allow the appreciation of the consequences" caused to the environment following the realisation of facilities or works. Thus, environmental impact studies must contain "an analysis of the initial state of the site and its environment, the study of the modifications that the project would generate and the measures envisaged to avoid, reduce and, if possible, compensate for the harmful consequences for the environment" (Law No. 76–629 – art. 2). A limitative list of works that are not subject to EIS is also established, although in principle EIS is mandatory when the project cost greater is than € 1.9 million. For years, the compensation was rarely applied. This was due to various reasons, including poor means for its implementation, monitoring and control of its application by the administration. This led to poorly completed environmental impact studies, a lack of process operationalisation and a small number of projects subject to environmental impact studies (Quétier, Regnery, and Levrel, 2014). However, things have changed since the 2000s. An easing of the strict protection of certain species in the framework of article 86 of law 2006–2011 of 5 January 2006 (application February 2007) makes it possible to bring harm to these so-called protected species by means of compensation in the framework of development projects. It then becomes important to frame and envisage tangible compensation measures. Since 2007, the number of requests for exemptions from the strict protection of protected species submitted to the Conseil National de Protection de la Nature (National Council for the Protection of Nature) increased from 44 requests in 2006 to 197 requests in 2011 (Regnery, 2013). In the context of the development projects and the EIS procedure, it is the requests for exemptions that have increased. Moreover, in 2012, impact study reform broadened the spectrum of projects that must comply with this obligation (Decree No. 2011–2019), leading to an increase in the requests for exemptions on protected species in the framework of the projects and therefore the needs in terms of compensation. In 2006, the law on water and aquatic environments specified that the impacts caused to wetlands by development projects should be evaluated in order to avoid, reduce and compensate for damage to bodies of water (Law No. 2006–1772). Similarly, the habitat directive requires Natura 2000 impact assessment dossiers to be filled out in order to be able to implement compensatory measures, if impacts can be neither reduced nor avoided (Directive 92/43/EEC, 1992).

The first clearing bank was set up in France in 2008, when the Caisse des Dépôts et Consignation (CDC) bought a former abandoned industrial orchard located near the Coussouls de Crau Nature Reserve. The purchase of the land cost 5.5 million euros. The goal is to conduct experiments on the compensation offer in France. Hence, the first French "Réserve d'Actifs Naturels" (Natural Asset Reserve, RAN)[1] was created. In 2011, the French government launched a call for expressions of interest (CGDD, 2011) to test this compensation mechanism more broadly through the offer. Three new operations emerged. The so-called PRM (the property developer is responsible for the implementation of the compensation, see Monin-Soyer, 2011) remains the most widely used system today. With the 2016 law for the recovery of biodiversity and landscapes, which finally included the possibility of creating NCS (Art L. 163-3), we can expect to see this system grow. The article stipulates: "operations for the restoration or development of elements of biodiversity, known as "natural compensation sites" (NCS), can be put in place by public or private persons, in order to implement the compensation measures of defined in I of article L. 163-1, in a manner that is both anticipated and shared. The natural compensation sites are subject to prior approval by the State, according to procedures defined by decree" (Law No. 2016–1087).

ECOLOGICAL COMPENSATION SHOULD NOT BE CONFUSED WITH

REFORESTATION

The compensation for clearing or reforestation has existed in France since 1952 under the Forest Code. Article 163 of the same code allows clearing in return for carrying out reforestation work. It is the State representative in the departments

---

1 This term was replaced in 2016 following the law for the restoration of biodiversity and landscapes (in particular, the law on "Natural Compensation Sites"(NCS) (Law No. 2016-1087) (Vaissière et al., 2017).

who authorises clearing through reforestation, with a ratio ranging from 1 to 5 depending on the economic, social and environmental value of the forestation (Art L341-6 of the Forest Code). This compensation is therefore intended to compensate for the losses of the forest heritage in these different dimensions, and especially with regard to the economic value associated with forestry production (Labat, 2015). This vocation is reinforced by the law dated 13 October 2014, the law on the Future of Agriculture, Food and Forestry (loi d'avenir pour l'agriculture, l'alimentation et la forêt, LAAF), which henceforth gives petitioners the possibility to fulfil their reforestation obligation for clearing by contributing to a "Strategic Forest and Wood Fund" instead of reforesting nature, which was previously the only option (Decree No. 2015–776, 2015, Law No. 2014–1170, 2014). The objective of this fund is to finance priority investment projects for forests and research and development actions defined as part of the "national forest and wood programme" and their regional variations (Labat, 2015).

Collective Agricultural Compensation

Article 28 of the same law provides for a preliminary study to assess the negative consequences of the developments made on the agricultural economy, and to consider avoidance, reduction and compensation measures in order to consolidate the agricultural economy of the territory (Law No. 2014–1170). Thus, since 2014, there have also been plans to introduce forms of collective agricultural compensation, which were initially planned to be in kind. The possibilities were broadened in the implementation decree dated 31 August 2016 (Decree No. 2016-1190), which allows the developer to contribute to agricultural investment funds intended to finance specific sectors and/or develop distribution channels, among other things (Labat, 2015).

"Territorial Compensation"

It is not subject to legislative or regulatory requirements. It is often negotiated at the time of signing contracts and can also provide a competitive advantage during a call for tenders. It can take different forms: preferential hiring of local residents to construct the infrastructure, improvement of the living environment, financing of training, support to local economic dynamism, establishment of foundations, etc. It is intended to repair the negative externalities produced by a development and/or to improve the territorial equity and/or to allow better acceptability of the project locally (Gobert, 2010).

Ecological compensation is intended to counteract damage to biodiversity from infrastructure developments, plans and programmes. The objective is to achieve no net loss of biodiversity (French General Commission for Sustainable Development, 2012, 2013; Law No. 2016–1096) in order not to worsen the state of conservation of the impacted species, to allow the good ecological state of bodies of water to be maintained (Law No. 2006–1772) and/or to preserve the integrity of the Natura 2000 network

(European Commission, 2007). A number of principles are recommended to this end, such as the integration of the project in a hierarchical approach to the avoidance, reduction and, ultimately, compensation of impacts to biodiversity; environmental equivalence; environmental additionality; integration of compensatory measures in a functional ecological whole; the effectiveness of the measures before the arrival of the impacts; the durability of the measures; the integration of stakeholders in the development of the measures; respect for fairness in the choice of measures; the monitoring and control of measures (French General Commission for Sustainable Development, 2012, 2013; Law No. 2016-1096, BBOP, 2012; Bull et al., 2013; Gonçalves et al., 2015; Gordon et al., 2015; IUCN, 2016; McKenney and Rana, 2010; Moreno-Mateos et al., 2015; Quétier and Lavorel, 2011).

The implementation of ecological compensation has tended to intensify over the past ten years. Many actors are mobilising in this field, thus creating systems of actions for implementing compensation. However, many uncertainties remain with regard to the operationalisation of this public policy and these effects on the anticipated objectives (Guillet and Semal, 2018; Levrel et al., 2018). Indeed, any new instrument (here ecological compensation) produces more or less expected and desired effects on the objective it has to achieve (Crozier and Friedberg, 1977; Lascoumes and Gales 2007). Here, it is thus a question of putting the compensation systems in France in perspective for the first time, in order to see to what extent these systems respond to the expected principles. The purpose of this chapter is also to attempt to better characterise the systems that are being established in France through various examples. We will see to what extent they refer to the PRM system or the clearing bank system (Monin-Soyer, 2011; Pilgrim, 2014).

In this paper, we will first discuss certain principles of ecological compensation and the advantages that each of the systems offers, in theory, with respect to each other, with regard to these major principles. We will then show the ecological compensation systems that are set up in France and, finally, we will discuss how they allow or do not allow the aforementioned principles to be met.

5.2 Materials and Methods

5.2.1 Definition of Two Compensation Systems

Several compensation systems exist around the world (Gelcich et al., 2016; Jacob et al., 2015; Monin-Soyer, 2011). Two major systems are generally mentioned:

- Clearing bank or natural compensation site (a term used since the 2016 law on the recovery of biodiversity and landscapes – Law No. 2016–1087): "a set of actions carried out on a site generating environmental gains that could be subject to transactions with developers to compensate for their authorised impacts. These environmental gains are centralised and managed by a third party (called an "operator" in France) and result from compensation actions" (Levrel et al., 2015, 291).
- Individual permit or PRM: "Environmental compensation is carried out directly by developers who have caused an authorised impact. In the French context, we also talk about compensation on demand" (Levrel et al., 2015, 293).

In the United States, in the first case, there is a transfer of responsibility for the success of the compensatory measures (Vaissière et al., 2015) from the petitioner to the compensation operator, which is not the case in France. Whatever the system adopted in France, it is the petitioner who remains responsible for the implementation and achievement of the results expected by law (French General Commission for Sustainable Development, 2012, 2013; Law No. 2016–1087).

The choice of one system or another will produce different effects (Crozier et al., 1977) on achieving the net loss of biodiversity.

Each of these two systems has advantages and disadvantages for achieving these principles and the goal of no net loss of biodiversity. The following table details which system best meets which principles, based on the functioning of clearing banks in the United States.

For each system, the (+) and (−) show which responds best or worst to the expected principles of compensation (Table 5.1).

5.2.2 Semi-Structured Interviews and Documentary Study

This study was made possible, thanks to approximately 15 semi-structured interviews that were conducted in France between June 2016 and October 2017. These interviews were held with petitioners, decentralised government departments, advisory bodies, local authorities, nature conservation associations and researchers. Particular emphasis was placed on the actors working on the implementation of compensation for the high-speed line project built between Tours and Bordeaux. The aim of these

Table 5.1 Principle of ecological compensation and systems

Compensation Principles	Criteria Admitted	Clearing Bank	PRM System
Ecological efficiency of the compensation	The restoration is more ecologically viable on large spaces in one unit	+	−
	It is easier to track and monitor the implementation of compensatory measures on a single site than across scattered sites and different owners	+	−
	Competences of the stakeholders	+	−
	Effectiveness of the measures before the start of the impacts, because it is necessary to have ecological results in order to get compensation credits	+	−
	Anticipation		
Durability of compensatory measures	Land acquisition and long-term protection status	+	−
Ecological equivalence	The compensatory measures must correspond to the specificities of the biodiversity impacted	−	+
Proximity	The measurements must be located as close to the impacts as possible	−	+

Adapted from Levrel et al. (2015, 53) and Vaissière et al. (2017)

semi-structured interviews (Kaufmann, 1996) was to understand the role of each of the actors involved in the implementation of compensation measures, by taking up the different aspects that are important within ecological compensation: ecological equivalence, durability of measures, landownership and proximity (McKenney and Kiesecker, 2010). A documentary study[2] focused on the regulatory expectations in terms of compensation (protected Species and Water Act issues, nature of the compensatory measures, date of implementation) and the concrete implementation on the ground (number of sites, surface, location).

5.3 Results: Examples of Ecological Compensation in France

5.3.1 Permittee Responsible Mitigation or Compensation on Demand: The Example of the Southern Europe Atlantic High-Speed Railway Line

The PRM is the compensation system that is used most. Indeed, there are no real alternatives to this system in France today (Dutoit et al., 2015). The compensation system was studied with PRM through the implementation of the compensation on the Southern Europe Atlantic High-Speed Railway Line (HSRL) project linking Tours to Bordeaux. This HSRL was built, thanks to the setting up of a public-private partnership signed between RFF (Réseau Ferré de France) and Lisea in 2011.

As shown in Figure 5.1, 3,500 ha was needed to be found to set up the compensation measures. One hundred and sixty-nine sites were validated by the State services. They represent an area of 1403 ha[2].

Their sizes vary from less than 1 ha to more than 70 ha. There is an average size of 8 ha per site, with 25% of sites having areas of less than 1.6 ha, 50% of sites having less than 6.58 ha and 75% of sites having less than 12 ha. It can therefore be seen that more than half of the sites are smaller than 10 ha. Moreno-Mateos et al. suggest that below 1 ha the recovery success of ecosystems is almost nil and that beyond 100 ha recovery success is almost certain (Moreno-Mateos et al., 2012). The multitude of sites with highly variable sizes also makes monitoring and control more complex and costly (Levrel et al., 2015). The final managers of the compensatory measures are either the Conservatoire des Espaces Naturels (CEN), local authorities, private actors who own land (but in a minority), foresters or farmers. The competences of these actors with regard to the conservation of natural environments can be discussed. CEN is a historic figure in the conservation of natural areas. Without being the final manager of the compensatory measures, many naturalist associations carry out initial diagnoses

---

2 SEA HSRL: Protected Species Decrees and Water Act (Ministry of Environment, Energy and Sea, 2017; Préfecture of Gironde et al., 2012; Préfecture of Charente, Préfecture of Charente-Maritime, & Préfecture of Gironde, 2012; Préfecture of Charente & Préfecture of Deux-Sèvres, 2012; Préfecture of Indre et Loire, 2012; Préfecture of Indre et Loire & Préfecture of Deux-Sèvres, 2012).

Natural Compensation Sites (NCS): the "commitment documents" for each of the operations were studied (CGDD, 2011, Departmental Council of Yvelines, 2014, Dervenn, 2014, Dutoit et al., 2015, EDF et al., 2014).

Specifications for compensatory measures (Lisea, 2014).

Compensatory measure database (MC2; Cosea, 2017).

of potential sites for compensatory measures and the ecological monitoring thereof, providing certain know-how for dealing with these compensation issues, especially when this concerns knowledge of species and natural environments. With regard to the anticipated implementation of compensation measures, as the timeline Figure 5.1 shows, it started at the same time as the works (first impacts on biodiversity). The effectiveness of the compensation measures was not fully achieved by the time the impacts took place.

Compensation measures are considered sustainable when the land on which they are located is acquired by the developer (Moreno-Mateos et al., 2015). We know that two systems of landownership coexist: out of the 3,500 ha requested for compensation, 700 ha must be acquired by the developer while 2,800 ha will be contracted. More than 225 protected species were impacted by the project, as well as more than 8,000 water-courses. Four types of environments were identified as being equivalent to the habitats impacted by the project. Thus, the search for compensatory measure sites focused on:

- agricultural plains, especially for lowland birds (little bustards, Eurasian stone curlews, butchers, Montagu's harriers, etc.);
- open mesophilic or calcicolous grasslands for birds (woodlarks, Dartford warbler, etc.), amphibians, plant species (umbel-flowered sun rose, Odontites jaubertianus) and insects (large blue butterflies);
- mature afforestations for small mammals (squirrels, hedgehogs, bats, etc.), birds (little owls, lanius, etc.), amphibians and insects (great capricorn beetle, Rosalia longicorn);
- wetlands for semi-aquatic mammals (European mink, etc.), birds (hen harriers, etc.), amphibians, insects (false ringlets, etc.) and some species of plants, such as bog-myrtle;

An initial inventory in terms of biodiversity was made on each potential site, and the restoration measures to be implemented according to the predefined specifications were also chosen. The types of species and the surfaces that will be counted on the site are then validated by the State services. It can be seen that ecological equivalence is relatively accurate.

Regarding proximity, it can be observed that 7% of the measuring sites (1,403 ha in surface area) are within 1 km of the line, 23% of the sites are between 2 and 5 km away, and 36% of the sites are between 5 and 10 km and 34% of the sites are more than 10 km from the line. The proximity of the sites to the place of impact ranges between a few hundred meters to more than 10 km for a significant part of the sites. This notion of proximity is therefore relative.

5.3.2 The Emergence of Compensation Banks

Table 5.2 summarises the main characteristics of these different operations.

A fairly large disparity can be observed in these operations at different levels. Indeed, none of these NCS consist of a single large terrain, as in the Cossure intervention. The other three NCS consist of several sites split up over a larger or smaller area and with a greater or lesser degree of "naturalness". The integration of measures within a functional landscape will have an effect on the recovery of the ecosystem following

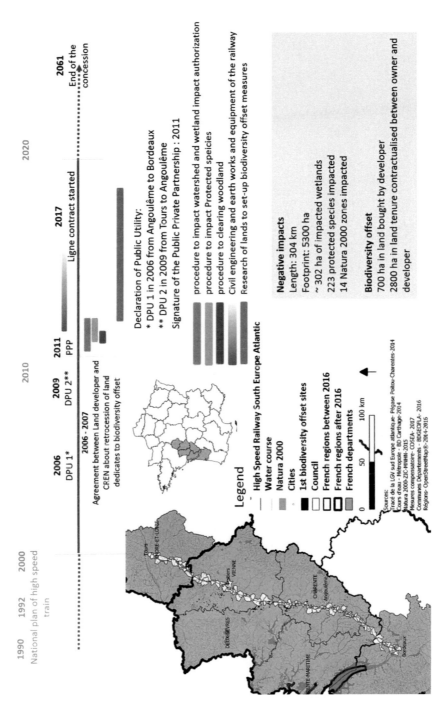

Figure 5.1 Spatial-temporal data of the SEA HSRL. (All figures concerning the number of sites, their sizes and distances are based on the first 1,403 ha of compensation sites implemented on this project.)

Table 5.2 Characteristics of NCS

	Cossure	Combe Madame	Yvelines	Vallée de l'Aff
Types of environments to be restored	Former industrial orchards, the restoration of which is intended to recover the Coussoul de Crau	Ecosystem of the sub-alpine and upper stages (alt. 1,200 m to 2,930 m) in the Belledonne massif, the restoration is intended to rehabilitate open areas; Wetlands; Forest environment	Mosaics of prairies and shrub habitats without human activity, 84% wooded areas	Sub-watershed of the Aff
Targeted element of biodiversity	Stone curlew Little Bustard Pin-tailed sandgrouse common wall lizard	Mountain Galliformes (black grouse, rock ptarmigan, hazel grouse, rock partridge), Lepidoptera, Chiroptera	Avifauna and Entomofauna in open and semi-open dry environments	Wetlands In agricultural and forestry environments
Surface area	357 ha in one unit	68 ha not adjacent, spread over a territory of 1,852 ha	100 ha not adjacent	1,480 ha not adjacent
Service area	600 km^2	25 km in every direction	Seine Aval estuary 750 ha	Vilaine watershed
Status of the compensation operator	CDC biodiversity Subsidiary of Caisse des Dépôts et Consignment – State-owned public bank	Combe Madame Biodiversity Initiative Association under the law of 1901 From EDF	Departmental Council of Yvelines	Private research organisation (Derven)
And other actors involved	CEN PACA[a], CA, Inra d'Avignon	CBNA, association Gentiana, ONCFS, fédération de chasse (Federation of Hunters), LPO, ONF, IRSTEA	CBNBP, MNHN, OPIE, association de naturalistes des Yvelines	16 farmers, CA, Forum des marais atlantiques, Bretagne Vivante, APPCB
Mobilisation of land	Purchase by the CDC (full acquisition)	Property of EDF, founder of the association (full acquisition)	Difficulties in convincing landowners to sell. 6 ha acquired at the end of 2016. (contractualisation acquisition)	Difficulties in convincing the owners to enter into this process (contractualisation acquisition)

(Continued)

Table 5.2 (Continued) Characteristics of NCS

	Cossure	Combe Madame	Yvelines	Vallée de l'Aff
Unit	1 ha = one unit	The "quality hectare" will be defined according to the method for calculating the ecological gains	The expected gains will be assessed both in terms of natural habitats and species habitats, functionalities (including the sites' contribution to the ecological continuity network) and ecosystem services	Compensation unit: "extended hectare" => an asset of 1ha, 1 ecosystemic production zone extending over 25 ha and additional actions targeting protected species
Stage of implementation	Tenth year of testing 155.57 units sold on 357 units (Dutoit et al., 2015)	2013–2015 initial diagnoses. Start of restoration work in 2016	Attempt to acquire land from communities	Seeking landowners to contract restoration measures
Commitment period	30 years minimum			

[a] CEN PACA, Conservatoire d'Espace Naturels Provence Alpes Côtes d'Azur; CA, Chambre d'Agriculture; ONCFS, Office National de la Chasse et de la Faune Sauvage; CBNA, Conservatoire Botanique National Alpin; LPO, Ligue pour la Protection des Oiseaux; ONF, Office Nationale de la Forêt; IRSTEA, Institut de Recherche en Sciences et Technologies pour l'Environnement et l'Agriculture; CBNBP, Conservatoire Botanique National du Bassin Parisien; MNHN, Museum National Histoire Naturelle; OPIE, office pour la protection des insectes et de leur environnement; APPCB, Assemblée Permanente des Présidents des commissions locales de l'eau Bretagne.

the implementation of restoration measures (BenDor et al., 2009; McKenney and Kiesecker, 2010). Different types of actors (public, private for-profit or not-for-profit) are responsible for setting up these NCS. However, none of these are directly specialised in biodiversity, even though two new structures have been created, namely CDC biodiversité and IBCM (Initiative Biodiversité Combe Madame – Combe Madame Biodiversity Initiative). Historically, no actor has been involved in the management of natural areas. Many are surrounded by specialists in the field. Three of the four interventions are not yet operational, meaning it is not possible to compensate for the project impacts that have occurred in the meantime. One of the main interests of the NCS or clearing bank system lies in the anticipation of restoration, but it is not yet operational in France. Land control seems to be a problem for the Vallée de l'Aff and Yvelines. Indeed, these two operators have almost no land and it is therefore difficult for them to start their operation. In the Vallée de l'Aff, the idea is not necessary to acquire land but to convince the owners to adopt a change of practices. The principle of durability through acquisition is therefore weakened. It is also observed that the elements of biodiversity in question are different. Three of the four systems currently being implemented focus on creating suitable sites for protected species, in order to argue this under protected species exemptions, whereas for the Aff catchment basin, the environments subject to restoration attempts are rivers and wetlands, which is more in line with the expectations of the water act. The specificity of the restored assets is not without consequences for the propensity of these natural compensation site operators to sell units (Calvet et al., 2015; Coggan et al., 2013; Scemama and Levrel, 2014; Vaissière et al., 2017). Indeed, if the assets are too specific to a given environment, they are less likely to be impacted because they are less present in the area. As such, the Cossure steppes are a good example of a very specific environment, for which the CDC has not found an outlet. However, many NCS focus on different taxa and different types of environments, which diversify the supply and can potentially affect a variety of markets, some of which may be linked to high specificity of biodiversity. As can be shown in Table 5.2, the service areas[3] are variable. Thus, the proximity between impacts and compensatory measures will be unequal from one NCS to another. Implementation Decree No. 2017-265 of 28 February 2017 concerning the NCS specifically requires the justification of financial and technical capacities for the implementation of measures compensating threats to biodiversity in an anticipated and shared manner, and justification of the rights to implement compensatory measures in the NCS field. It is specified that the approval cannot be valid for less than 30 years. The four operations presented earlier have until 1 July 2019 to be approved (Decree No. 2017-264). However, it is not clear whether all these operations will be. Indeed, questions can be asked about the aspects concerning land control and sharing. It is also not clear whether local authorities can systematically enter into these approvals, as the financial terms are not specified. The Cossure and Combes Madame operations seem to fulfil the expected conditions.

However, the approval is not a prerequisite for implementing anticipated and shared compensation measures. Other approaches are now being considered.

---

3 The space within which a developer generating an impact can purchase compensation credit for a NCS (Vaissière et al., 2015).

5.3.3 Effects of French Systems on Ecological Compensation Principles

Table 5.3 summarises the effects of the two examples of PRM studied and those related to the NCS presented previously, in the light of the expected compensation principles. This allows us to reconsider: the principles, which system allows them to be met and, therefore, what the difference is between the systems.

On the question of surfaces (EfEC A), we see that two NCS (Yvelines and Vallée de l'Aff) out of four do not have large areas of continuous compensatory measures. Small spaces make the recovery of restorations slower than large spaces (Moreno-Mateos et al., 2012). We could see that some SEA sites had an area of nearly 70 ha, which is larger than some sites belonging to Yvelines and Vallée de l'Aff NCS, and even Combe Madame. It would be interesting to study the integration of all these sites (NCS or SEA) in a landscape framework that is functional from the point of view of biodiversity, in order to better understand the effects of the environment on the recovery of restored ecosystems. Proper networking of the compensatory sites will contribute to the recovery of restored ecosystems (McKenney and Kiesecker, 2010).

For the durability (P) of the compensatory measures, three temporalities can be taken into account: the commitment period (30 years minimum for NCS and 50 years for the concession on the SEA HSRL); the type of landownership (acquisition or contracting) and the duration of the contracts with the final managers. In terms of land control, we saw that the Cossure NCS acquired 357 ha while 700 ha will be acquired for the SEA HSRL, for example. These acquired areas should be reduced to the total area impacted by the project, in order to correctly compare whether, in some cases, PRM may lead to a more sustainable land situation than NCS. Similarly, two NCS currently being studied have as much, if not less, land acquisition capacity than the PRM system. In terms of durability and competence (EfEC C), the final managers are a State-owned public bank and, on the contrary, the Conservatoires d'Espaces Naturels for the 700 ha acquired for SEA. Indeed, the legitimacy of the actors is quite questionable. There are still few specialists in ecological compensation today, or rather there is a multitude of actors trying to grasp the subject. This latter is vast and requires many

Table 5.3 Effects of compensation systems on the principles of ecological compensation

Compensation Principles	Clearing Bank in the United States	NCS System				PRM System	
		Cossure	Combe Madame	Yvelines	Vallée de l'Aff	SEA HSRL	Expected Effects
(EfEC) A	+	+	+	−	−	−	−
B	+	+	−	−	−	−	−
C	+	+	+	+	+	+	−
D	+	+	+	?	?	−	−
(P)	+	+	+	−	−	+	+/−
(Ee)	−	−	−	−	−	+	+
(Prox)	−	−	−	+	+	−	+

(EfEC) Effectiveness of compensation measures: (A) Large spaces, (B) Facilitated monitoring and control (C) Competences, (D) Anticipation; (P) Durability; (Ee) Ecological equivalence; (Prox) Proximity naturalist skills, and for the moment, none of the compensation operators or petitioners (SEA) can quite declare themselves to be more competent than another. All these actors are surrounded by biodiversity specialists (naturalists and ecologists) and, historically, by some natural area management organisations, as we have seen.

On the anticipation (EfEC D) of the measures with respect to impacts, it has been seen that few NCS have been set up since 2011, and, of those that have been created, some are not yet operational. Thus, for the Yvelines NCS and that of the Vallée de l'Aff, land control does not seem easy to obtain. Perhaps more measures would have been anticipated had these systems been put in place. Similarly, in some PRM projects, the implementation of a certain amount of measures could have been more anticipated (more than 20 years elapsed between the project idea and its realisation). This could be improved on projects whose implementation lasts from 10 to 20 years or even longer, as may have been the case for the very large infrastructures of the HSRLs. In addition, in the case of clearing banks, the first units are sold before they have recovered a satisfactory ecological level at the time of the impacts, and the total recovery of the ecosystem is effective several years after the implementation of the clearing bank in the United States. This is also true of NCS if the same terms for releasing the compensation units apply. The first units sold would then not be really anticipated in relation to the arrival of the impacts, and the ecological recovery level of the ecosystem would not yet be truly restored. With regard to the proximity (Prox), we saw that some measurements on SEA were located more than 10 km from the impact sites, while in the case of the Yvelines NCS, the service area does not extend beyond 7.7 km².

No examples fulfil all the expected principles. Other approaches may be able to respond to these principles.

5.3.4 "Territorial Approaches": A New Way for Ecological Compensation?

Other initiatives are flourishing all over France. In many cases, these initiatives are led by public authorities seeking to organise the implementation of ecological compensation in their territory. This is especially true when these communities have expertise in terms of management. This territorialisation can be based on several aspects of the territory: its biophysical component, its legal and administrative component, its actors and its inhabitants (Moine, 2006). Different types of initiatives are developing, here is a brief overview.

5.3.4.1 Organisation of the Network of Actors

The Landes Department is considering becoming a compensation operator. As this approach is in-line with the framework of the new 2016 law on biodiversity, the department did not respond to the call for the expressions of interest in 2011. It is above all seeking to organise ecological compensation within its territory, in order to streamline initiatives and have an overview of what is happening. It also wants to combine this departmental operator approach to ecological compensation with other forms of compensation, such as reforestation under the Forest Code and collective agricultural compensation in the framework of the Loi d'Avenir pour l'Agriculture l'Alimentation et la Forêt (Law for the Future of Agriculture, Food and the Forest) from 2014. It wants to create a dedicated departmental organisation. The Services Directorate is considered the relevant actors to be involved in such an approach, in connection with the Environment Department.

5.3.4.2 Integration of Compensation in Contractual Arrangements

The Comité Intersyndical d'Aménagement du Lac du Bourget (Lac du Bourget Inter-Union Management Committee, CISALB) has put in place a Plan d'Action en Faveur des Zones Humides (Wetland Action Plan, PAFZH) to guide efforts to restore the wetlands that have the most need. This action plan provides for these wetlands to be restored both with funding from the water agency and with funding from project owners, with obligations to implement ecological compensation measures. The CISALB proposes each Public Corporation for Inter-Municipality Cooperation (Etablissement Public de Coopération Intercommunal, EPCI) forming part of the Lac du Bourget catchment area to join this approach.

Within the framework of the expectations of the Masterplan for Water Development and Management (Schéma Directeur d'Aménagement et de Gestion des Eaux, SDAGE), a departmental inventory of the territory's wetlands was carried out, which made it possible to identify close to 300 wetlands representing approximately 3,300 ha. Of these 3,300 ha, it was identified that 330 ha was already managed by an organisation, that 5 ha of the wetlands disappears on average each year, that nearly 150 ha of the wetlands requires restoration work and that 25 ha of the wetlands will be very heavily degraded or potentially disappear, following development work. The territory is therefore faced with four challenges: halting the loss of the wetlands, restoring the degraded wetlands, controlling the impact of development on wetlands and finding wetlands on which to carry out restoration and maintenance measures, as ecological compensation. Thus, the implementation of the PAFZH forms a part of the Lac du Bourget catchment area contract passed with the water agency. This PAFZH consists of restoring, maintaining and limiting the degradation of the wetlands, and guaranteeing their preservation through their inclusion in planning documents. This can be done either voluntarily within the framework of the financing foreseen by the water agency, or in a compulsory way as ecological compensation, in the event that the petitioner finances and delegates all the restoration work and maintenance of the wetlands identified as requiring priority restoration to the competent EPCI. The petitioner remains responsible for the implementation and the expected results of the ecological compensation operations. It can clearly be seen here that the choice of wetland sites to be restored was pre-identified and is part of a process to integrate the compensation device and even the avoid, reduce and compensate sequence on a territorial level, thanks to the coupling with other devices that have actions directed at the same objectives. The willingness to register the wetlands in urban planning documents even allows territorialisation thereof (CISALB, 2012), i.e. the appropriation of the wetland issues and the establishment of different actions by the institutional actors in the territory. More broadly, compensation under the Water Act is incorporated into the provisions made at the level of the Masterplan for Water Development and Management (Schémas d'Aménagements et de Gestion des Eaux, SDAGE).

5.3.4.3 Pre-identification of Sites Based on Ecological Criteria

Several projects for identifying interesting sites for accommodating compensatory measures have emerged. One example of this is the project to identify sites for compensatory measures in the Grenoble basin, allowing them to be mapped. The ESNET[4] (Ecosystem Services NETworks) project carried out by the Alpine Ecology Laboratory, among others, develops prospective scenarios for changing land use by 2040. Future impacts on biodiversity and wetlands have been identified based on the guidelines of the Territorial Coherence Scheme (Schéma de Cohérence Territoriale, SCoT). Two main types of compensation scenarios have been considered (surface compensation only with variable ratios, and surface and functional compensation). A cartographic simulation was constructed in order to find as many available wetland surfaces as there were destroyed wetland surfaces. In the second scenario, the functional dimension was integrated by assigning a wetland functionality index to each cultivation practice. The importance of the functional impact has thus been identified and matched with plots with functional interests equivalent to the degradation. This made it possible to identify many sites that seem interesting from an ecological point of view at first glance. A field inventory will help expand on this pre-identification (Vaissière et al., 2016). Other criteria may be included in the model, such as land availability or integration within the "Trame verte et bleue" (blue-green infrastructure). This method makes it possible to guide and organise ecological compensation on a territorial scale.

In conjunction with research institutes, the Montpellier urban area and the Occitanie region are considered the construction of indicators for assessing ecological pressures and issues at the territorial scale, in order to assess the impacts of projects, plans and programmes on biodiversity. The identification of ecological issues and potential impacts makes it possible to better organise ecological compensation upstream.

All these initiatives aim to organise the compensation before the arrival of the projects, in order to allow a certain control of the subject on the part of the public power and also to standardise the initiatives and make them more effective, as well as to integrate them within a more coherent biodiversity conservation policy. Once again, the identification of the problem areas (Witté and Touroult, 2014) makes it possible to act by first avoiding the most interesting areas in terms of biodiversity (Bigard et al., 2017), identifying sites that are damaged and that require to be restored or rehabilitated and which could be in the context of ecological compensation. Finally, construction could be oriented towards areas that have no interest in terms of biodiversity, which are exempt. This would make it possible to integrate projects better within an avoid, reduce and compensate approach, and to better understand the cumulative effects of projects and apply the avoid, reduce and compensate sequence on a territorial scale, which makes sense in terms of the functioning of biodiversity (Bezombes, 2017; Whitehead et al., 2017).

The territorialised approaches are still too recent for us to really understand their effects. This is an interesting avenue of the development that could help to fulfil some of the compensation principles.

---

4 http://www.projet-esnet.org/

5.4 Discussion and Conclusion

We have seen that the strength of the clearing bank system in the United States in particular does not seem to constitute a strength in the French NCS. Indeed, various examples have shown that the compensation principles met by the American clearing banks – such as having a large unitary space, compulsory acquisition, the effectiveness of the controls, the competence of the operators, anticipation of compensation measures before the arrival of the impacts and the durability of the compensation measures – are not consistently found in the French NCS. Several aspects may be involved in this: the acquisition of land is complex and the organisation of access to land is not simple in France (Levesque, 2013). It is necessary to be able to have long-term control of the land over time, which can be tantamount to investing in the purchase of land. Even though land management may involve land contracting, the anticipation of measures requires the ability to invest a minimum upstream to restore all or part of the site before being able to sell the units (Vaissière et al., 2015). Regarding the benefits of NCS monitoring and control, this is only true if the NCS consist of a single site. Otherwise, time will need to be spent monitoring and controlling all the sites. We have seen that some NCS are composed of many sites. It still seems difficult to say whether compensation operators are more qualified to carry out compensation than a petitioner who carries out their compensation using PRM. Indeed, the examples have shown that they are all surrounded by specialists in the field of ecology and or historical actors engaged in the protection of natural areas. With regard to the durability of the compensatory measures, the NCS system (30 years in the examples presented) is not more sustainable than the PRM system (50 years of concession and retrocession of 700 ha to the CEN).

Finally, these examples show that the distinction between these two systems (NCS and PRM) is actually not very clear. Some of the principles met by each of the two systems seem to blend together, posing the question, what really characterises them? The number of sites and their sizes (a large contiguous site or a multitude of smaller sites)? The type of measures (measures with non-specific biodiversity and another with very particular biodiversity)? The anticipation of the implementation of the compensatory measures? The financing arrangements? The land control (acquisition, contractual agreement)? The durability of the measures (30, 50 years, in perpetuity)? The nature of the operator (private, public, for-profit, non-profit)? The ecological equivalence and proximity (what spatial scale is being sought)? Responsibility for implementation and results? A better distinction between systems would provide a clear characterisation of them, and therefore, the principles which they can meet. The new law for biodiversity advocates anticipation and sharing of measures, and the fact that an operator can be public or private (Law No. 2016-1087). It should be clarified what is meant by sharing (the grouping of measures on an adjoining site?). Similarly, the anticipation methods are not specified, and anticipation probably requires a significant investment capacity that small businesses and local authorities do not have. The advantages of the clearing banks present in the United States do not seem to be clearly reflected in the French NCS. It is regrettable that no real assessment was conducted on the NCS before including this apparatus in the law (Vaissière et al., 2017). We have also seen that the Vallée de l'Aff and Yvelines NCS can be related to the territorial approaches described. Indeed, the choice of sites for Yvelines followed a reflection on the development of the Seine Aval territory.

We are therefore seeing the emergence of initiatives geared towards territorial reflection, both at the level of governance and the integration into already existing conservation systems (particularly with regard to monitoring and administrative management), or a reflection on the spatial or cartographic scale. These reflections could also fulfil a number of expected principles of ecological compensation, such as better integration of the avoid, reduce and compensate sequence with the land-use choices. The anticipation of land control on the sites of interest for compensation and integration of sites into a functional framework would allow faster, more viable recovery (McKenney and Kiesecker, 2010). The durability of the compensatory measure sites could be reinforced through inclusion within the urban planning documents for these areas. The Agence Française de la Biodiversité (French Agency for Biodiversity) has also planned to identify the areas relevant to compensation at the national level (Law No. 2016-1087). The PRM seems more precise in terms of ecological equivalence. The territorial approach could make it possible to rationalise this equivalence with a satisfactory territorial scale. The difficulty would lie in defining the relevant scale of action (Bezombes et al., 2017; Bigard et al., 2017). The development project territory seems too small to meet all the expected principles (Vaissière, Levrel et al., 2017), and the administrative territory only seems to correspond to a small part of the actual ecological environments. So perhaps we could adapt more to these later.

Acknowledgments

The authors would like to thank all those who agreed to share their experiences on the subject of compensation by giving us their time. We would also like to thank the Eco-design Chair of Buildings and Infrastructures for financing the engineer contract study by Pauline Delforge and the doctoral grant of Julie Latune.

Bibliography

BBOP. (2012). Standard on Biodiversity Offsets. Retrieved 11 October 2016, from http://www.forest-trends.org/documents/files/doc_3078.pdf.x.

BenDor, T., Sholtes, J., & Doyle, M. W. (2009). Landscape characteristics of a stream and wetland mitigation banking program. *Ecological Applications,* 19(8), 2078–2092.

Bezombes L. (2017). *Développement d'un Cadre Méthodologique pour l'Evaluation de l'Equivalence Ecologique: Application dans le Contexte de la Séquence « Eviter, Réduire, Compenser » en France*, PhD, Université Grenoble Alpes.

Bigard C., Pioch S., & Thompson J. D. (2017). The inclusion of biodiversity in environmental impact assessment: Policy-related progress limited by gaps and semantic confusion. *Environmental Management*, 200:35–45.

Bull, J. W., Suttle, K. B., Gordon, A., Singh, N. J., & Milner-Gulland, E. J. (2013). Biodiversity offsets in theory and practice. *Oryx*, 47(03), 369–380. https://doi.org/10.1017/S003060531200172X.

Calvet, C., Levrel, H., Napoléone, C., & Dutoit, T. (2015). La réserve d'actifs naturels, chapter 12. In *retaurer la nature ppour atténuer les impacts du développement, analyse des mesures compensatoires pour la biodiversité*. (Quae, 314).

CGDD. (2012). DOCTRINE relative à la séquence éviter, réduire et compenser les impacts sur le milieu naturel. Ministère de l'écologie du développemet Durable des trasports et du Logement.

CGDD. (2013). Lignes directrices nationales sur la séquence éviter, réduire et compenser les impacts sur les milieux naturels. References.

CISALB Comité InterSyndical d'Aménagement du Lac du Bourget. (2012). Plan d'action en faveur des zones humides (PAFZH) sur le bassin versant du lac du Bourget.

Code forestier (nouveau) - Article L341-6, L341-6 Code forestier (nouveau) §.

Coggan, A., Buitelaar, E., Whitten, S., & Bennett, J. (2013). Factors that influence transaction costs in development offsets: Who bears what and why? *Ecological Economics*, 88, 222–231. https://doi.org/10.1016/j.écolecon.2012.12.007.

Commissariat général au développement durable. (2011). Appel a projet d'opérations expéri-mentales d'offre de compensation. *Ministère de l'écologie, du développement durable, des transports et du logement.*

Conseil Départemental des Yvelines. (2014). Engagement opération yvelinoise. Cosea. (2017, March 14). Sites géographiques des mesures compensatoires. Shapefile.

Crozier, M., & Friedberg, E. (1977). L'acteur et le système (Seuil).

Décret n° 2011-2019 du 29 décembre 2011 portant réforme des études d'impact des projets de travaux, d'ouvrages ou d'aménagements, 2011–2019 § (2011).

Décret n° 2015-776 du 29 juin 2015 relatif à la gouvernance du fonds stratégique de la forêt et du bois et aux règles d'éligibilité à son financement, 2015–776 § (2015).

Décret n° 2017-264 du 28 février 2017 relatif à l'agrément des sites naturels de compensation | Legifrance.

Dervenn. (2014). Engagement Sous-bassin versan de l'Aff.

Directive 92/43/EEC. (1992, May 21). Council Directive 92/43/EEC of 21 May 1992 on the conservation of natural habitats and of wild fauna and flora.

Dutoit, T., Calvet, C., Jaunatre, R., Alignan, J.-F., Bulot, A., & Buisson, É. (2015). Première expéri-mentation de compensation par l'offre: bilan et perspective, (16). Retrieved from http://www.set-revue.fr/premiere-   experimentation-de-compensation-par-loffre-bilan-et-perspective/text.

EDF, & Initiative Biodiversté Combe Madame. (2014). Engagement relatif à l'offre de compen-sation Combe Madame.

European Commission. (2007). Clarification des concepts de: solutions alternatives, raisons impératives d'intérêt public majeur, mesures compensatoires, cohérence globale, avis de la commission Document d'orientation concernant l'article 6, paragraphe 4, de la directive "Habitats".

Gelcich, S., Vargas, C., Carreras, M. J., Castilla, J. C., & Donlan, C. J. (2016). Achieving bio-diversity benefits with offsets: Research gaps, challenges, and needs. *Ambio*, 1–6. https://doi.org/10.1007/s13280-016-0810-9.

Gobert, J. (2010). Éthique environnementale, remédiation écologique et compensations terri-toriales: entre antinomie et correspondances. *VertigO - la revue électronique en sciences de l'environnement*, 10(1). https://doi.org/10.4000/ vertigo.9535.

Gonçalves, B., Marques, A., Soares, A. M. V. D. M., & Pereira, H. M. (2015). Biodiversity offsets: From current challenges to harmonized metrics. *Current Opinion in Environmental Sustainability*, 14, 61–67. https://doi.org/10.1016/j.cosust.2015.03.008.

Gordon, A., Bull, J. W., Wilcox, C., & Maron, M. (2015). FORUM: Perverse incentives risk undermining biodiversity offset policies. *Journal of Applied Ecology*, 52(2), 532–537. https://doi.org/10.1111/1365–2664.12398.

Guillet, F., & Semal, L. (2018). Policy flaws of biodiversity offsetting as a conservation strategy. *Biological Conservation*, 221, 86–90. https://doi.org/10.1016/j.biocon.2018.03.001.

IUCN. (2016). IUCN policy on biodiversity offsets. Retrieved from http://cmsdata.iucn.org/downloads/iucn_biodiversity_offsets_policy_jan_29_2016.pdf.

Jacob, C., Quétier, F., Aronson, J., Pioch, S., & Levrel, H. (2015). Vers une politique française de compensation des impacts sur la biodiversité plus efficace: défis et perspectives. *VertigO - la revue électronique en sciences de l'environnement*, 14(3). https://doi.org/10.4000/ vertigo.15385.

Journal Officiel de la République Française. LOI no 2016-1087 du 8 août 2016 pour la reconquête de la biodiversité, de la nature et des paysages (2016).

Kaufmann, J.C. (1996). *L'entretien compréhensif.* Paris: Nathan Université.

Labat B. (2015). COMPENSATION FORESTIÈRE OU COMPENSATION EN FORÊT? Enjeux et ambiguïtés de la compensation écologique dans le contexte forestier. Study carried out in the framework of an agreement between the Association Humanité et Biodiversité and the Bureau de la gestion durable de la forêt et du bois du Ministère de l'Agriculture (DGPE-Minagri) - Humanité Biodiversité.

Lascoumes, P., & Gales, P.L. (2007). Introduction: Understanding public policy through its instruments—from the nature of instruments to the sociology of public policy instrumentation. *Governance,* 20(1), 1–21. https://doi.org/10.1111/j.1468-0491.2007.00342.x.

Levesque, R. (2013). Les SAFER. D'un opérateur foncier agricole à un opérateur rural. *Pour,* (220), 185–192. https://doi.org/10.3917/pour.220.0185.

Levrel, H., Guillet, F., Latune, J., Delforge, P., Frascaria, N. (2018). Application de la séquence éviter-réduire-compenser en France: le principe d'additionnalité mis à mal par 5 dérives. VertigO la revue électronique de l'environnement.

Levrel, H., Frascaria-Lacoste, N., Hay, J., Martin, G., & Pioch, S. (2015). Restaurer la nature pour atténuer les impacts du développement: Analyse des mesures compensatoires pour la biodiversité. Editions Quae.

Lisea Ligne SEA Tours Bordeaux. (2014, February 14). Mesures compensatoires procédure de mise en oeuvre.

Loi no. 76-629 du 10 juillet 1976 relative à la protection de la nature - Article 2. Loi no. 2006-1772 du 30 décembre 2006 sur l'eau et les milieux aquatiques.

Loi no. 2014-1170 du 13 octobre 2014 d'avenir pour l'agriculture, l'alimentation et la forêt, 2014–1170 § (2014).

Loi no. 2016-1087 du 8 août 2016 pour la reconquête de la biodiversité, de la nature et des paysages, 2016–1087 § (2016).

McKenney, B. A., & Kiesecker, J. M. (2010). Policy development for biodiversity offsets: A review of offset frameworks. *Environmental Management,* 45(1), 165–176. https://doi.org/10.1007/s00267-009-9396-3.

Ministère de l'Environnement de l'Energie et de la Mer. Arrêté ministériel portant dérogation aux interdictions portant sur un certains nombre d'éspèces protégées pour la réamlisation des travaux de construction de la Ligne à Grande Vitesse Sud Europe Atlantique entre Tours et Bordeaux. (2017).

Moine, A. (2006). Le territoire comme un système complexe: un concept opératoire pour l'aménagement et la géographie. *L'Espace géographique,* 35(2), 115–132.

Monin-Soyer, H. (2011). *La compensation écologique état des lieux & recommandations.* Paris: IUCN France.

Moreno-Mateos, D., Maris, V., Béchet, A., & Curran, M. (2015). The true loss caused by biodiversity offsets. *Biological Conservation,* 192, 552–559. https://doi.org/10.1016/j.biocon.2015.08.016.

Moreno-Mateos, D., Power, M. E., Comín, F. A., & Yockteng, R. (2012). Structural and functional loss in restored wetland ecosystems. *PLoS Biology,* 10(1), e1001247. https://doi.org/10.1371/journal.pbio.1001247.

Pilgrim, J.D. (2014). *Technical Conditions for Positive Outcomes from Biodiversity Offsets. An Input Paper for the IUCN Technical Study Group on Biodiversity Offsets.* Gland: IUCN, 46.

Préfecture de la gironde, préfecture de la v ienne, p réfecture de la c harente, p réfecture des Deux -Sèvres, p réfecture de la c harente -m aritime, & p réfecture de l 'i ndre - et -l oire. Arrêté portant dérogation à l'interdiction de destruction d'espèces et d'habitats d'espèces animales protégées et de destruction d'espèces végétales protégées, Pub. L. No. Arrêté no. 20120.59–0013, 86 (2012).

Préfecture de Charente, Préfecture de Charente-Maritime, & Préfecture de Gironde. Réalisation de la Ligne à Grande Vitesse Sud Europe Atlantique entre Tours et Bordeaux Bassin Versssant de la Dordogne, Arrêté no. 2012–02–23/23, 97 (2012).

Préfecture de Charente, & Préfecture des Deux-Sèvres. Autorisation des installations de la Ligne à Grande Vitesse Sud- Europe Atlantique au titre de la loi sur l'eau Bassin versant de la Charente, Arrêté n °2012363–0002 § (2012).

Préfecture d'Indre et Loire. Réalisation de la Ligne à Grande Vitesse Sud europe Atlantique entre Tours et Bordeaux Bassin versant de l'Indre, Pub. L. No. Arrêté no. 12–E–11, 72 (2012).

Préfecture d'Indre et Loire, & Préfecture des Deux-Sèvres. Réalisation de la Ligne à Grande Vitesse Sud europe Atlantique entre Tours et Bordeaux Bassin Versant de la Vienne, Pub. L. No. 2012/DDT/847, 59 (2012).

Quétier, F., & Lavorel, S. (2011). Assessing ecological equivalence in biodiversity offset schemes: Key issues and solutions. *Biological Conservation*, 144(12), 2991–2999. https://doi.org/10.1016/j.biocon.2011.09.002.

Quétier, F., Regnery, B., & Levrel, H. (2014). No net loss of biodiversity or paper offsets? A critical review of the French no net loss policy. *Environmental Science & Policy*, 38, 120–131. https://doi.org/10.1016/j.envsci.2013.11.009.

Regnery, B. (2013). *Les mesures compensatoires pour la biodiversité Conception et perspectives d'application*, PhD, Université Pierre et Marie Curie.

Scemama, P., & Levrel, H., (2014), L'émergence du marché de la compensation des zones humides aux États-Unis: impacts sur les modes d'organisation et les caractéristiques des transactions. *Revue d'Economie Politique*, 123(6): 893–924.

Vaissière, A.C., Bierry, A., & Quétier, F. (2016). Mieux compenser les impacts sur les zones humides: modélisation de différentes approches dans la région de grenoble. *Sciences Eaux & Territoires IRSTEA*, (21), 64–69.

Vaissière, A.-C., Levrel, H., & Pioch, S. (2017). Wetland mitigation banking: Negotiations with stakeholders in a zone of ecological-economic viability. *Land Use Policy*, 69, 512–518. https://doi.org/10.1016/j.landusepol.2017.09.049.

Vaissière A.C., Levrel H., Scemama P. (2015). « Les banques de compensation aux Etats-Unis : une nouvelle forme organisationnelle et institutionnelle pour la conservation basée sur le murché ? », in Levrel H., Frascaria-Lacoste N., Hay J., Martin G., Pioch S. (eds.), *Restaurer la nature pour atténuer les impacts du développement*. Analyse des mesures compensatoires pour la biodiversité, Éditions QUAE, pp. 116–127.

Vaissière A.C., Quétier F., Levrel H. (2017). « Le nouveau dispositif des sites naturels de compensation : est-ce trop tôt ? » in Cans, C., Cizel, O., (dir.) *La loi biodiversité. Ce qui change en pratique*, Éditions Législatives, pp. 126–137.

Whitehead A.L., Kujala H., & Wintle B.A. (2017). Dealing with cumulative biodiversity impacts in strategic environmental assessment: A new frontier for conservation planning. *Conservation Letters*, 10(2).

Witté, I., & Touroult, J. (2014). Répartition de la biodiversité en France métropolitaine: une synthèse des Atlas faunistiques. *VertigO - la revue électronique en sciences de l'environnement*, 14 (1). https://doi.org/10.4000/vertigo.14645.

World Bank Group. (2016). *Biodiversity Offsets: A User Guide*, Profor eds, 60.

Chapter 6

Biodi(V)strict®

A Tool for Incorporating Biodiversity within Development

Angevine Masson and Nathalie Frascaria-Lacoste

Université Paris-Saclay, AgroParisTech, ESE Lab

6.1 Introduction

Today, more than half of the world's population lives in urban areas. According to the latest United Nations report on urbanisation prospects published in 2014, these areas are likely to absorb most of the population growth. Thus, while currently 54% of the world's population lives in urban areas, this proportion is projected to increase to 66% by 2050 (United Nations, 2014). This increase in population density in urban areas poses economic and demographic problems that must be solved quickly, but it also presents numerous ecological consequences.

Indeed, the destruction and fragmentation of habitats, among others due to the urban sprawl, are now considered one of the major causes of the erosion of biodiversity and a threat to the services rendered thereby (Millennium Ecosystem Assessment, 2005).

Yet, urban ecosystems are liable to provide a number of services and promote the well-being of urban dwellers, particularly if the former are well managed (McGranahan et al., 2005). Indeed, while overly rapid and poorly organised urban growth poses a threat to sustainable development (United Nations, 2014), the negative effects of urbanisation on biodiversity and on the well-being of urban dwellers can be mitigated by well-thought-out urban projects (Whiford, Ennos, & Handley, 2001).

Urban biodiversity is recognised by all actors involved in urban planning as an essential component. However, despite this recognition and appropriation of the subject by the actors, effective integration of biodiversity into project design remains challenging.

The literature identifies three main obstacles to this phenomenon:

- The very mixed compilation of ecological data makes them difficult to access and also causes confusion between scientists and developers with regard to terminology (Löfvenhaft, Björn, & Ihse, 2002).
- The time scales for project design and ecological studies are significantly different. Although detailed studies of urban biodiversity are necessary for the generation of scientific knowledge, they are time-consuming and difficult for the actors to understand and exploit directly (Yli-Pelkonen & Niemelä, 2005). These detailed studies are therefore too complex for the rapid assessment of biodiversity required for many of the projects (Tzoulas & James, 2010).
- The number of stakeholders involved in developing a project complicates exchange and dilutes information (Löfvenhaft, Björn, & Ihse, 2002).

Therefore, to facilitate the integration of biodiversity in projects, the actors need effective tools and methods that can be understood by non-ecologists and allow for a precise description of habitats (Tzoulas & James, 2010).

One example is life cycle assessment (LCA), which is widely used in project design as a tool to support decision-making. However, in this method, which involves assessing the overall environmental impact of a project throughout its lifetime, biodiversity is incorporated in such a way that it makes it difficult for stakeholders to understand. Indeed, LCA calculations remain approximate from the point of view of biodiversity, and the frequent use of this tool could lead to a homogenisation of the arrangements put in place to promote biodiversity. As a consequence, this would impoverish biodiversity and alter the functioning of urban ecosystems in a systemic manner (Henry & Frascaria-Lacoste, 2012). To take biodiversity into account effectively, however, the tools must highlight the specificity of each project and not tend towards the homogenisation of the solutions. Thus, the competence of an ecologist is necessary for any development and construction project that aims to integrate this inseparable dimension of urban resilience. However, effective integration of biodiversity also requires tools that are easy to use and understandable by non-ecologists in order to allow for a dialogue between the actors involved in the planning and construction and the ecologists.

In response to these challenges, Alexandre Henry developed a first version of Biodi(V)strict®, a decision-support tool for better integrating biodiversity in projects, as part of his thesis "Development of Eco-neighbourhoods and Biodiversity", defended in December 2012. This tool has since been taken up again and improved and will be presented later in this chapter.

6.2 Presentation of Biodi(v)strict®

6.2.1 Construction of the Tool

Biodi(V)strict® is a diagnostic and decision-making tool for improving the "biodiversity potential" of urban and suburban development projects. It takes the form of an IT tool and a methodology formalised in four main stages.

The approach is based on the principles of Ecological Land Use Complementation (Colding, 2007) and focuses on the concept of habitat and its potential for hosting local biodiversity, without targeting a particular species. The identification of habitats is based on a predefined list, where each habitat corresponds to a weighting coefficient reflecting its potential for hosting biodiversity and a coefficient of soil-to-water permeability.

6.2.1.1 The "Ecological Land-Use Complementation" Approach

Ecological land-use complementation (ELC) is based on the idea that urban green spaces can support greater biodiversity when they are brought closer together in different combinations. This comes from the theoretical concept of the landscape complementation developed by Dunning in 1992. In a very heterogeneous landscape made up of very diverse patches, as in urban landscapes, a species needs to move between these different habitats to have enough resources to fulfil its life cycle (feeding, reproduction, rearing, dispersion, etc.). Landscape complementation implies that a species that needs at least two types of resources from different habitats during a life cycle can move between these habitats, and therefore these habitats are not too far away in the matrix not favourable to its movement.

Thus, it is the availability and composition of habitats and their configuration in the urban landscape that will influence the survival of individuals, populations, and communities in a highly fragmented environment.

ELC applies the notion of landscape complementation at the scale of a city and shows that different urban green spaces can interact in synergy to support biodiversity when they are arranged in different combinations.

By ensuring better access and use of resources, the functioning of the ecosystem will be improved by creating new ecological niches, allowing the arrival of new species that can support new ecological functions. Thus, ELC should aim to (i) assemble habitats that support different functions and (ii) create combinations that support new functions. This will then promote a variety of functional traits, and therefore complementarity of the niches. However, diverse functions lead to diverse responses.

Ecosystems built in this way adapt better to environmental changes and are therefore more resilient. *Ultimately*, they favour functional biodiversity as a whole.

ELC allows us to address the subject of the functionality and resilience of an urban landscape. For the construction of Biodi(V)strict®, it was assumed that the principle of ELC can be applied to scales smaller than that of cities, such as neighbourhoods, or smaller still, such as construction projects.

6.2.1.2 Selection of Biodiversity Potential Measurement Indicators

The Biodi(V)strict® indicators were created to respond to strong ecological issues in urban areas and with the aim of promoting better ecological functioning in the site. These indicators are the proportion of green spaces, the diversity of habitats, the diversity of strata, soil-to-water permeability and connectivity. The general reasoning was carried out according to the ELC principle (Colding, 2007) and the notion of habitats, integrating several constraints including the speed of acquiring the necessary data and the facilitated interpretation of results (Tzoulas & James, 2010).

First, the basic data for calculating the indicators must be able to be collected quickly. The species are therefore not taken into account directly for the calculation of Biodi(V)strict® indicators. The time resources needed to conduct wildlife inventories in the field have been identified as a real barrier to the rapid assessment required for the majority of urban projects.

The habitat scale was initially selected because habitat diversity is an inherent component of the definition of biodiversity and is often considered and used as an indicator of species diversity (Hermy & Cornelis, 2000). Secondly, habitat identification can largely be carried out on the basis of aerial photography, significantly reducing the time spent on the ground.

Then, in order for the indicators to be easily interpretable, they must have a defined range of variation (for example, between 0 and 1), independent of the landscape analysed, to facilitate the interpretation of the results. (Saura & Pascual-Hortal, 2007).

The issues and calculations for each indicator are described in the following table:

Indicator	Issue	Calculation
Proportion of green spaces	Vegetation is the basis of most terrestrial ecosystems. Maintaining a minimal proportion of green spaces in the city is a prerequisite for creating spaces conducive to biodiversity (Natureparif, 2012)	Proportions of green spaces on the site = $\dfrac{\text{Surface of green spaces}}{\text{Total site area}}$
Permeability of the soil to water	Soils fulfil indispensable functions: they play a role in the storage and purification of water, provide a physical support for plants, constitute water and nutrient reserves for the vegetation and transform organic matter, thus taking an active part in the carbon cycle (Natureparif, 2012) Urbanisation leads to the destruction and waterproofing of soils and thus to the disappearance of vegetation. The amount of flowing water increases as a result of reduced soil infiltration and interception by the plant cover. This has negative effects, such as increased flood peaks for rivers, bank erosion and flooding (Whiford, Ennos, & Handley, 2001).	Permeability of the site = $$\sum_{i=n}^{n}\dfrac{S_i \times c_i}{\text{Total site area}}\times100$$ where • n is the number of habitats on the site studied, • S_i is the surface area of the habitat i, • c_i is the infiltration coefficient related to the habitat i,
Habitat diversity	The correlation between habitat diversity at a study site and species diversity is most often positive (Tews et al., 2004). Thus, habitat composition and heterogeneity are good indicators of the overall biodiversity of a site (Hermy & Cornelis, 2000; Tzoulas & James, 2010). Habitat diversity is easier to apprehend than species diversity, which requires a protocol and identification studies in the field, which can be both time-consuming and expensive. The different habitats that can be found in cities or suburban areas do not have the same potential to host biodiversity, nor play the same role in the functioning of ecosystems. This is why each habitat has been associated with a weighting factor reflecting its potential to host biodiversity. These coefficients were assigned according to the degree of naturalness and management associated with each habitat.	$$H=-\sum_{i=1}^{n}\dfrac{n_i \times cpi}{N}\times \ln\left(\dfrac{n_i \times cpi}{N}\right)$$ where • n is the total number of habitats present on the site • N is the total surface area of the site • n_i is the surface area of habitat i • cpi is the weighting coefficient of the habitat i Habitat diversity indicator = $\dfrac{H}{H_{max}}\times100$ where • H is the Shannon diversity index for the site studied • H_{max} is the maximum diversity index (Continued)

Indicator	Issue	Calculation
Diversity of plant layers	The structure of vegetation is also a good indicator of biodiversity (Tzoulas & James, 2010; Madre, Vergnes, Machon & Clergeau, 2013). Various biotic and abiotic conditions develop through plant layers, giving species a greater variety of ecological niches. Thus, an environment with a higher number of layers would favour more species and thus support a more complex ecosystem.	Diversity of layers = $-\sum_{i=1}^{n} \frac{x_i \times nsi}{N} \times \ln\left(\frac{n_i \times nsi}{N}\right)$
	To describe the structure of the vegetation, we use a classification of classical layers in plant ecology used by Madre et al. (Madre, Machon, Vergnes, & Clergeau, 2014) to describe the different types of green roofs. This classification is appropriate for all types of ecosystems, whether natural or artificial and is as follows:	where
	Mossy layer: composed of mosses (bryophytes), lichen, mushrooms and small herbaceous plants or sedums. This layer also includes grassy lawns that are frequently mowed.	• n is the total number of habitats present on the site
	Herbaceous layer: dominated by non-woody herbaceous plants such as succulents or flowering plants not exceeding one metre at maturity.	• n_i is the surface area of habitat i
	Shrubby layer: shrubs, bushes and young trees.	• nsi is the number of layers of habitat i
	Tree layer: composed of mature trees.	• N is the total surface area of the site
Connectivity	The connectivity of a landscape is considered a key point for maintaining the stability and integrity of ecosystems (Taylor, Fahrig, Henein, & Merriam, 1993). It has been shown that small urban green spaces, such as business gardens, are important for maintaining the overall connectivity of urban green spaces and that these spaces thus contribute to the functioning of the urban ecosystem (Serret, 2014). Considering this, many other urban green spaces (woods, parks, gardens, etc.) of varying sizes can also play an important role in keeping urban biodiversity functional.	Intrasite connectivity indicator = $\dfrac{\sum_{i=1}^{n}\sum_{j=1}^{n} cp_i a_i \times cp_j a_j \times p_{ij}}{A^2}$
	Cities are characterised, in particular, by a very fragmented landscape. The few green spaces present are very often isolated in a matrix of constructions, roads and other artificialised and impermeable surfaces. It is therefore difficult for species to move between different green spaces in cities.	where
	The establishment of ecological corridors within cities would reduce the isolation of green spaces and allow better functioning of urban biodiversity. By maximising the greening of projects and encouraging the connectivity of green spaces within the same project, we could finally promote the creation of "stepping stone" corridors within the city, and therefore participate both in good local functioning of biodiversity (at the project level) and also in good overall functioning of the urban ecosystem (at city level) in the long term.	• PC is the probability of connectivity • a_i and a_j are the respective areas of patches i and j • cpi and cpj are the respective weighting coefficients of habitats i and j to the total area of the site studied p_{ij} is the probability of dispersion, i.e. the probability of a displacement between the patch i and the patch j. $p_{ij} = e^{-d_{ij}}$, with d_{ij} being the distance between i and j

6.3 Construction of the Habitat Database

Databases presenting habitat typologies may have the limitation of being too specific to particular locations to be applied elsewhere or too general to capture the heterogeneity of urban habitats (Tzoulas & James, 2010). This is why a specific list was developed for use with Biodi(V)strict® by combining a list of habitats from the literature (Hermy & Cornelis, 2000) with the EUNIS databases (Louvel, Gaudillat, & Poncet, 2013) and Corine Land Cover (European Environment Agency, 2012). This habitat list is intended for urban and suburban sites in France and elsewhere in Europe.

The list of habitats is structured in four levels of definition, ranging from a very general characterisation (level 1) to the management of each habitat in detail (level 4). Each habitat is characterised by an identification number, a weighting coefficient reflecting its potential for hosting biodiversity and a permeability coefficient, and is categorised as blue (aquatic or wet), green (vegetated) or grey (artificial surfaces).

The current habitat list therefore makes it possible to take into account the capacity of each habitat considered and the associated management method (for example, a regularly mown lawn will have a lower weighting coefficient than a lawn with differentiated management) in the calculation of indicators.

6.3.1 Presentation of the Method

The analysis of the biodiversity potential of a project is based on the comparison of a site with different states, for example in the initial state and in the projected state after construction. It is therefore a comparative method based on an iterative reasoning that highlights a project's impacts on urban and suburban ecosystems and the scope for improvement.

The methodology for conducting a Biodi(V)strict® study can be divided into four stages:

- The "site diagnosis", which consists of identifying the habitats present on the site before and after the project from aerial photographs, a visit to the site planned for the development and the project's site plans.
- The mapping of the habitats showing the initial site (before completion of the development) and the projected state.
- Measuring the five Biodi(V)strict® indicators
- The biodiversity potential is assessed by identifying the pressures and/or opportunities that the project will involve on the site, based on the graphical and cartographic results of the tool and the proposal for developments promoting biodiversity, leading to a variant of the project.

The proposed developments supporting a more biodiversity-friendly project variant will support the following elements:

- Increasing the proportion of green spaces on the site, as the greater the surface available for the species, the richer the biodiversity will be

- Preserving soil permeability to encourage restoration of the water cycle and protection of the soil, the foundation of any terrestrial ecosystem
- Tending towards a diversity of habitats, as a heterogeneous environment provides more refuge, feeding and breeding grounds for local species and thus promotes a greater diversity of species
- Increasing the diversity of plant layers (mossy, herbaceous, shrubby, tree), as a complex vertical tier diversifies the environment a little more and provides more resources for biodiversity.
- Bringing habitats closer together and creating connections between site spaces to ensure that the species can move

6.4 Application to Cité Descartes

Cité Descartes is a research and teaching centre spanning 123 ha, located in the municipalities of Champs sur Marne (77) and Noisy le Grand (93). This site is the subject of an urban expansion project coordinated by the EPAMARNE (Etablissement Public d'Aménagement de Marne-la-Vallée, Marne-la-Vallée Public Planning Establishment), which aims to make it an eco-district. This project involves restructuring the site and creating new buildings. An analysis of the site has been conducted with Biodi(V)strict® to identify the pressures on biodiversity that will be caused by the project to expand the Cité Descartes site.

Mapping of the habitats in Cité Descartes in the initial site (Figure 6.1) and project (Figure 6.2) was carried out on QGIS. These habitat maps made it possible to calculate the Biodi(V)strict® indicators (Table 6.1) and analyse the project's impacts on the biodiversity of the initial site.

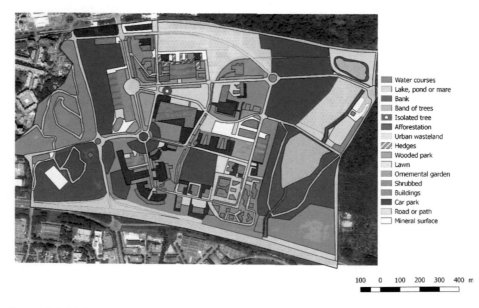

Figure 6.1 Habitat mapping of the initial site (preliminary project) of Cité Descartes.

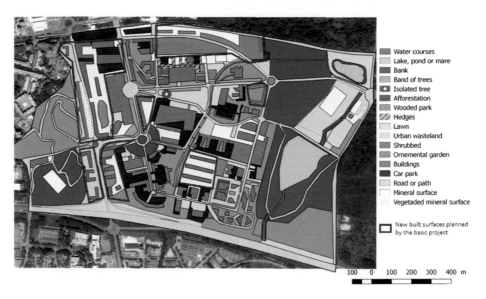

Figure 6.2 Project habitat mapping.

Table 6.1 Biodi(V)strict® Indicator Values for Different States of the Cité Descartes Project (Initial Site and Project)

Project Status	Proportion of Green Spaces	Habitat Diversity	Diversity of Layers	Connectivity	Permeability of Soils to Water
Initial site	0.66	0.32	0.39	0.23	0.6
Project	0.51	0.28	0.29	0.14	0.49

The urban expansion project has a negative effect on the biodiversity potential indicators. This is not surprising, given the number of natural areas impacted, such as afforestation, woodlands and wastelands. The construction project will convert 22.7% of the green spaces present on the initial site into artificial surfaces. The basic project will negatively affect all indicators with a decrease of 12.5% for habitat biodiversity, 25.6% for plant layer diversity, 39% for connectivity and 18% for soil and water permeability.

The greening of roofs is often a solution put forward by the design team to mitigate the impacts of new constructions on natural areas. In the interests of presenting the results simply and quickly, we chose to test a variant of the project that only includes the installation of green roofs on the new buildings. The results of the indicators for this variant are presented in Table 6.2. We could have imagined other variants, including moving some buildings to preserve existing natural areas, such as wooded areas, implementing differentiated management measures across the campus, etc.

Table 6.2 Biodi (V)strict® Indicator Values for the "Green Roof" Variants of the Project

Project Status	Proportion of Green Spaces	Habitat Diversity	Diversity of Layers	Connectivity	Permeability of Soils to Water
Variant with green roofs	0.57	0.31	0.32	0.17	0.51

6.5 Discussion and Conclusion

Despite the creation of new tools and methods to facilitate the integration of biodiversity in urban development projects (Henry & Frascaria-Lacoste, 2012; Tzoulas & James, 2010), professionals still struggle to actually integrate this issue within their projects. This observation could certainly be explained by the fact that these tools and methods were created to help operational actors to integrate biodiversity into their projects, without working with them to really identify their needs.

This is why we have focused on producing an easy-to-use decision tool for non-specialists, who could integrate it within their eco-design process and which could also improve the dialogue between actors and ecologists.

We have worked with a high level of interaction with urban design professionals to ensure that Biodi(V)strict® can be included in the eco-design process and that these actors can effectively familiarise themselves with it.

Thus, the analysis of a project's biodiversity potential with Biodi(V)strict® is based on the comparison of a site with different states, for example the initial state and the projected state after construction. This approach concentrates on habitats without focusing on species and saves considerable time by avoiding field inventories. The input data needed to calculate biodiversity potential indicators can be acquired very quickly by identifying the habitats of a site from aerial photographs, confirmed by a quick visit to the site. The factor determining the duration of a Biodi(V)strict® study is the time required to map a site starting from the predefined list of habitats. On the basis of several studies carried out in urban or suburban contexts and on projects of different sizes, we discovered that the duration of this mapping depended on both the area of the site studied and the fragmentation of the mapped entities. Overall, the larger the site under study and the higher the heterogeneity of small entities, the more substantial the mapping time, although it still remains acceptable to the operational actors. The calculation of the indicators is automatic and takes a few minutes once the GIS data are entered into the computer.

In addition, the indicators measuring the biodiversity potential were chosen both because they make it possible to understand the quality and ecological functioning of a site and because they represent a major ecological challenge in urban areas. These strong questions stand out and are well understood by architects, planners, decision-makers and project teams. Thus, communication around the Biodi(V)strict® indicators facilitates a dialogue between all the stakeholders involved in a project.

The results of the tool are quantitative. This has the advantage of simplifying relatively complex ecological concepts and allowing the actors to incorporate the subject in their design processes.

The example of the application to Cité Descartes presented in this chapter aims to illustrate the results that could be obtained with Biodi(V)strict®. These results are both visual thanks to the maps and quantified thanks to the indicators, and they allow the design teams to quickly understand the issues and results in terms of biodiversity, which was hitherto barely taken into account. In our example, we see that the installation of green roofs on new constructions is not enough to have biodiversity potential that is at least as much as the initial site. This result could be a point of entry into the discussion with the project's designers: biodiversity potential is not considerably increased because all the roofs are green. Further improvements should be considered in order to arrive at biodiversity potential that is as close as possible to the initial state or to go beyond it. Once the numerical and comprehensible results support the ecologists' discourse on the need to think about the design of the project in terms of the ecological functionality of the site, the discussions are facilitated and we can hope to move towards a more functional consideration of biodiversity.

The methodological stance taken by Biodi(V)strict® is to not consider species. This choice results from a compromise between the desire to correctly describe a "potential state of biodiversity" and to remain highly operational. However, this choice is justified because the nature and diversity of habitats is an inherent component of the definition of biodiversity and is generally considered as an indicator of species diversity (Hermy & Cornelis, 2000). In addition, habitat identification can be carried out by aerial photography, significantly reducing identification time in the field.

It should be noted here that this method is not intended to replace more detailed ecological studies. It does, however, allow for effective assessment of a potential state of biodiversity in a neighbourhood, building or infrastructure.

In conclusion, Biodi(V)strict® is a tool based on scientific principles that are understandable for urban and suburban project design teams. We therefore believe that this method makes it possible to find the right compromise between comprehension and efficiency to face the complex problem of considering biodiversity within the design of the cities of tomorrow.

However, it is important to keep in mind that Biodi(V)strict® is only part of the response to the effective integration of biodiversity within development and construction projects. Biodiversity is often not yet systematically integrated into the heart of projects and even less so as an issue of interest (Vandevelde, Penone, Kerbiriou & Le Viol, 2012). In fact, the need to preserve biodiversity often appears to be a poorly understood regulatory constraint, as actors often perceive this as a "disproportionate additional cost" (Tourjansky-Cabart & Galtier, 2006). In addition, biodiversity is often placed behind other goals, such as economic development, transportation or energy performance (Ahern, 2013), and very often it is considered too late in the project development process for the projects to integrate it as effectively as possible.

Therefore, the effective consideration of biodiversity in projects will depend heavily on the involvement of operational figures (Vandevelde, Penone, Kerbiriou, & Le Viol, 2012). This will inevitably imply a paradigm shift by the actors involved in constructing the cities of tomorrow, who must stop viewing consideration of biodiversity as a constraint and instead see it as an opportunity for renaturation and action for better sustainability of cities.

Bibliography

Ahern, J. (2013). Urban landscape sustainability and resilience: the promise and challenges of integrating ecology with urban planning and design. *Landscape Ecology 28*, pp. 1203–1212.

Colding, J. (2007). Ecological land use complementation for building resilience in urban ecosystems. *Landscape and Urban Planning 81*, pp. 46–55.

European Environment Agency. (2012). Retrieved from http://www.statistiques.developpement-durable.gouv.fr/donnees-ligne/t/nomenclature-standard.html.

Henry, A., & Frascaria-Lacoste, N. (2012). Comparing green structures using life cycle assessment: a potential risk for urban biodiversity homogenization. *Journal of Life Cycle Assess 17*, pp. 949–950.

Hermy, M., & Cornelis, J. (2000). Towards a monitoring method and a number of multifaceted and hierarchical biodiversity indicators for urban and suburban parks. *Landscape and Urban Planning 49*, pp. 149–162.

Löfvenhaft, K., Björn, C., & Ihse, M. (2002). Biotope patterns in urban areas: a conceptual model integrating biodiversity issues in spatial planning. *Landscape and Urban Planning 58*, pp. 223–240.

Louvel, J., Gaudillat, V., & Poncet, L. (2013). *EUNIS, European Nature Information System, Système d'information européen sur la nature. Classification des habitats. French translation. Habitats terrestres et d'eau douce.* Paris: MNHN-DIREV-SPN, MEDDE.

Madre, F., Machon, N., Vergnes, A., & Clergeau, P. (2014). Green roof as habitats for wild plant species in urban landscapes: first insights from a large scale sampling. *Landscape and urban Planning 122*, pp. 100–107.

Madre, F., Vergnes, A., Machon, N., & Clergeau, P. (2013). A comparison of 3 types of green roof as habitats for arthropods. *Ecological Engineering 57*, pp. 109–117.

McGranahan, G., et al. (2005). *Urban Systems.* Washington, DC: Island Press.

Millennium Ecosystem Assessment. (2005). *Ecosystems and Human Well-being: Synthesis.* Washington, DC: Island Press.

Natureparif. (2012). Bâtir en favorisant la biodiversité. Un guide collectif à l'usage des professionnels publics et privés de la filière du bâtiment. Victoires Editions.

Saura, S., & Pascual-Hortal, L. (2007). A new habitat avaibility index to integrate connectivity in landscape conservation planning: comparison with existing indices and application to a case study. *Landscape and Urban Planning 83*, pp. 91–103.

Serret, H. (2014). Espaces verts d'entreprise en Ile-de-France: quels enjeux pour la biodiversité urbaine? Thesis at the Muséum National d'Histoire Naturelle.

Taylor, P., Fahrig, L., Henein, K., & Merriam, G. (1993). Connectivity is a vital element of landscape structure. *Oikos 68*, pp. 571–573.

Tews, J., Brose, U., Grimm, V., Tielbörger, K., Wichmann, M., Schwager, M., & Jeltsch, F. (2004). Animal species diversity driven by habitat heterogeneity/diversity: the importance of keystone structures. *Journal of Biogeography 31*, pp. 79–92.

Tourjansky-Cabart, L., & Galtier, B. (2006). La biodiversité dans les projets d'amé. nagement. Evaluation environnementale et socio-économique. *Responsabilité & Environnement 44*, pp. 57–64.

Tzoulas, K., & James, P. (2010). Making biodiversity measures accessible to non-specialits: an innovative method for rapid assessment of urban biodiversity. *Urban Ecosystem 13*, pp. 113–127.

United Nations. (2014). *Report on urbanization prospects.* Retrieved from http://www.un.org/en/development/desa/population.

United Nations, P. D. (2014). *World Urbanization Prospects: The 2014 Revision*, Department of Economic and Social Affairs Population Division, New York, 517.

Vandevelde, J.-C., Penone, C., Kerbiriou, C., & Le Viol, I. (2012). Grandes infrastrucutres de transport et biodiversité: quelle prise en compte? In C. Fleury, & A.-C. Prévot-Julliard, *L'exigence de la réconcialiation: biodiversité et sociétés* (pp. 195–209). Paris: Fayard (Le temps des sciences).

Whiford, V., Ennos, A., & Handley, J. (2001). "City form and natural process" - indicators for the ecological performance of urban areas and their application to Merseyside, UK. *Landscape and Urban Planning 57*, pp. 91–103.

Yli-Pelkonen, V., & Niemelä, J. (2005). Linking ecological and social systems in cities: Urban planning in Finland as a case. *Biodiversity and Conservation 14*, pp. 1947–1967.

Part 2

Incorporating the Human Factor within Eco-design

Chapter 7

Social Practices and Ways of Living

Christophe Beslay

Bureau d'Études Sociologiques Christophe Beslay (BESCB), Toulouse, France

7.1 Introduction

There is no need to recall the energy and environmental challenges faced by the building sector in terms of achieving the objectives set by Law No. 2015-992 of 17 August 2015 on the energy transition for green growth. Under the impetus of the regulations (RT 2012) and numerous performance labels (BBC, Effinergie, HQE, etc.), a new constructive model is becoming standardised to achieve these objectives (Beslay, Gournet, and Zélem, 2015). This results in the implementation of an approach that relies especially on natural thermal inputs, insulation and airtightness of the building, the inertia provided by the materials, reduced heating production recovered and distributed by a mechanical ventilation system (MVS), the production of solar energy (thermal and/or photovoltaic), geothermal energy (heat pump and underground heat exchangers) and, more rarely, wind power as well as automatic regulation.

Energy performance is not just a technical matter, but rather it results from a constant "dialogue" between the occupants and the equipment available to them. Technologies do not exempt people from acting. Energy performance, like comfort, remains a sociotechnical co-production involving humans and technical objects. In buildings, it is as much about the structural characteristics of buildings and equipment as it is about occupant practices and lifestyles. Lifestyles are expressed by what people do, own and use and by the meaning they give to their social practices, which reflect their identity (relationship to self and relationship to others) (Le Gallic et al., 2014). Housing facilities are thus partially constitutive of the lifestyles they help to structure. Conversely, equipment choices, its uses and energy consumption refer directly to lifestyles. There is therefore a dual determination of lifestyles and domestic technical objects.

Beyond the problems of design and construction, occupant behaviour thus emerges as a determining component of the actual performance of buildings. "Without occupants, buildings do not consume energy" (Thellier, 2015). Their arrival in a building is always a source of uncertainty. The challenge of eco-design is thus to create a harmony or a balance between the needs for comfort and "usability", the practices of using the economical energy and technical systems in order to obtain high-quality comfort and energy consumption in line with performance objectives. In this perspective, behaviours and techniques must be understood in a complementary way, as part of an exchange process that leads the occupants to acquire new domestic knowledge and new routines, or even to redefine certain uses. Energy performance can only be achieved with the help of the occupants, whose expectations, needs and skills should be anticipated as much as possible, especially as the energy issue is not generally their main concern.

7.2 Energy Performance, a Sociotechnical Production

The relationship that individuals have to housing or the premises they occupy is a social construct that can be analysed as sociotechnical action systems. Energy performance is thus built within the interaction between the occupants/users and the technical objects at their disposal, in contexts that are always particular. Similarly, energy practices – rather the practices of using energy services, provided that energy is not consumed directly, but consumed by the use of equipment for heating, domestic hot water, cooking, lighting, cooling, washing, etc. – do not refer to social affiliations alone, even if they contribute to them, but rather relate to sociotechnical systems (Figure 7.1) that refer to a combination of interacting and interdependent elements: occupants/users (decision-makers, designers, users, etc. with their social properties, beliefs, imaginations and routines), techniques (considered as nonhuman actors), the social dynamics in which individuals participate (life cycles, social norms, network games, learning processes, information flows, etc.), environments (climatic, energy, political) and configurations (organisational, institutional, family, etc.).

Housing and the workplace are specific sociotechnical configurations that can be considered as "heterogeneous arrangements that not only mix individual and collective actors, but also techniques, procedures and rules, which enter into the configuration alongside humans" (Callon, 2001). In this perspective, techniques (building, equipment)

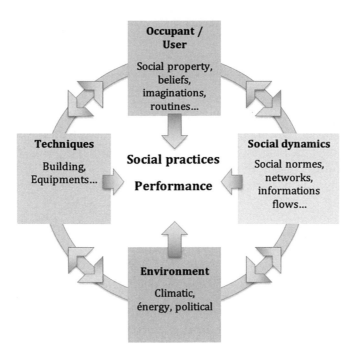

Figure 7.1 **A sociotechnical system.**

interact with humans. Their activities are interdependent in space and time. This relationship is marked by use and action rules, forms of negotiation and interpersonal coordination, and also strategies for control or seizure of power. This gives rise to room for manoeuvre (strengths/constraints), the outlines and evolutions of which can typically be punctuated by the composition/recomposition of the household or work group or by the nature of the domestic or professional activities.

Each equipment or types of equipment can be considered as a specific sociotechnical microsystem. These microsystems operate in a relatively autonomous way, as long as the technical devices, knowledge and habits, imaginations, collective dynamics and market logic are specific to each one. In fact, little consistency can be seen between the different social practices of energy regulation or use. The energy attention and comfort standards can vary significantly according to the posts and uses. They are economical for heating but expensive for lighting, stringent for thermal comfort but not for visual or acoustic comfort, simple for interior comfort but very energy consuming for eating practices or leisure habits, etc. This would make it difficult to describe households unequivocally with regard to the use of energy services.

7.3 The Social Part of Energy Performance

The copious feedback we now have from "high-performing" buildings coincides in finding a gap between the expected performance and actual performance (Illouz and Catarina, 2009, Carassus, 2011, Assegond et al., 2016).

The monitoring of low-energy model buildings carried out by CEREMA from 2012 to 2016 (CEREMA, ADEME, 2017) made it possible to compare the expected consumption (Th-CE 2005 regulatory calculation method) and the actual consumptions measured for 24 residential and non-residential operations (Figure 7.2).

In one out of two cases, the measured consumption is higher than the expected consumption by 10 kWcp/m².year. For a quarter of operations, the difference is greater than 35 kWhep/m².year. It may double. Conversely, in some operations, the measured consumption is lower than the expected consumption. However, in almost half of the cases, the consumption measured corresponds fairly well to the expected consumption. Significant differences therefore only relate to one out of two operations.

The authors of the CEREMA study note that the differences observed between the expected and measured performance are due to multiple factors. Some are technical, including the performance of the building envelope and the efficiency of the heat generators. Others relate to occupant practices, such as solar gain management and facility control. In the residential sector, difficulties in learning and appropriating the systems on the part of the occupants have been observed on a fairly recurring basis.

This feedback highlights the social part of building energy performance, which is expressed by the adaptive actions taken by occupants to obtain a level of comfort corresponding to their aspirations and by their manner of appropriating the equipment and housing. Thermal comfort, ventilation and, more broadly, the use of systems are the issues that cause the most tension and the most interventions and workarounds on the part of the occupants (Figure 7.2).

User practices that could hinder energy performance generally come from devices that are unsuitable for their way of life or work, the difficulty of controlling or maintaining the building's equipment or the discomfort they feel, causing them to modify

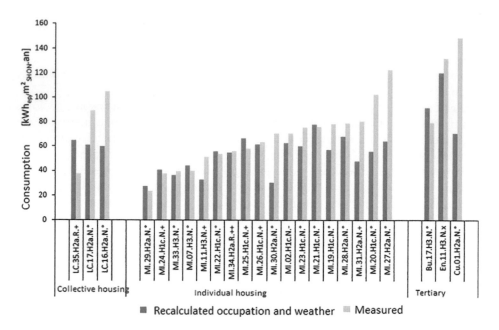

Figure 7.2 Expected consumption and measured consumption of regulated posts.

them. Performance is not limited to energy consumption alone. For the occupant, comfort and the quality of use more broadly are important elements, which can be very energy intensive.

7.3.1 Thermal Comfort

In high-performance buildings, winter comfort is generally not lacking, with few problems and a high level of satisfaction. However, the comfort temperature is rarely consistent with the temperature of 19°C taken into account in regulatory calculations. The most thorough surveys carried out on representative samples of the French housing stock (Crédoc 2010, 2012, SOeS-Ipsos 2010) show that the average heating temperature in the living room amounts to 21°C. About 30% of households are heated to 21°C and 75% to more than 19°C. The temperature distribution is more spread out in the rooms, with just over 50% of rooms heated to above 19°C.

All other things being equal, the more recent the housing, the higher the heating temperature, approaching the standard of comfort at 22°C, as in the most developed countries, the United States in particular. When the temperature set point is fixed to the standard of 19°C in a binding manner, occupants tend to equip themselves with additional heating. The gap between the technical and regulatory standard and the social norm contributes to a direct rebound effect. Efficient equipment is used more intensively and does not produce the expected energy savings. Pointing out that the average temperature of dwellings increased from 19°C to 21°C between 1986 and 2003, ADEME itself notes that "the quest for comfort continues to outweigh the desire for financial savings" (ADEME, 2008).

For occupants, management of winter thermal comfort in a high-performance building involves new practices, ranging from a logic of production to a logic of heat conservation and recovery. The occupant's job thus consists of not leaving the windows open and taking advantage of solar gains.

Summer comfort is an element that is often underestimated in design. There is not always enough protection (canopies, blinds, or living fences). Summer discomfort is found to be excessive in the surveys, with premises sometimes described as "uninhabitable". From the occupants' point of view, summer and off-season comfort brings about other practices than just heating management, which require them to be more active or even proactive in preventing heat build-up: closing windows, curtains and shutters during the day, ventilate at night in summer, and dressing better in the off-seasons. These practices refer to cultural habits and knowledge developed in the southern regions to protect against the heat, but are little known in the northern regions, where inappropriate practices can explain situations of greater discomfort. In these regions, learning needs to take place and new habits still need to be acquired to guard against this discomfort, such as accepting a certain amount of darkness inside the home during the day, not opening windows to enjoy the sun and heat, etc. Other factors may constrain these adaptive practices, such as the risk of intrusion or noise pollution resulting in a lack of night ventilation.

Thermal comfort is a complex problem that is difficult to grasp from a strictly technical or regulatory point of view. It is only one aspect of residential comfort, which is multidimensional and therefore thermal, olfactory, visual and acoustic at the same time. It is linked to the location, the space, the interior fittings (the distribution of the rooms), the available volumes, the environment, accessibility, emotional attachment, etc. Dissatisfaction with one aspect may render other elements unacceptable and, conversely, certain benefits may make sources of discomfort acceptable. Discussions of thermal comfort can thus only be interpreted in relation to all these elements.

The physiological dimension of thermal comfort is part of a socio-historical construction that evolves according to the available techniques, knowledge of use and social norms (Brisepierre, 2015). In the hygienist design of the late nineteenth century, it was recommended not to exceed 16°C inside homes, as higher temperatures were considered harmful to health, probably in relation to the wood or coal heating systems used. Although dependent on the objective temperature, thermal well-being also concerns perception and the subjective impressions from the senses. A fairly weak relationship between thermal comfort and indoor temperature can be observed. Many factors come into play in the perception of temperature: humidity, airflow, colours of materials, furniture and light, clothing, food, activities, etc.

There are strong individual variations in the perception and satisfaction of thermal comfort, depending on gender, age, sociocultural affiliations, modes of socialisation, etc. The sociotechnical construction of comfort is part of the collective space of the household or the work team. It results in social games of negotiation, regulation and power: the "war of the buttons" over lighting, the "quest for fire" over heating (Desjeux, 1996) and, now, a third front over the control of sunshine, brightness and ventilation through the manipulation of openings, shutters or curtains. Daily negotiations and therefore compromises are established regarding the level of comfort expected and the efforts needed to produce it. There is no guarantee that these compromises will always be compatible with the expected performance objectives.

There is thus a significant gap between the measurements and the discussions by the occupants, especially since less than half of the dwellings have a reliable instrument for measuring temperature. In fact, there are multiple "temperatures" that rarely correspond to each other: the set temperature, the measured temperature, the perceived temperature, the comfort temperature, the acceptable temperature and the regulatory temperature. This multiplicity of referents confuses the analyses and is a source of controversy within the groups of occupants and with collective housing and tertiary sector managers.

Thermal comfort is therefore not a stable state but a permanent process of striving for resolution, over which the occupants wish to retain control. It is not the "zero state" of physiological feeling, "the absence of disturbance" or "the optimum of climatic conditions". "It is a dynamic notion that, without doubt, oscillates between freedom of movement and the feeling of its reliability, and cannot be reduced to standard values [...]". In common experience, the notions of comfort and aesthetics share this quality of being unstable, polysemic and fluctuating" (Dard, 1986). Jean Fourastier (1973) defined comfort as "the control of one's inner environment, the ability to adjust the parameters according to external conditions, activities and forms of occupation". A permanent and homogeneous temperature does not guarantee a good level of satisfaction among the occupants.

7.3.2 Controlled Mechanical Ventilation (CMV)

There is a real lack of knowledge of the MVS, especially dual flow, which appears like a strange object that the occupants have trouble taming, when they know what it is. Indeed, the occupants are sometimes not even aware of its existence or function. In the tertiary sector, it is often confused with "air-conditioning". We also observe a lot of confusion between ventilation, aeration and air-conditioning, as well as a lack of knowledge of the primary purpose of ventilation, which refers to the health issue of indoor air quality.

Ventilation vents can be perceived as a source of draughts. Some people obstruct them, and others turn the grill upward to redirect the airflow or even change the layout of the room. The noise of the MVS is also a recurring problem. Inhabitants seek to reduce it by tinkering with the sheath, while others go as far as cutting the ventilation. Conversely, in the summer, the strategy may be to increase the airflow during the day with the intention of cooling the dwellings, although this brings in more warm air and raises the temperature of the incoming air. However, by creating the air circulation in this way, the occupants think they are doing the right thing. Dual flow ventilation introduces new uncertainty in relation to the question of whether or not to open windows to ventilate the premises in winter.

The maintenance of the CMV is still not taken into account in the management of either residential or non-residential buildings. Occupants are not particularly aware of the maintenance that needs to be carried out. Filters are changed fairly arbitrarily, and some occupants even believe that it is not necessary to change them or that they are not responsible for doing it. In this case, very dirty filters are observed, which reduce the flow rates. The impact is felt in terms of both energy consumption and indoor air quality. The manual conception of aeration remains predominant and is a routine inherited from the hygienist policies of the nineteenth century, transmitted from generation to generation and relayed by the media (with instructions to air 5 to 10 minutes per day).

Unknown by the occupants, the CMV is also "mistreated" by professionals (GC Magazine, 2017). Cerema's analysis of 373 reports on 1287 building compliance checks reveals that two out of every three ventilation systems installed in new homes are non-compliant, compared to just under half in the non-residential sector. According to Pierre Deroubaix from ADEME, "with time, nothing works any more. The situation is not worse today than in the past, but the health consequences of a failing CMV in an airtight RT2012 building are more significant" (GC Magazine, 2017). The malfunctions affect the air intake modules and outlets (46%), the performance of the exhaust air flows (27%), the system configuration (10%), the extraction unit (9%) and circulation and air transfer (8%).

7.3.3 Increasingly Sophisticated Equipment

Feedback (CEREMA, 2017) shows that poor energy performance is linked to the use of new high-performance equipment: the difficulties of use, as well as the desire to neutralise systems suspected of producing discomfort, explain at least part of the perceived poor performance, especially for systems or feelings of discomfort. Occupants experience recurring difficulties in learning and grasping systems in a residential context. The example of ventilation is undoubtedly emblematic of the difficulties in aligning energy issues and quality of use issues.

There is a tendency towards the hyper-sophistication of system management equipment that tends to keep users away by multiplying automatic regulation and/or the injunction to "touch nothing", at the risk of disrupting the systems. The mode of producing comfort is then marked by ambiguity concerning the place of humans and techniques. The injunction made to the occupants remains paradoxical: it is up to you to ensure your comfort, to adopt behaviours that meet the performance and energy saving requirements, but let the technology manage it for you! A balance still needs to be found, as it is not always very clear for both the designers and the users.

In the process of spreading and socially integrating the innovations described by Scardigli (1996), acceptability is a first step that refers to the ability of the techniques to find their place in the everyday world and to fulfil the function(s) for which they were designed. Acceptability therefore depends on the coherence between the design logic, the lifestyles and the usage logic. Appropriation of the techniques truly marks their social integration into the users' lifestyles and universes of meaning. It manifests itself in forms of misuse, DIY and transformation, which rely on mastery of the technical object and the often-misappropriated uses. In this sense, the logic of use outweighs the logic of design at this stage. The technical object takes place in everyday life, and somehow escapes its designers to adapt to the social world of its users.

The occupants seek to tame the multiple interfaces that will allow them to adjust the interior comfort through thermostats, buttons regulating the boiler or water heater, shutters, etc. In a way, manipulation of the control commands gives them experience and concrete knowledge of the relationship between the operation of the equipment and the phenomena generated within the premises, such as airflow, phase shift or thermal radiation. As such, with a greater or lesser degree of success and ease, they manage to find the "suitable actions" (Boltanski and Thévenot, 2006), i.e. the gestures that form part of "an intimate arrangement of things and people who live together, sanctioned by the absence of surprise in the course of the intervention". Appropriation can involve the adaptation of people or things to new uses of technology or an evolution of living patterns, such as an air-conditioned environment, for example. This process of appropriation involves experimenting with systems on a daily basis. This experience leads to an emphasis on the importance of practical knowledge of system management. It emphasises the need to transmit this knowledge to the occupants, in the form of "instructions" and to go through a learning phase in relation to the housing technical systems.

Thus, arrival in a new dwelling constitutes a "test", and this is even more the case in buildings that fall under new design models. New housing is always a new technical system that challenges past habits. It involves new ways of employing the equipment that require learning new usage rules (Beslay, 2017). Instructions for use, where they exist, are insufficient when they are not explained concretely, in a usage situation. Poor mastery of equipment management and control tools can result in high consumption. In a recent study on households in fuel poverty (Beslay and Gournet, 2018), arrival in a new home was shown to be an event that could tip fragile households into energy poverty and unpaid energy bills, caused by failure to master the equipment for managing comfort. It takes two to three years to appropriate a new home, including the time needed to experiment with and master the management of interior comfort over the four seasons. In some cases, the return of energy-consuming practices in terms of bills only becomes apparent after one year, during the regularisation of the monthly payments.

Appropriation depends primarily on the complexity of the systems and the quality of the interfaces. If they are too complex, they become unusable by the occupants, who give up on making use of them. It also relates to the occupants' previous experiences and skills, and their technical knowledge with regard to understanding the operation of the building and appliances. Thus, a household that has always lived in homes with collective heating will potentially find it more difficult to deal with the need to intervene in their heating. The new instructions for efficient buildings (do not open the windows in winter and let the CMV manage the ventilation) can overcome the previous practical knowledge. Finally, appropriation refers to the perception of the role played by one system or another in the production of discomfort and the occupant's position within the technical management (what is the responsibility of the occupants and what is the responsibility of the landlord or maintenance technicians).

7.4 Should We Model Social Practices and Ways of Inhabiting?

The consideration of social practices and ways of inhabiting within eco-design partly depends on the integration of social practices and living arrangements in the models used to simulate energy consumption and energy performance. It also involves other forms of associating occupants with eco-design: working groups, sociological surveys, support, etc.

The modelling of practices and ways of inhabiting is tripped up by the social complexity and remains controversial in social sciences, notably among sociologists. For some, "society cannot be entered into an equation" (Jensen, 2018). Unlike the natural sciences, the social sciences cannot construct "islands of stability" (Jensen, 2018) or "things that hold" (Desrosières, 2010), that is to say, things that resist scientific criticism, which build coherent knowledge and allow agreement within the discipline. For Jensen, "there are four essential factors that make simulations of society qualitatively more difficult than those of matter: the heterogeneity of humans; the lack of stability of anything; the many relationships to be considered, both temporally and spatially; the reflexivity of the humans who react to the models made of their activities". The difficulty also lies in quantitatively mobilising parameters that are generally evaluated qualitatively.

For other authors, this position is an ideological stance, even intellectual laziness (Roggero and Sibertin-Blanc, 2008). By contrast, they believe there is an "urgent need to model in sociology" (Manzo, 2007), as far as sociology should be able to follow the model that states "as they become perfected, the sciences tend to become quantitative" (Pareto, 1968). Several avenues have been explored: a micro approach by the " agents", a macro approach by the social determinants of the practices and a descriptive probabilistic approach.

7.4.1 The Multi-Agent Approach

The multi-agent approach has been the subject of numerous methodological and epistemological reflections and studies in the social sciences and seems to open up a field that is potentially very fertile in terms of social simulation. Its language (or grammar) can echo certain concepts and paradigms of sociology. The principle of this approach is to "directly manipulate digital entities representing people, instead of having to stick to aggregated terms such as averages or statistical distributions" (Manzo, 2014). The " agents" are represented by attributes, such as gender, age, marital status, level of education, and type of employment. They have resources and are caught in contexts of interaction governed by rules and standards, etc.

This type of approach to individual behaviours in interaction contexts allows the formalisation of configurations that take the form of concrete action systems and mobilises the concepts of the sociology of organised action (Friedberg, 1993), as proposed by Roggero and Sibertin-Blanc (2008). For Manzo (2007), this approach is based on "structural individualism" or " complex methodological individualism". It is " a form of non-reductionist individualism because the analytic primacy of the people coexists with the acceptance of the importance of social 'structures' upstream, and systems of direct and indirect interdependence continuously created between the actors downstream".

7.4.2 Seeking Out Practice Determinants

Other studies that are more forward-looking seek to identify the determinants of social practices and lifestyles. In his thesis on science and information technologies and communication, Le Gallic (2017) proposes a modelling approach to "think about our future lifestyles in terms of energy forecasting approaches". He identifies a set of determinants for planning choices, mobility and ways of living: the size and composition of the household, the type of habitat (size, employment, location), the use of time (activities carried out and the time spent on them), consumer behaviour (diet, the structure of consumption of goods and services) and the relationship to time, space, self and others.

In a prospective study on the cooling and washing equipment in a building in 2030 (Beslay et al., 2014), we identified a set of a dozen elements to determine lifestyles and their evolutions, which structure domestic practices of equipment usage, the forms of their social integration and their appropriability by households: urban and habitat forms, family structures (configuration, size and composition of households, social roles and responsibility for domestic work), systems of activity (employment, working time, activities outside of work), food supply and consumption patterns, beliefs and social representations (energy, refrigeration and freezing, food risks, responsibilities), social norms (individualisation of practices, hygiene standards, forms of sociability), knowledge of use, energy and technical culture, relationships to techniques, energy and environmental awareness, the logic of the market, climate and political contexts and levels of confidence in technology, innovation, public authorities and market regulation. This perspective is also that adopted in the numerous works of social psychology on "consumer behaviour" (Polizzi di Sorrentino, Woelbert and Sala, 2016) or on the obstacles and levers of change in social practices (Martin and Gaspard, 2016).

The difficulty of this type of approach by practice determinants lies in segmenting or "taking aside" each of these dimensions without considering their interrelations or coevolutions. However, as Jensen (2018) remarks, " social life cannot be explained by the addition of some important variables playing a uniform role, but rather by specific conjunctions and events".

7.4.3 The Stochastic Approach

This approach consists of "observing reality and reproducing it without explaining the cause-and-effect relationships between the different phenomena" (Vorger, 2014). It is based on calculating the probability that phenomena or social practices will occur in a given environment, for example, the open or closed state of windows according to the temperature of a room. It is also based on the formulation of a set of hypotheses about the players and their practices. The stochastic option of "understanding the behaviour of the occupants as a highly random phenomenon" (Vorger, 2014), without considering their logic or their motivations, may, at first glance, offend sociologists' sensibility. The intention "to imitate a visible or measurable behaviour, but not the processes or the mechanisms at the origin of this behaviour" (Varenne, 2010) is opposed to the sociological approach, which, by contrast, aims to elucidate the social logic or sociotechnical methods by which practices are constructed and acquire meaning. Their variety does not refer to "their highly random nature", but to individual and collective histories, to the diversity of social experiences and to the specificity of the sociotechnical systems, which are difficult to reduce to just a few standard forms.

However, the main obstacle to the development of a stochastic model is that it must be based on representative and substantial data, so that the calculation of probabilities can be generalised. In the sociological studies on social practices, the approaches remain very qualitative, seeking to highlight the sources or the logic of the action, leading, at best, to ideal typical qualitative typologies that cannot be used for modelling. The few existing quantitative surveys are often confined to the grey literature of study reports and opinion polling institutes. All are based on declarative data. In surveys, whatever the form (face-to-face, self-declaration, telephone or internet), actors do not always say what they do and do not always do what they say. This is even more the case when the subject of the study has social connotations, such as energy practices where the economical "proper actions" are overvalued, while expensive practices tend to be stigmatised. In surveys, there is usually a more or less significant discrepancy between declarations, observations and measurements. Thus, in our own field surveys on energy practices, as far as possible, we pair the collection of words with commentated observations on housing and equipment. Discussions outside situations are quite often different from discussions during situations of use (in front of the equipment, for example, or by analysing the results of the instrumented measures). More generally, practices fall within the sphere of observation, whereas discussion provides access to the meaning that the actors give to their practices and not their practices in themselves. Observation also makes it possible to identify all the "micro practices", the ad hoc, DIY, improvised adjustments made on the basis of implicit practical knowledge and experiences, which are rarely verbalised in the interviews but which constitute the backbone of the ways of doing things.

Data taken from measurements (instrumented or administrative) are considered *a priori* more reliable than the declarative data by the world of engineers, yet they have a significant degree of uncertainty, sometimes of a magnitude equal to that of the declarative data. Instrumented measurement campaigns often involve small numbers (a few dozen dwellings, rarely more than a hundred), if they are not limited to the premises of the researchers themselves. Finally, we cannot rule out possible Hawthorne effects, whereby the occupants who know that they are being observed spontaneously adopt practices that conform to what they perceive to be socially valued, even if they temporarily abandon some less acceptable habits. These uncertainties not taken into account in the measurements are undoubtedly difficult to evaluate *a posteriori*.

The methodological and practical difficulties of social observation, the multiple possible biases in data collection and the acute awareness that any measure is socially constructed, from the point of view of both the researcher and the people surveyed, encourage social scientists to assume a certain modesty and place themselves in the "likely" category (Dubet, 1994) rather than claiming the accuracy of the measurement to be useful for modelling. This obstacle is emphasised by most of the authors who implement this approach (Page et al., 2008, Polizzi di Sorrentino et al., 2016).

In the end, it is questionable whether the stochastic approach really needs the social sciences. Above all, it needs data on social practices and their context of effectiveness. In this perspective, the widespread roll-out of Linky and Gazpar smart meters, and the proliferation of connected objects open wide perspectives for knowledge of energy consumption. Ultimately, it should be possible to have very accurate knowledge of most of the uses and practices and their impact on overall energy consumption, in a very short time and according to different consumer profiles. There is reason to think that Big Data will revolutionise energy simulation tools for understanding the social logics of energy service use.

In the framework of the Gazpar experiments (Paulou et al., 2017), specific work was carried out by the Énergies Demain research department to foresee these forms of using and processing consumption data. Smart meters trace a total volume of consumption on a daily basis. This overall volume of consumption can be broken down by usage to assist the user in controlling their consumption, or to facilitate other analysis applications, such as the creation of consumption profiles or consumption forecasts. Several means can be used to break down consumption by use. A supervised learning algorithm can be implemented, assuming that distribution data are available to drive the model. In the absence of data histories, it is also possible to train the learning algorithm using a standard load curve per use. Some information may help improve the reliability of the algorithm. Hourly consumption data will make it easier to identify load profiles for each use. An understanding of household or housing composition will also be useful for more accurately sizing the overall need per use. Household consumption history can be used to drive predictive models using supervised learning techniques. By comparing a household's consumption histories with weather history and forecast data, it is possible to assign a consumption profile to it, and thus predict future consumption. Consumptions are modelled based on the user's consumption history, and weather forecast files.

7.5 Concluding Points

The practices prove to be a key lever for achieving the expected energy performances, whether they may be actors involved in the occupation and management or those of the realisation and design: the whole acting system is affected. The study of real performance and the identification of levers of action to reform energy use practices require a cross analysis of the building's technical operation and the behaviours and interplay of the actors involved.

It is often forgotten that occupants are primarily users of the building and not involved in its performance. They live in/occupy a building with their own motivations and awareness of energy, but, in general, energy performance is not the predominant criterion in choosing their places or way of inhabiting them. Indoor comfort remains a strong preoccupation with energy goals that seem very abstract. In this sense, energy performance remains a concern for engineers and politicians, quite outside the social logic of the occupants.

In fact, there are multiple conceptions of performance. We have identified at least three, which apply to different social levels and engage different configurations of players. Energy performance first and foremost applies to technical systems (a car, an electrical appliance or a building) and their uses. It affects designers and managers, and, to a lesser extent, the users, who are mainly concerned with the quality of energy services. At a more macro level, environmental or societal performance takes the entire life of the technical system into account and integrates socio-environmental impacts. It involves designers and policies. At the micro level of individuals, housing performance is assessed in terms of the comfort level obtained, the adequacy of the building (residential or non-residential) for the activities and the costs generated. It falls within the category concerned with occupants, but is of interest to designers and managers. Without doubt, when they talk about "performance", the contributors are not necessarily thinking of the same thing. Moreover, the very term *performance*, the "new slogan of Western societies" (Heulbrunn, 2004), is not without ideological implications and is the bearer of a certain model of society.

From our point of view, one of the challenges of eco-design lies in the establishment of a true technical democracy, in which technical and architectural solutions are integrated and reflect the lifestyles, uses and energy practices for which they act as a kind of "spokespersons". In this perspective, the aim is to promote energy performance through the adoption of eco-efficient design that assists users, and which considers the users' needs (practical and symbolic) and their socio-demographic, economic and cultural characteristics, in order to provide them with ergonomic products and/or technologies that are adapted to their usage, usable and, therefore, acceptable and appropriable. The challenge is also to develop "sensitive" technical solutions that are able to enter into dialogue with users, to understand their needs and to adapt to different contexts of action.

Such an approach has the advantage of placing the occupants of the buildings in the centre. Above all, it enriches the design of technical solutions through the experience and practical knowledge of users, and it promotes the socialisation of techniques, that is to say their integration into the occupants' lifestyles and culture.

Although imperfect, modelling is a way to better take the occupants into account in eco-design. The widespread deployment of smart meters and connected objects should help to refine knowledge of social practices relating to energy use in buildings. However, in order to produce a level of performance that can meet the needs of environmental transition, occupant aspirations and economic and territorial constraints, it remains necessary for occupants to take an active part in design collectives and the consideration of their opinions, according to various modalities (surveys, workshops, etc.) adapted to the operation types (construction, renovation, residential, non-residential).

Bibliography

ADEME, 2008, Press kit "Faisons vite, ça chauffe".

Assegond, C., Beslay, C., Brisepierre, G., Gournet, R., 2016, Études de cas dans le cadre des suivis de bâtiments démonstrateurs à basse consommation d'énergie, ADEME.

Barbat, M., Gournet, R., Beslay, C., 2011, Confort intérieur des bâtiments résidentiels à basse consommation, COSTIC, BESCB, DHUP.

Beslay, C., Attali, S., Filliard, B., Gournet, R., Hainaut, H., Radanne, P., Schaeffer, F., Waide, P., Zélem, M-C., 2014, Étude prospective sur les équipements de froid et de lavage dans le bâtiment de 2030, BESCB, Futur Facteur 4, SoWatt, Waide Stratégic Efficiency, ADEME.

Beslay, C., Gournet, R., Zélem, M-C., 2015, Le "bâtiment économe". Utopie technicienne et résistance des usages, - in Boissonade, J. (Dir.), *La ville durable controversée. Les dynamiques urbaines dans le mouvement critique, Ed. Pétra, col. Pragmatismes.*

Beslay, C., 2017, Les occupants, acteurs de la performance des travaux de rénovation thermique et énergétique, CVC La revue des climaticiens, n°897.

Beslay, Christophe & Romain Gournet, 2018: Parcours et pratiques des ménages en précarité énergétique : enquête auprès de 30 ménages. BESCB. Toulouse, France.

Boltanski, Luc & Laurent Thevenot, 2006: On Justification. Economies of Worth. Princeton University Press, Princeton and London.

Brisepierre, G., 2015, Les ménages français choisissent-ils réellement leur température de chauffage? La norme de 19°C en question, in Zélem, M-C., Beslay, C., *Sociologie de l'énergie. Gouvernance et pratiques sociales*, CNRS Éditions.

Callon, M., 1997, *Concevoir, modèle hiérarchique et modèle négocié, in Collectif, Élaboration des projets urbains et architecturaux.* Paris: PUCA.

Carassus, J., 2011, *Les immeubles de bureaux "verts" tiennent-ils leurs promesses? Performances réelles, valeur immobilière et certification "HQE exploitation".* Paris: CSTB/ CERTIVEA.

CEREMA, 2017, Bâtiments démonstrateurs à basse consommation d'énergie. Enseignements opérationnels tirés de 119 constructions et rénovations du programme PREBAT 2012-2016, ADEME.

CREDOC, 2010, Enquête Consommation d'énergie 2009–2010.

CREDOC, 2012, Enquête Consommation d'énergie 2011–2012.

Dard, P., 1986, Quand l'énergie se domestique, Ed. Plan construction.

Desjeux, D. (Dir.), 1996, Anthropologie de l'électricité. Les objets électriques dans la vie quotidienne en France, Paris, L'Harmattan.

Desrosières, A., 2010, La politique des grands nombres. Histoire de la raison statistique, Ed. La Découverte.

Dubet, 1994, Sociologie de l'expérience, Ed. Seuil.

Génie Climatique Magazine, 2017, Arrêtons de maltraiter la ventilation, octobre.

Heilbrunn, R., (Dir.), 2004, La performance, une nouvelle idéologie? La découverte, col. Critique et enjeux.

Illouz, S., Catarina, O., (2009), *Retour d'expérience de bâtiments de bureaux certifiés HQE*. Paris: ICADE/CSTB.

Jensen, P., 2018, *Pourquoi la société ne se laisse pas mettre en équation*, Seuil.

Kaufmann, J-C., 2004, *L'invention de soi. Une théorie de l'identité*, Armand Colin.

Le Gallic, T., Assoumou, E., Maïzi, N., Strosser, P., 2014, Les exercices de prospective énergétique à l'épreuve des mutations des modes de vie, *VertigO*, 14, no. 3.

Le Gallic, T., 2017, Penser nos futurs modes de vie dans les démarches de prospective énergétique: proposition d'une approche par la modélisation, Thèse de doctorat en sciences et technologies de l'information et de la communication, Mines Paris-Tech.

Manzo, P., 2007, Progrès et "urgence" de la modélisation en sociologie. Du concept de "modèle générateur" et de sa mise en œuvre, L'Année sociologique, 57.

Manzo, G. (Dir.), 2014, La simulation multi-agents: principes et applications aux phénomènes sociaux, Revue Française de sociologie, no. 55–4.

Martin, S., Gaspard, A., 2016, *Changer les comportements, faire évoluer les pratiques sociales vers plus de durabilité*, ADEME.

Martuccelli, D., 2010, *La société singulariste*, Armand Colin.

Page, J., Robinson, D., Morel, N., Scartezzini, J-L., 2008, A generalised stochastic model for the simulation of occupant presence, *Energy and Buildings*, 40(2):83–98.

Pareto, V., 1968, *Traité de sociologie générale (1917–1919)*, Droz.

Paulou, J., Suaud, C., Beslay, C., Gournet, R., Biau, J-B., Souchu, R., Viandon, A., 2017, *Potentiel de maîtrise de l'énergie des compteurs communicants gaz. Les compteurs communicants peuvent-ils changer les pratiques de consommation énergétique? Retours d'expérience sur le pilote de déploiement du compteur Gazpar, Compteurs communicants gaz, pratiques des ménages et économies d'énergie*, GRDF, ADEME.

Polizzi di Sorrentino, E., Woelbert, E., Sala, S., 2016, Consumers and their behavior: State of the art in behavioral science supporting use phase modelling in LCA and ecodesign, The International Journal of Life Cycle Assessment 21 (2), 237–251.

Roggero, P., Sibertin-Blanc, C., 2008, Quand des sociologues rencontrent des informaticiens: essai de formalisation, méta-modélisation, modélisation et simulation des systèmes d'action concrets, *Nouvelles perspectives en sciences sociales*, 2, no. 2.

Scardigli, V., 1992, *Les sens de la technique*, Paris: PUF.

SOeS-IPSOS, 2010, Enquête sur les pratiques environnementales des Français, CGDD.

Thellier, F., 2015, Sans occupant, les bâtiments ne consomment pas d'énergie ! in Zélem, M.-C., Beslay, C., *Sociologie de l'énergie. Gouvernance et pratiques sociales*, CNRS Éditions.

Varenne, F., 2010, Les simulations computationnelles dans les sciences sociales, *Nouvelles perspectives en science sociales*, 5, no. 2.

Vorger, E., 2014, Étude de l'influence du comportement des occupants sur la performance énergétique des bâtiments, PhD thesis, École nationale supérieure des mines de Paris.

Zélem, M-C., Beslay, C., (Dir.), 2015, Sociologie de l'énergie. Gouvernance et pratiques sociales, CNRS éditions, col. Alpha.

Chapter 8

Modelling the Energy Consumption of Buildings

Jean-Pierre Lévy and Fateh Belaïd

École des Ponts ParisTech, Marne-la-Vallée, France

8.1 Introduction

There is now worldwide acknowledgement of the need for energy transition to reduce the harmful effects of consumption, the impacts of which are considered to be devastating for the climate. In this context, the building sector is identified as one of the major factors in overall energy consumption and greenhouse gas (GHG) emissions. For example, this sector is estimated to be responsible for 36% of GHG emissions (BPIE, 2013) at the European level alone. In response to this, international incentive directives are being put in place to ensure that states take adequate measures to renovate dilapidated buildings and construct buildings that consume little energy. At the European level, this is the case of the Directive on the Energy Performance of Buildings (DEPB) from 2002 to 2010, or the Directive on Energy Efficiency from 2012. These directives have been transposed into the national laws of the Member States. As a result, most EU countries (and beyond) are now adopting financial incentives or binding regulations to achieve substantial energy savings and significant GHG reductions over time horizons of 30 or even 50 years (BPIE, 2013).

In France, observation of changes in consumption per sector shows an upward trend in energy consumption in the building sector over the past 30 years, despite stagnation in the 2010s. It increased by almost 20% between 1990 and 2012. It represented 45% of the country's total energy consumption in 2015. Residential and non-residential buildings account for 20% of GHG emissions within all the national emissions from 2014 (MEEM, 2017). Thus, the European Directives have been translated into a regulatory framework by the new law on energy transition, which aims to reduce GHG emissions by 40% compared to their 1990 level by 2050. France has also committed to reducing its total energy consumption by 50% compared to the 2013 level by 2030 (MEEM, 2016). To achieve these objectives, standards and labels are decreed to improve building energy performance, such as energy positive buildings (Bâtiments à Energie Positive, BEPOS), high environmental quality (Haute Qualité Environnementale, HQE) or low-energy buildings (Bâtiment Basse Consommation, BBC).

The development of these constructive measures generally relies on models integrating variables related to materials, orientations and sometimes the life cycle of buildings (Peuportier et al., 2013). In this logic, technical systems are supposed to solve measured and modelled environmental problems. However, despite the success of these models in the operational environment, the introduction of binding standards, the development of alternative energies and the investment in innovating energy efficiency technologies are still insufficient for solving the problems related to decreasing the energy consumption of buildings. Thus, several studies on French or international data show a gap in energy performance, that is a difference between the energy consumption observed and that modelled (Cayla, 2011, Majcen et al., 2013). The source of this discrepancy may be poor model estimates, such as climatic conditions or the thermal characteristics of the dwellings. However, the main argument is that of the standardisation of housing energy use in models that does not correspond to the diversity found in reality. Nevertheless, although they are difficult to estimate, the knowledge, understanding and integration of household energy behaviours in global models could optimise the effectiveness of the directives and the resulting norms (Ouyang and Hokao, 2009, Abrahamse et al., 2007). Indeed, in this sphere, it seems difficult to ignore the behaviour of households as they are active users whose consumption level can be measured, while buildings are static energy-consuming objects. The law on energy transition also takes this aspect into account by developing information and support measures to raise awareness of energy savings among households, by developing economic and fiscal incentives (MEEM, 2016).

In this context, this chapter looks at different models and approaches concerning the energy consumption of households and buildings. Our goal is to establish a global framework (technical, economic, social, etc.), to shed light on their construction logic, even though it seems difficult to carry out a complete scan of the scientific literature in this area, given how copious it is.

8.2 Meaning and Utility of Consumption Modelling

Conventional belief generally associates a model with quantitative approaches. However, its vocation is not so much to provide algorithms as to constitute a tool of knowledge that is likely to provide a simplified overview of the reality. It serves as a sort of reproducible reading chart for interpreting the complex processes that we study. The theory is therefore never far from the construction of a model (Varenne, 2010), and it even acts as a mediator between the theory and the "real world" (Morgan and Morrison, 1999, Sergi, 2014). Any model therefore relies *a priori* on theoretical constructs reflecting a level of knowledge of processes that will be validated *via* the model and, at the same time, will also be confronted with their own simplification.

There is therefore a heuristic scope in the model, but also a practical and utilitarian one. Hence, there is an ambiguity depending on the purpose attributed to its use. A model can act as a chart for reading the causal sequences of the process we are trying to understand in terms of its functioning. In this case, it can reach a level of complexity where an attempt will not necessarily be made to reduce it to indicators or to digitise it at the risk of reducing its explanatory scope. It can also be a tool for mediating between knowledge and power (Sergi, 2014). It will then be made available to an actual project. Digitisation will be unavoidable as the model will have to provide indicators to adjust the operational choices. The technician or policy in charge of the operation will use these indicators to help design and predict project performance, evaluate it once the project is completed, and measure the discrepancies between the simulated performance and the actual performance.

The modelling of building energy consumption falls completely within this context. It can be a description of causal sequences that associate a practice with a material container (the building). While the impact of the practice on the final building consumption will be measurable *a priori* on the one hand, it will be more difficult to reduce a social behaviour to an indicator. Hence, there is a difficulty of digitally modelling energy behaviours, the knowledge of which is important for understanding the consumption process. In this sense, and given the lack of digital models, the question of experimentation with different practices remains open.

On the other hand, when the model relies on technical elements like building components (materials, openings, windows and glazing, orientation, etc.), the impact on the building's energy performance will be easier to evaluate. This impact will then be directly measurable based on the indicators produced by the model. In this context, it will constitute a useful decision-making tool, which can be used for determining construction choices or managing the operators' strategies, for example.

In a complete theoretical construction of building energy consumption, it is therefore not easy to associate the constructive aspects and the domestic uses – with regard to the residential field – or the office practices – with regard to the non-residential field. Few models associate these two dimensions, and the approaches are generally segmented according to the authors' disciplines. However, in economic approaches in particular, consumption models sometimes associate social variables with morphological variables without taking into account their use, considered as processes.

The field of building energy consumption modelling is therefore vast, diverse and involves many disciplines spanning the social sciences, engineering, physics, economics and ecology, to name just a few. In this subject, each discipline finds a support for improving knowledge of a dimension of its field: a social fact, a resistance of materials and their insulation, an architectural construction, a relationship between nature and the built environment, etc. All of them enter the environmental field with the aim of reducing consumption through the dimension studied. However, the models produced do not necessarily fit into the same perspectives. Some have an explanatory scope and others are more experimental, while models associating experimentation and explanation, uses and conceptions or practices and techniques are more rare.

In order to attempt to shed light on the way in which these models are constructed, it is possible to try typologies that are known to necessarily be reductive. At first, the models can be classified according to the apparent methodological oppositions.

- Deductive modelling approaches are based on hypotheses constructed from the authors' experiences or accumulated knowledge in the field.
- Inductive modelling approaches are built on observations or field surveys to highlight causality chains.

The former are generally used to construct quantitative, mathematical or statistical models, while the latter make it possible to describe qualitative processes that do not necessarily lead to digitisation (they are also described as deterministic by physicists or engineers, meaning that the approach is more descriptive than experimental).

This first typology remains very reductive and too dichotomous, however. We can try to overcome it by focusing on gaining more precise understanding of how quantitative or qualitative models are constructed. In this framework, the scientific literature on energy consumption mobilises more quantitative approaches than qualitative approaches. Thus, a study conducted by Benjamin K. Sovacool in 2014 on the basis of

4,444 articles on this topic, published between 1999 and 2013 by 9,549 authors in three leading journals specialised in energy (*Energy Policy*, *Electricity Journal* and *The Electricity Journal*), shows that only 19.6% of the authors have a background in the social sciences. In addition, 12.6% of articles in the whole corpus use qualitative methods, and less than 5% of the citations refer to journals in the social sciences and humanities.

8.3 Quantitative Models

Quantitative approaches therefore constitute the bulk of scientific production on the question of energy consumption. The vast majority of them aim to build models that simulate and experiment with reducing the intensity of residential and tertiary energy consumption, without necessarily providing the explanatory elements of the process. We will focus here more on the domestic dimension, which makes it possible to question the validity of building consumption models, with or without taking the uses in these approaches into account.

8.3.1 Statistical Approaches

The first major type of quantitative approach relies on statistical calculations. Generally, they estimate the relationship between end use and energy consumption to determine the building energy consumption representative of building stock, which is most often residential. The statistical approach attempts to establish a causal relationship between the energy consumption of buildings, users (individuals or households) and different contextual variables, such as climatic conditions or urban density. We can distinguish three calculation methods.

*   *Regression*: This makes it possible to determine the influence of the input parameters on final energy consumption, most often of a dwelling or a residential building. Input variables can range from socio-economic information about occupant households to behavioural data and other more technical features concerning the housing or building (Belaïd, 2016, Tonn and White, 1988). The estimation of the model's validity is assessed based on the quality of the adjustment.
*   *Conditional demand analysis*: This model uses the regression of energy consumption on potential influential variables, but involves conditioning with the presence of energy equipment (Nesbakken, 1999).
*   *Neural network*: This rare technique uses a simplified mathematical model to represent the interconnections between different energy uses (Issa et al., 2001), in the manner of biological neural networks. Similar to regression, neural network models minimise estimation errors and also make it possible to capture the nonlinearity of relationships. So far, this method has been used to forecast the electrical load variations of energy services.

These different models are widely used in econometric studies on building energy consumption and agree on several variables that explain household energy behaviours, such as the age of the reference person and the income or the size of the household.

Beyond the characteristics of households, economists are also interested in the deterrent role played by energy prices. While these results may appear consistent, statistical methods, nevertheless, have limitations. They are intended to highlight the explanatory variables of mechanisms for which an attempt is made to give reading grids. They result in algorithms in which coefficients indicate the effects of the variables affecting energy consumption. All the variables used in the regression must be quantitative, including those for households, which excludes any behavioural indicator (even if binary variables can sometimes be used). The algorithm will therefore translate the impact of each of the variables into coefficients, without necessarily giving the elements that make it possible to understand the reasons for which they play a role. This means that the purpose of the approach is not so much to explain the consumption process as it is to model it. We thus remain at the deductive process stage, insofar as the choice of the explanatory variables rests on hypotheses pertaining to constructed theories or experiences. In other words, the variables are misleadingly described as explanatory since they are based solely on hypothetical-deductive methods, which, by definition, are not demonstrative. This is why results sometimes appear contradictory, as there are no means for understanding the mechanisms by which close variables can act in the opposite direction. For example, this is the case for results showing that income has a positive effect on energy consumption, while employment regressions show that people with the highest occupational status rank among those who consume the least amount of energy (Table 8.1).

8.3.2 Engineering Approaches

The other major type of quantitative model is usually developed by engineers. These are mathematical approaches aimed at modelling consumption from algorithms developed using deductive approaches. Quantification is necessary here, and the indicators are based on acknowledged theories or hypotheses. These models estimate consumption according to the technical infrastructure of the building, housing and household equipment, especially household appliances and the physical environment (outdoor temperature, sunshine, etc.). When taking behaviours into account, hypotheses most often concern residence times, temperatures, lighting or window openings, for example, in order to integrate the impacts of uses on the variations of the building's thermal environment within the model.

We can distinguish three methods for constructing these models:

- *Distributions*: This technique mainly concerns the calculation of the final energy consumption of domestic equipment. It relies on estimates of their distribution and penetration within dwellings. The calculation uses common efficiency coefficients (Kadian et al., 2007).
- *Archetypes*: This method is based on housing stock characteristics (date of construction, type of building, size, etc.). Within it, the energy consumption is modelled for each housing or housing group archetype. The model is then used to extrapolate consumption to an entire building stock (at a specific scale: local, regional, national). However, it is constrained by the number of archetypes defined ex ante, and the estimates lead to limited representativeness of the housing stock (Huang and Brodrick, 2000).
- *Sampling*: This approach is based on a representative sample of dwellings at the scale of the perimeter studied. It aims to estimate the variations in consumptions

Table 8.1 Some Determinants of Household Energy Consumption (Conditional and Regression Analyses)

Variables	Effects on Energy Consumption	Sources
Occupation state	Although theories indicate that tenant's energy consumption should lie under that of owners, no consensus has been reached based on current empirical evidence.	Charlier (2015), Jones et al. (2015), Yohanis (2012)
Income	Income has a positive effect on energy consumption (in general, the prices elasticity lies under 0.15).	Cayla et al. (2011), Nesbakken (2001), Belaïd (2016), Labandeira et al. (2006), Santamouris et al. (2007)
Number of occupants	The more household occupants, the higher the energy consumption.	Leahy and Lyons (2010), Vaage (2000)
Age of the reference household occupant	Energy consumption is the highest at households with a reference occupant between 45 and 65 of age − relative to other households.	Brounen and Kok (2011); Brounen et al. (2013)
Employment type	Energy consumption is inversely proportional to social and professional statuses.	McLoughlin et al. (2012)
Urban area type	There is no agreement on the effect of urban density on energy consumption.	Kaza (2010)
Energy price	Higher energy prices reduce energy consumption (price elasticity −0.20 à −1.6).	Belaïd (2016), Dubin and McFadden (1984), Halvorsen and Larsen (2001), Labandeira et al. (2006), Larsen and Nesbakken (2004), Nesbakken (1999), Risch and Salmon (2017)
Comfort preferences	Individual preferences about comfort have a positive impact on energy consumption. Potential energy savings could range from 1.1% to over 29%.	Estiri (2015), Jones et al. (2015), Lopes et al. (2012)

between the housing within the corpus. The energy consumption of the housing stock can be estimated after extrapolation, subject to the strength of the sample (Swan and Ugursal, 2009).

8.3.3 A Transversal Vision of Quantitative Models

Statistical and mathematical models of energy consumption are not compartmentalised even though the former are mainly developed by economists and the latter by engineers (climatologists, building physicists, ecologists, etc.). In particular, they coincide in terms of their input entries.

A first approach to quantitative models is to use global or aggregated indicators concerning energy prices, household incomes, possession of equipment on a national scale or even the main construction types according to the materials used, the climatic contexts or building orientations. These models, which can be described as "top-down" or "general", are based on the results of major surveys or national public data generally provided by the technical services of a particular Ministry (finance, environment, etc.). The quality of the data gives them reliability and objectivity in that they can be updated as the input data evolves. "Top-down" models are useful on a macro-scale (economic or structural) and can influence national policies to reduce energy consumption. On the other hand, their application is more uncertain on a micro-scale. They are therefore little used in the context of eco-neighbourhood projects and especially in estimates of localised building consumption.

Another approach, called "bottom-up", is the reverse of the previous approach. It does not focus on overall energy consumption but on that of housing or building types. The input data are disaggregated and the projection of the micro-model to a larger set is intended to provide an estimate of overall consumption, under the condition of taking the diversities peculiar to each individual or local situation into account. This condition disregards the fact that the totality is not equal to the sum of peculiarities, which limits the scope of these models to a macro-scale.

Quantitative models also try to integrate user behaviour into the input variables because their energy consumption contributes to that of the building, if only in terms of the heat loss. The statistical models focus more on the socio-demographic characteristics of the buildings' residents, and the mathematical models introduce data relating to thermal environment uses. Whatever their specificity, these behavioural data must imperatively be quantitative in order to be introduced into a mathematical model or to constitute a variable of an algorithm resulting from a statistical regression. The most striking result of this work is that of having demonstrated the existence of a rebound effect, which corresponds to the increase in energy consumption following the introduction of new technologies in housing (Brookes, 1990, Wirl, 1997, Schipper and Grubb, 2000, Orea et al., 2015). This effect can be defined as a behavioural response emanating from an improvement in the energy efficiency of housing and buildings (Binswanger, 2001). However, the behavioural variables are discretised and are not associated to reconstruct a series of uses that would make it possible to approach behaviours as a logical process, specific to socially and demographically differentiated households, located in a specific environment.

Comparing the input modes of the mathematical and statistical approaches shows that, ultimately, a quantitative model of energy consumption is valid at its level of its development, but that it is difficult to conceive of it as a nesting of scales. It also appears that the probabilities assigned to deductive model variables, including global variables, often rely on local or national survey processing. In other words, these models rely on algorithms whose coefficients are estimated based on empirical analyses. However, this approach does not necessarily give an inductive character to the methods, insofar as, before resorting to the survey results, the choice of the variable to be integrated in the algorithm in itself constitutes a deductive hypothesis of the energy consumption process.

Variables that allow mathematical or statistical models to be constructed are therefore fairly close. It includes the property and housing features, and often also estimated equipment rates and their category according to their energy consumption, indicators of housing and building insulation, as well as the socio-demographic characteristics of the occupants or behavioural parameters. Even if these determinants are not associated according to the same logic and if the mathematical models pay more attention to the construction materials, they coincide in terms of their input modes and their limits.

8.4 Qualitative Models

The polarisation of approaches around quantitative methods underlines the dominance of deductive, statistical or mathematical models in the field of building energy consumption. It probably leads to a reduction in the role of energy behaviours in constructing these models, reinforcing their unambiguous character. However, qualitative studies within the social sciences show the interest of integrating the complexity of social facts within the comprehensive knowledge of consumption.

8.4.1 Multiform Behaviours

Based on the qualitative work of his corpus, Sovacool (2014) highlights the way in which the technical, political, economic and social factors interact on the process of household energy consumption. In this framework, the effects of the context as well as the individual and collective dimensions are the essential elements of the process. From the point of view of individuals and groups, the consumption mechanism thus mobilises personal factors like the predispositions inherited from experiences, but also more structural elements such as household size or incomes.

Still, many studies on behaviours emphasise the relationship to the domestic comfort of groups and individuals. Sometimes associated with cleanliness (Shove, 2003), when it comes to water consumption, comfort usually relates to room occupancy and housing temperatures, and in this sense, households are adapting to the domestic context by developing "energy intelligence" (Subremon, 2012). Several authors underline the paradox between ecological commitment and the level of household consumption, which refers to evolving definitions of comfort that are specific to each domestic situation. Thus, for households, comfort appears to be "at the heart of an ideological paradox, between the intentionality of their practices presented as eco-responsible but sometimes anecdotal, and a significant propensity to develop housing equipment that actually becomes increasingly energy-consuming" (Roudil and Flamand in Lévy et al., 2014, p. 52). In other words, the quest for comfort is the origin of domestic behaviours that vary according to the family situations and the type of housing occupied. Beyond the temperature and the occupancy of the rooms, it would induce a capacity for adaptation that mobilises equipment that consumes energy. Thus, there are as many types of energy behaviour as there are household capacities to adapt to produce a comfortable domestic universe, regardless of the energy cost involved.

8.4.2 A Social Practice

How, therefore, can we provide a comprehensive and coherent reading grid for such multifaceted behaviours? To overcome this heterogeneity, Shove et al. (2012) have proposed shifting the paradigm of behaviour towards that of social practice. This would involve three types of elements: materials, meanings and skills (Figure 8.1).

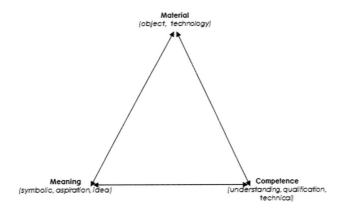

Figure 8.1 Elements of social practices according to Shove et al. (2012).

In this configuration, individuals would develop practices by mobilising the material goods to which they attribute symbolic meanings, according to their ability to make use of them. The interest of this proposal does not lie in affirming the role of these three poles, but rather in their relationship and in the fact that the importance of each of them would evolve in step with the practice. Thus, the authors try to show that as the material object becomes more technical, the skills needed to use it become less significant in practice. At the same time, the symbolic dimension of the object becomes central. They support their demonstration on the evolution of car use, but the model could equally be extended to the use of other material objects, like the computer.

It is also interesting to apply the model to energy consumption as a social practice in its own right. This transfer acquires all the more meaning since energy is an intangible and abstract good that requires a material vector for its user to become aware of its value (Bonnin, 2016). This vector is generally a type of equipment, which can range from a light fitting to an appliance, among others. Each of these devices does not require any particular technical skill to be used and, if we follow Shove et al., its acquisition therefore refers to its symbolic value. For example, this process has been very well described by Margot Pellegrino, through the use of air conditioners in Calcutta, for which "energy behaviours (…) are determined by social aspiration, which aims to demonstrate belonging to a precise and identifiable social status through the possession of an object (the air-conditioner) and the modality of its use (independent of a real and perceived necessity)" (Pellegrino, 2013).

Qualitative approaches therefore highlight the existence of consumption processes that are part of complex social logics. They reflect a diverse range of adaptations to the domestic space according to individual and collective experiences, perceptions of comfort, sensitivities to the environmental question or domestic and family situations. These adaptations mobilise pieces of material equipment, which are vectors of consumption of an immaterial source but which therefore acquire a value that is all the more symbolic since they are commonly used.

Qualitative approaches therefore reveal the limits of a simplified representation of energy behaviours, which cannot be reduced to discrete household characteristics, quantifiable energy intensity and energy costs. However, this poses the question of how to construct generalisable models based on reduced observation corpora. This is especially the case as their experimentation – and consequently their validation – is made difficult by the lack of indicators.

8.5 Three Examples of a Transversal Approach

The above shows that digital models have the advantage of allowing experimentation with the materials used, insulation modes and orientations, or indeed all the technical elements for which the energy efficiency is measurable. Sometimes, variables aiming to integrate the energy behaviours of households are taken into account in the algorithms, but on the whole these models remain very techno-centric and fail to represent socially and demographically differentiated series of use. For their part, qualitative models seek to reproduce household energy behaviours as social processes, but ignore the technical constraints of uses. Even if they succeed in translating the diversity and complexity of the individuals' and groups' relationships to energy, the chains of causality highlighted cannot be reduced to indicators and are difficult to model digitally. As a result, they cannot be verified and tested, and are not taken into account in the construction standards.

Although the two types of models at first appear difficult to reconcile, each of them gives us part of the chart for reading the energy consumption processes of buildings. It is also revealing that most of the reviews of the work carried out on consumption lead to the conclusion that transversality is necessary within the approaches: between the study of the behaviour of individuals and that of the life cycle of equipment (Polizzi di Sorrentino et al., 2016), between economic, environmental and urban inputs (Tu, 2018), or even by an association between the constructive or architectural aspects of buildings, the urban or geographical context, the sociological structure of the inhabitants and the environmental psychology (Pellegrino and Musy, 2017). However, this transversality is not self-evident. Below, by way of an indication and not an exhaustive overview, we present three research studies (two of which we took part in) to illustrate the diversity of methods for modelling the energy consumption of buildings, those of users and the difficulty of bringing them together within a common approach.

8.5.1 Modelling Behaviours to Measure Building Energy Performance

The first type of work concerns attempts to model the impact of behaviour on the energy performance of buildings using probabilistic approaches. Éric Vorger's thesis (2014) is a good example. By extending Page (2007)'s approaches to office building occupancy, he proposes applying stochastic modelling based on inhomogeneous Markov chains to offices and dwellings. These probabilistic calculations are based on the assumption that changes in user behaviour are not constant. Vorger uses results from the statistical processing of national databases (time-use survey, census, housing survey) or local statistics to determine the probabilities of presences in dwellings and offices, the number of times windows are opened and closed, the use of electrical appliances, room lighting or heating control. In addition, the model takes the diversity of behaviours into account according to the socio-demographic characteristics of users and the types of housing they occupy.

This probabilistic modelling of building energy performance based on occupant behaviour is undoubtedly one of the most successful. It does, however, have several limitations inherent to quantitative approaches that lie at the intersection of inductive and deductive logics. For example, the model's role in terms of representing the totality of behaviours obliges its author to rely on very heterogeneous works. For example, to study windows being opened by residential building occupants, Vorger uses the Haldi and Robinson model (2009), which was developed for office buildings located in Switzerland and validated in Austria (Vorger, 2014, p. 190). From this point of view, projecting behaviours displayed in offices into housing contexts raises questions, as does applying models constructed in different climatic conditions within France. Moreover, the need to know the occupancy of dwellings to determine residents' activities leads to the construction of uncertain probabilities based on mechanical associations between the surface area of the dwellings and the size of the households (p. 144). Finally, the model was used to simulate the consumption of heating and hot water in a residential building of 16 dwellings that had undergone energy renovation. The results show significant differences between the model and the observed situations with regard to the number of occupants, the heating temperatures, the energy consumption and, in particular, the estimates of the gains expected from the renovation (p. 404) (Figure 8.2).

Here, we see the difficulty of producing a deductive model that encompasses all energy behaviours. The necessary data, which are often missing, require the use of survey results from different contexts and scales to calibrate the behavioural probabilities. This diversity of sources introduces important biases in the model, which cannot be used at refined scales and in specific situations. The use of inductive approaches (via surveys) to construct probabilistic hypotheses must therefore be based on binding protocols that reduce the possibilities of developing generic deductive models that are valid at different scales and in different contexts.

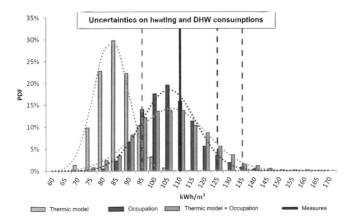

Figure 8.2 Differences between estimated consumption (ECS and heating) and actual consumption after work on a 16-unit building located in Feyzin (69) according to Éric Vorger's model. The dashed vertical lines correspond to the guaranteed thresholds (risk of exceeding 2.5%). The consumption measured during the year following the works is represented in the red continuous line (110 kWh/m²). (Vorger, 2014, p. 403.)

8.5.2 Life Cycle of Households and Energy Consumption

The second type of work does not concern modelling approaches, but rather inductive quantitative methods designed to describe the behavioural logics of energy consumption. They are based on ad hoc surveys or national databases that indicate household consumption. As an example, we can present the work we have carried out on the links between the life cycle sequences of households and the intensity of household consumption (Lévy and Belaïd, 2018).

The effect of family structures and their evolution was highlighted in 1976 by a longitudinal study in the metropolitan area of Lansing (Michigan) by Morrison and Gladhart. Numerous quantitative studies linked the life cycle to energy consumption during the 1980s (Fritzsche, 1981, Frey and LaBay, 1983), but they were based on a variety of different samples and methods, and did not lead to convergent results. Since then, apart from some qualitative research (Garabuau-Moussaoui, 2009), work on household energy behaviours has shifted away from the life cycle approach to household demographic characteristics, such as age or size, considered as specific explanatory variables in static consumption models (O'Neill and Chen, 2002, Brounen et al., 2012, Valenzuela et al., 2014). However, there was no research showing the links between the combination of household energies used, the demographic characteristics of the users, their locations and the type of habitat they occupied.

Our study mobilises the National Housing Surveys by INSEE (French National Institute of Statistics, 2002 and 2006). In a first step, we performed factor analyses and hierarchical classifications in order to construct ten types of energy consumers (Table 8.2), associating the modes and energy intensities used (electricity, gas, fuel, LPG, wood, district heating and charcoal), household characteristics (family status, age, size, socio-professional category and income), characteristics of the type of dwelling occupied (occupation status, type of dwelling, number of rooms, area, date of construction of the building) and the residential context (urban, suburban, rural). These types show that domestic energy modes are markers of dwelling characteristics and occupant profiles.

In a second step, we verified the temporal stability of these relations between 2002 and 2006. Our analyses show divergences between the factors affecting overall consumption, per m^2 and per person according to the different types of consumers constructed. Finally, we tried to identify the causes of changes in consumption specific to each type, by reasoning according to groups of homogeneous consumers, that is to say, by neutralising the effects of the type of housing occupied, its location and the energy modes used. This method made it possible to fully explore the concept of efficiency and identify households that could regulate their consumption according to their social and demographic characteristics (Figure 8.3).

The results show that for identical energy modes, housing type and urban contexts, consumption intensities per person vary according to positions in the life cycle of households, while consumption per m^2 does not change or changes little. This means that the energy behaviours of stable households in their homes do not adapt to changes in the size of the family over the course of the life cycle.

Ultimately, this research has highlighted the need to combine consumption measures per m^2 (which are very dependent on housing surface areas) with those of consumption per person (which are very dependent on changes in family structures), whereas incentive policies to reduce domestic energy consumption are generally only based on

Table 8.2 Dominant Profiles of Households and Dwellings of the ten Types of French Energy Consumers in 2002

Types of Consumers	%	Energy Types	Dominant Profiles
Type 1	4.6	E–F–P (DH/C+)	Childless couples or young or older single people belonging to the working classes with medium or low incomes; tenants of an apartment with less than forums and smaller than 100 m² in a building constructed between 1950 and 1974 in a rural or urban area
Type 2	8.5	E–FP (W)	Childless couples or single retired people belonging to the working classes; owner occupiers of a house larger than 100 m² built before 1975 in a rural or periurban area
Type 3	26.3	E+ P W (DH/C–)	Middle-income families aged 40–59; owner occupiers of a house larger than 70 m² built before 1948 or after 1975 in a rural or periurban area
Type 4	1.5	E+F+P W	Families aged over 40, upper or middle management levels (including retired); owner occupiers of an old house larger than 100 m² in a rural or periurban area
Type 5	13.2	E–G– (DH/C+)	Young working-class middle-income or low-income childless couples or single people (sometimes precarious); tenants in social or private housing with four rooms or less built between 1949 and 1974 in an urban area
Type 6	1.1	E–G–F–	Childless couples or single retired people, management or working-class category; tenants in social or private housing with 4 rooms or more built between 1949 and 1974 in an urban area
Type 7	22.2	E+	Childless young couples or single people; tenants in social or private housing with one or two rooms, smaller than 70 m², built after 1975 in an urban area
Type 8	7.2	E–P (DH/C F+)	Low-income, working-class childless couples or retired single people; tenants in a house with four rooms or more and smaller than 70 m², built before 1975 in a rural or periurban area
Type 9	9.7	E–G+	Management-level or working-class childless couples or retired single people; tenants in social or private housing with 3 or 4 rooms, between 70 and 100 m² built between 1949 and 1974 in an urban area
Type 10	5.5	E+G+(W–)	High-income upper or middle management families aged 40–59; owner occupiers of a house larger than 100 m² built before 1975 in an urban area

Example of interpretation: the consumption modes not in brackets represent a minimum of 60% of the type. The consumption modes in brackets are in a minority and are represented at least 1.5 times more within the type than in the total population.

Sources: Secondary processing of the 2002 National Housing Survey (INSEE).

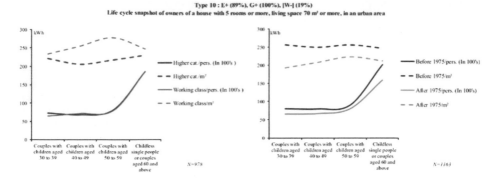

Figure 8.3 Average annual consumption per m² and per person according to the current life cycle for owners of a 5-room or more house measuring 70 m² and over located in urban areas in France (ENL, 2006.)

the former measurement. By showing that energy consumption is an evolving social practice, this dynamic approach is in line with most qualitative models. This is important because building consumption estimates that take occupancy characteristics into account do not anticipate the impacts of ageing in stable households, nor the change in the occupancy of buildings as the residents are replaced. These results thus question the robustness of inductive models that, with the exception of life cycle assessment (LCA) approaches, address energy consumption as a static fact.

On the other hand, while these results open up prospects for digitisation, they do not constitute models that can be mobilised by operators. In addition, they largely erase the impact of the constructive aspects of buildings and do not address energy practices in all their dimensions, as Eric Vorger was able to do, for example. Nevertheless, they encourage the construction of complex models that combine the impact of constructive aspects with consumption per m² and per person. They also invite the evolution of energy practices during the life cycle of households to be integrated within energy consumption models of dynamic buildings, according to their stability or their residential mobility.

8.5.3 Mapping Energy Behaviours

The third types of work circumvent the difficulty of modelling uses from consumption, based on a synthetic approach to domestic energy behaviours. We present here a research project in which we participated (Bourgeois et al., 2017), the objective of which was to produce a digital model of uses acting on the level of consumption, in other words, to model a social behaviour that is generally addressed by qualitative studies.

As a first step, we constructed three synthetic indicators of energy behaviours (regulation of consumption, equipment rate, intensity of large equipment usage) developed from a survey of 1950 households in the Paris region. In a second step, these indicators were modelled statistically (logistic or multiple regression) using the survey data, but care was taken to ensure only variables present in the population census were mobilised (fuels, households, dwellings). This precaution was intended to allow the model to be used at large scales for which we do not have precise data on the energy behaviours of households.

The three models show that energy lifestyles are essentially constrained by the mode of heating and the fuels used. But they also depend on household characteristics such as the age of the reference person or the size of the family. Like qualitative models of use, this quantitative approach confirms the existence of temporal flexibility in energy behaviours.

The third step was to mobilise the models to map these behaviours at the IRIS scale of Ile de France, using the 2011 census.

As a broad outline, the maps show the cumulative effects of the urbanisation process of the region (type of housing and suburbanisation) and the demographic distribution of the population on the allocation of energy behaviours in space.

This approach has a triple heuristic scope: better knowledge of the determinants of uses, anticipation of domestic energy consumption, production of a dynamic model to simulate behavioural variations in time and space. In the context of energy transition, it opens up prospects for development policies aimed at monitoring or anticipating energy consumption and GHG emissions, by using the model to test the effects of a variation in population, housing construction or types of buildings on a fine scale.

Again, however, building performance variables are much less accurate than those used in engineering models, as opposed to synthetic behaviour variables that encompass dozens of forms of use. The approach raises the question of modelling complex behaviours from simple and accessible variables found in large national databases. This model was used within research to improve the Météo-France urban heat island TEB model (Schoetter et al., 2017).

However, it is symptomatic that the synthetic indicators were retained from the sole point of view of measurable variables, like the heating temperature of housing, without taking the all-encompassing and dynamic nature of the behaviours into account (Table 8.3).

Each of these three types of work has limitations. However, we can also consider that each one of them makes it possible to provide a partial response to the limitations of the other two. In this sense, they appear to be complementary. Associating

Table 8.3 Ownership of Large Household Appliances Model

Variable	Level	Coefficient	Odd_ratio
(Intercept)		−1.22	0.30
Type of housing	Individual		
	collective	−0.73	0.48
Surface		0.55	1.72
Number of persons	1		
	2	0.23	1.26
	3 or more	0.57	1.78
Couple	yes		
	no	−0.52	0.60
Occupancy	owner		
	others	−0.40	0.67

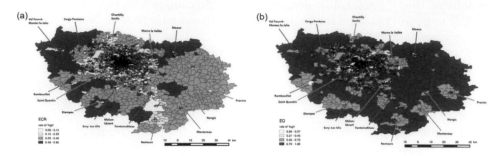

Figure 8.4 Maps of strong energy regulations (a) and high rates of household electrical appliances (b) in Île de France (scale Iris RGP 2011). (Bourgeois et al., 2017, pp. 189–190.)

them could open interesting avenues to produce new concepts, but also to model the energy consumption of buildings from a dynamic point of view in terms of the diversity of their design and their uses, taking the effects of scale and context into account (Figure 8.4).

8.6 Conclusion

At a time of energy transition, building consumption modelling constitutes an important mediator of expertise for operational actors. It is an essential element in the implementation of construction and rehabilitation policies that achieve factor 4. At the same time, this consumption falls within the constructive constituents of the buildings and social behaviours of the buildings' occupants. However, the numerical modelling of building consumption and that of energy behaviours do not belong to the same approaches.

On the one hand, since multifaceted behaviour processes cannot be digitised, quantitative models reduce domestic energy use to a few discrete indicators that do not reflect the overall logic of social practices. This results in a significant difference between the estimation of consumption by energy models and the actual performances of buildings. On the other hand, while the qualitative studies on household energy behaviours manage to describe processes in their complexity and comprehensiveness, they are difficult to model digitally. When they rely on synthetic quantitative indicators, behavioural models struggle to introduce the effects of the construction components of buildings.

However, in order to anticipate or estimate the energy consumption of a project, operators must integrate the energy efficiency of construction or renovation, its economic balance and also the behaviour and expectations of users. In this regard, to constitute real decision support, models of building energy consumption must be able to meet this triple requirement. However, the linking of inductive and deductive approaches, deterministic and stochastic models, mathematics and statistics and even all these approaches between them is not self-evident since they mobilise the social and economic sciences, urban ecology and urban engineering in the broad sense. However, in the field of energy consumption, there is no truly hermetic compartmentalisation between mathematical models, statistical models and qualitative models, as the empirical data is often useful for calibrating or quantitatively estimating the assumptions of deductive approaches, whereas the production of digital and dynamic indicators that introduce the life cycle, the comprehensiveness and the spatiality of social processes is not an insurmountable dimension of qualitative approaches to energy behaviours. Methods that make it possible to combine the technical, economic and social dimensions to achieve more robust models of building consumption and come closer to the complexity of the "real world" still remain to be found. The co-construction of new concepts and methods integrating causality into numerical modelling, to theorise, simplify and experiment with models of building energy consumption is a field that is more necessary than ever, but which is still largely open.

Bibliography

Abrahamse W., Steg L., Vlek C., & Rothengatter T. (2007) "The effect of tailored information, goal setting, and tailored feedback on household energy use, energy-related behaviors, and behavioral antecedents", *Journal of Environmental Psychology*, 27(4), 265–276.

Belaïd F. (2016) "Understanding the spectrum of domestic energy consumption: Empirical evidence from France", *Energy Policy*, 92, 220–233.

Binswanger M. (2001) "Technological progress and sustainable development: What about the rebound effect?", *Ecological Economics*, 36(1), 119–132.

Bonnin M. (2016) *Habitable et confortable: modèles culturels, pratiques de l'habitat et pratiques de consommation d'énergie en logement social et copropriétés*, PhD Thesis on Architecture, Université Paris-Nanterre.

Bourgeois A., Pellegrino M., & Lévy J.-P. (2017) "Modeling and mapping domestic energy behavior: Insights from a consumer survey in France", *Energy Research & Social Science*, 32, 180–192.

Brookes L. (1990) "The greenhouse effect: The fallacies in the energy efficiency solution", *Energy Policy*, 18(2), 199–201.

Brounen D., & Kok N. (2011) "On the economics of energy labels in the housing market", *Journal of Environmental Economics and Management*, 62 (2), 166–79.

Brounen D., Kok N., & Quigley J. M. (2012) "Residential energy use and conservation: Economics and demographics", *European Economic Review*, 56, 931–945.

Brounen D., Kok N., & Quigley J. M. (2013) "Energy literacy, awareness, and conservation behavior of residential households", *Energy Economics*, 38, 42–50.

Buildings Performance Institute Europe (BPIE) (2013) *Stimuler la rénovation des bâtiments: un aperçu des bonnes pratiques*, Bruxelles: BPIE. http://www.planbatimentdurable.fr/IMG/pdf/150_EX_Stimuler_la_renovation_des_batiments_bonnes_pratiques_BPIE_2013.pdf

Cayla J.-M. (2011) *Les ménages sous la contrainte carbone*. PhD thesis, École des Mines.

Cayla J.-M., Maizi N., & Marchand C. (2011) "The role of income in energy consumption behaviour: Evidence from French households data", *Energy Policy*, 39 (12), 7874–83.

Charlier D. (2015) "Energy efficiency investments in the context of split incentives among French households", *Energy Policy*, 87, 465–479.

Dubin J. A. & McFadden D. L. (1984) "An econometric analysis of residential electric appliance holdings and consumption", *Econometrica*, 52 (2), 345–62.

Estiri H. (2015) "The indirect role of households in shaping US residential energy demand patterns", *Energy Policy*, 86, 585–94.

Frey C. J. & LaBay D. G. (1983) "A comparative study of energy consumption and conservation AC Ross family life cycle", in Bagozzi R. P., Tybout A. N., & Abor A., (eds), *Advances in Consumer Research*, 10, 641–646.

Fritzsche D. J. (1981) "An analysis of energy consumption patterns by stage of family life cycle", *Journal of Marketing Research*, 18 (2), 227–232.

Garabuau-Moussaoui I. (2009) "Behaviours, transmissions, generations: Why is energy efficiency not enough?", *ECEEE Summer Study Proceedings*, 33–43.

Haldi F. & Robinson D. (2009) "Interactions with window openings by office occupants", *Building and Environment*, 44, 2378–95.

Halvorsen B. & Larsen B. M. (2001) "The flexibility of household electricity demand over time", *Resource and Energy Economics,* 23 (1), 1–18.

Huang Y. J. & Brodrick J. (2000) *A Bottom-up Engineering Estimate of the Aggregate Heating Andcooling Loads of the Entire U.S. Building Stock*. LBNL--46303.Berkeley, CA: Ernest Orlando Lawrence Berkeley National Laboratory.

Issa R. R. A., Flood I., & Asmus M. (2001) "Development of a neural network to predict residential energy consumption", In *Proceedings of the Sixth International Conference on Application of Artificial Intelligence to Civil & Structural Engineering*, ICAAICSE' 01. Stirling: Civil-Comp Press, 65–66.

Jones R. V., Fuertes A., & Lomas K. J. (2015) "The socio-economic, dwelling and appliance related factors affecting electricity consumption in domestic buildings", *Renewable and Sustainable Energy Reviews,* 43, 901–17.

Kadian R., Dahiya R. P, & Garg H. P. (2007) "Energy-related emissions and mitigation opportunities from the household sector in Delhi", *Energy Policy*, 35 (12), 6195–6211. doi:10.1016/j.enpol.2007.07.014.

Kaza N. (2010) "Understanding the spectrum of residential energy consumption: A quantile regression approach", *Energy Policy,* 38 (11), 6574–85.

Labandeira X., Labeaga J.-M., & Rodríguez M. (2006) "A residential energy demand system for Spain", *The Energy Journal*, 87–111.

Larsen B. M. & Nesbakken R. (2004) "Household electricity end-use consumption: results from econometric and engineering models", *Energy Economics,* 26 (2), 179–200.

Leahy E. & Lyons S. (2010) "Energy use and appliance ownership in Ireland", *Energy Policy,* 38 (8), 4265–79.

Lévy J.-P., Roudil N., Flamand A., & Belaïd F. (2014) "Les déterminants de la consommation énergétique domestique", *Flux*, 96, June, 40–54.

Lévy J.-P. & Belaïd F. (2018) "The determinants of domestic energy consumption in France: Energy modes, habitat, households and life cycles", *Renewable and Sustainable Energy Reviews*, 81, 2104–14.

Lopes M. A. R., Antunes C. H., & Martins N. (2012) "Energy behaviours as promoters of energy efficiency: A 21st century review", *Renewable and Sustainable Energy Reviews,* 16 (6), 4095–104.

Majcen D., Itard L. C. M., & Visscher H. (2013) "Theoretical vs. actual energy consumption of labelled dwellings in the Netherlands: Discrepancies and policy implications", *Energy policy*, 54, 125–36.

McLoughlin F., Duffy A., & Conlon M. (2012) "Characterising domestic electricity consumption patterns by dwelling and occupant socio-economic variables: An Irish case study", *Energy and Buildings,* 48, 240–48.

MEEM. (2016) *Chiffres clés de l'énergie*, Paris: Service de l'Observation et des Statistiques (SOeS) Édition.

MEEM. (2017) Indicateurs et indices. Chiffres clés de l'environnement. Partie 2 Gestion et utilisation des ressources naturelles. http://www.statistiques.developpement-durable.gouv.fr/indicateurs-indices/li/partie-2-gestion-utilisation-rcssources-naturelles.html.

Morgan M. S. & Morrison, M. (1999) "Models as mediating instruments" in Morgan M. & M. Morrison (dir.) *Models as Mediators: Perspectives on Natural and Social Science, Ideas in Context* (Vol. 52), Cambridge: Cambridge University Press, 10–37.

Morrison B. M. & Gladhart P. M. (1976) "Energy and families: The crisis and the response", *Journal of Home Economics*, 68, 15–18.

Nesbakken R. (1999) "Price sensitivity of residential energy consumption in Norway", *Energy Economics*, 21(6), 493–515.

O'Neill B. C. & Chen B. S. (2002) "Demographic determinants of household energy use in the United States", *Population and Development Review*, 28, 53–88.

Orea L., Llorca M., & Filippini M. (2015) "A new approach to measuring the rebound effect associated to energy efficiency improvements: An application to the US residential energy demand", *Energy Economics*, 49, 599–609.

Ouyang J. & Hokao K. (2009) "Energy-saving potential by improving occupants behavior in urban residential sector in Hangzhou City, China", *Energy and Buildings*, 41(7), 711–720.

Page J. (2007) Simulating occupant presence and behaviour in buildings, PhD thesis at l'École Polytechnique Fédérale de Lausanne.

Pellegrino M. (2013) "La consommation énergétique à Calcutta (Inde): du confort thermique aux statuts sociaux", *VertigO - la revue électronique en sciences de l'environnement*, 13(1) April, URL: http://vertigo.revues.org/13395 ; DOI: 10.4000/ vertigo.13395.

Pellegrino M. & Musy M. (2017) "Seven questions around interdisciplinarity in energy research", *Energy Research & Social Science*, 32, 1–12.

Peuportier B., Thiers S., & Guiavarch A. (2013) "Eco-design of buildings using thermal simulation and life cycle assessment", *Journal of Cleaner Production*, Elsevier, 39, 73–78.

Polizzi di Sorrentino E., Woelbert E., & Sala S. (2016) "Consumers and their behavior: state of the art in behavioral science supporting use phase modeling in LCA and ecodesign", *International Journal of Life Cycle Assess*, 21, 237–251, DOI 10.1007/s11367-015-1016-2.

Risch A. & Salmon C. (2017) "What matters in residential energy consumption: evidence from France", *International Journal of Global Energy Issues*, 40 (1–2), 79–115.

Santamouris M., Kapsis K., Korres D., Livada I., Pavlou C., & Assimakopoulos M. N. (2007) "On the relation between the energy and social characteristics of the residential sector", *Energy and Buildings*, 39 (8), 893–905.

Sergi F. (2014) Quelle méthodologie pour une étude des modèles DSGE? Suggestions à partir d'un état des lieux des recherches sur la modélisation, *Work papers from the Centre d'Economie de la Sorbonne*, 67.

Schipper L. & Grubb M. (2000) "On the rebound? Feedback between energy intensities and energy uses in IEA countries", *Energy policy*, 28(6), 367–88.

Schoetter R., Masson V., Bourgeois A., Pellegrino M., & Lévy J.-P. (2017) "Parametrisation of the variety of human behavior related to building energy consumption in the Town Energy Balance (SURFEX-TEB v.8.2)", *Geoscientific Model Development*, (10), https://doi.org/10.5194/gmd-10-2801-2017.

Shove E. (2003) "Converging conventions of comfort, cleanliness and convenience", *Journal of Consumer Policy*, 26 (4), 395–418.

Shove E., Pantzar M., & Watson M. (2012) *The Dynamics of Social Practice. Everyday Life and How it Changes*, London: Sage.

Sovacool B. K. (2014) "What are we doing here? Analyzing fifteen years of energy scholarship and proposing a social science research agenda", *Energy Research & Social Science*, 1, 1–29.

Subremon H. (2012) "Pour une intelligence énergétique: ou comment se libérer de l'emprise de la technique sur les usages du logement", *Métropolitiques*, 7 November. URL: https://www.metropolitiques.eu/Pour-une-intelligence-energetique.html.

Swan L. G. & Ugursal V. I. (2009) "Modeling of end-use energy consumption in the residential sector: A review of modeling techniques", *Renewable and sustainable energy reviews*, 13(8), 1819–35.

Tonn B. E., White D. L. (1988) "Residential electricity use, wood use, and indoor temperature; An econometric model", *Energy Systems and Policy*, 12(3), 151–65.

Tu Y. (2018) "Urban debates for climate change after the Kyoto Protocol", *Urban Studies*, 55(1), 3–18.

Vaage K. (2000) "Heating technology and energy use: A discrete/continuous choice approach to Norwegian household", *Energy Economics*, 22(6), 649–69.

Valenzuela C., Valencia A., White S., Jordan J. A., Cano S., Keating J., Nagorski J., & Potter L. B. (2014) "An analysis of monthly household energy consumption among single-family residences in Texas, 2010", *Energy Policy*, 69, 263–72.

Varenne F. (2010) "Les simulations computationnelles dans les sciences sociales. Nouvelles perspectives en sciences sociales (Online)", *Prise de parole (Ontario, Canada)*, Special issue on the topic of simulation, 5(2), 17–49. http://www.erudit.org/revue/npss/2010/v5/n2/index.html. 10.7202/044073ar.

Vorger E. (2014) *Étude de l'influence du comportement des habitants sur la performance énergétique du bâtiment*, PhD thesis in Energy and Process, Paris: École Nationale Supérieure des Mines de Paris.

Wirl F. (1997) "Economic analysis of energy conservation." In Wirl F. *The Economics of Conservation Programs*. Boston, MA: Springer, 13–59.

Yohanis Y. G. (2012) "Domestic energy use and householders energy behaviour", *Energy Policy*, 41, 654–65002E.

Chapter 9

Reconstruction of Building Occupation by Machine Learning

Eric Vorger and Maxime Robillart

Kocliko, Bordeaux, France

9.1 Introduction

The lack of knowledge regarding building uses currently constitutes a scientific barrier when it comes to predicting the energy consumption and comfort of buildings. Indeed, users have a strong and complex influence on the energy consumption and comfort of buildings (through their presence and activities, the management of heating instructions, opening/closing windows, managing shading devices, lighting, the use of artificial lighting and electrical appliances and pumping domestic hot water). The short-term predictions used for the real-time control (around 24 hours) – and also the medium-term predictions (one year) that are useful for sizing renovation operations – are tainted by very significant uncertainties (often above 50%) due to this lack of knowledge regarding uses.

One promising method for identifying uses in the existing buildings is to set up a measurement protocol associated with machine learning methods. Indeed, the development of these methods is favoured by the rise of Internet of Things (IoT), which provides metering and measurement data at controlled costs, as well as the emergence of platforms for connected buildings.

9.2 Latest Developments in Occupancy Modelling

9.2.1 Stochastic Models Representative of the Population

In recent years, the scientific community working in the field of building energy simulation has developed many models in order to replace the conventional occupancy scenarios used as inputs for simulations. Particular efforts have been made with stochastic models that make it possible to account for the diversity of human behaviour and its random nature.

For example, stochastic models have been developed and implemented in the Pleiades software suite (Vorger 2014). Thus, Pleiades now offers a refined modelling of behaviours for residential and office buildings *via* the Amapola extension.

The construction of models draws on a large number of data taken from measurement campaigns, sociological surveys (e.g. the population census, the time-use survey or the household equipment survey by INSEE) and scientific literature.

Stochastic models make it possible to traverse the field of possibilities virtually by associating them with a probability. They are calibrated on statistical data representative of the population, with a possibility of refinement based on contextual elements like location (e.g. the number of inhabitants in a T2 follows a different probability in Paris than in a small French town). This contextualisation is limited, however. The occupancy scenarios generated by the models do not represent the "exact" occupancy of the building studied. It is difficult to do better in the design stage of new buildings, or to overcome the lack of information on occupancy in the case of existing buildings. In many situations, however, uncertainty regarding occupancy could be reduced through measurement.

9.2.2 Integration of Information Measured In Situ

The presence of users is the aspect of occupancy whose evaluation has attracted the most interest in the industry, as well as in the academic world. It has been found that questionnaires are time-consuming and unreliable, motion detectors have a limited detection radius, they do not provide information on the number of occupants and disregard people with low mobility (office work, reading, television, sleep, etc.), and geolocation *via* smartphones is insufficiently precise and intrusive, and also significantly increases battery consumption.

Thanks to the rise of IoT, work is being undertaken to simultaneously exploit data from different sources (questionnaires, expert knowledge and above all measurements of temperature, CO_2, humidity, electricity consumption, motion detectors, etc.) *via* models built by machine learning (Amayri et al. 2016a). This promising work is based on restricted samples, which hinders their validation and generalisation. It is therefore interesting to pursue them on the basis of larger measurement campaigns. In addition, similar methods can be used for other aspects of occupancy, such as the use of thermostats, windows or concealment devices.

9.2.3 Occupancy Identification by Machine Learning

Machine learning refers to a set of methods and algorithms for extracting information from the data or learning a behaviour by observing a phenomenon. It is divided into two main branches: supervised learning and unsupervised learning. In the supervised case, the objective is to determine a new output Y from a new input X, given a learning database $\{(X_1, Y_1),..., (X_n, Y_n)\}$. We are talking about a classification problem when the Y_i take discrete values, and a regression problem when the Y_i are at real values. In unsupervised learning, there is no output, Y. The automatic learning algorithm only applies in this case to find the similarities and distinctions within the observations $(X_1,..., X_n)$ and to build a model to represent them better.

In the context of applying machine learning methods to estimate occupancy, the input data X are conventionally constituted by numerous observations of different predictors (or features) such as temperature, sound or CO_2 concentration. Y outputs correspond to occupancy or occupancy levels.

9.2.3.1 Supervised Learning Methods

In the literature, many supervised learning methods have been used to estimate occupancy in buildings. For example, to detect the number of occupants, Yang et al.

(2012) developed a model using a set of measurements, including temperature, humidity, CO_2, brightness, sound and motion. The model was based on a neural network and had an accuracy of 88% in relation to the case study and 65% when the model was applied to another room. Similar studies have also used neural networks with an accuracy ranging from 75% to 84.5% (Ekwevugbe et al. 2013).

Decision trees were also used to detect occupancy in real time (Hailemariam et al. 2011). Using measurements of brightness, sound, CO_2, motion and power consumption, accuracy of 97.9% is forecast.

More recently, Candanedo and Feldheim (2016) evaluated three classification methods (linear discriminant analysis, decision trees and random forests) by considering measurements of brightness, temperature and CO_2 concentration. The results obtained (between 95% and 99% accuracy) showed that the accuracy of the classification models strongly depended on the sensors considered and their association.

Amayri et al. (2016a) proposed a general approach for determining which sensors to use to estimate the number of occupants in an office. Measurements included motion detection, energy consumption, CO_2 concentration, noise, and the position of doors and windows. In this study, the authors only considered decision trees as they have the advantage of providing human-readable decision rules, corresponding to rules of the "if-then-else" type and where the thresholds can be adjusted according to the living spaces considered.

The main difficulty in using supervised learning methods is that training the models requires labelled training data, that is data for which the true classification is known. Thus, in order to train a model to predict the "occupied/unoccupied" state of a room, this state must be provided to all the training data in advance, and the same applies for predicting the number of occupants. These training data are generally obtained using cameras, but these approaches raise confidentiality issues. One of the solutions for avoiding the use of cameras is to use unsupervised learning methods.

9.2.3.2 Unsupervised Learning Methods

Many unsupervised learning algorithms have been used in the literature to identify building occupancy. For example, Lam et al. (2009) developed a hidden Markov model (HMM), with 80% accuracy in detecting the number of occupants using CO_2 sensors (indoor and outdoor), acoustic sensors and passive infrared sensors. A similar approach was also adopted by Dong et al. (2010) with an average accuracy of 73%. In this study, it was noted that these models were unable to detect sudden changes in occupancy over short periods.

Electricity consumption data for five houses were also used to detect occupancy (Kleiminger et al. 2013). The authors used HMMs and the k-nearest neighbours algorithm, and obtained an average accuracy of more than 80%. For example, the accuracy of HMMs varied from 81% to 87% for summer and from 70% to 87% for winter.

An auto-regressive HMM or ARHMM was introduced by Ai et al. (2014) and compared to HMM to identify the number of occupants in a room (number between 0 and 6). Many sensors were used in this study (passive infrared sensor, CO_2, temperature, relative humidity, air velocity). In the case of an uncommon variation in occupancy, ARHMM and HMM showed an accuracy of 84% and 76.2%, respectively.

A general approach has also been proposed by Amayri et al. (2016b) to determine occupancy in a room-based sensor data and knowledge from both observations and questionnaires. The model was based on a Bayesian network and used data relating to motion detection, sound (sound pressure) and energy consumption. This approach has been applied to an office with an average accuracy of 90%.

HMMs can also be used to establish occupancy profiles from the measurements made. This method has been used by Candanedo et al. (2017) using temperature and humidity measurements, for example. Similarly, D'Oca and Hong (2014) proposed a learning method for identifying occupancy and use scenarios for an office building (16 offices). This method thus provided information on the occupancy profiles of the building.

9.3 Case Study

9.3.1 Description of the Case Study

An office of the Centre efficacité Energétique des Systèmes (CES, Energy Efficiency Systems Centre) of MINES ParisTech was instrumented to test different machine learning algorithms. The office has five workstations and can frequently accommodate visitors. A sensor network has been installed, including:

- A camera connected to a Raspberry Pi to record occupancy and activities (manual labelling of occupancy data);
- Connected WeMo Insight outlets to evaluate electricity consumption;
- A Netatmo station and a GreenMe cube allowing continuous measurement of temperature, humidity, lighting, noise and air exchange (CO_2 concentration).

A measurement campaign was conducted from September to November 2017. These data were subsequently processed (manual labelling of the occupancy) and averaged over a period of 30 minutes to take the non-instantaneous evolution of the CO_2 concentration into account (Amayri et al. 2016a).

The data thus obtained made it possible to evaluate the prediction capacities of the different machine learning algorithms on two application cases, namely to predict the periods of occupancy (absence or presence) of the instrumented office, on the one hand, and to estimate the number of people present, on the other hand. Supervised or unsupervised learning techniques have been tested on these classification problems.

9.3.2 Evaluation Criteria

To measure the quality of a classification system, a confusion matrix is conventionally used. Table 9.1 thus presents a confusion matrix in the case of a binary classification. Each row of the matrix represents the number of occurrences of an estimated class, while each column represents the number of occurrences of a real (or reference) class. The interest of the confusion matrix lies in the fact that it can show the performances of the classification system studied quickly.

Several performance criteria can be derived from the confusion matrix. In the case of a binary classification, for example, mention may be made of the recall corresponding

Table 9.1 Confusion Matrix in the Case of a Binary Classification

		Actual Classification	
		Unoccupied (−)	Occupied (+)
Predicted class	Unoccupied (−)	True negatives (TN)	False negatives (FN)
	Occupied (+)	False positives (FP)	True positives (TP)

to the proportion of positives that the system has correctly identified (rate of true positives (TP)), that is to say, the model's capacity to detect all occupancy periods.

$$\text{Recall} = \frac{TP}{TP + FN}. \qquad (9.1)$$

There can also be interest in the precision, that is the proportion of correct predictions among the points predicted as positive (corresponding to the office's occupancy periods).

$$\text{Precision} = \frac{TP}{TP + FP}. \qquad (9.2)$$

Finally, we also often look at specificity, which is the rate of true negatives (TN), that is the ability to detect all the offices' vacancy periods. This is a complementary measure to the recall.

$$\text{Specificity} = \frac{TN}{TN + FP}. \qquad (9.3)$$

These indicators are easily generalised in cases where several classes are to be predicted. It is then sufficient to gather the un-analysed classes together in a single class.

9.4 Results

9.4.1 Supervised Learning

Following the latest developments, learning by decision tree appears to be the technique conventionally used to estimate occupancy. However, this method presents risks of overfitting. An alternative is to use the random forest to limit these risks.

Random forests are a set of decision trees generated individually on slightly different subsets of data.

The learning process for a random forest is as follows (Figure 9.1):

- Creation of new learning sets by a double sampling process;
  - Selection of the learning sample by bootstrap (corresponding to a random sampling with replacement with the same number of observations as the original data);

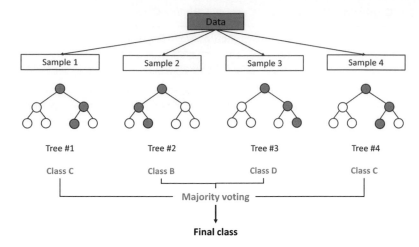

Figure 9.1 Illustration of how a random forest works.

- Random selection of p predictors (usually $p \ll P$) from P predictors of input data;
- Generation of decision trees from each training set created.

Subsequently, the prediction of a new value consists of carrying out a classification for each tree and choosing the value with the most votes among all the trees (majority class).

9.4.2 Training and Test Data

In supervised learning, it is necessary to divide the measured data into two subsets: the training/validation data and the test data. The learning model (in this case, the random forest) is generated based on the first data subset. During this step, the internal parameters (hyperparameters) are adjusted (e.g. the depth of the decision trees). Cross-validation with three folds was considered to avoid any risk of overfitting.

Cross-validation consists of splitting the dataset into k parts (folds). In turn, each of the k parts is used as a test game. The rest (the union of $k-1$ other parts) is used for training. Figure 9.2 illustrates how the method works for fivefold cross-validation.

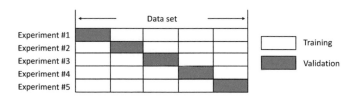

Figure 9.2 Illustration of how fivefold cross-validation works.

In the event of a classification problem, an attempt is made to create *k* folds so that they contain roughly the same proportions of examples of each class as the complete dataset.

Finally, the performance of the learning model (generalisation error) is evaluated on the test data. The measurement periods considered in this study are summarised in Table 9.2.

9.4.2.1 Estimation of Occupancy Periods (Binary Classification)

Figure 9.3 shows the performance obtained by the random forest on the test data, where the real occupancy profile is plotted in blue and the estimated profile is shown in the green dotted line. It can be seen that the algorithm is able to accurately reproduce (96%) the real occupation profile. It also shows a 96% recall, meaning that 96% of the occupancy periods were detected (Table 9.3). Similarly, 94% of the vacancy periods were identified (specificity). However, we note that the algorithm is unable to recognise rapid changes in the occupancy and has a slight latency during transitions.

Table 9.2 Training and Test Data

	Dates of Measurements
Training data	From 15/09/2017 to 19/10/2017
Test data	From 21/10/2017 to 12/11/2017

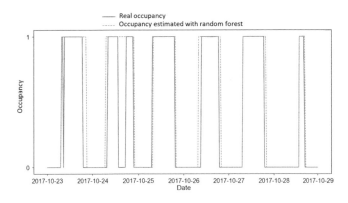

Figure 9.3 Occupancy estimation (binary classification) by the random forest (test data from 23 to 29 October 2017).

Table 9.3 Confusion Matrix (Binary Classification) on Test Data (Random Forest)

		Real Class	
		Unoccupied	Occupied
Predicted class	Unoccupied	726	2
	Occupied	43	271

The analysis of the various decision trees constituting the forest shows that electricity consumption, noise level and CO_2 concentration are the most used and therefore most relevant predictors for identifying and predicting occupancy in offices. This finding is in agreement with the results obtained by Amayri et al. (2016a).

9.4.3 Estimated Number of People Present

It is also interesting to be able to test the capabilities of the random forest to identify and predict occupancy levels in the office. However, since some occupancy levels have low occurrences (e.g. six occupants), the algorithm will have difficulty in being able to identify and predict all of them accurately. It is therefore essential to gather different levels of occupancy to improve the performance of the algorithm.

In this case study, three occupancy levels were considered, with the same training and test data:

- Level 0: absence;
- Level 1: occupancy between one and three people;
- Level 2: occupancy strictly greater than three people.

Figure 9.4 shows the performances obtained by the algorithm on the test data (the real occupation profile is given in blue and the estimated profile in the green dotted line). Note that the performance of the algorithm has deteriorated. Although it has an overall accuracy of 93%, the performance of the algorithm is very disparate depending on the occupancy levels to be predicted (Table 9.4). Thus, for occupation levels 0 and 1, it has 99% and 78% accuracy, respectively. By contrast, only 44% accuracy was achieved for level 2 occupancy. This poor performance is explained by the low occurrence of this level of occupancy in the training data (28 occurrences). The algorithm then has difficulty identifying the relevant rules to predict it.

The random forest offers very good performance for predicting occupancy (binary classification) and, to a lesser extent, for estimating the people present. However, this

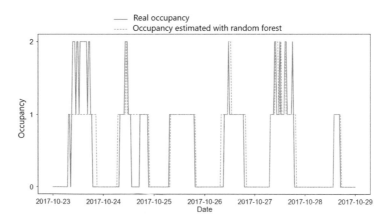

Figure 9.4 Occupancy estimation by the random forest (test data from 23 to 29 October 2017).

Table 9.4 Confusion Matrix on Test Data (Random Forest)

		Real Class		
		Class 0	Class 1	Class 2
Predicted class	Class 0	726	2	0
	Class 1	43	238	24
	Class 2	0	5	4

technique needs training data that require the use of invasive devices (e.g. cameras) to label the data beforehand. To overcome this limitation, unsupervised learning methods may be considered, like HMMs.

9.4.4 Unsupervised Learning

HMM is a statistical model in which the modelled system is assumed to be a Markovian process (Markov chain) with unknown parameters.

Traditionally, Markov chains are used to characterise transition probabilities between different states (e.g. absence and presence in our case study). These transition probabilities thus materialise the possibility of passing from one state to another (e.g. absence => presence).

Unlike classical Markov chains, where transitions are unknown but where the process states are known, in HMM the process states are also unknown. We cannot observe the sequence of states directly: the states are hidden. However, each state emits "observations" that are observable. The work is therefore not carried out directly on the sequence of states but on the sequence of observations generated by the states (Figure 9.5).

Although HMMs do not require training data, only the test data were considered in this study to evaluate its performance and to be able to compare it with that obtained by the random forest.

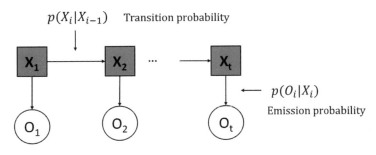

$$p(X_i|X_{i-1}) \quad \text{Transition probability}$$

$$p(O_i|X_i) \quad \text{Emission probability}$$

X_i : *hidden state at time i*
O_i : *observations at time i*

Figure 9.5 Illustration of how HMM works.

9.4.5 Estimation of Occupancy Periods (Binary Classification)

In order to reduce the size of our dataset (dimension equal to the number of predictors), a principal component analysis was performed. The purpose of this unsupervised method is to find a new orthonormal basis (principal components) on which to represent our data so that the variance of the data along these new axes is maximised. Classically, to reduce the size of the data, we focus on the proportion of variance explained by each of the principle components (ranked in descending order). The number of principle components is then chosen to explain a percentage (set by the user) of the variance. In this study, the percentage was set at 80%.

Figure 9.6 shows the performance obtained by HMM on the test data, where the real occupancy profile is plotted in blue and the estimated profile is shown in green dotted lines. The performance is substantially equivalent to that obtained by the random forest, with 96% overall accuracy, 96% recall and 95% specificity (Table 9.5). Similarly, HMM is unable to identify rapid changes in occupancy.

HMMs therefore show very good performance for the estimation of occupancy periods (binary classification) and have the advantage of not requiring training data to function.

9.4.6 Estimated Number of People Present

As with random forests, tests were also carried out on the HMM's ability to predict and identify occupancy levels. Figure 9.7 shows the performances obtained by

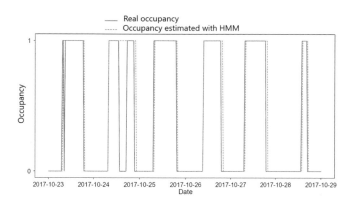

Figure 9.6 Occupancy estimation (binary classification) by HMM (test data from 23 to 29 October 2017).

Table 9.5 Confusion Matrix (Binary Classification) on Test Data (Hidden Markov Model)

		Real Class	
		Unoccupied	Occupied
Predicted class	Unoccupied	732	9
	Occupied	37	264

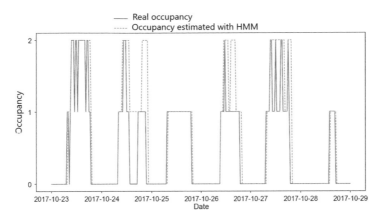

Figure 9.7 Occupancy estimation by HMM (test data from 23 to 29 October 2017).

Table 9.6 Confusion Matrix on Test Data (Hidden Markov Model)

		Real class		
		Class 0	Class 1	Class 2
Predicted class	Class 0	732	9	0
	Class 1	28	147	0
	Class 2	9	89	28

the algorithm on the test data (the real occupation profile is shown in blue, and the estimated profile in the green dotted line). HMM has an equivalent performance to random forests with very good overall accuracy (93%), but uneven performance depending on the occupancy levels to be predicted (Table 9.6). Thus, for occupancy levels 0 and 1, the algorithm has a precision of 99% and 84%, respectively, whereas it is only 22% (with 100% recall) for level 2 occupancy. As with the random forest, HMM is unable to identify reliable emission and transition probabilities to predict occupancy level 2 given the low occurrence of this level in the data.

9.5 Conclusion

This study has shown the ability of machine learning algorithms to identify and reconstruct occupancy from measurements. Supervised (random forest) and non-supervised (HMM) learning methods have been tested and show similar performance. For the prediction of the periods of occupation (binary classification), these algorithms obtain very good performances with an accuracy of 96% in both cases. Both algorithms show a deterioration in performance in terms of the prediction of different occupancy levels, with great difficulty in identifying high occupancy levels (occupancy greater than three people) due to the low occurrence of this situation in the dataset.

The labelling of learning data is time-consuming and requires expensive and intrusive systems like cameras. It is therefore interesting to privilege methods of unsupervised learning, but these require additional knowledge to identify classes *a posteriori*. Hybrid methods allowing reduced labelling could provide an appropriate response. These methods consist of labelling the data only when they provide a significant added value to the model. This method, which was initiated by Amayri et al. (2016c), relies on interaction with the user to ask a question (*via* a mobile application or a web interface, for example) when the response would provide information that could significantly improve the quality of the model. According to this method, it would be possible to improve the prediction capabilities of the algorithms on the occupancy levels with a low occurrence.

Bibliography

Ai, B., Z. Fan, and R. X. Gao. 2014. "Occupancy estimation for smart buildings by an autoregressive hidden Markov model". In *2014 American Control Conference*, 2234–39. https://doi.org/10.1109/ACC.2014.6859372.

Amayri, M., A. Arora, S. Ploix, S. Bandhyopadyay, Q. D. Ngo, and V. R. Badarla. 2016a. "Estimating occupancy in heterogeneous sensor environment". *Energy and Buildings* 129 (October): 46–58. https://doi.org/10.1016/j.enbuild.2016.07.026.

Amayri M., Q. D. Ngo, and S. Ploix, 2016b. "Estimating occupancy from measurement and knowledge with Bayesian Networks". *International Conference on Computational Science and Computational Intelligence (CSCI)*, Las Vegas, NV, 508–13. doi: 10.1109/CSCI.2016.0102.

Amayri M., S. Ploix, P. Reignier, and S. Bandyopadhyav, 2016c. "Towards interactive learning for occupancy estimation". *ICAI'6 - International Conference on Artificial Intelligence (as part of WORLDCOMP'16 – World Congress in Computer Science, Computer Engineering and Applied Computing)*, Las Vegas, NV.

Candanedo, L.M. and V. Feldheim. 2016. "Accurate occupancy detection of an office room from light, temperature, humidity and CO_2 measurements using statistical learning models". *Energy and Buildings* 112 (January): 28–39. https://doi. org/10.1016/j.enbuild.2015.11.071.

Candanedo, L.M., V. Feldheim, and D. Deramaix. 2017. "A methodology based on Hidden Markov Models for occupancy detection and a case study in a low energy residential building". *Energy and Buildings* 148 (August): 327–41. https://doi.org/10.1016/j.enbuild.2017.05.031.

D'Oca, S. and T. Hong. 2014. "A data-mining approach to discover patterns of window opening and closing behavior in offices". *Building and Environment* 82 (December): 726–39. https://doi.org/10.1016/j.buildenv.2014.10.021.

Dong, B., B. Andrews, K. P. Lam, M. Höynck, R. Zhang, Y.-S. Chiou, and D. Benitez. 2010. "An information technology enabled sustainability test-bed (ITEST) for occupancy detection through an environmental sensing network". *Energy and Buildings* 42 (7): 1038–46.

Ekwevugbe, T., N. Brown, V. Pakka, and D. Fan. 2013. "Real-time building occupancy sensing using neural-network based sensor network". In *2013 7th IEEE International Conference on Digital Ecosystems and Technologies (DEST)*, 114–19. https://doi.org/10.1109/DEST.2013.6611339.

Hailemariam, E., R. Goldstein, R. Attar, and A. Khan. 2011. "Real-time occupancy detection using decision trees with multiple sensor types". In *Proceedings of the 2011 Symposium on Simulation for Architecture and Urban Design*, 141–8. Society for Computer Simulation International.

Kleiminger, W., C. Beckel, T. Staake, and S. Santini. 2013. "Occupancy detection from electricity consumption data". In *Proceedings of the 5th ACM Workshop on Embedded Systems for Energy-Efficient Buildings*, 10:1–10:8. BuildSys'13. New York, NY: ACM. https://doi.org/10.1145/2528282.2528295.

Lam, K.P., M. Höynck, B. Dong, B. Andrews, Y.-S. Chiou, R. Zhang, D. Benitez, and J. Choi. 2009. "Occupancy detection through an extensive environmental sensor network in an open-plan office building". *IBPSA Building Simulation* 145: 1452–59.

Vorger, E. 2014. "Étude de l'influence du comportement des habitants sur la performance énergétique du bâtiment". Ph.D. thesis, École Nationale Supérieure des Mines de Paris, 474 p.

Yang, Z., N. Li, B. Becerik-Gerber, and M. Orosz. 2012. "A multi-sensor based occupancy estimation model for supporting demand driven HVAC operations". In *Proceedings of the 2012 Symposium on Simulation for Architecture and Urban Design*, 2:1–2:8. SimAUD'12. San Diego, CA: Society for Computer Simulation International. http://dl.acm.org/citation.cfm?id=2339453.2339455.

Chapter 10

What Future for Nature in the City?

Isabelle Richard

ENVIRONNONS

10.1 Introduction

10.1.1 Major Issues Concerning Nature in the City and Current Problems

The growth of the urban population is obvious: more than 50% of the world population resides in cities, whereas in France 3/4 of the population is now urban. This will lead to even greater environmental and social issues in future. These different issues lead us to define challenges for making urban life liveable, a source of well-being and able to cope with climate change.

In a time of climate change, nature in the city offers obvious and significant ecosystemic services for responding to environmental challenges. Indeed, it not only improves air quality but also has a role in reducing heat islands and in the regulation of rainwater. In this way, real ecosystems are recreated within cities. In addition, nature in the city affects the sensation of well-being experienced by the users of the city. Some studies even show that it has therapeutic effects (Ulrich, 1984). Urban settings can be sites of concurrent sources of environmental stress for city dwellers. The presence of nature in these same settings therefore has a resource function for urban inhabitants if they have access to nature nearby that enables this pacification. Finally, nature and biodiversity in the city also offer the possibility of reconnection, for several reasons. It allows reconnection to others through the social bond it promotes (especially in the context of shared gardens and all the city greening activities carried out by urban collectives more broadly), the reconnection to nature itself through contact with and experience of it and, finally, reconnection to the self, in other words, reconnection to life and its functioning in the broad sense. All these functions obviously respond to the challenges of environmental health and adaptation to climate change.

Nature can also have economic functions. Firstly, in the context of urban agriculture, it can generate an economic model, jobs and profits but also budget savings in the context of the reduction of health problems, which can sometimes be very expensive for a society.

Nature in the city is a concept that can be interpreted in very different ways in terms of both its creations and its representations. The literature on this subject tells us that there is a gap between what urban trends offer to citizens (such as natural gardens in which vegetation grows without much anthropic intervention) and the representation of horticultural gardens, which are still very much rooted and very popular with city dwellers. This discrepancy can also be seen when we compare the knowledge of developers and city dwellers on certain terms such as notions of the green and blue belt ("trame verte et bleue") or biodiversity. Indeed, while these terms are part of the current vocabulary of developers, they are often unknown or even understood differently among the general public (Cormier et al., 2012).

However, as previously seen, the environment, and more precisely the landscape, has a definite impact on the quality of life and well-being of an urban population. As such, Hinds and Sparks (2009, 2011) study the relationships between the natural environment, identity, emotional well-being and its meaning. These authors show that individuals who grew up in a rural area would have a stronger environmental identity, a more developed environmental experience and a greater sense of well-being than people who grew up in an urban or suburban setting.

City dwellers sense this need, which is now growing and widely expressed. As such, White and Gatersleben (2011) conducted two studies to measure the impact of a green roof on the perception of a building. The authors show that green buildings are more highly appreciated. They are considered more aesthetic, more relaxing and have more affective quality compared to buildings without vegetation. These results are consistent with other research and suggest integrating vegetation with construction could be a solution for preserving nature within urban settings. Long and Tonini (2012), for their part, refer back to these observations and show that the greening of the urban landscape is due in particular to the demand for nature in the city on the part of citizens. According to the authors, there are three main motivations for city dwellers to frequent a green space: (i) to escape, hide, forget the city and relax, (ii) to frequent a sociable space, and (iii) to use places for leisure and sports practices.

The desire for nature is therefore important to city dwellers, and Bourdeau-Lepage (2013) uses the term "homo qualitus" to define the person who not only seeks to satisfy their material and immaterial well-being, but also their desire for nature and the preservation of its environment as an element of their well-being.

The type of nature proposed in cities today tends to be diverse, and this renaturation ranges from a logic of creating greenery (green spaces, green walls), in which city dwellers are simple spectators, to a logic of involving residents through the advent and roll-out of urban agriculture and shared gardens, among other things. We thus see expressions appearing such as "local food network", "urban agriculture", "nature in the city", "greening" and "green and blue belt" in the discourse concerning the city and its management. The development of urban agriculture is taking place in local and resident-driven initiatives, as well as in larger-scale political initiatives. There is also a growing enthusiasm for these new forms of consumption among city dwellers.

However, despite this quest for urban nature, city dwellers appear not to see the potential "disadvantages" of nature in the city, and the longing for nature is supposedly driven more by a desire for aesthetics or comfort in urban forms. Nature as it is idealised and represented in the collective consciousness does not always seem to coincide with the nature encountered in everyday reality. Renauld's research (2012) reflects this well and shows that, after initial enthusiasm about having some greenery on their balconies, residents perceive certain inconveniences related to this type of installation (the presence of insects, for example) and prefer to cut off the water that feeds the plants or to use pesticides. The author concludes that greening would be accepted provided that it does not encroach on the private area. This corroborates the findings of Long and Tonini's research (2012), according to which the lack of vegetation management by inhabitants is rejected, with spontaneous vegetation being seen as a failure on the part of the city's technical services. This goes against the rise of the natural garden model, in which human control is minimal. Finally, in relation to this question, Blanc (2004) describes the fact that the negative perception of animal species can often be related to the proliferation of these species or to their harmful and/or invasive nature in certain cases.

What is this desire to have nature in the city? Do city dwellers have a unified relationship to this nature in the city? To what extent do proposals for nature in the city such as those offered in new urban developments (green walls and roofs, vegetable gardens, green urban furniture, street pots, etc.) respond to urban demand? Do they influence or inhibit this demand? What are the constraints hindering integration of nature in the city today? What is the perception of nature in the city of tomorrow according to experts and citizens?

We propose to give answers to these questions in this chapter by questioning both experts and users on this topic.

10.1.2 The ReCreA Green Vil' Project – Objectives and Methodologies

The ReCreA Green vil' project (representation, creation and appropriation of nature in the city) has been funded within the framework of the "Eco-design of Buildings and Infrastructure" Chair of Entrepreneurship, which is linked to the VINCI group and three engineering schools (MINES ParisTech, Ponts ParisTech and AgroParisTech).

This project aimed to gather representations, expectations and difficulties around the question of integrating nature in the city, both from a sample of experts and from also users. Through different methodologies, we tried to understand the factors that facilitate or inhibit the appropriation of nature in the city.

To build our reflection, we first carried out a series of semi-structured interviews with experts involved in the integration of nature in the city. These interviews consisted of establishing feedback for existing projects integrating nature, to identify levers and obstacles relating to these integrations. We carried out 23 interviews in total, with the sample including experts from different fields (project leaders, scientific and technical experts, local actors).

The results of stage 1 inspired us to create stage 2, which consisted of developing a comprehensive workshop around the concept of nature in the city with users/inhabitants. The idea was to make a list, to look into the relationship to meaning in nature, but also to bring up project evaluations and imagine future steps for integration. The comprehensive workshop provides better understanding of the dynamics involved in the representations of the theme. The idea was to take inspiration from the results of this phase to reflect on new lines of research. A total of 10 inhabitants participated in this first comprehensive workshop.

The results of stage 2 allowed us to build stage 3, which proposed gathering experts/developers and city dwellers together in a creative workshop to discuss and imagine solutions on the question of integrating nature in the city. The creative workshop invited participants to imagine new ways of integrating nature in the city. In total, 12 inhabitants and five experts took part in this third stage.

10.1.3 Contribution of the Workshops to the Prospective Approach

We thought it would be interesting to develop the two workshops described above as part of a prospective approach. Indeed, not only did they allow us to shed light on the topic through different perspectives, they also enabled a dynamic exchange that proved highly pertinent for outlining the contours of nature in the city of tomorrow.

Moreover, we found it interesting to think about research on nature in the city through the lens of participatory sciences for three essential reasons: (i) to produce knowledge that meets the needs of the inhabitants while guaranteeing a scientific interest; (ii) to establish a real dialogue between two worlds that are often separated, namely the scientific world and society; and (iii) to offer inhabitants a form of empowerment by shifting from the status of users of the city to that of stakeholders.

Finally, in human sciences, collective workshops are often mobilised as a methodological scientific approach and are more commonly called "focus groups". However, these methodologies remain very focused on traditional types of questioning, and often favour stereotypical responses from the participants. More active methods, like facilitation (proposed in our workshops), make it possible to go beyond these issues and to obtain more original and more detailed results on the desired theme. Our work is therefore innovative from this point of view, in that it makes it possible to test the contribution of new methodologies in the scientific field.

10.2 What Are the Levers and Difficulties for Nature in the City Today?

10.2.1 How Do the Experts Perceive Current Projects?

During the interviews carried out in stage 1 of the research, the experts were invited to cite some examples of the integration of nature in the city, which they perceived as either successful or unsuccessful. The exemplification of projects allowed us to identify the main difficulties for their implementation, but also to pinpoint their levers. Thus, we tried to collect data on their perceptions concerning the different spaces mentioned, and at the same time also to come back to the experience of these same spaces of nature in the city.

We analysed the data on this topic and attempted to make a quantified summary in the graphs below. We have expressed the number of occurrences for each item based on the overall number of occurrences provided. We would like to point out that these are not questions that participants were asked to answer, but rather the grouping of terms spontaneously mentioned and related to the levers for integrating nature in the city. The spontaneous nature of the answers explains the low percentages for the different items mentioned.

We counted 71 occurrences regarding difficulties related to integrating nature in the city. Fewer levers were mentioned, with a total of 53. Note the pre-eminence of the difficulties noted in comparison with the levers. This is surprising at first as nature in the city is overwhelmingly supported by city dwellers. However, the desire for nature as it is imagined often comes into conflict with the reality of the devices and constraints associated with nature in the city, making the integration of this nature complex at times (Figure 10.1).

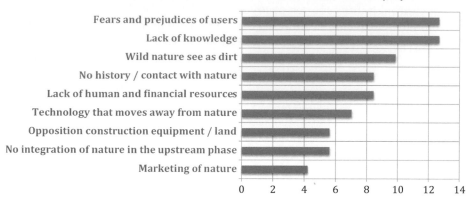

Figure 10.1 Perceived difficulties relating to the integration of nature in the city.

The difficulty that arose most when the participants evoked projects relating to nature in cities is the fear and prejudices of the users. Yet nature is sought after by the majority of city dwellers. In other words, the imagination of nature sometimes comes into conflict with the embarrassing, even anxiety-provoking reality of nature present in the city.

This fear of a desired nature is paradoxical. However, if we delve into the idea of disconnection, we see an initial explanation concerning this "anxious desire". This ambivalent fear/desire of nature can indeed be explained by the city dwellers' lack of experience with nature in the city. Indeed, although cities tend to transform by becoming more "green", the supply of green space remains insufficient. This is due to land reasons because it limits the construction of housing and because it does not offer enough square metres of nature per city dweller as urban spaces become increasingly densely populated.

The demographic growth of cities is not the only cause of this disconnect, however.

There is also the idea of a loss of contact with nature, which would accentuate the appearance of prejudice in relation to it. In psychology, this phenomenon refers to the contact hypothesis (Allport, 1954). This theory, first developed in the United States to reduce ethnic inequalities, states that improving inter-group relations is facilitated when they are brought into contact. However, according to the author, in order for this change of attitude to be operative, it is necessary to meet four conditions:

1. To have sufficient social and institutional support;
2. To have regular and sufficiently long contact that tends towards a certain proximity (what the author calls proximity potential);
3. To have equal status;
4. To be in a position of cooperation with regard to an objective.

If we transpose this to the theme of relationships to nature, we realise that the conditions are not optimal for favouring these positive attitudes (as described in the contact hypothesis) towards nature and for each of the four conditions mentioned above. Indeed, the content of the interviews shows that nature is not yet placed upstream of urban projects; rather, it is only thought at the end. Institutional support is therefore lacking. Moreover, given their way of life, city dwellers are not required to maintain daily relations with a natural setting, meaning the potential for proximity is low. It can also be noted that the economic and societal changes of the "30 glorious years" in France (development of pesticides, the rise of individual cars, globalisation, etc.) did not favour an egalitarian positioning between people and nature, making the latter increasingly vulnerable and relegating it to a lower position than that of humans. They therefore do not have equal status. As a result, cooperation has been and is still difficult today, and relates to Hardin's famous "Tragedy of Commons" (1968), which states that humanity can either privilege its personal interests for immediate profit to the detriment of the planet in the future, or alternatively privilege common interests and future pleasures by renouncing its immediate desires. It is the first option that dominates, rendering cooperation insufficient.

A third explanation for this paradox can be found in the scientific literature relating to climate change. In the populations that are least connected to nature, there is no systemic vision of the climate (Shepardson et al., 2012). In more concrete terms, people who are physically closer to nature have a systemic vision of climate change, while more urban populations have an analytical view centred on one element at a time. This is known as "cognitive vulnerability... which leads to cognitive conflicts, difficulties in understanding, as well as the failure of problem-solving strategies" (Lammel et al., 2012). Indeed, if individuals consider themselves a part of the "environment system", they are more sensitive to their physical and social environment, since they are part of it. They are therefore stakeholders within the system. If they extract themselves from this system environment, they become spectators observing a distant physical world, which they use but of which they are not a part. This feeds the paradoxical fear of a desired nature, as what we do not know does not reassure us. If individuals do not understand how the environment system works, they only perceive a part of the prism of this nature. In other words, they consider nature to be the way they idealise it. However, when faced with a more spontaneous form of nature, they encounter species and spaces that they have not envisaged and which are therefore potentially anxiety-provoking, since they do not know how this nature works, except by the stereotypes that these spaces convey.

Some respondents link this disconnect to the fear that this nature in the city will become commercial and fall under the influence of large economic groups. The fear expressed here is that the management of projects concerning nature in the city will go from being a system made by and for city dwellers (community gardens or the appropriation of wastelands, for example) to a more liberal system within the classical economy (e.g. intensive production of urban agriculture without involving citizens, who are only perceived as consumers).

In addition, some of the respondents come back to the notion of communication and support for users in the process of greening the city. Recalling the misunderstandings of city dwellers with regard to certain management decisions concerning nature in the city (including differentiated management), they refer back to the need to communicate the type of nature being integrated in the city and the explanations for these choices. They also stress the importance of informing and raising public awareness about biodiversity and the very functioning of nature.

According to the experts, this lack of communication and information becomes a source of dissatisfaction and misunderstanding on the part of the inhabitants with regard to the management of green spaces in the city. Indeed, the notion of dirtiness often came up in respondents' discussions, especially when describing city dwellers' perception of differentiated management.

The discussions also brought up terms that contrast the economic pressure of land with city dwellers' need for natural spaces.

The experts also spoke about the levers for integrating nature in the city, the main results of which are presented here (Figure 10.2).

The first lever mentioned by the respondents was that of the acculturation of experts in charge of integrating nature with regard to the functioning of the territory, both from a physical point of view and also in terms of uses. A good project would consist of thinking about and taking into account the endemic species, to leave room for spontaneous growth while proposing a form of nature that is adapted to the uses and desires of the inhabitants.

However, the experts also referenced the question of support and user involvement, which seem to guarantee the sustainability of projects integrating nature in the city.

This support and this involvement should be made possible by city dwellers' own experimentations with nature. In other words, the sensoriality and experience of nature should serve as a gateway for promoting reconnection and therefore also the involvement of the inhabitants in developing and maintaining spaces of nature. Underlying these notions is also the idea of rediscovering nature, its functioning and respect for it.

Figure 10.2 Perceived levers for integrating nature in the city.

One of the ways to improve this reconnection with nature can be found in the narrative and poetic aspect of nature. Indeed, some respondents evoked the idea of reconnecting populations by narrating stories and experiences of nature. The theme of stories about nature was particularly present in the whole body of responses. As previously seen when it was discussed as an obstacle preventing the appropriation of nature in the city, it was mentioned with regard to our lack of memories involving this kind of nature. However, it is also brought up as a lever that could foster reconnection.

The appropriation and integration of nature in the city also seems to depend on a collective desire and shared efforts. Through nature in the city, we can perceive the will for a paradigm shift in the current governance of our societies, which should evolve in the direction of other more community-based and participatory paradigms (Table 10.1).

In total, 43 occurrences referred to projects judged exemplary by the experts. We can note the predominance of urban agriculture projects, which seemed to inspire enthusiasm among experts as they promote sociability, reintegration, circular economy and also participation. Projects that arise from the needs of users and the territory were also put forward. These are projects that promote the involvement of users or are designed and built by them. These projects are often related to the importance of knowledge of the territory and its resources. Finally, urban transformation projects were also cited as projects that are very well appropriated by urban dwellers and often go beyond their original expectations. The most mentioned projects concern those intended to renovate the existing offer of "soft links" or walks. Territorial and bottom-up project types were regularly associated to define what represents a good project for the respondent.

10.2.2 How Do Inhabitants Perceive the Nature Offered in the City?

We wanted to question the inhabitants about their perception of nature in the city as they know it today during a comprehensive workshop. A few days before it took place, the participants were invited to send us some photos that, for them, mobilised the different senses found in nature. We therefore asked them to send us photos of kinds of nature that were pleasant and/or unpleasant to (i) touch, (ii) see, (iii) feel, (iv) hear and (v) taste. Then, during the workshop, all the participants were invited to select the nature photos they liked and explain their choices.

Through this activity, we tried to gain a better understanding of the representations of nature and also relationships to the senses in nature. We made the deliberate choice not to specify what kind of nature we were referring to, and the participants were free to send us photos of "wild" nature or nature in the city. This allowed us to expand the scope of the definition of nature, in order to be able to better return to the subject of nature in the city in a second time. Indeed, we were interested in going beyond the traditionally offered patterns of nature in the city to analyse what elements are now absent from nature in the city as it is designed, but which it would be interesting to reintegrate into the urban space. This allowed us to gather some initial information on how to think about the senses in nature, and also to be able to put these representations up for debate during the "photo-language" (photo-elicitation) session.

The photos centred around a visual notion of nature were selected 13 times, with the idea of a horizon dominating the explanations. The photos referring to smell centred

Table 10.1 Summary of Projects Deemed Positive for the Appropriation of Nature in the City

Types	Urban Agriculture Projects	Bottom-Up Projects	Territory Projects	Urban Transformation Projects
Assessment criteria	Projects advocating reintegration, sociability, circular economy, participation	Projects promoting citizen involvement, leaving free spaces for user investment	Projects that use the territory's resources and advocate diversity of forms and functions	Renovation projects enhancing the existing offer of soft links for walking
Number of occurrences/43	23	7	7	6
Examples cited	AAA Colombes, Agrocité, Recyclerie, Vergers Urbains, Santropol roulant (2), Lufafarm, Rurban, Poteaux Beaux, Toits tout verts, Veni Verdi (4), Topager (2), Community gardens (4), Le paysan urbain, Ferme de Gally, V'île fertile	Saint Brice Courcelle neighbourhood, Vauban neighbourhood in Freiburg, Frontage in England, Noaille neighbourhood in Marseille, Ville de Saintes, German architecture, Charte main verte	École de la biodiversité Boulogne Billancourt (3), Île de Nantes (2), Ville d'Angers, Jardins des plantes in Paris	Reopening of the Bievre Coulée verte (2), Cheonggyecheon in Seoul, Jardins des Etangs Gobert, Promenade du Paillon in Nice

on preferences relating the smells of edible nature and were chosen 12 times. There were 11 occurrences of touch, which mainly refer to the preference for floral elements. With eight occurrences, taste was mostly related to fruits and vegetables. Finally, also with eight occurrences, hearing was mainly associated with preferences for sounds related to the relationship to water and the calming effect of these spaces. Finally, several positively rated photos of nature referred to feelings of astonishment and surprise on the part of walkers.

These first exchanges naturally led the participants to debate the question of the senses within nature in the city. The participants agreed that the senses were not really appealed to in nature in the city, and, when they were, it was not quite in the way they would like in the presence of nature. They evoked the idea of a kind of nature in the city that is suspicious, dirty and not really positive, despite the fact that they like seeing it. Two results thus appear: on the one hand, an absence of sensoriality in nature in the city and, on the other hand, a rather negative relationship to the sensorial when the senses are evoked.

> *First meeting between experts/users*: During the interviews, the experts told us that one of the levers for the appropriation of nature in the city was experience, and putting the user in contact with it. However, the users explained to us that it is not only the presence of nature that guarantees sensoriality, but also the context and the different atmospheres of the place (sound, climate, cleanliness, etc.). This leaves us to think that building a natural space in the city also requires thinking about the context of its implementation, to make it more pleasant and therefore to reflect its integration more holistically.

In a second part of the workshop, using the brainstorming methodology, the users were asked to give as many examples as possible of projects of an exemplary nature that they thought promoted appropriation by the inhabitants (Table 10.2).

The participants identified 18 locations that they felt referred to a relevant/pleasant city space. There were repetitions among the quotations, such as the "incredible edible" projects, the "coulée verte" (green corridor) or "Parc de Bercy". The various comments given by the participants allowed us to draw up a list of criteria that can promote the appropriation of nature in the city.

Thus, we note that the typo-morphology of the project plays an important role in the appreciation of a natural space in the city (see Coulée verte, Parc de Bercy, Parc de Belleville). Indeed, spaces with a "gradient" seem to be appreciated in that they offer viewpoints, and they give users a feeling of openness or of a horizon in particular. Everything happens as if people had the feeling of taking their place in the city – their city – through this type of morphology of nature in the city whereby they feel like they are above thing and able to see them from afar.

Projects favouring the integration of city dwellers were also put forward. Thus, the "incredible edible" approaches and the permission to green the bases of trees were appreciated by the respondents, although they expressed concern for the sustainability of each of these approaches.

Table 10.2 Exemplary Nature Projects Favouring the Appropriation of Inhabitants According to the Participants

Places Mentioned	Criteria
Incredible Edibles (2)	Creations of synergies by and for the inhabitants/ bottom-up approach
Coulée verte (2)	Domination of viewpoints, overviews
Parc de Bercy (2)	Place where there is activity with diversified spaces, interesting elevations
Greening tree bases	Good grassroots initiative for the bottom-up approach but the sustainability of the maintenance is problematic
Grands Voisins	Many activities and stakeholders who mingle
Transfo garden parking in Nice	The idea of going from something hostile to something beautiful
Parc de la Villette	Place that offers a variety of uses and attracts a diversity of profiles
Peach wall in Montreuil	Idea of resistance on a space
ZAC in Batignolles	Quiet and preserved, wastelands
Parc de Belleville	Gradients and horizon
Citroën Boulogne factory renovation	Good ratio between green spaces and buildings, positive transformation of a previously seedy place
Green space in Lyon, Scoop	Pooling tools and facilitators and training residents
Shared habitat project, including a green space	These are people who take care of the land and greening
Sheep in the parks	Sensation of astonishment
Recycling with hens	Sensation of astonishment

The projects offering a variety of uses and users were also appreciated by respondents, who referred us to the Grands Voisins projects, as well as Parc de Bercy or Parc de la Villette.

Transformation and rehabilitation projects were also cited as examples, in that they allow for reinvestments in the existing projects and the transformation of places that were essentially not very attractive and were not designed for hedonic uses.

Finally, the notion of astonishment seems to be an interesting criterion to take into account, and was generally related to fauna settled in urban contexts (hens, sheep, etc.). This naturally brings us back to the question of the connection to nature. Fauna provokes astonishment when it is found in a place that seems closed to any other form of life than the human being within the collective representations. This can be reinforced by the fact that wildlife is often absent in the discourse on nature, and this is also the case in our sample. When fauna is evoked, it provokes astonishment because city dwellers are no longer used to rubbing shoulders with it, and this is especially the case for those who are not able to leave the city for the weekend or holidays.

Following this, we asked the participants to carry out the same exercise but this time to designate the nature in the city projects that seemed least successful to them. Unlike the previous question, few examples were mentioned.

Only four types of projects were cited. However, all the projects mentioned were the subject of a consensus among the group, and several people mentioned the criteria that, according to them, stopped this natural space in the city from meeting their expectations.

Table 10.3 Nature Projects Deemed Least Successful According to the Participants

Places Mentioned	Justification
Les Halles (Paris Canopy)	Disappointment regarding what was announced compared to the result, spaces without green energy, children's games are not better than before, feeling of loss of heritage
Removal of allotment gardens around Paris	Idea of individualism that no longer works today because of the return to more community-based things, feeling of waste.
Defence	No ecological ambition, very mineral
Projects that do not include the main users	We have to adapt to people's lifestyles and ask them questions before installing things that do not suit them and which they will not maintain.

They essentially bring up two fundamental criteria in the construction of a project concerning nature in the city, namely (i) the consideration of environmental issues and (ii) the participatory nature of the projects (Table 10.3).

> *Second meeting between experts/users*: If we compare the data between inhabitants and experts on projects that seem relevant to them, we see several things. Firstly, and contrary to the experts, urban agriculture projects were not cited as a priority by the inhabitants. Secondly, both the experts and the inhabitants prefer projects favouring the appropriation of nature in the city (bottom-up projects). Finally, diversified projects are also put in the spotlight by the inhabitants, as were urban transformation projects, which was also the case for the experts.

The last exchange proposed in the workshop consisted of working together in small groups to imagine a "user pathway" within a neighbourhood that would allow a city dweller to connect to nature. The idea was to ask them to fill out a form previously developed by ENVIRONNONS and ECO-WORK. Several places were proposed to the participants (business district, suburban district, hyper-dense urban district), as well as several types of people (family with child, couple without children, single person, child, elderly person). The game consisted of imagining a plan of action specifying the persona and the chosen place. We asked them to highlight three key elements of their production and three essential experiences (which they had to rank) for a successful natural construction in the city or the suburbs. The main objective of this activity was to analyse the types of transformation envisaged to introduce nature to the city, in order to better understand the representations of how nature in the city should be.

The first group proposed the pathway of a woman actively working in defence. In this user pathway, we find three interesting experiences that combine technology with equipment that promotes the transformation of energy that is already present, the desire for more sustainable and more socially responsible food and finally greening, which offers perspectives that allow users to remove themselves from a very mineral environment.

The second user pathway proposed was that of a little boy in a residential area. It draws our attention to the fact that housing estates deserve to be thought of more as places of collective life. Indeed, there seems to be a contradiction between the fact that houses are gathered in one place and yet these are not thought of as community places.

In addition, the question of security is also posed in these urban forms. In fact, housing estates are often crossed by roads allowing residents to return home or leave. If children have no collective space or playing area except in individual gardens, it becomes impossible if not dangerous to play together in these spaces. Thus, the transformation proposals concern the development of these spaces with pedestrian zones in the centre of blocks of houses, coupled with play areas and vegetation.

The last pathway imagined by the third team is that of a young skateboarder in a very dense urban space (Place de la République). This user pathway is a little different from the other two in that it invites us to see nature differently. It does not contrast greening and minerality, but rather unites them to develop skating activities in a festive atmosphere. The idea is to leave room for pedestrians while respecting the existing objects (including trees) and lowering energy consumption (lighting).

10.2.3 Cross-Perceptions of Nature in the City

In this research, we strive to understand the different representations of nature in the city in order to formulate forward-looking avenues for the integration and appropriation of new forms of nature in the city.

Starting from this, the aim of this second workshop is to put the different points of view on nature in the city from both experts and inhabitants into perspective, in order to imagine prototypes of nature in the city that respond to environmental preservation issues as well as human well-being/health.

In the first stage of the workshop, the idea was to organise a period of expression based around a photo library. We then created a sheet with the aim of getting the participants to work in small groups to define, prioritise and debate nature projects deemed to be positive. Finally, in a third step, we gathered the participants in a large group to discuss the stages, difficulties and levers concerning the integration of nature in a fictional city.

The objective of the photo library was to highlight the different representations and reactions relating to the notion of nature in the city. To build this photo library, we drew inspiration from the results of the two previous stages (interviews and inhabitant workshop). Thus, the proposed urban representations of nature were those previously mentioned by the experts and the inhabitants. More than 60 photos, classified according to different categories of nature in the city, were presented to the participants. From these, participants were asked to choose an image that best represents nature in the city and an image that is least representative of their idea of nature in the city. Following this, they had to explain the reasons that led them to these choices. If these photos did not inspire them, they could also take a post-it, add their own elements representing nature in the city and explain them later.

Several types of nature were thus put forward without the different themes being mentioned. The photos were simply hanging on the wall at random. We have grouped them here and established categories.

With regard to the photos representing forms of nature in the city appreciated by the participants, 13 people spoke out and 4 abstained. The selection of photos best representing nature in the city shows a broad consensus around projects of a spontaneous nature. Indeed, the majority of participants chose photos relating to spontaneous vegetation and urban wastelands (9/13). According to the participants who chose this type of photo, this type of nature offers a stronger sense of naturalness because it is less controlled by humans. It also reveals the power of nature, which can grow in places that are not primarily dedicated to its development. Three other people selected urban nature photos focusing on urban agriculture. They justified this by the fact that they are participatory approaches with strong potential for sociability. According to them, these types of projects offer the possibility to reconnect to a rich kind of nature.

Eleven participants selected a picture that, in their opinion, represents the nature in the city least. In contrast to the previous task of selecting photos that best represent nature in the city, there was less consensus among the participants in selecting the photos least representative of nature in the city.

Among the selected items, five referred to photos of wildlife (a pigeon was chosen twice, a rat once and a fox twice). People who selected the pigeon and the rat referred to the fact that it is the city that makes them ugly, by devaluing them. Given the sheer number of them to be found, these animals also evoke questions concerning the lack of biodiversity among the fauna in the city according to the participants. The choices concerning the photo of the fox mainly referred to the destruction of its habitat, and thus its necessity to migrate towards the city ("it is not normal" to see a fox in the urban environment). It should be noted that photos relating to fauna were all chosen by the inhabitants and not by the experts.

Following the photo library, the participants were divided into small groups to develop collective reflections on urban projects integrating nature. We were careful to distribute them so that there was a mixture of experts and inhabitants in each group. This activity had multiple aims. One intention was to start a dialogue between the experts and inhabitants on different projects they chose and judged positively in small groups. Thus, four small groups were formed and were given the task of exchanging their preferred places of nature in city. They then had to agree on a project, explain why they picked it by indicating their main criteria of choice and, finally, identify any points of debate on the project.

The first project chosen by the first group was that of "coulée verte" (green course). The criteria for its choice were the idea of greening the city, the importance of a pathway and the sensation of height that offers walkers on the horizon. The main point of debate rested on whether or not this kind of space has a socialising nature.

The second project described was that of a wild garden, the Jardin Saint Vincent, located in the 18th arrondissement of Paris. This garden is not very accessible to the inhabitants as it is only open one day a month and allows biodiversity to return to the city. The criteria for its choice were those of biodiversity and its preservation, the awakening of the senses and a feeling of discovery.

The debate then centred on the problem between letting nature grow and thus redeveloping non-anthropised environments, and that of city dwellers' uses and desire for natural spaces. The third project turned to the choice of the îlot d'Amaranthes in Lyon, which is a shared garden created in an infill on the initiative of the inhabitants. The main criteria of choice were the social link and the contribution to social diversity, the appropriation by the inhabitants and the fact that this kind of project generates conviviality.

Finally, the last project picked by the last group concerned the Veni Verdi urban agriculture community project. The participants were prompted to choose it because of the community and educational aspects, as well as its usefulness for biodiversity. The main point of debate centred on the anthropised aspect of nature in the city as opposed to a desire for more spontaneous management of it (Figures 10.3–10.6).

Figure 10.3 Coulée Verte – Paris 12th arrondissement.

Figure 10.4 Jardin Sauvage Saint Vincent – Paris 18th arrondissement.

Figure 10.5 Ilôt d'Amaranthes – Lyon. (Brin d'Guill.)

Figure 10.6 Association Veni Verdi – Paris (credit Veni Verdi).

10.3 Conclusion: Tomorrow... Nature in the City

10.3.1 Summary of Main Findings

This exploratory work allowed us to highlight the different representations of nature in the city, and it enables us to sketch the first outlines of nature in the city as it is imagined by both inhabitants and experts. Below, we have summarised the main data collected on the perception of nature in the city today and expectations for tomorrow.

Regarding the different results, there are duplications in the discussions between the different stages of the research. There therefore seems to be some consensus between experts and non-experts around the representation of and desire for nature in the city.

Indeed, there is little difference in the discourse other than the relationship to the economy of the project and the preponderance of urban agriculture in the discussions by the experts compared to the users. In the light of these results, we propose a prospective study on nature in the city by 2040, which is given below. We will formulate this in the form of an essay because foresight is never produced alone, and we feel it would be interesting to start a debate around the ideas laid out below, and to enrich the foresight by organising a new workshop bringing together experts in nature in city, economists, developers and users. This foresight implicitly outlines a number of avenues that can promote the appropriation and integration of nature in the city.

The limits of nature in the city today are as follows.

- A lack of contact with and experience of nature on the part of city dwellers;
- An almost total sensory absence in nature in the city;
- User fears/doubts concerning fauna and flora;
- A lack of communication and information between managers and users of green spaces;
- Strong land pressure in cities does not favour its development;
- Differentiated management not yet well perceived/understood by users.

The expectations of nature in the city of tomorrow are as follows.

- An experiential form of nature that promotes return to the senses in a positive way;
- Nature designed by and for users in collaboration with experts;
- A coherent form of nature in line with developments aimed at reducing the carbon footprint of cities;
- A more spontaneous form of nature that is accompanied by awareness of its users;
- Diversified nature in terms of both uses and forms;
- A form of nature that promotes surprises and discovery by creating a poetic dimension;
- A form of nature that allows a return to the self and one's own humanity;
- A form of nature that takes the notion of climate change into account in its integration.

10.3.2 A Prospective Study on Nature in the City by 2040

The scenario proposed below is based on the hypothesis that cities will densify, with overcrowding and a high concentration of individuals per square metre.

10.3.2.1 Nature and Relationship to Land

At first, the densification of cities will lead to a dilemma over whether to allocate land dedicated to housing, offices and shops, or land dedicated to natural areas. In other words, spaces of nature in the city will be redesigned to be able to offer comfort in urban areas without hindering the development of buildings to accommodate new-comers and activities dedicated to the functioning of the city. This land issue and this demand for natural areas will therefore lead developers to think differently about spaces of nature. Thus, we will witness the gradual disappearance of urban park construction in favour of integrating nature more widely throughout the city. There will be more nature but it will be less concentrated in one place. Indeed, the de-waterproofing of soil will be accentuated to allow a more continuous form of nature to develop, which will gradually break down the split between nature and city. Nature will no longer be confined between the gates of a park but will take its place on the public roads as such (pavements, integration in building materials, city lighting, on bridges, etc.). Nature will then be a component of the city, and it will be conceived at the same time as transport infrastructure and buildings.

If the spaces dedicated to nature change, then the forms of nature will also change and will tend towards a more differentiated kind of nature that is more wild, depending on the zones of its growth, and more diversified. The diversity of forms of this nature may lead to a diversity of uses thereof. For example, the de-waterproofing of pavements will allow for more appropriation on the part of the inhabitants, thus creating open green areas that give free rein to various uses (agriculture, horticulture, open spaces, play areas, etc.). This would be closer to the idea of "Frontage" already present in some countries.

There will also be an increase in green and blue belts linking the existing green spaces and wildlife corridors will be created. This will lead to a proliferation of animal species other than pigeons and rats.

Users will be both more used to seeing nature in differentiated management and richer fauna in the city. As a result, the representations of nature in the city will gradually change. It will no longer appear as a means for getting out of the city but will be considered a real infrastructure in the same way as transport or buildings. Nature in the city will therefore be normalised. This normalisation and this constant relationship to nature, including both flora and fauna, will have the effect of reducing city dwellers' fears, as well as that of reconnecting them to the natural systems. This will result in a stronger awareness of nature and its preservation, and will therefore lead to less energy-consuming and more reasoned consumption practices. Hedonic activities will also be affected, as will relationships to work. By being closer to nature, city dwellers will rethink their life priorities and move towards more peaceful and simplified lifestyles.

10.3.2.2 Nature, Energy Needs and Innovations in Construction

In addition, the densification of cities will lead to a strong need for green energy. Indeed, fossil fuels will no longer be the dominant energy model in 2040 and other alternatives will meet this need. However, this will not be to the detriment of the space reserved for construction, especially green spaces. One of the solutions provided in this context will be to systematise the waste sorting system, in particular that of organic waste, and to develop methanation on a large scale. The roofs of cities will also share the hosting of green energy versus natural installations.

Changes are also expected in building methods. Indeed, builders will develop externalities (balconies, terraces, hanging gardens) in new housing by co-building green space with the future inhabitants once they have signed the promise of sale (just as the flooring finishes and other elements are chosen in future home at present). Thus, we will witness a diversity of forms of nature at the scale of buildings, which will promote the enrichment of biodiversity. These developments will not only be co-constructed between residents and builders, but an ecologist will also be there not only to guide the various choices and, above all, to inform the inhabitants about the benefits and possible risks of a given kind of natural development. Thus, individualised nature at the scale of housing will develop and inhabitants will also be able to transform their natural projects as soon as they wish (as with the adaptation of spaces).

In addition, the ambiances of the city will be reworked by the builders and architects, who will integrate more sensory elements in their projects. They will create natural spaces in their projects and/or buildings that will adapt to the season and its light by taking endemic elements from both fauna and flora. This again refers to the collaboration between constructors, urban planners and ecologists.

Finally, we will witness the rise of biomimicry, with builders who will be sensitive to the functioning of nature and who will seek to emulate it by creating materials that reflect their functions. As a result, planning professionals, like users, will have renewed representations of urban projects.

10.3.2.3 Nature, Sustainable Food and Urban Agriculture

In 2040, city dwellers will be more involved in health issues because of media pressure and the various health scandals on the topic that have occurred in recent years. This will lead to a change in household consumption. Thus, the demand for sustainable food will be more consistent and households will prefer organic products. In addition, climate change issues will lead households to move closer to short supply chains and consequently to buy locally and without packaging. This will lead to a strong development in urban agriculture. The lack of space will not meet the demand, and urban agriculture will be broken down into several modes of production and management. Community gardens will multiply in the infills of urban projects and will promote forms of agriculture that are more social than concerned with producing food. Vegetable gardens on roofs will achieve higher yields, which will be placed on the market. Permaculture will be favoured in both forms of agriculture, and users will be more aware of this kind of practice. Users may also be involved in the production, or at least the formulation, of projects around rooftop agriculture (which does not necessarily mean gardening, but it could involve informing gardeners about their consumption preferences, for example, or to having access to this natural space to relax). However, these roof spaces will be restricted by the need for spaces for installing green energy systems. To achieve a yield that can feed the city, vertical greenhouses will be developed in suburban areas. Unlike the other two types of agriculture, vertical greenhouses will not be located in city centres because they bring less direct benefits to the well-being of urban dwellers than visible natural spaces. On the other hand, to promote reconnection to and therefore understanding of the functioning of nature, it is necessary to bring humanity and nature into contact. However, this system does not allow it. Moreover, this would profoundly transform representations of the functioning of nature. In fact, green skyscrapers make it possible to produce non-endemic species and off-season species, which is very different to the classical functioning of nature. Spaces of hedonic and experiential nature will therefore be privileged in city centres. Thus, users will be invested in the construction and management of spaces dedicated to urban agriculture, and also to the construction of urban spaces in all its forms more broadly. Finally, we will see the spreading of the practice of agro-ecology on more extensive perimeters, especially in suburban and rural areas.

10.3.2.4 Nature, Mobility and Urban Ambiences

By 2040, the urban context and urban environments will have evolved and transformed. Green mobility will be the dominant model, resulting in much quieter sound in the city. Thus, natural spaces become more widespread, and the feeling of well-being in the city will increase. Residential mobility will not evolve in that city dwellers will inexorably go through the same stages of the life cycle. Thus, they will begin to inhabit small spaces in order to change them during the advent of family transformations. This turnover will be taken into account in the development of nature in the city, especially in cases where the inhabitants are involved in its management. A social economy of projects involving nature in the city will be considered with the creation of a system of governance at the level of blocks of buildings. Representatives will be elected and will meet to redefine the expectations of their natural space. Indeed, this turnover entails changes in representations of and desires for nature, as well as changes in the availability of household times. Also, consultations will be carried out regularly to modify and/or preserve the natural space (e.g. to transition from a kitchen garden to a wasteland, from a wasteland to a horticultural space, from a horticultural space to a space with multiple or recreational functions). Thus, this governance, which will be renewed according to the relocations, will make it possible to organise, implement and sustain nature projects.

10.3.2.5 Nature and Regulations

Finally, regulatory changes will be imposed by the challenges of climate change requiring cities to systematically introduce nature into their new projects. Much like thermal regulations today, biodiversity regulations will be imposed in major cities and for each new construction project.

To conclude, in environmental psychology, the term "behavioural" site refers to a "complex system of human-environment interrelations whose regulation determines and deploys behaviours, which may or may not be deployed, or which should not be deployed" (Moser & Weiss, 2003, p. 247). This means, for example, that certain behaviours are expected in a given place, while others are proscribed. For example, in a library, screaming is not allowed and a quiet atmosphere is expected. With its form and its uses, the city is also a behavioural site, where city dwellers expect to see certain types of behaviour take place, while others are not even considered, because they too far removed from the expectations of the site. The transformation and growth of nature in the city will transform the behavioural site of the city, and therefore the behaviours and representations of the city. Thus, de-waterproofing, biomimicry, agriculture in the city, the continuity of natural spaces, the advent of a less controlled for of nature and a renewed governance of spaces will lead to a change of mentality and changes in the uses of the city.

Bibliography

Allport G.W. (1954). *The Nature of Prejudice*. Cambridge, MA: Addison-Wesley.

Blanc N. (2004). De l'écologique dans la ville. *Ethnologie française*, Vol. 34(4), 601–607. doi:10.3917/ethn.044.0601.

Bourdeau-Lepage L. (2013). Nature(s) en ville, Métropolitiques, 21 February 2013. http://www.metropolitiques.eu/Nature-s-en-ville.html.

Cormier L., Joliet F., & Carcaud N. (2012, July). La biodiversité est-elle un enjeu pour les habitants? *Développement durable et territoires [online]*, Vol. 3(2), posted on 8 January 2013.

Hardin G. (2009). The tragedy of the commons. *Journal of Natural Resources Policy Research*, 1(3), 243–253.

Hinds J., & Sparks P. (2009). Investigating environmental identity, well-being, and meaning. *Ecopsychology*, Vol. 1(4), 181–186.

Hinds J., & Sparks P. (2011). The affective quality of human-natural environment relationships. *Evolutionary Psychology*, Vol. 9(3), 451–469.

Lammel A., Dugas E., & Guillen-Gutierrez E. (2012). L'apport de la psychologie cognitive à l'étude de l'adaptation aux changements climatiques: la notion de vulnérabilité cognitive. *VertigO*, 12(1). http://vertigo.revues.org/11915; DOI: 10.4000/ vertigo.11915.

Long N., & Tonini B. (2012). Les espaces verts urbains: étude exploratoire des pratiques et du ressenti des usagers. *VertigO-la revue électronique en sciences de l'environnement*, 12(2).

Moser G., Weiss K. (dir), 2003, *Espaces de vies. Aspects de la relation homme-environnement*. Paris. A. Colin. Collection Sociétales, 396.

Pol E. (2011). Les causes de l'incertitude environnementale. Acte du 4ème colloque ARPENV: l'individu et la société face à l'incertitude environnementale, Ifsttar, Lyon, Bron, 6–8 June 2011.

Renauld V. (2012). Les conceptions techniques innovantes face aux règles d'usage des habitants: enquête sur un bâtiment écologique emblématique de l'écoquartier De Bonne à Grenoble. *Contribution scientifique et technique sur la notion d'appropriation dans les opérations d'aménagements urbains durables*, 34–38.

Shepardson D.P., Niyogi D., Roychoudhury A. & Hirsch A. (2012). Conceptualizing climate change in the context of a climate system: Implications for climate and environmental education. *Environmental Education Research*, 18(3), 323–352.

Ulrich R.S. (1984). View through a window may influence recovery from surgery. *Science*, 224(46–47), 420–421.

White E.V., & Gatersleben B. (2011). Greenery on residential buildings: Does it affect preferences and perceptions of beauty? *Journal of Environmental Psychology*, 31(1), 89–98.

Chapter 11

Housing Demand

Residential Paths and Inequalities in Comfort and Access to Property

Vincent Lasserre-Bigorry, Fabien Leurent, and Nicolas Coulombel

Ecole des Ponts ParisTech, Laboratoire Ville Mobilité Transport

11.1 Introduction

11.1.1 The Issue of Housing through the Prism of Sustainable Development

Housing is a sphere that carries important social issues. For a human person, housing is the privileged space of domestic life and the intimate setting for the household to which they belong. Housing shelters people and facilitates the basic biological needs (sleep, meals, washing), as well as communal living, if the household includes several people. By its construction, housing is sustainable equipment, with a support, walls and a roof, devices providing water and energy, as well as telecommunications and physical access, and various domestic appliances (heating, kitchen, computer, etc.) in addition to furniture.

The layout of the housing and its quality in terms of surface, hygiene, tranquillity, brightness and thermal comfort are the important factors for the well-being of the people who live there. An extrinsic quality is added to this intrinsic quality. The location of the housing determines the accessibility of urban amenities (shops, services, leisure), employment and social relations (neighbourhood, district, family and friends in the urban area and beyond). In other words, the location of housing is an important factor in the household's relationship to the city.

The services associated with housing are provided at the price of a significant economic contribution. Current usage generates costs based on the associated consumption (water, energy, telecom) and local charges (collective ownership fees where necessary, waste collection and other levies and taxes). The cost also includes the provision of accommodation per period, according to the occupancy status: the rent if the household is a tenant, or the investment cost in the case of ownership. The amount of these expenses in relation to the household income constitutes the household's effort ratio for its housing. In the city, this value can easily reach 20% or 30%, which makes housing the number one item of expenditure in the household budget.

The economic challenge is therefore also present, especially since the urban area is larger and more populated. As households bid to secure housing, property values include not only the cost of construction but also the land rent. Although housing seems more expensive in the city centre, living in the suburbs requires more time and money to be spent for mobility. This is the fundamental principle of property market equilibrium in urban economic theory.

In fact, environmental issues are just as important as social and economic issues. The quality of construction determines the energy needs required to ensure thermal comfort, and the quality of equipment determines domestic consumption as well as the gestures of use. Added to these current consumptions is a "grey area" as consumptions of energy and materials during the construction or renovation phase can be assigned to the current system on taking the technical lifetime into account. We also count consumption outside housing: access to activities and jobs leads to greater energy and material consumption as housing is further away from the city centre.

All in all, the question of housing is important for the three pillars of social, economic and environmental sustainability.

11.1.2 Eco-Design of Housing Systems: A Management Problem Requiring Tools

Individual dwellings or collective residences are durable objects that must be planned in order to ensure quality of use, technical robustness and environmental performance in both the construction and use phases. A particular requirement is to limit the end-user expenditure related to property holding use charges.

The complexity of the planning problem is obviously multiplied at the city scale as it must also integrate the technical infrastructures of urban planning and mobility, as well as transport services for access to amenities, the presence of nature and associated conditions. Additionally, the system of social and economic activities needs to be planned, with their own property and infrastructural needs. In the end, overall planning presents a high level of complexity in terms of combining the number of premises to include in the space; questions concerning the respective configuration of residences, jobs, services, etc.; anticipation of performance over time; and therefore the long-term preparation.

Overall optimisation therefore seems very difficult to plan a priori. Moreover, each city is a long-term socio-technical organisation, so the managerial problem therefore affects the progressive evolution of the system. However, this evolution depends on a plurality of decision-makers that include not only the planning authorities but also all the households that freely decide on their residential establishment, as well as all the firms that decide where to locate their activities.

This is why public policies focus on privileged targets. On the demand side, this includes housing assistance for households, in cash or in kind (social housing), while on the supply side, it involves the facilitation of construction and energy renovation, as well as specific urban planning and mobility policies.

This collective management of the housing system still remains to be equipped with tools according to its different components, by means of observation and simulation tools. In an urban area where housing is plentiful and inexpensive in relation to household incomes, it suffices to ensure the condition of the most modest households, as well as the overall environmental performance. Several indicators at the disaggregated level of households are enough to examine the lower part of the statistical distribution of incomes and the upper part of the statistical distribution of environmental footprints. These statistics motivate household housing surveys conducted nationally in France and other countries, as well as specific sections in the general population censuses. However, in an urban area where housing is strained and expensive, the system must be managed in economic terms of quantity and price. As such, the model of interaction between transport and urbanisation called the "land use–transport interaction" (LUTI) model is appropriate (see the other chapter on housing), and makes it possible to simulate the interaction between housing supply and demand.

In an LUTI model, demand for residential establishment is represented as the household population. Each household is described by characteristics such as the number of people, income and location(s) of employment. Their choice of housing is modelled as a residential establishment decision in terms of floor area, location and occupancy status (tenant or owner). In a "static" LUTI model, all the households make their residential establishment choices simultaneously in each scenario relating to the state of the system. Their prior state is disregarded. In a dynamic LUTI model, the co-evolution of supply and demand is represented in successive periods, typically year by year. The previous situation can be taken into account in a given period. Often, however, in order to simplify the model, the distinction between tenant and owner is blurred, as well as that of the occupancy status in the prior state.

However, sociology of the family has established the notion of the "residential path" for individuals, whereby housing is adjusted gradually (by mutations) according to the stages of their life and the composition of the household. One of the crucial steps in this individual path is home ownership.

11.1.3 Our Knowledge Objectives

The purpose of our research is to contribute to an improved representation of housing demand and use, in a form that can be integrated into a planning model. The contribution is threefold:

- Firstly, we shall propose indicators to represent the objective conditions of the demand and the use of housing, specifically a surface indicator per person to express the "surface comfort" and another to measure the effort ratio, as well as equity indicators among households;
- Secondly, we will model household behaviours, including both residential paths, which constitute demographic behaviours, and the decision to become an owner, which is an economic behaviour;
- Thirdly, we will analyse the overall evolutions of the housing system in the case of the urban area of Paris through a diachronic analysis spanning half a century. This analysis will consider the development of the population and the housing stock, its expansion in space, generational renewal and the respective residential paths of the "demographic cohorts".

11.1.4 Method

Our method combines social sciences and statistical modelling. The constitution of indicators, the analysis of values and user costs, and the revelation of microeconomic behaviours fall within the sphere of econometric modelling. The analysis of disparities between households in the light of equity indicators is as much social as it is econometric. Finally, the highlighting of residential paths and generational effects is based on demographic modelling.

The econometric apparatus thus constituted is applied to the Ile-de-France region, that is essentially the Parisian urban area, during a period spanning from 1968 to 2012. We studied the population and housing system from two sources of statistical data: population censuses and national housing surveys.

11.1.5 Outline of This Chapter

The rest of this chapter is organised into four main sections followed by a conclusion. We will begin by describing the population and housing system in Paris region over course of half a century and at two levels: the macroscopic level of the whole and the microscopic level of the household, according to its particular conditions in terms of age, generation and residential path. Next, we will model the housing service as it is received by households. We will relate the surface comfort indicator to household characteristics, including the number of people and income, and we will analyse the disparities between households in order to establish a diagnosis from the perspective of equity.

After these two sections dominated by demographic analysis, the next two sections will be devoted to economic analysis. We will model housing use costs as well as effort rates for each household in the general population, before focusing on the households likely to have access to the property in order to model the decision to make a first-time house purchase.

In the conclusion, we will summarise the main demographic and economic findings from the long-term analysis of the Paris system, and we will outline the adaptability of our results for LUTI modelling.

11.2 Diachronic, Demographic and Geographical Analysis of Households in the Paris Region

11.2.1 Definitions

As a socio-economic service based on tangible property, the "demand" for housing, in the economic sense of the term, consists of a set of "households". The individuals who make up the population are effectively divided into "households" according to the housing they occupy with one or more people, generally with family ties between parents and children and between spouses. Through a mirror effect, a blended family is defined according to the housing that it shares.

Within an urban area, the notion of a resident household is equivalent to the notion of a principal residence. We can therefore analyse the demand and the use of housing in the urban area by considering the composition of the resident population in households.

11.2.2 Population and Housing: Half a Century of Households in the Paris Region

A particular urban area is transformed over the long term. Its population is renewed, and it evolves in number, distribution in space and in "demographic behaviours" understood in terms of lifespan, fertility rate and formation of households.

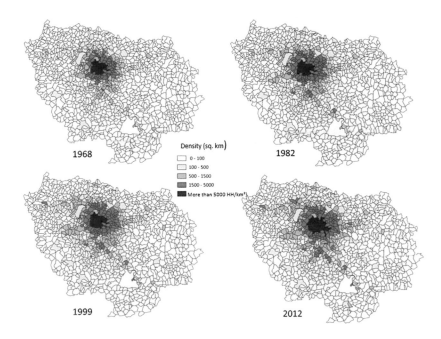

Figure 11.1 Paris region from 1968 to 2012: the progress of urbanisation.

For the Paris region, the half-century from 1968 to 2012 was marked by a significant increase in population: from 8.9 million inhabitants to 11.7 million, that is 30% growth. At the same time, the number of households increased from 3.3 million to 5.0 million, or +53%. This means housing stock evolved significantly faster than the population counted as the number of individuals. These growths were inscribed in the space through the signification expansion of urbanisation (see Figure 11.1).

This evolution differential was due to the transformation of household compositions. The average number of persons per household decreased from 2.74 in 1968 to 2.33 in 2012, that is by 15% (see Figure 11.2). Depending on the number of people in the household, there was a high degree of stability in the number of households with three or more persons. It was the households with 1 or 2 persons whose numbers increased (see Figure 11.3).

Therefore, there were fewer people per household. At the same time, there were more rooms per household, however: +22%, from the ratio of 2.76 in 1968 to that of 3.37 in 2012. These two developments contributed to an even greater rise in the number of rooms per person in a dwelling: from 1.23 to 1.77 between 1968 and 2012, that is +44%. This is a considerable increase, but still less than the transition from 1.37 to 2.17, or +59% in metropolitan France (see Figures 11.4 and 11.5).

11.2.3 Demographic Aspects: The Decline of Cohabitation

Several demographic factors are at play in the structural evolution of the population in terms of numbers of inhabitants and households. They should be considered both

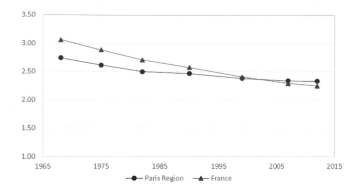

Figure 11.2 Evolution of the average household size: Paris region and metropolitan France.

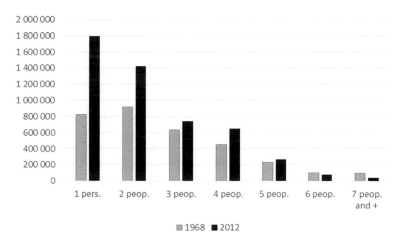

Figure 11.3 Evolution of the sizes of households in the Paris region between 1968 and 2012.

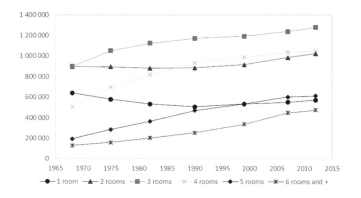

Figure 11.4 Structure of the Paris housing stock according to the number of rooms.

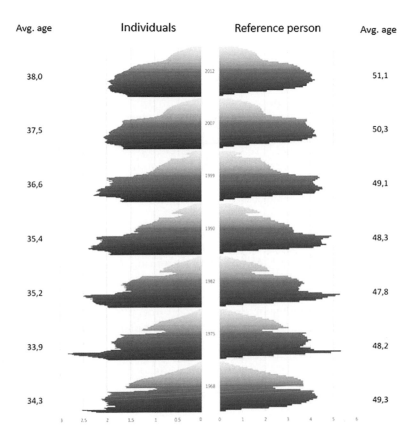

| Avg. age | Individuals | Reference person | Avg. age |

Avg. age			Avg. age
38,0			51,1
37,5			50,3
36,6			49,1
35,4			48,3
35,2			47,8
33,9			48,2
34,3			49,3

Figure 11.5 Age pyramids in the Paris region for individuals and households according to the reference person.

in absolute numbers and in relative proportions. First of all, the general ageing observed for the whole of metropolitan France barely affected the Paris region. Older age groups saw their numbers increase slightly, but at a rate that was significantly lower than the national average. The reason is that a large proportion of households leave the Paris region for another region during the transition to retirement. As a result, the population of the Paris region remained relatively young during the last half-century, and the contrast with the rest of France is growing.

The number of children per household slowly decreased. The ages of parents at the birth of their children increased on average, and their statistical distribution widened, making the presence of children more statistically persistent on a longer range of household lifespans, confusing the signs of a slow decline in the birth rate.

The most salient demographic fact is the considerable increase in people living alone or in couples, in both absolute and relative terms. The share of one-person households increased from 31% to 39% over the period from 1968 to 2012. Factors include an increase in single life, a decline in cohabitation with elderly parents, an increase in separations and single parenthood, lower birth rates and a higher age for having a first child. Only one factor is working in the opposite direction: young adults are leaving their parental home later.

11.2.4 Life Pathway and Residential Pathway

In demographic analysis, we separate the respective roles played by age and generation (group of individuals born at the same time) by studying social and economic phenomena per demographic cohort. We trace the trajectory of socio-economic variables as a function of age, per generation. Since households are formed in a shifting manner, cohorts of households refer to a reference person, namely the oldest person if they are single, or the oldest actively employed man in the case of multiple adults. We will consider pseudo-cohorts in the rest of this chapter, as we identify generations from moments in time over successive surveys, instead of progressively observing the trajectories of individuals.

Cohort analysis of the number of persons per household reveals the role of age first of all: growth from 20 to 45 or 50 years, then a decrease (see Figure 11.6). This applies to each generation. Moreover, cohort analysis among successive generations reveals a gradual downturn towards a uniformly low trajectory, which confirms the overall decline in household size in detail.

With regard to the number of rooms in the dwellings, the cohort analysis reveals rising residential paths. Each cohort has a growing trajectory that reaches a plateau at

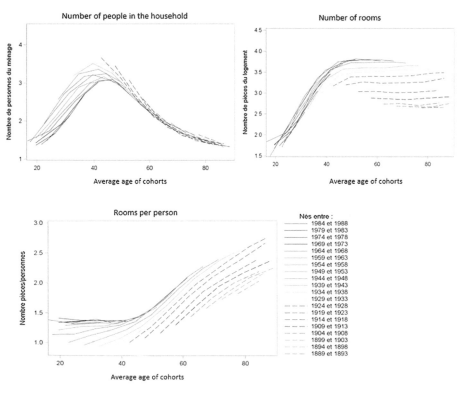

Figure 11.6 Sizes of households and dwellings by demographic cohorts.

around 45–50 years and then remains consistent. The levels rise between the successive generations. Although not very perceptible in the part from 20 to 40 years, it is much more marked between the levels reached from 50 years onwards. Older households maintain their housing consumption, while the number of people in the household decreases. We will come back to this "fossilisation" of an increasing part of the housing stock in the next section.

Regarding the number of rooms per person in the household, the trajectory by cohort also rises. The increase in the number of rooms according to age is faster than that of the number of people among the "junior" households. For the "seniors", the stability of the numerator combined with the decrease in the denominator ensures the growth of the ratio. The trajectories of successive generations only converge in the most recent cases. The dispersion of the "junior" parts results from the decrease in the number of persons per household, while the variations in the "senior" parts result from the increase in the level of room numbers obtained according to the generation.

11.2.5 Spatial Aspects: Expansion of Urbanisation and Development of Individual Housing and Property

The increase in the number of rooms available per person in the Paris region took place through a significant growth in the number of dwellings with three or more rooms, while the number of one- and two-room dwellings varied little (see Figure 11.4). The development of the number and size housing took place in the suburbs through an expansion of urbanisation in the inner suburbs (+44%). This took place on even larger scale in the outer suburbs (+136%), while the central city (Paris) changed little (+2%).

Figure 11.7 shows the respective evolutions of these three spaces in terms of the number of inhabitants and the number of households, as well as the types of individual or collective housing. The phenomenon of suburbanisation occurred in close interaction with the transformation of travel conditions. On average, the number of daily trips per individual remained relatively stable, but the distances travelled grew. The car mode is used more, as is public transport for crossing distances of more than one kilometre.

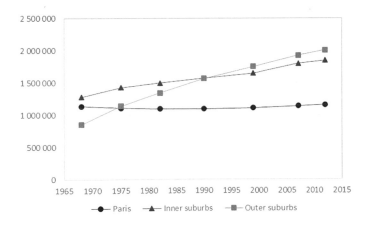

Figure 11.7 Distribution of inhabitants and households in three concentric spaces.

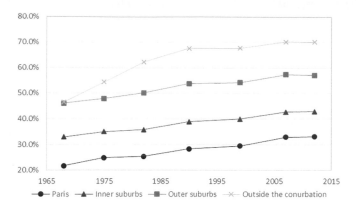

Figure 11.8 Progression of occupying property, Paris region.

There was an increase in the time spent, energy consumption, greenhouse gas emissions and air pollutants, despite the concomitant improvement in motor vehicle performance. The health effects are all the more amplified as the accumulated impacts affect the population, which has itself become larger.

The development of the housing stock benefited individual housing in particular. The stock of single-family houses increased by 60% between 1968 and 1990, and by a further 22% from 1990 to 2012, reaching a quarter of the total stock. In a correlated manner, the proportion of households owning their homes rose from 32% to 47% during the period (see Figure 11.8). The 15 point increase is identical to that observed for the whole of France, with the regional level remaining 10 points lower than the national level.

11.3 Housing Size by Household: Factors and Disparities

Several factors influence the size of the housing available to a household: the number of people, the age of the reference person, the generation to which this person belongs, the location in the territory by large area, socio-professional category (SPC) related to income and occupancy status as either owner or tenant. So far, the observation has remained macroscopic although we have detailed it according to the demographic cohorts.

In this section, we will analyse the size of housing according to household characteristics, using a disaggregated model at the individual level of each household. Methodologically, we specified this disaggregated model (with several variants, in fact), and we associated it with a statistical estimation method, based on population census data (Section 11.3.1). Empirically, we applied the model to the Paris region for the period from 1968 to 2012. The results obtained have high statistical reliability (Section 11.3.2). Finally, we investigated the factual situation of the housing system on a third analysis plan, this one of a social kind, as the disparity of the housing conditions among households calls the fairness of this service within society into question. In order to shed light on this question, we used the relative dispersion indicator known as Hoover or Pietra–Ricci (Section 11.3.3), and we applied it to the Paris region to measure the equity of the housing service, which constitutes a social issue (Section 11.3.4).

11.3.1 Econometric Modelling of Factors and Influences

Let i be a particular household among the population I, n_i the number of people (or units of housing consumption) in the household and p_i the number of rooms available for the housing. The number p_i is a variable that quantifies the housing service for the household. The surface s_i would also be appropriate, and we have considered it in another model.

In the model presented here, p_i is the "dependent" variable, the one we want to "explain" by the factors. Foremost among the factors, we consider the number n_i of people in the household. There are several possible options for taking into account the characteristics of individuals (adult or child, age of children) and interpersonal relationships (e.g. adults in couples), in order to express a number of housing consumption units that correspond to the minimum number of rooms suitable for the household, with regard to the definition of overpopulation accepted by INSEE (the French National Institute of Statistics) since 1970. Thus, this variable takes the composition of the household into account beyond the gross number of people (number of children, age of children, the presence of a couple). More broadly, Xi denotes the vector of household characteristics. In the model presented below, Xi incorporates ni as well as

- The age of the reference person, which may indicate the capacity for material or financial accumulation available to the household;
- The demographic generation of this person, which indicates the particular historical conditions that the household has encountered in its life, notably the precedence of access to the housing stock and certain price and mortgage conditions;
- SPC classification of the reference person, which loosely indicates the household's level of income and therefore its financial capacity in relation to the economic conditions of the property market;
- The residential location in the urban area: a location that is a greater distance from the centre (also loosely) shows a certain relaxation of property prices, as well as more spacious housing and the facilitation of individual housing;
- The occupation status, between an owner occupying their own dwelling, a tenant on the private market or a social housing tenant.

We modelled the number of rooms minus 1, that is p_i-1, as a random Poisson variable with integer values starting from zero, the parameter of which is an exponential function of a linear combination of the different factors. Formally, the model consists of the following influence function, in which β denotes the parameter vector:

$$E[p_i \mid X_i] = 1 + \exp(\beta \cdot X_i).$$

The notation E for the expected value means that the influence is modelled on average for households with the same characteristics X_i, and that the residential characteristics that are not explained as factors are the source of the individual variations relative to this average.

II.3.2 Application in the Paris Region

Figure 11.9 presents the model estimation results for all the households observed in the general population censuses conducted between 1968 and 2012. The degree of statistical significance is very high. The influence of a particular factor is understood as follows. Take the example of occupation status, indicated by a binary variable with a value of 0 (tenant) or 1 (owner): all things being equal, the estimated value of 0.251 for the multiplying coefficient of this variable implies a number of rooms that is 29%=exp (0.251) larger on average for homeowners versus tenants.

Statistical regression reveals the following influences, factor by factor:

- The influence of the number of people is strong: the value153 for the transition from 2 to 3 people corresponds to 16% more rooms;
- The age of the reference person influences the number of rooms in an increasing manner, at a fast pace up to 38–40 years, and after that with a softened gradient that still remains distinctly positive;
- The generation effect is very perceptible: there is a growth up to generation N born between 1949 and 1953, thus at the height of the baby boom, then stagnation beyond this point. This corroborates the convergence of the residential pathways observed in the previous section;
- The SPC influences the size of the housing, all things being equal. Except for farmers, who are spaciously housed in rural areas, managers have the largest dwellings, while employees and workers occupy smaller dwellings;
- Regarding the residential location, an establishment in the outer suburbs can have 25% (= exp (.2)) more rooms compared to a residential establishment in inner Paris;
- The occupation status is important: a homeowner has a quantity that is 29% more than that of a private tenant, all things being equal. A social housing tenant is relatively more generously housed than private tenants, with a difference of +10% (= exp (.08)).

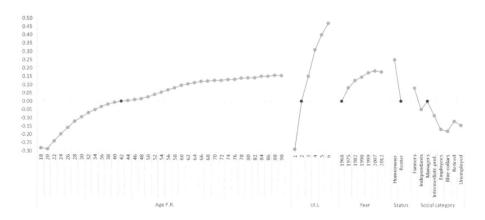

Figure 11.9 Parameters of the regression model, according to the age of the reference person, the number of housing consumption units (HCUs) and the SPC.

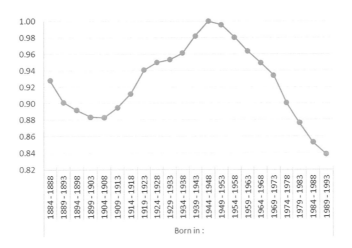

Figure 11.10 Parameters of model with generations.

Thus, statistical regression according to multiple factors makes it possible to discern the respective influences. The model presented here juxtaposes the factors but does not explain the underlying relationships that would combine them. Other models have been specified to study more complex interactions, in particular a cohort model to better discern the respective effects of age and generation per every five years for birth years (see Figure 11.10). Generation *M*, born between 1944 and 1948, obtained the most favourable housing conditions, which were better than those of the generations that preceded or followed it. This second regression brought our attention to the most recent generations, born after 1979. Their coefficients are the lowest, and they decrease with later arrivals. This attests to the deterioration of the conditions relating to access to housing for young generations in the Paris region.

11.3.3 Disparity Analysis Method

We have just observed significant disparities among the household population in terms of housing, according to several axes of analysis, taking into account the number of people in the household. Age, generation, SPC and location within the city are all sources of disparities.

In order to assess the disparities in a synthetic manner, there are different "statistical summaries". To study the economic development and to analyse income, the Gini index is used, among other indicators that share similar principles. We prefer to use the relative mean dispersion (RMD) indicator mobilised by many authors in the scientific literature (Pietra, Ricci, Hoover). We again note i as a given household composed of n_i persons, with p_i as the number of rooms in its dwelling. We study the fairness of the distribution of the number of rooms per person, p_i/n_i, among the households. For each household, the algebraic difference $p_i - n_i - \bar{p}$ between the individual value and the fair value $\bar{p}$ (mean of the distribution) affects n_i persons, so the difference to consider is $p_i-n_i-\bar{p}$, since it is the number of people who would remain stable in the household if we redistributed the housing between households.

The sign of the difference indicates the direction of the transfer to be achieved in order to improve the fairness: if $p_i < n_i < \bar{p}$, the household has a shortage of rooms, whereas if $p_i > n_i > \bar{p}$, the household has a surplus.

To summarise all the individual disparities, we must add them in absolute value: the total quantity $\Sigma_{i \in I} \mid p_i - n_i \bar{p} \mid$ can be divided by 2, because a transfer from the richest to the poorest (in the style of Robin Hood) would affect two households at a time. It is also related to the total number of rooms noted as P in order to normalise the result. The final indicator is the "relative mean dispersion", or RMD:

$$R = \frac{\Sigma_{i \in I} \mid p_i - n_i \bar{p} \mid}{2P}. \tag{11.1}$$

Its value is between 0 and 1: a lower value indicates greater equity, while a value close to 1 indicates very high inequity.

In order to compare categories of households k, we define a relative deviation indicator; by noting N_k the number of persons in class k households and P_k the total number of rooms in their dwellings, their overall deviation from the average, in relation to their number of rooms, is

$$s_k = \frac{P_k - N_k \bar{p}}{P_k}. \tag{11.2}$$

This is a surplus since a positive value marks an excess over a perfectly equitable distribution. By denoting $\bar{\bar{p}} = P_k/N_k$ as the average number of rooms per person in this category, the relative surplus shows that

$$s_k = 1 - \frac{\bar{P}_k}{\bar{\bar{P}}}. \tag{11.3}$$

In the population divided according to the categories $k \in K$, we still note that

$$\Sigma_{k \in K} \frac{P_k}{P} \mid s_k \mid \le 2R.$$

Literally, the weighted average of the relative surplus is less than $2R$. The weighting is based on the number of rooms obtained by the category of households, and not on the number of households or persons.

11.3.4 A Summary of Disparities in Housing for the Paris Region

We calculated the RMD indicator in the Paris region for each population census between 1968 and 2012. Figure 11.11 shows the values around 20%, which grow at a moderate pace over time. The figure also shows some theoretical references: on the one hand, the minimum disparity that would inevitably remain after the most advanced redistribution, because both the number of rooms per dwelling and the number of persons per dwelling are integers; on the other hand, the dispersion that would result remain after a totally random distribution. The actual distribution is clearly situated between the two theoretical references, a little closer to the second than to the first.

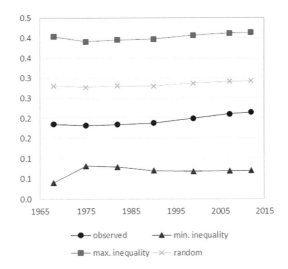

Figure 11.11 The RMD index for the Paris region between 1968 and 2012.

The analysis of relative surpluses by age group clearly shows the increasing influence of age on the surplus at each census (see Figure 11.12a). The analysis of the relative surpluses according to the SPC establishes an opposition between the retired households on the one hand and the active households on the other. The retired households have a positive surplus that grows over time, whereas all the categories of active workers have a negative surplus, with the exception of managers, whose surplus approached zero in 2012 (Figure 11.12b).

Overall, the summary indicators show a strong inequality between generations (in the current sense of the term), after or before retirement, respectively.

We must bear in mind that this observation is static and does not integrate the life trajectory of the households. In fact, it is dynamic, as is the residential path. Throughout its existence, each household will pass through the two active and retirement phases, respectively. On the two phases as a whole, therefore dynamically, the balance per household will probably be much more equitable.

11.4 Spending and Housing Costs by Household and Its Strategy

Let us turn to the monetary counterpart of the housing service. For the households that benefit from it, the service has a cost that, up to its sign, reveals the associated value. It is necessary to distinguish between the concepts of expenditure and cost: an expenditure is effective and directly observable, whereas a cost is a theoretical notion intended to evaluate an option (a strategy) in a decision problem.

We will begin by describing the housing system prices and market conditions before defining the notion of expenditure and quantifying it in absolute terms. In the following, we will focus on the cost of housing occupation, net of charges, to relate it to the income and thus measure the budgetary effort ratio consented by a household in order

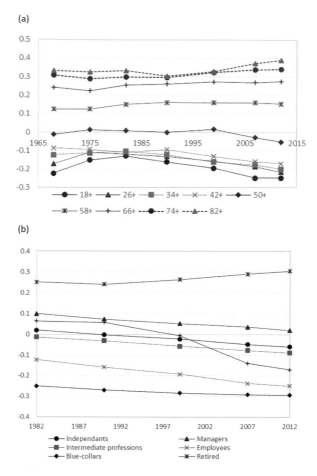

Figure 11.12 **Relative differences (a) according to age, (b) according to SPC.**

to have access to housing. To understand the evolution of effort ratios over time, we will define an effort ratio index per purchase price unit. Finally, we will give a theoretical definition of the cost of housing for a household that plans to become owners. This cost is directly comparable to the price of rent for an equivalent property during the same period.

N.B. In this part and in the following, we rely on national housing surveys (enquêtes nationales de logement, ENL) that have been carried out in France every five years since 1973, the last being in 2013. These surveys describe housing in terms of quality and quantity, including household characteristics, surface area of the property, rent or mortgage repayment. We also use Insee-Notaires indices for the prices of purchased property and *Clameur* for rental.

11.4.1 Prices and Market Conditions

The statistical population of dwellings changes over time. Depending on the year of observation, the distribution of dwellings by surface or space varies. In order to describe the prices of housing goods in a way that is both synthetic and comparable over time, it is necessary to define a "price index" that relates the price of a good to the associated service quantity.

Let us now denote P as the price of a dwelling at the time of purchase (or L as the price of rent) and Q as the quantity of the good, which can be interpreted as the "floor area". The price index denoted as I_P (or I_L for rent) verifies that

$$P = Q \cdot I_P \, (\text{Or } L = Q \cdot I_L). \tag{11.4}$$

Thus, I_p can be thought of as a price per unit area (m^2) at the time of buying the property, and I_L as the rent per unit area and per period of time, typically a month. Between the different periods, we must also take the evolution of prices (inflation phenomenon) into account. In the following, we consider "constant price" indices after adjusting for the general rate of inflation.

From 1973 to 2013, housing purchase price index varied across several phases (see Figure 11.13). There was a significant increase until 1980, then a decline until 1985,

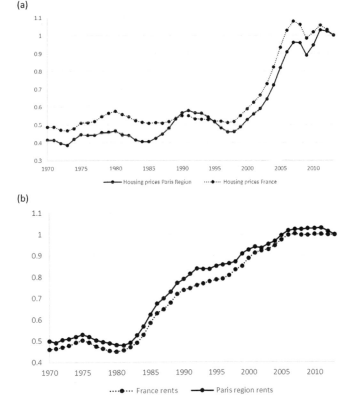

Figure 11.13 Property price index (a) for ownership, (b) for rental [ref. 2001 = 1].

followed by a stagnation until the end of the 1990s for metropolitan France. In the case of the Paris region, there was a large housing bubble between 1985 and 1992 and a corresponding decline until 1997. There was a sharp rise from 1998 to 2008. The price doubled in ten years, then stabilised at the highest level.

The temporal variations of the purchase price are faster and less monotonous than the demographic and rent variations. In fact, in the same period, the price index for rents increased fairly regularly (see Figure 11.13b), and the price index for the whole of France doubled between the end of the 1970s and the 2000s.

The purchase price is not the only parameter of a decision to buy. Borrowing possibilities also come into play. On the demand side, this concerns the ability to repay credit, whereas the supply side for this issue is constituted by banking institutions, mortgage rates and the loan terms granted.

Interest rates on home loans declined steadily during the period under review, remaining at a net inflation rate of around 2% during the 2000s (see Figure 11.14a). The complementary indicator of loan duration was fairly stable from 1970 to 1985, then decreased until 1997 and since then rose sharply from 14 to 20 years (see Figure 11.14b).

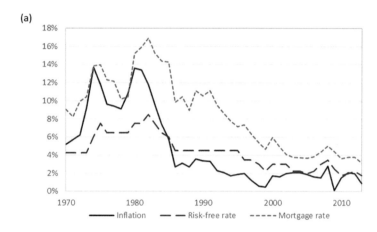

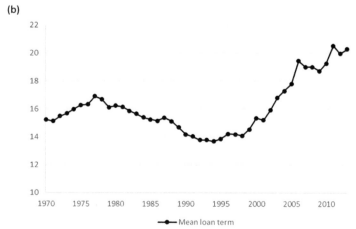

Figure 11.14 Variations in mortgage terms (a) interest rate, (b) repayment term.

In other words, people who access property through credit go into debt for a significant period – a quarter of their life expectancy, a third of their adult life. The sum of the duration of credit and the age at which the residential pathway is stabilised, around 40 years, is now close to retirement age, which in most cases is associated with a significant decrease in household income.

11.4.2 On Housing Expenditure

According to INSEE,[1] "the effort ratio is (...) the ratio between the sum of the expenses related to the main house and the household income. Expenses for homeowners include loan repayments for the purchase of housing, property tax and co-ownership fees. They exclude the cost of fixed capital and therefore differ from the user cost of housing. For tenants, they include rents and rental charges. For all households, they include the housing tax and the water and energy costs associated with housing".

These expenses represent the household's gross expenditure and are subtracted from the housing allowances to obtain the "net" expenses. As the expenditures are roughly proportional to the area of the property, among other factors, they therefore vary widely among the household population. Figure 11.15a shows the net monthly expenditure for households in the Paris region, by housing occupation status. "Recent" buyers (less than four years) have higher expenditures than the first-time buyers overall (loans in repayment), and they spend more than tenants in private housing, while tenants in social housing have the lowest expenses, due to both affordable rents and the benefit of the allowances. The level reached by recent buyers in 2013 represents more than a quarter of the average income of a household in the Paris region at that date (€ 4180 per month). The level of expenditure reached in the Paris region is much higher than in the rest of France: between 30% and 80% more depending on the occupation status (Figure 11.15b), although the dwellings are more spacious in the rest of France.

11.4.3 The Relationship to Income: The Effort Ratio

To retain its housing for a certain period, say a month, the household i spends rent L_i or repays an amount M_i for its mortgage. Relative to its income Y_i during the same period, the effort ratio denoted μ_i is defined as

$$\mu_i = \frac{L_i}{Y_i} \text{ or } \mu_i = \frac{M_i}{Y_i}. \tag{11.5}$$

This effort ratio does not take into account the charges for water and energy supplies, nor the expenses of co-ownership, etc.

For the Paris region, there was a gradual increase in effort ratios from 1973 to 2013, whatever the housing occupation status (Figure 11.16a). The levels of effort ratios in the most affluent categories (the first-time buyers) are lower than for the more modest categories (the social housing tenants).

---

1 Accessed on https://www.insee.fr/fr/statistiques/2492224 in April 2017.

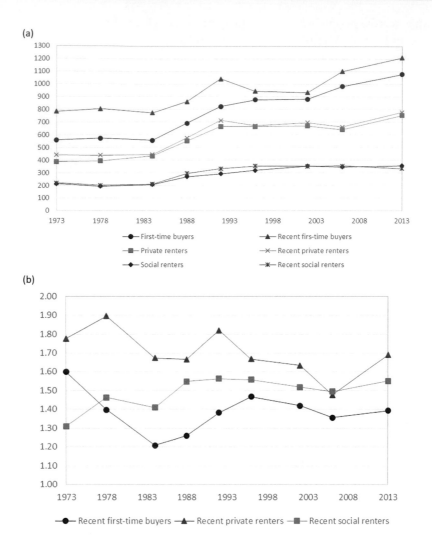

Figure 11.15 (a) Net monthly expenditure according to occupation status in the Paris region, (b) relationship between the respective expenses in the Paris region and in the provinces.

These are empirical effort ratios, as observed in household surveys. For purchasers, the effort rate for the owner option can also theoretically be modelled. The purchase price of the dwelling, P, is broken down into borrowed capital K and personal contribution P-K: the rate of loan borrowing is denoted $\tau = K/P$.

Depending on the interest rate of the loan r and the duration d (the number of repayment periods), the amount of the periodic repayment is proportional to the amount borrowed:

$$M = F(r,d)K, \text{with } F(r,d) = \frac{r}{1-(1+r)^{-d}}. \tag{11.6}$$

Also, depending on the quantity of housing Q and the price index $I_p = P/Q$, it comes to

$$\mu = \frac{M}{Y} = F \cdot \frac{K}{Y} = F \cdot \tau \cdot \frac{P}{Y} = F \cdot \tau \cdot I_p \frac{Q}{Y}. \tag{11.7}$$

This theoretical rate was calculated for each year with the prevailing conditions for F, I_p and Y, as well as $\tau = 40\%$. The results for the Paris region and the rest of metropolitan France are illustrated in Figure 11.16b with empirical effort ratios. The respective evolutions show that the theoretical rate is about 5% points lower than the empirical rate: this difference can be attributed to the increase in the quality of housing demanded by buyers.

(a)

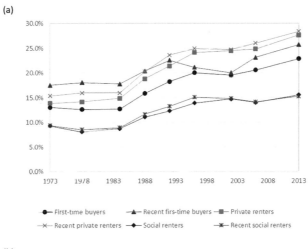

(b)

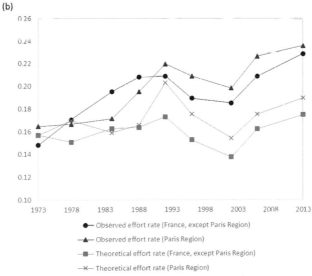

Figure 11.16 (a) Net effort ratio in the Paris region, (b) relationship between different notions of the purchasers' effort ratio.

11.4.4 The Effort Ratio Index

The effort ratio refers to the concept of household solvency: acquisition of a property that would require an excessive effort ratio is not sustainable for the household budget. Per unit of housing quantity (according to the Q variable scale), the effort ratio index is defined as follows:

$$I_\mu = \frac{\mu}{Q} = F \cdot \tau \cdot \frac{I_P}{Y}.$$ (11.8)

By normalising to $\tau = 1$, comes $I_\mu = F \cdot \frac{I_P}{Y}$. It is a demand solvency index. Since $F = M/K$, the index verifies that

$$I_\mu = \frac{I_P / Y}{K / M}.$$ (11.9)

It is therefore the ratio between the price level relative to the household income, in the numerator, and the repayable capital at a given repayment amount, K/M, which can be evaluated in years, in the denominator.

Figure 11.17 shows the variations of the two components under a convention of equality in 2013. The evolutions are largely comparable, corresponding to the stability of the ratio and therefore of the index I_μ.

11.4.5 Becoming an Owner: An Option and Its "User Cost"

A household looking for housing on the private market has two options: either to rent a home from a landlord or to buy a home to occupy it. These two occupation options each have advantages and disadvantages. By restricting the analysis to the financial aspects of the occupation, in the current system, the cost of the "Tenant" option per period is simply the price of the rent. The cost of the "Homeowner" option includes

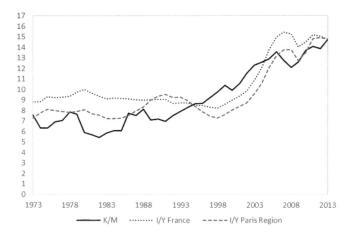

Figure 11.17 Joint variations of the two components of the theoretical effort ratio index.

several parameters. For a housing purchase price P bought by borrowing capital K at the interest rate r, there is first the amount for the repayment of the loan for the period of use, that is $r.K$. Added to this is the opportunity cost of the capital for the initial contribution, $P-K$, at the interest rate r', that is $r'(P-K)$. Property taxes for the period, assumed to be proportional to the purchase price and denoted $p \cdot P$, as well as the depreciation of the property at the rate δ, that is $\delta \cdot P$, must also be taken into account. Finally, the property market will evolve with an algebraic growth rate of π, so the property will be appreciated by $\pi \cdot P$. In total, the user cost of the "Homeowner" option per period is

$$UC = r \cdot K + r' \cdot (P - K) + (p + \delta - \pi)P. \tag{11.10}$$

By relating this cost to the housing quantity $Q = P/I_p$ and noting again $\tau = K/P$, we obtain the unit cost

$$\frac{UC}{Q} = (r\tau + r' \cdot (1 - \tau) + p + \delta - \pi) \cdot I_P. \tag{11.11}$$

This unit cost is directly comparable to the rent price index I_L up to one slight difference: capital and income taxation. We denote v as the marginal income tax rate for the household in question. The opportunity cost of the initial contribution is reduced to $(1-v) \cdot r'(1-\cdot\tau)$. If, moreover, mortgage loan repayments are deductible from income, as is the case in the United States, then the cost of the capital borrowed is reduced to $(1-v) \cdot r \cdot \tau$. The revised expression can be denoted by $UC \cdot v / Q$.

This cost formula remains incomplete. There is a lack of terms for transaction costs, such as moving expenses, property transfer taxes, notary fees *"in the strictest sense"* and agency fees. The same is true for the risks affecting the appreciation of the property by the market (uncertainty on π). There are still some advantages of being both owner and occupant at the same time: the satisfaction of being "master of one's own home", the possibility of carrying out part of the maintenance by oneself, the economy of a relationship with a landlord and faster interventions in the event of disturbances (failures), etc. These additional costs, insofar as they are monetisable, can be fully integrated with the user cost. As for the "Tenant" option, there is also a lack of transaction costs, in particular moving expenses, which are proportionally greater because they have to be depreciated over a period of occupation that is generally less than for the alternative option.

We measured the respective costs UC/Q and I_L of the two options between 1970 and 2013 for metropolitan France and Ile-de-France in particular (Figure 11.18). The cost of the "Homeowner" option has varied widely over time, with growth until 1982, decline until 1997, when it reached a historically low level, followed by a rebound in the 2000s due to higher prices (index I_P).

11.5 Access to Property: A Decision-Making Problem

We have highlighted the diversity of individual household situations (in Sections 11.2 and 11.3) and modelled the economic value of occupation options as homeowner or tenant (Section 11.4). The next step is to study the choice of a housing option at the

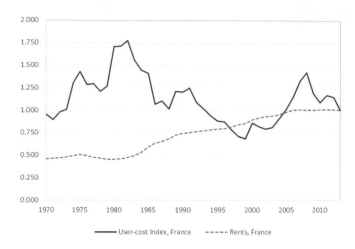

Figure 11.18 The respective cost indices of the two occupation options.

individual level of a household, to model the economic behaviour of choice and to econometrically estimate the model thus constituted for Ile-de-France. We will start with the problem of choice and by composing a modelling structure. We will then specify a specific econometric method, based on the French national housing surveys. Following this, we will state the status choice model and estimate its parameters. Similarly, we will state the quantity of service models before estimating the parameters. We will end with a summary of the overall model in order to outline its outreach and limitations.

11.5.1 The Problem of Choice and Modelling Structures

Over the course of its life, a particular household is caused to make decisions regarding its residential establishment. The issue for the household is twofold, and not only concerns the service rendered for everyday life, but also concerns the constitution of assets. Since the chosen housing has an economic cost, the user cost is an outgoing flow of money from the "balance sheet". Thus, the choice of residential establishment constitutes a decision problem for the household, which arbitrates between the quality of the service (in particular, the housing floor area, which is quantitative) and the economic cost. The household leverages its freedom of choice among the housing options. At a given moment, it usually occupies a dwelling. One reference option is to stay in the occupied dwelling, while alternative options require a move – or simply a change of occupation status.

We are interested here in the problem of the residential transition, changing from a reference option to one or another of the various alternative options. We classify these alternative options in priority according to the occupancy status, which are described in three categories of Owner Occupant (Propriétaire Occupant, PO), Private Market Tenant (Locataire du marché Privé, LP) and Social Housing Tenant (Locataire du parc Social, LS). Each modality covers a subset of concrete dwellings, which present a variety of intrinsic characteristics (surface area, type of individual or collective habitat, quality of construction, environmental performance, etc.) and extrinsic features (the position in relation to amenities and the configuration of urban places, i.e. the location).

We reduce the diversity of concrete options to a single summary characteristic: the quantity of housing service, per period of occupation. This notion refers to the floor area of the housing, the quality of the type and the economic valuation of the location within the property market in question (specific to the occupation status and the urban area considered), etc., bringing them together within a single quantitative indicator of an economic nature measured on a monetary scale.

Thus, our model of choice is constituted and structured in two stages: a first stage to choose the occupancy status and a second stage to choose the quantity of service. This model of housing demand is microeconomic since it involves a decisional unit – the elementary household – and by the principle of economic rationality in decision-making. However, it is also synthetic, and therefore aggregated, with regard to the housing options.

11.5.2 Econometric Method

We availed ourselves of data from seven national housing surveys (ENL) carried out between 1980 and 2013. The questionnaire completed by a household describes the main characteristics of this household (number of persons, gender and age of the reference person, SPC), the characteristics of the dwelling occupied during the survey (including the initial year of the occupation), as well as the characteristics of the dwelling occupied four years before the survey, and finally the number of residential changes that occurred in the meantime.

For each of the surveys, we only consider the households based in Ile-de-France during the survey, and out of these, those who occupied a rented home on the private market in the situation four years prior to the survey. We thus observe the subpopulation of private renter households four years before the survey.

For each of these households, the observation reveals whether at least one change has occurred, and if so, to what occupation status and for what quantity of housing service. Per household i observed, we denote A_i the year of survey and A_i' the initial year in which the dwelling was occupied. If $A_i' < A_i - 4$, then no change has occurred in the observed interval, and we define $A_i'' \equiv A_i$. If $A_i - 4 \leq A_i' \leq A_i$, then at least one change has occurred. In cases with a change, we focus on the last one made and we set down $A_i'' \equiv A_i'$.

For any year m between $A_i - 4$ and A''_i, the household i provides a point of observation (i, m), to which we associate a description of the household characteristics, through backwards projection from the situation declared in the year A_i, especially for the age of the reference person and their income, according to the SPC and the annual rates of change successively recorded at the regional level. We postulate that there is statistical independence between the points (i, m) of the same household. This interdependence postulate is simplifying. We completed the data set by associating the "market conditions" for both the year and the household with each observation point. These are indices I_P of purchase price and I_L of rent and user costs. For the ownership option, we took into account the borrowing rate and the loan interest rate, which are specific to the household.

II.5.3 The Choice of Occupation Status: Transition Model

Per household i as a private sector tenant, we consider a year m to constitute a choice between staying in the housing occupied, which is the reference option, and changing housing by making a transition to one of the three options associated with the occupation statuses: Owner Occupant, Private Tenant or Social Tenant. The realm of choice thus consists of four options k.

We postulate (i) that the household evaluates each option k by a "random utility" function, denoted $U_{ki}^{(m)}$, which incorporates the characteristics of the household and those of the option, and (ii) that the household chooses the option that presents the maximum utility for it. This second postulate is an axiom of rational preferences for the microeconomic behaviour of the household.

In this kind of discrete choice model, each option is chosen with a certain proportion, which is the probability of offering the best utility:

$$\Pr\{k\,|\,i,m\} = \Pr\{U_{ki}^{(m)} \geq U_{k'i}^{(m)} : \forall k' \in K\}. \tag{11.12}$$

As an initial approach, we favoured a simple specification for random utility functions. The function $U_{ki}^{(m)}$ is the sum of a deterministic part $V_{ki}^{(m)}$ and a random term $\eta_{ki}^{(m)}$:

$$U_{ki}^{(m)} = V_{ki}^{(m)} + \eta_{ki}^{(m)}. \tag{11.13}$$

Assuming that variables $\eta_{ki}^{(m)}$ are independent and identically distributed according to a Gumbel law, the model specification constitutes a multinomial logit model, such that equation (11.12) is solved by the following formula:

$$\Pr\{k\,|\,i,m\} = \frac{\exp\left(V_{ki}^{(m)}\right)}{\Sigma_{k' \in K}\exp\left(V_{k'i}^{(m)}\right)}. \tag{11.14}$$

We set the deterministic utility of the reference option to zero. For the other options, we formulated the deterministic utility as the following function:

$$V_{ki}^{(m)} = \theta_{ki}^{(m)} + \alpha_k \ln P_{ki}^{(m)} + \alpha_k \ln Y_i^{(m)} + \gamma_k \ln n_i^{(m)}, \tag{11.15}$$

where $\theta_{ki}^{(m)}$ is a "constant" parameter that may depend on the age of the reference person in i to m, α is the coefficient of the option for a price $P_{ki}^{(m)}$, β_k is the option coefficient that measures the sensitivity to income $\gamma_i^{(m)}$ (to its logarithm, in fact), and Y is the coefficient of sensitivity to the log of the number $n_i^{(m)}$ of consumption units in the household (i.e. the equivalent number of adults). As price variable $P_i^{(m)}$, we selected 1 for the LS option (hence $\alpha_{LS} = 0$), the rent price index $I_L^{(m)}$ for the LP option and, for the PO option, the user cost product for the purchase price index $I_P^{(m)}$.

Thus, the dwellings that materialise an occupation status option are synthetically described by the price variable: neither the area nor the location is modelled.

Table 11.1 Estimation Results of the Transition Model

	Transition Towards Private Rental		Buying Property		Transition Towards Social Rental	
Permanent income (log)	0.45	***	0.95	***	−0.77	***
Rent prices (log)	−0.29	***				
User cost of purchase (log)			−0.85	***		
Age of the reference person						
under 25	0	Ref.	0	Ref.	0	Ref.
[25; 30]	−0.66	***	0.21	n.s.	−0.37	*
[30; 35]	−1.18	***	0.11	n.s.	−0.75	***
[35; 40]	−1.37	***	−0.28	n.s.	−0.94	***
[40; 45]	−1.54	***	−0.62	**	−1.10	***
[45; 50]	−1.69	***	−0.90	***	−1.41	***
[50; 55]	−2.01	***	−1.09	***	−1.05	***
[55; 60]	−2.22	***	−1.70	***	−1.65	***
[60; 65]	−2.40	***	−2.48	***	−1.83	***
65 and over	−2.95	***	−1.72	***	−1.81	***
Household size (HCU)	−0.16	**	−0.19	*	0.16	*
Marital status						
Couple with child(ren)	0	Ref.	0	Ref.	0	Ref.
Couple without children	0.15	n.s.	−0.25	n.s.	0.07	n.s.
Single person	0.10	n.s.	−1.23	***	−0.55	**
Single parent family	0.37	n.s.	−0.62	***	−0.07	n.s.
Constant	−1.99	***	−2.73	***	−0.06	n.s.

$N = 36{,}457$ Significance, ***$p < 0.001$, ** $p < 0.01$, * $p < 0.05$, n.s. not significant.

11.5.4 Estimation of the Transition Choice Model

Table 11.1 reports the estimated values for the coefficients of the status choice model for Paris region households observed between 1980 and 2013. It is retained that

- Income has a positive influence on the utility of private options, but has a negative influence on the LS option;
- The negative influence of the price variable for private options;
- The size of the household has a positive influence on the social option but is negative on private options, which reflects a greater inertia towards housing change;
- The age of the reference person exerts a negative influence and becomes increasingly strong, which also reflects the inertia in the face of change.

Figure 11.19 illustrates the probabilities of choosing each status option for a certain household category according to income. Beyond very low incomes (for which households actually benefit from unreported external support), income appears to be an important and growing factor for access to the occupying property, and, conversely, a decreasing factor in the transition to social housing rental.

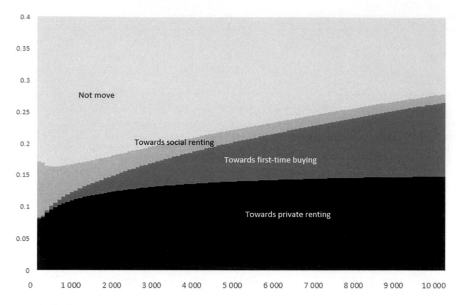

Figure 11.19 Occupancy status transition probabilities, by income, for a couple from the Paris region with no children, whose reference person is between 30 and 35 years old, 2012 situation.

11.5.5 Consumption-Level Models

Above the transition choice model, the second stage of the demand model deals with the quantity demanded by the household, that is its level of housing service consumption. We focused on private options, with a quantity model for the PO option and another for the LS option.

For each of the two options, we modelled the quantity demanded Q according to the following formula:

$$\ln Q_{ki}^{(m)} = \tilde{\theta}_{ki}^{(m)} + \tilde{\alpha}_k \ln P_{ki}^{(m)} + \tilde{\beta}_k \ln Y_i^{(m)} + \tilde{\gamma}_k \ln n_i^{(m)} \,, \tag{11.16}$$

where $\tilde{\theta}_{ki}^{(m)}$ is a parameter called "constant", $\tilde{\alpha}_k$ is the elasticity parameter of the quantity at the price, $\tilde{\beta}_k$ is the income elasticity and $\tilde{\gamma}_k$ is the elasticity household size, counted in number $n_i^{(m)}$ of consumption units in the household.

The quantity variable is determined as the ratio of the property price on the market to the market price index to the move-in year, that is P/I_P for PO or L/I_L for LP. As a price variable, the index is used for the LP option, and the product of I_P and the user cost for the PO option.

11.5.6 Estimation of Housing Quantity Models

Table 11.2 presents the estimation results of the two consumption-level models, for households in the Paris region from the private rental stock who moved in the four years preceding the observation by the survey. The indicators of the elasticities are

Table 11.2 Estimation Results of the Consumption Quantity Models

Quantity of Service Model	Home Owners	Private Renters
"Constant" coefficient	10.901	5.099
Price elasticity	−0.486	−0.257
Income elasticity	0.437	0.339
Household-size elasticity	0.246	0.287

The price elasticity is higher in absolute value for the "Homeowner" option than for the "Private Renter" option, and similarly, the income elasticity is higher for the PO option, while the elasticities to the household size are close between the two options.

in line with the theory: negative for price, positive for income and the number of consumption units.

Figure 11.20 shows the average price predicted for the property occupied according to each occupation option, per household category and per income.

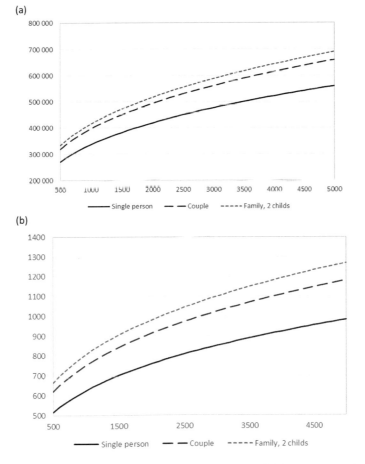

Figure 11.20 Property price by income, (a) Homeowner, (b) Private rental.

11.5.7 Summary of the Demand Model

Our two-level model of housing demand refers to a specific category of households: private sector tenants, the most likely to go on to own their own home. The observation by the ENLs is both accurate and limited. It is accurate for the characteristics of the household and those of the occupied dwelling (and also of the previous dwelling), but restricted to the dwellings occupied by the household. The choice of the household can be carried out in relation to many concrete options, however! We have additionally modelled the price conditions in each private housing market, but not the detail of the properties offered in these markets.

This is why our model is primarily concerned with access to home ownership. The estimation results obtained are reasonable, both for the propensity to change status and for the demand for quantity of floor area. The factor coefficients are consistent with economic theory, especially for price, income and household-size elasticities for the two quantity models. In addition, between the demand functions of the two private markets, the respective elasticities are arranged in a certain order according to the influence factor: income has a bigger effect for a property that is purchased than for a dwelling that is rented, as with the price.

11.6 In Conclusion: Mobilising the Social Sciences for the Analysis and Eco-Design of the Property System

11.6.1 A Socio-Economic History Informed by Demography

Reviewing over a half-century of housing history in metropolitan France and the Paris region, we have highlighted a long-term growth in "area comfort", measured in the number of rooms per person. This macroscopic observation is reinforced by the analysis of "residential pathways" in terms of living space: the demographic cohorts show that the individual residential pathways are ascending according to the age and that pathways have been increasingly higher between the successive generations, reaching an asymptotic trajectory over the past 20 years.

In parallel, the analysis of the economic conditions of housing reveals that the prices of property have increased sharply over this recent period. The temporal conjunction of the two bullish trajectories – for the quantities, on the one hand, and the prices, on the other – could be interpreted as an enrichment of the households that own the properties. It could also be expressed as the progressive increase of the expenditures made by households to achieve a certain social standard as regards housing.

In this "social game", the younger generations are at a disadvantage compared to previous generations. They are under pressure from the property market, even though they are also suffering from the labour market. In the French case, this explains the progressive development of housing subsidies, some of which specifically target young adults.

11.6.2 Socio-Economic Indicators for Housing Inequalities

We modelled the size of the dwelling in terms of the number of rooms and according to the characteristics of the household by a Poisson regression, which reveals the respective influences of the composition of the household, the income, age and SPC of the reference person, as well as the type of individual or collective housing and the residential location (with respect to the centre of the urban area). This model accounts for a large part of the diversity of housing conditions among households.

In a simplified way, we also applied inequality indicators to the household population in keeping with the Pietra–Ricci–Hoover method. We noticed inequalities that are obvious in a given period, and which tend to grow over time. This socio-economic indicator reveals the effect of age in particular, showing a significant advantage for the older category, the one of retired households.

These "cross-sectional" indications should be cross-referenced with the demographic analysis according to cohorts. Housing equity between households should be conceived across the life cycle, taking the entire residential pathway into account and going beyond one particular stage.

II.6.3 Contributions to the Economic Modelling of the Housing System

In the introduction to this chapter, we noted the importance of LUTI models for understanding the economics of the housing system in an urban area. Our econometric models are preparatory contributions. The revelation of residential pathways and generation effects can be used to segment and profile the demand for housing in a dynamic model of the housing system in the Paris region. Furthermore, the specification and estimation of the two-level demand model yielded results (in terms of decision structure, formulas of functions and parameter values) that could be used to simulate demanders' behaviours.

However, we limited ourselves to a very synthetic description of the property options. Our contribution falls more in the field of housing economics than in the field of urban economics. The main lesson to be learned for an LUTI model is the highlighting of the "patrimonial fact" and the financial efforts that households devote to access property.

II.6.4 A Largely Perfectible Information System

To diagnose the situation of households in the housing system, as demanders, we used the well-established statistical databases, namely the population census and the national housing surveys, as well as observatories of price and, more broadly, transaction conditions. These various bases are the components of an information system that still remains to be integrated, to be constituted in the form of a powerful technical device. Even when considered together, the bases reported poorly reflect the interaction between housing supply and demand. The housing stock is revealed by the population census.

It would be even better described (at the French level) through tax files, but they do not indicate the market value of the properties, nor the willingness to sell, and are limited to the status of occupation or vacancy. With regard to demand, the population census describes the characteristics of households and the use of occupied housing, but not the prices, nor the aspirations and property plans of the households. As for usage, the census and the ENL describe concrete use, but without indicating the potentialities, whereas to understand an economic market, it is necessary to identify the actors concerned and, for each of them, the opportunities available to them and which constitute their domain of action and decision-making.

To produce the information that we have signalled as being of interest, the solution certainly requires modelling and simulation. LUTI models produce virtualities and constitute a particular solution. There is another solution that appears to be more powerful for knowing the states reached. In the digital age, property transactions are all recorded somewhere and their computerised descriptions could be centralised. This kind of an observation system would make it possible to reveal and synthesise the completed transactions, and also to find out about property offers and transaction circumstances, for example the duration of an offer, the progression of a negotiation conducted with or without success, the evolution of the price and the participation of one or more estate agents in the processes

The web platform of an estate agent network certainly resembles this type of centralised system. The possibility of an all-encompassing system to serve collective management of housing system issues deserves to be studied, considering factors such as development and location of the housing stock, price levels, adequacy to the needs and aspirations of the population and environmental performance. With regard to this last point, it would be necessary to add specific questions to the census and also the ENL.

11.6.5 An Operating System Whose Management Remains to be Empowered

The technological potentiality of a centralised information system on the housing system challenges public authorities, as an opportunity to improve their social, economic and environmental performance. Indeed, the public authorities directly control important levers of action: land use and urban development, the development and management of social housing, as well as a wide range of financial incentives – fiscal instruments and financial support of a social character. They indirectly influence other levers of action, notably property development and urban construction and renewal initiatives.

The digital era lends itself to the development of a centralising system, not only for information but also for management, to prepare for the medium- and long-term future of the housing system. This is the founding ambition of the LUTI models. It is now possible, and certainly opportune, to develop a 3.0 generation of LUTI models, beyond generation 1.0 (static LUTI models) and generation 2.0 (dynamic LUTI models with high geographic information content), by adding strong behavioural content and enhancing the coherence of specific policies.

Bibliography

Accardo, J., Bugeja, F., 2009. Le poids des dépenses de logement depuis 20 ans. Cinquante ans de consommation en France, INSEE.

Arrondel, L., Lefebvre, B., 2001. Consumption and investment motives in housing wealth accumulation: a French study. *Journal of Urban Economics* 50, 112–137.

Batten, D.C., 1999. The mismatch argument: the construction of a housing orthodoxy in Australia. *Urban Studies* 36, 137–151.

Berger, M., 2016. *Les périurbains de Paris: De la ville dense à la métropole éclatée?*, *Anthropologie*. CNRS Éditions, Paris.

Berger, M., Rougé, L., Thomann, S., Thouzellier, C., 2010. Vieillir en pavillon: mobilités et ancrages des personnes âgées dans les espaces périurbains d'aires métropolitaines (Toulouse, Paris, Marseille). *Espace populations sociétés. Space populations societies* 53–67.

Bonvalet, C., Bringé, A., 2010. Les trajectoires socio-spatiales des Franciliens depuis leur départ de chez les parents. Temporalités.

Bourassa, S.C., 1996. Measuring the affordability of home-ownership. *Urban Studies* 33, 1867–1877.

Brueckner, J.K., 1986. The downpayment constraint and housing tenure choice: a simplified exposition. *Regional Science and Urban Economics* 16, 519–525.

Bugeja-Bloch, F., 2013. Logement, la spirale des inégalités: une nouvelle dimension de la fracture sociale et générationnelle. PUF.

Díaz, A., Luengo-Prado, M.J., 2008. On the user cost and homeownership. *Review of Economic Dynamics* 11, 584–613.

Frosini, B.V., 2012. Approximation and decomposition of Gini, Pietra–Ricci and Theil inequality measures. *Empirical Economics* 43, 175–197.

Goodman, A.C., 1988. An econometric model of housing price, permanent income, tenure choice, and housing demand. *Journal of Urban Economics* 23, 327–353.

Jacquot, A., 2006. Cinquante ans d'évolution des conditions de logement des ménages. *Données sociales, la société française (édition 2006)*, INSEE Références 467–473.

Kain, J.F., Quigley, J.M., 1972. Housing market discrimination, home-ownership, and savings behavior. *The American Economic Review* 263–277.

Laferrère, A., 2008. L'impact du vieillissement de la population sur les marchés immobiliers. Recherches et prévisions.

McFadden, D.L., 1973. Conditional logit analysis of qualitative choice behavior, in: Zarembka, P. (Éd.), *Frontiers in Econometrics*. Wiley, New York.

Mulder, C.H., 2006. Population and housing: A two-sided relationship. *Demographic Research* 15, 401–412.

Robinson, R., O'Sullivan, T., Grand, J.L., 1985. Inequality and housing. *Urban Studies* 22, 249–256.

Timbeau, X., 2013. Les bulles « robustes »: Pourquoi il faut construire des logements en région parisienne. *Revue de l'OFCE* 128, 277.

Tunstall, B., 2015. Relative housing space inequality in England and Wales, and its recent rapid resurgence. *International Journal of Housing Policy* 15, 105–126.

Part 3

Simulation at the Service of Eco-design

Chapter 12

DREAM

An Urban Equilibrium Model

Fabien Leurent, Nicolas Coulombel, and Alexis Poulhès

École des Ponts ParisTech

12.1 Introduction

12.1.1 Context

Two socio-economic systems are of particular interest for urban planning: the land-use system (which includes the real estate system) and the transport system. These two systems are strongly intertwined, as established by body of scientific contributions (e.g. Levinson 2007, and King 2011 for recent contributions, and Lowry, 1964, for a more historic one). The relationship between the two is not symmetrical however, with their respective urban change processes working on different time scales (Simmonds et al., 2011). Land use–transport interaction (LUTI) models, the aim of which is to model the urban change processes linking these two systems, are therefore especially useful for the design and ex ante evaluation of urban development scenarios.

The typology proposed by DSC et al. (1999) distinguishes two main families of LUTI models: static models such as RELU-TRAN (Anas and Liu, 2007) or Pirandello (Delons et al., 2009) that focus on the long-term supply–demand equilibrium, on the one hand, and quasi-dynamic models through the micro-simulation of agents and decisions, such as UrbanSim (Waddell et al., 2003) or ILUTE (Salvini and Miller, 2005) on the other.

Static models are simpler and constitute the first generation of models. They are therefore the natural choice for simulating a territory and informing the planning process, before using a more detailed and more complex dynamic model if this proves necessary.

It is above all vital for the model's hypotheses to be realistic and to be verifiable by an analyst. Most static models assume that the total floor area of the dwellings in each zone in the area can be influenced by the demands of households and their aggregation, and also that each household is a perennial microeconomic agent.[1] The dynamic framework removes these limitations, but at the cost of an increase in the number of hypotheses whose interactions are not only more difficult to check but also more difficult to identify qualitatively.

---

1 *i.e.* does not change over time, without any notion of stages in the life cycle.

12.1.2 Objective

We present here a static model for the equilibrium between supply and demand on an urban housing market. This model disaggregates demand according to household size, place of work, category of activity and income. It is designed for the private rented property market, so social housing and home purchase are not dealt with at this stage but may be covered by subsequent extensions. The size of each dwelling is fixed, and the number of dwellings, broken down on a qualitative basis (including size) and with respect to location, is elastic to price according to a given specification.

We shall provide an efficient mathematical formulation for this model: starting from primal–dual equilibrium conditions, a variational inequality problem will be formulated, as well as a concave maximisation programme. This has enabled us to demonstrate the existence and uniqueness of the equilibrium and propose a robust computation method.

The model is intended for simple applications, providing the first-round results in the context of an operational study or for teaching purposes.

12.1.3 Approach

Overall, an economic model is specified for which a mathematical and algorithmic treatment is provided: the paper is situated at the crossroads between urban economics and operations research.

The mathematical treatment is derived from traffic assignment on a transport network, in particular, dual criteria assignment with distributed cost–time trade-offs. Demand will be disaggregated discretely according to the employment area and household size, and continuously according to income. For each household, we will model the discrete choice between the housing options, according to zone and quality (which includes size), assuming that the household is sensitive to the price of housing and the cost of transport between home and work. Transport between home and work has a monetary cost and a temporal cost, which determine budgetary affordability and affect utility in relation to income. Thus, for each employment area and discrete type of household, the choice of residential zone and quality of dwelling is analogous to the choice of a route on an origin–destination pair on the basis of cost and time, for a population of users who are differentiated on the basis of the cost–time trade-off. We will adapt our previous treatment of the dual criteria assignment of traffic onto a network (Leurent, 1993) to the microeconomic choice of housing: this adaptation is straightforward for a simple utility function, for example Cobb–Douglas or similar.

12.1.4 Outline of This Chapter

The rest of this chapter consists of four parts followed by a conclusion. We begin by stating the hypotheses of the model (Section 12.2). From these, we derive a "grouped" demand function by aggregating microeconomic agents without restricting their individual behaviour (Section 12.3). Next, we will formulate the supply–demand equilibrium, give mathematical formulations, establish the properties of existence and uniqueness, and provide a resolution algorithm (Section 12.4). Last, we present the first results from an application to the Paris region, including the analysis of a test scenario of increased fuel prices (Section 12.5). The conclusion will lay out some possibilities for extension (Section 12.6).

12.2 Modelling Hypotheses

The model's hypotheses relate to job supply (Section 12.2.1), housing supply (Section 12.2.2), transport supply (Section 12.2.3) and demand for housing and transport (Section 12.2.4).

12.2.1 Job Supply

Let us assume that the number of jobs per zone for each category of activity (which can include the nature of the activity and the job category) and for each level of salary is exogenous in the medium term. Let $W = J \times K$ be the set of pairs $w = (j, k)$ of an employment $j \in J$ and a category of activity $k \in K$. The number of jobs in the category is denoted by Q_w, and the wages are distributed according to the distribution function H_w. Thus, economic activity in the area does not depend on the economic conditions that apply to the housing and transport markets.

12.2.2 Housing Supply

Let us assume that housing is distributed according to quality and locational zone. In particular, the quality includes the floor area of the dwelling, the quality of construction, the quality of the neighbourhood and the available local amenities. R denotes all the types of property, s_r denotes the average floor area per dwelling of type r, S_r the quantity supplied and p_r the price per unit of floor area.

Assume also that in the medium term, supply is elastic to the market price according to the function:

$$S_r = O_r(p_r). \tag{12.1}$$

In general, we assume that O_r is an increasing function. Let $O_r^{(-1)}$ the reciprocal function also be increasing.

In principle, these hypotheses are valid for the private rented property sector. The social rented property sector operates differently, with a supply function that depends on both the production cost and private sector prices, while demand depends on both the private sector price and the waiting period. The owner-occupier sector bears a closer resemblance to the private rented sector than the social sector, but the transaction costs depend on the length of occupation and the price for the demander depends on the cost of credit and the initial deposit (therefore capital as well as income).

12.2.3 Transport Supply

It is assumed that transport services are available between the different types of residential areas (by zone and housing mode) $i \in I$ and the types of employment $w \in W$ (making a distinction according to the zone and, if necessary, the category of activity which can generate specific modal requirements such as car access). Servicing the pair (i, w) involves a travel time $\tilde{t}_{i,w}$ and a monetary cost $t_{i,w}$.

These terms are taken as exogenous, which is independent of network flows and the income derived from the job held.

12.2.4 Demand for Housing and Transport

Household demand is the keystone of the system as it links jobs, housing and transport. We have assumed that demand consists of a set of households indexed by m, each of which has one job and a size v_m that is expressed as the number of individuals or consumption units. Household employment is fixed in terms of location, category of activity k_m and salary ρ_m.

For the household, an r-type dwelling is a residential option with floor area s_r and quality q_r, but it has housing expenditure of $p_r.s_r$ and travel expenditure ($t_{r,j}$ in money and $\tilde{t}_{r,j}$ in time). These expenditures reduce the amount of the household's income that is available for all types of consumption apart from housing and transport. These types of consumption are known as the "numeraire good" which is monetised and denoted by z.

We have made the assumption that the household desires space in its dwelling, quality (apart from the floor area) denoted by q, the numéraire good z and free time τ, and that its preferences for a "bundle" (s, q, z, τ) are represented by utility function $U_m(s, q, z, \tau)$. This function models the household's interest in the bundle. Each housing option constitutes a specific bundle. The household is assumed to be micro-economically rational, meaning that it selects the option with the maximum utility from those that are available to it on the market.

A residential option must be considered in relation to two budgetary constraints, one of which is monetary and the other temporal. We use $c_r(q_r, s_r, v_m)$ to denote the dwelling's running cost in each period. The monetary budgetary constraint requires the household's expenditure on housing, transport and the numéraire good to be compatible with income:

$$p_r s_r + c_r(q_r, s_r, v_m) + t_{r,j} + z \leq \rho_m \tag{12.2}$$

The temporal budgetary constraint compares the time at work $\tilde{\rho}_m$, travel time $\tilde{t}_{r,j}$ and the residual time τ to the available time θ_m:

$$\tilde{\rho}_m + \tilde{t}_{r,j} + \tau \leq \theta_m. \tag{12.3}$$

Let R_m denote all the housing options which are within the household's budgetary reach, that is which satisfy (12.2) and (12.3). An option of this type leaves the household with an amount of numéraire good and free time formulated as follows:

$$\tilde{z}_{m,r} = \rho_m - p_r \cdot s_r - c_r(q_r, s_r, v_m) - t_{r,j} \tag{12.4}$$

$$\tilde{\tau}_{m,r} = \theta_m - \tilde{\rho}_m - \tilde{t}_{r,j}. \tag{12.5}$$

Its utility for the household is therefore

$$\tilde{U}_m(r) = U_m\left(s_r, q_r, \tilde{z}_{m,r}, \tilde{\tau}_{m,r}\right). \tag{12.6}$$

The microeconomic behaviour of the household as a demander consists of selecting the best available option, that is

$$\text{Finding } r^* \in R_m, \forall r \in R_m : \tilde{U}_m(r^*) \geq \tilde{U}_m(r). \tag{12.7}$$

12.3 The Demand Function for Each Class (Segment)

We have specified the microeconomic model for an individual demander of housing and transport. This model is disaggregated on the basis of job location j, activity category k, income level ρ and household size v. Disaggregation is performed discretely for (j, k, v) and continuously for ρ. A demand class is defined by the triple $\sigma = j, k, v$ and a distribution function H_σ for income. Income is thus the central parameter in our analysis. By considering its distribution, we treat a class collectively in order to formulate its overall demand for a given set of supplied options.

12.3.1 Residual Consumption and Residential Utility

The relationship between the residual amount of the numéraire good and income is affine: denoting $P_{\sigma,r} = p_r.s_r + c_r\left(q_r, s_r, v_{m(\sigma)}\right) + t_{r,j(\sigma)}$ equation (12.4) yields that

$$\tilde{z}_{\sigma,r} = \rho - P_{\sigma,r}. \tag{12.8}$$

This is an increasing function of income.

The amount of free time $\tilde{\tau}_{m,r}$ does not depend directly on income ρ and can be denoted by $\tilde{\tau}_{\sigma,r}$.

The residential utility $\tilde{U}_{m,r}$ of a residential option r for a demander $m = (\sigma, \rho)$ is the utility function $U_m\left(s_r, q_r, \tilde{z}_{m,r}, \tilde{\tau}_{m,r}\right)$, on condition the option is compatible with the temporal and monetary budgetary constraints. This compatibility depends simultaneously on r, σ and ρ. When this compatibility exists (each constraint must be examined in relation to ρ), under given r and σ, the residential utility is an increasing function of the residual consumption $\tilde{z}_{\sigma,r}$, and therefore increases with income as $\tilde{z}_{\sigma,r}$ increases with ρ.

Let $\hat{U}_{\sigma,r}$ denote the utility of the option r in relation to income ρ, within segment σ. This function is increasing and continuous if the initial function U_m is continuous.

Let us also define the maximum residential utility function out of all the options that are feasible in terms of the budget,

$$U_{R\sigma}^*(\rho) = \max\left\{\hat{U}_{r\sigma}(\rho) : r \in R_{\sigma\rho}\right\}. \tag{12.9}$$

As an increase in income makes it easier to satisfy a budgetary constraint, a higher income makes it possible to access more residential options. A higher income therefore provides higher maximum residential utility for two reasons: firstly because the utility of each accessible option is higher and secondly because the choice set is larger.

12.3.2 Efficiency Domain of a Housing Option

For a given option r, we shall define the feasibility domain $F_{\sigma r} = \{\rho : r \in R_{\sigma \rho}\}$ and the efficiency domain, which is the set of incomes for which the option provides the maximum utility:

$$E_{or} = \left\{\rho : r \in R_{\sigma \rho} \text{ and } \hat{U}_{\sigma \rho}(\rho) = U^*_{R_{\sigma \rho}}(\rho)\right\}. \tag{12.10}$$

The assignment of each demander to their preferred option amounts to specifying the efficiency domains of the various options. If the utility functions have a simple form, then the efficiency domains do too, as they are intervals that are bounded by special values, the "break-off points". Let us make the specific hypothesis that there is an increasing function φ_σ such that the compound function $u_{\sigma,r} = \varphi_\sigma \circ \hat{U}_{\sigma,r}$ is affine in relation to income. As φ_σ is an increasing function, $\rho \to u_{\sigma,r}(\rho)$ so is, which is a utility function that ultimately depends on s_r, q_r, $\tilde{z}_{\sigma r}$ and $\tilde{\tau}_{\sigma r}$. Let's define the affinity parameters a and b according to income, such that

$$u_{\sigma r}(\rho) = a_{\sigma r} + b_{\sigma r}\rho = b_{\sigma r}(\rho - \zeta_{\sigma r}). \tag{12.11}$$

The efficiency domain of an option r is the intersection between its relative efficiency domains compared to any other option s, $E_{\sigma,r}(\rho) = \bigcap_{s \in R_\rho} E_{\sigma,r|s}$, for the relative domain

$$E_{\sigma,r|s} = \left\{\rho \in F_{\sigma r} : u_{\sigma,r}(\rho) \geq u_{\sigma,s}(\rho) \text{ if } \rho \in F_{\sigma s}\right\}. \tag{12.12}$$

Under the hypothesis of affinity, we can re-express the condition of superiority thus:

$$u_{\sigma r}(\rho) > u_{\sigma s} \Leftrightarrow b_{\sigma r}(\rho - \xi_{\sigma r}) \geq b_{\sigma s}(\rho - \xi_{\sigma s})$$

$$\Leftrightarrow (b_{\sigma r} - b_{\sigma s})\rho \geq b_{\sigma s}\xi_{\sigma s}.$$

If $b_{\sigma r} \neq b_{\sigma s}$, both sides of the previous line are equal at $\rho^*_{rs} = \dfrac{\left[b_{\sigma r}\xi_{\sigma r} - b_{\sigma s}\xi_{\sigma s}\right]}{b_{\sigma r} - b_{\sigma s}}$, and

$$u_{\sigma r}(\rho) - u_{\sigma s}(\rho) = (b_{\sigma r} - b_{\sigma s}) \cdot (\rho - \rho^*_{rs}).$$

So if $b_{\sigma r} > b_{\sigma s}$, then r is preferred when $\rho \geq \rho^*_{rs}$ and s when $\rho \leq \rho^*_{rs}$. If $b_{\sigma r} < b_{\sigma s}$, then s is better above ρ^*_{rs} and r is better below. In addition, it is necessary to include the admissibility conditions for each option: for r, the financial constraint is $\tilde{z}_{\sigma r} \geq 0$ so $\rho \geq \rho_{\sigma r}$, and the temporal constraint is $\tilde{\tau}_{mr} \geq 0$ so $\theta_{\sigma r} - \tilde{\rho}_m \geq \tilde{\iota}_{rj}$ (see Figure 12.1).

Finally, denoting by $D_{\sigma,r|s}$ the efficiency interval in the absence of admissibility constraints, it holds that

$$E_{\sigma,r|s} = F_{\sigma r} \setminus \left(D_{\sigma,r|s} \cap F_{\sigma s}\right) = D_{\sigma,r|s} \cap F_{\sigma s} \cap F_{\sigma r} \setminus \left(F_{\sigma r} \setminus F_{\sigma s}\right). \tag{12.13}$$

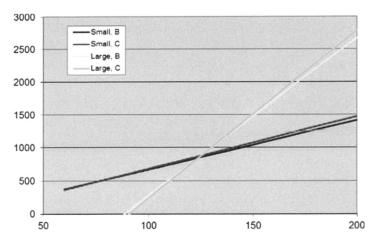

Figure 12.1 Reduced household utility function, by housing option, based on daily income.

The efficiency domains $E_{\sigma,r}$ of the options can be determined in a comprehensive manner, integrating the options in the order of increasing values of $b_{\sigma,r}$. Including a new option conserves or reduces the domain of each of the previous options, but it retains its character as an interval.

Ultimately, $E_{\sigma,r} = \left]\underline{\rho}_{\sigma r}, \bar{\rho}_{\sigma r}\right[$, perhaps with closed brackets and/or $\bar{\rho}_{\sigma r} = +\infty$. Set $E_{\sigma,r}$ may be empty, which we consider to be equivalent to $\underline{\rho}_{\sigma r} = \bar{\rho}_{\sigma r}$, if the distribution H_σ is continuous.

The intervals are arranged in increasing order of $b_{\sigma,r}$, so between two consecutive non-punctual intervals $s < r$, $\bar{\rho}_{\sigma s} = \underline{\rho}_{\sigma r} = \rho_{sr}^*$.

12.3.3 The Demand Function for Each Class

Under these conditions, the option r is optimal on $E_{\sigma,r}$, and selected for every value of $\rho \in E_{\sigma,r}$ (save possibly for the break-off points). Therefore, it attracts a number of customers

$$n_{\sigma r} = Q_\sigma Pr\{\rho \in E_{\sigma r}\} = Q_\sigma \left[H_\sigma\left(\bar{\rho}_{\sigma r}\right) - H_\sigma\left(\underline{\rho}_{\sigma r}\right)\right]. \tag{12.14}$$

The quantities $[n_{\sigma r} : r \in R_\sigma]$ make up the demand function of segment σ for residential options. They depend on the supply conditions.

12.3.4 Quasi-Cobb–Douglas Utility Function

The framework of hypotheses is particularly appropriate for the following utility function "à la" Stone–Geary:

$$U_{\sigma r} = (s, q, z, \tau) = K_\sigma\left((s - \upsilon_\sigma s_0)^+\right)^\alpha q^\beta z^\gamma \tau^\delta \tag{12.15}$$

This is of the Cobb–Douglas type with parameters $(\alpha, \beta, \gamma, \delta)$ that may themselves depend on σ, which is modified by requiring a minimum floor area, denoted by s_0 for each household member. The utility of an option with a floor area that is too small is reduced to zero.

The function $\varphi_\sigma : x \to x^{1/\gamma}$, applied to $U_{\sigma r}$, gives the following affine function:

$$u_{\sigma r}(\sigma) = \left[K_\sigma \left((s - v_\sigma s_0)^+ \right)^\alpha q^\beta z^\gamma \tilde{\tau}^\delta \right]^{1/\gamma} \tilde{z}_{\sigma r}(\rho) = b_{\sigma r}.(\rho - P_{\sigma r}). \tag{12.16}$$

Therefore,

$$b_{\sigma r}(\sigma) = \left[K_\sigma \left((s - v_\sigma s_0)^+ \right)^\alpha q^\beta z^\gamma \tilde{\tau}^\delta \right]^{1/\gamma}, \ \xi_{\sigma r} = P_{\sigma r} \text{ and } a_{\sigma r} = -b_{\sigma r}\xi_{\sigma r} = -b_{\sigma r}P_{\sigma r}.$$

The order of $b_{\sigma r}$ values depends on the characteristics q_r and s_r as well as the travel time $\tilde{\tau}_{rj}$ between home r and zone of work j for segment σ, via $\tilde{\tau}_{\sigma j} = \theta_m - \tilde{\rho}_m - \tilde{\tau}_{rj}$. A longer travel time reduces the $b_{\sigma r}$, and therefore acts in the opposite direction to a higher floor area s_r. *All other* things being equal, options which are identical in all ways except for travel time are arranged in order of decreasing travel time so, *a priori*, the efficiency domain of high incomes favours short home-to-work distances.

12.4 Mathematical Analysis

12.4.1 Primal–Dual Equilibrium Conditions

For option r and the segment σ, we shall associate the sum $N_{\sigma r} = \sum_{\ell=1}^{r} n_{\sigma\ell}$ and the attractiveness function as follows:

$$I_{\sigma r} = a_{\sigma r} + \sum_{\ell=r}^{R_\sigma - 1} (b_{\sigma\ell} - b_{\sigma,\ell+1}) H_\sigma^{(-1)}(N_{\sigma\ell}/Q_\sigma). \tag{12.17}$$

Let us assume that each demander is assigned to their optimum option. So between options s and r, which are used consecutively,

$$I_{\sigma r} - I_{\sigma s} = a_{\sigma r} - a_{\sigma s} - \sum_{\ell=s}^{r-1} (b_{\sigma\ell} - b_{\sigma,\ell+1}) H_\sigma^{(-1)}(N_{\sigma\ell}/Q_\sigma)$$

$$= a_{\sigma r} - a_{\sigma s} + (b_{\sigma r} - b_{\sigma s}) p_{sr}^*$$

Which is zero by definition of ρ_{sr}^*.

For an unused option t, the values $\rho_{\sigma t}$ and $\bar{\rho}_{\sigma t}$ are fixed as the break-off value between the two used options s and r, which are, respectively, lower and higher in the order of increasing values of $b_{\sigma r}$, therefore

$$I_{\sigma r} - I_{\sigma t} = a_{\sigma r} - a_{\sigma t} + (b_{\sigma r} - b_{\sigma t})\rho_{sr}^* \geq 0 \text{ since } u_{\sigma r}(\rho_{sr}^*) \geq u_{\sigma t}(\rho_{sr}^*).$$

Thus, the attractiveness function is at a maximum for any option that is used, and an option with a value that is strictly lower than the maximum value cannot be efficient (i.e. have a non-trivial efficiency domain). We can therefore define the equilibrium conditions with a dual variable μ_σ, as follows:

$$n_{\sigma r} \geq 0 \text{ and } \sum_{r=1}^{R\sigma} n_{\sigma r} = Q_\sigma \tag{12.18a, b}$$

$$I_{\sigma r} - \mu_{r\sigma} \leq 0 \text{ and } n_{\sigma r} \cdot (I_{\sigma r} - \mu_\sigma) = 0 \tag{12.18c, d}$$

It can easily be shown that these conditions are sufficient to characterise a local equilibrium for the segment σ. Coordination between segments is achieved by adjusting the price and quantity of the supply: $a_{\sigma r}$ depends on the price p_r, which in turn depends on $S_r = \Sigma_\sigma n_{\sigma r}$. This provides the basis for the general characterisation that follows.

12.4.2 Variational Inequality and Property of Existence

In vector terms, let us denote $\mathbf{n}_r = [n_{\sigma r} : r \in R_\sigma]$ and $\mathbf{I}_\sigma = [I_{\sigma r} : r \in R_\sigma]$ with components that are arranged in ascending order of $b_{\sigma r}$ values.

Theorem 12.1

The residential equilibrium resolves the following variational inequality, denoted by $\mathbf{n}_S = [\mathbf{n}_\sigma : r \in S]$ and $\mathbf{I}_S = [\mathbf{I}_\sigma : \sigma \in S]$: Find $\mathbf{n}_S^*$ such that $\forall \mathbf{n}_S$ that are admissible,

$$\mathbf{I}_S\left(\mathbf{n}_S^*\right) \cdot \left(\mathbf{n}_S - \mathbf{n}_S^*\right) \leq 0. \tag{12.19}$$

If the functions O_r and H_σ are continuous, then $\mathbf{I}_S$ is too, which guarantees the existence of a solution if the problem is feasible (in particular, the total supply must not be lower than the total demand). This demonstrates the existence of a residential equilibrium.

12.4.3 Extremal Formulation and Property of Uniqueness

$$J_\sigma = \sum_{r=1}^{R\sigma} \int_{}^{n_{\sigma r}} a_{\sigma r}(\theta)\,d\theta + Q_\sigma \sum_{r=1}^{R_\sigma} b_{\sigma r}\left(\eta_\sigma\left(\frac{N_{\sigma r}}{Q_\sigma}\right) - \eta_\sigma\left(\frac{N_{\sigma,r-1}}{Q_\sigma}\right)\right),$$

with $\eta_\sigma(x) = \int_{}^{x} H_\sigma^{(-1)}(\theta)\,d\theta.$

In fact, the function J_σ has the following partial derivative-s:

$$\frac{\partial J_\sigma}{\partial n_{\sigma r}} = a_{\sigma r}(S_r) + \sum_{\ell=1}^{R_\sigma} b_{\sigma,\ell}\left(H_\sigma^{(-1)}\left(\frac{N_{\sigma\ell}}{Q_\sigma}\right)1_{\{R_\sigma > \ell \geq r\}} - H_\sigma^{(-1)}\left(\frac{N_{\sigma,\ell-1}}{Q_\sigma}\right)1_{\{\ell > r\}}\right)$$

$$= a_{\sigma r} + \sum_{\ell=r}^{R_\sigma - 1} b_{\sigma,\ell} H_\sigma^{(-1)}\left(\frac{N_{\sigma\ell}}{Q_\sigma}\right) - \sum_{\ell=r+1}^{R_\sigma} b_{\sigma,\ell} H_\sigma^{(-1)}\left(\frac{N_{\sigma,\ell-1}}{Q_\sigma}\right)$$

$$= a_{\sigma r} + \sum_{\ell=r}^{R_\sigma - 1} \left(b_{\sigma,\ell} - b_{\sigma,\ell+1} \right) H_\sigma^{(-1)} \left(\frac{N_{\sigma\ell}}{Q_\sigma} \right).$$

Furthermore,

$$\frac{\partial^2 J_\sigma}{\partial n_{\sigma r}\, \partial n_{\sigma s}} = 1_{\{r=s\}} \frac{\partial a_{\sigma r}}{\partial n_{\sigma r}} + Q_\sigma^{-1} \sum_{\ell=1}^{R_\sigma - 1} \left(b_{\sigma,\ell} - b_{\sigma,\ell+1} \right) 1_{\{\ell \geq r\}} 1_{\{\ell \geq s\}} \dot{H}_\sigma^{(-1)} \left(\frac{N_{\sigma,\ell}}{Q_\sigma} \right).$$

Thus, $\displaystyle \sum_{r,s} \frac{\partial^2 J_\sigma}{\partial n_{\sigma r}\, \partial n_{\sigma s}} x_s x_r = \sum_r \frac{\partial a_{\sigma r}}{\partial n_{\sigma r}} x_r^2 + Q_\sigma^{-1}$

$$\sum_{\ell=1}^{R_\sigma - 1} \left(b_{\sigma,\ell} - b_{\sigma,\ell+1} \right) \dot{H}_\sigma^{(-1)} \left(\frac{N_{\sigma,\ell}}{Q_\sigma} \right) X_\ell^2 \text{ for } X_\ell = \sum_{r=1}^{\ell} x_r.$$

All the terms in the sum are below zero because $\Delta b \leq 0$, H_σ is an increasing function, so its reciprocal is increasing too and it has a non-negative derivative $\dot{H}_\sigma^{(-1)}$, while $\partial a_{\sigma r} / \partial n_{\sigma,r} = -s_r \dot{O}_r^{(-1)}$, which is non-positive since the inverse supply function is increasing and hence its derivative is non-negative. J_σ is therefore concave. This entails the uniqueness of the residential equilibrium (with regard to the components whose diagonal coefficient in the Hessian matrix $\left[\dfrac{\partial^2 J_\sigma}{\partial n_{\sigma r}\, \partial n_{\sigma s}} : r, s \in R_\sigma \right]$ is strictly negative).

In the case where there are several demand segments, the coefficients $[b_{\sigma r} : \sigma \in S]$ for a given option r are heterogeneous, so the term $a_{\sigma r}$ in the attractiveness function $I_{\sigma r}$ can no longer be included in an objective function. The reason for this is that $\partial a_{\sigma r} / \partial n_{\sigma' r} = -b_{\sigma r} s_r \dot{O}_r^{(-1)}$ so that if $b_{\sigma r} \neq b_{\sigma' r}$, then $\partial a_{\sigma r} / \partial n_{\sigma' r} \neq \partial a_{\sigma' r} / \partial n_{\sigma r}$.

The heterogeneity of the demand segments in the case of residential assignment is analogous to that of different classes of traffic in the case of assignment on a transport network (in particular passenger cars and trucks), in which case there is a journey time for each class and network arc that depends on the volumes of the different classes of vehicles, converted into passenger car units by means of specific coefficients of equivalence. From our practical experience of these models, we know that the heterogeneity within each segment (see income distribution) makes it possible to compute a satisfactory equilibrium (Leurent, 1997), while a variety of solutions is obtained with multiclass assignment models with homogeneous segments (Wynter, 1995).

12.4.4 Resolution Algorithm

Residential equilibrium can be calculated by using an algorithm that is taken from the assignment of traffic on a network. A natural first approach is to apply the "historical" algorithm developed by Beckmann et al. (1956), that is the method of successive averages (MSA), as it has been adapted for the dual criteria assignment model (Leurent, 1995).

The MSA algorithm performs successive iterations, which are given the index k. The price of the options is updated during each iteration on the basis of the clienteles $\left[S_r^{(k)} : r \in R \right]$. Each segment $\sigma \in S$ is then processed as follows:

- Determination of the coefficients $b_{\sigma r}$ and $a_{\sigma r}$;
- Determination of the efficiency intervals $\left[\underline{\rho}_{\sigma r}^{(k)}, \bar{\rho}_{\sigma r}^{(k)} \right]$. From this, an auxiliary state for the clientele is derived: $n_{\sigma r}^{\prime(k)} = Q_\sigma \left[H_\sigma \left(\bar{\rho}_{\sigma r}^{(k)} \right) - H_\sigma \left(\underline{\rho}_{\sigma r}^{(k)} \right) \right]$.

Once all the segments have been processed, the next state is determined by a convex combination of the present state of order k and the auxiliary state; the combination coefficient of which $\zeta_k \in \,]0,1[$ is predetermined:

$$\forall \sigma \in S, \; \forall r \in R_\sigma : \; n_{\sigma r}^{(k+1)} = \left(1 - \zeta_k \right) n_{\sigma r}^{(k)} + \zeta_k n_{\sigma r}^{\prime(k)},$$

$$\forall r \in R, \; S_r^{(k+1)} = \sum_{\sigma : R_\sigma \ni r} n_{\sigma r}^{(k+1)} = \left(1 - \zeta_k \right) S_{\sigma r}^{(k)} + \zeta_k S_{\sigma r}^{\prime(k)}.$$

Convergence towards equilibrium is assessed on the basis of the duality gap of the variational inequality, $DG^{(k)} = \sum_{\sigma \in S} \sum_{r \in R_\sigma} I_{\sigma r}^{(k)} \cdot \left(n_{\sigma r}^{\prime(k+1)} - n_{\sigma r}^{(k+1)} \right)$. This criterion is non-negative by construction, which is only cancelled at one equilibrium, and by continuity, a very small value indicates that an equilibrium state has almost been reached.

If the criterion is sufficiently small, the algorithm is stopped; otherwise, another iteration is performed.

For the initial state at $k=0$, it is possible to fix the values of $n_{\sigma r}^{(0)}$ as a uniform distribution $n_{\sigma r}^{(0)} = Q_\sigma / \mathrm{Card}(R_\sigma)$ to generate the values of $S_r^{(0)}$, and therefore the prices $p_r^{(0)}$ and attractiveness values $I_{\sigma r}^{(1)}$.

The sequence ζ_k must decrease towards zero, but not too rapidly in order to redistribute demanders efficiently between the options at each iteration.

12.5 Application to the Paris Region

We applied the DREAM (Disaggregate Residential Equilibrium Assignment) to a study of the Paris region.

12.5.1 Application Settings

Our study area is the "Région Île-de-France", the administrative region where the city of Paris and its urban suburbs are located. It covers 12,500 km². In 1999, its population amounted to 10 million inhabitants, or 18% of the population of metropolitan France.

Zone system: Following the DRIEA system (La Corte, 2006), we divided the region into 36 zones, each of which includes smaller transportation analysis zones. These 36 zones are relatively homogeneous in terms of population. The size of the zones decreases with density, hence resulting in smaller zones in the centre and larger zones in the outer ring.

Household segments: For the sake of simplification, this first application is limited to working households, that is those where the head of the household is employed.

In 1999 (our reference year), these households represented around 80% of the regional population.

Working households are segmented according to the number of members. The following categories have been distinguished (Table 12.1).

Households are also regrouped according to the head of the household's employment zone. There is therefore a total of $36 \times 5 = 180$ demand segments. We specify a shifted lognormal distribution of incomes by segment, from the 2001 to 2002 Household Mobility Survey (EGT).

Figure 12.2 shows the number of jobs per zone (symbolised by a proportional area disc), as well as the ratio between the number of jobs and the number of dwellings. Surplus jobs can be found in the areas of central Paris, Roissy around the airport and the

Table 12.1 Household Segments in the Metropolitan Area

	1 Person	*2 People*	*3 People*	*4 People*	*5 People and More*
Number of households	773,760	852,525	651,109	609,409	352,173

Source: 1999 population census.

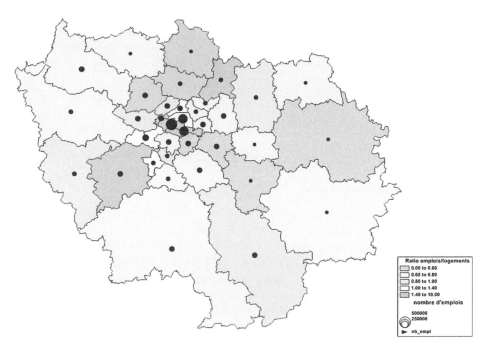

Figure 12.2 Number of jobs and job/population ratio by zone in 1999.

Table 12.2 Number of Dwellings Occupied by Active Households, and Average Surface Area, Ile-de-France 1999

	1 Room	2 Rooms	3 Rooms	4 Rooms	5 Rooms +
Number of dwellings	332,460	613,460	810,980	718,240	687,520
Average floor area (m²)	29.7	46.8	68.7	86.5	140

Source: 1999 population census.

scientific cluster of Paris–Saclay in the south-west. The residential function dominates in the peripheral zones.

Housing categories: We distinguished the housing according to zone and by the number of rooms (Table 12.2).

Transport costs: For this first application, we restricted ourselves to the automobile mode on the road network. For each origin–destination pair, monetary costs and travel times are computed for the morning rush hour using a road assignment model. This model, implemented in TransCAD, uses data from DRIEA for the situation in 2004.

Housing parameters: Ground rent at the edge of the urbanisation is set at €0.25/m² per day. The user cost is set at €3 per day per dwelling. Concerning zones, we set the elasticity of the housing supply at the market price at 0.1 in the centre (inner Paris), 0.2 in the inner ring and 0.3 in the outer ring. In relation to the stocks observed, the volumes are reduced by 20% in the centre, kept constant for the inner ring and raised by 20% for the outer ring.

Income distribution: For all household classes, the minimum income (shift) for the log-normal distribution is set to €10/working day in accordance with observed data in the EGT (as a way of comparison, the official minimum wage was actually €48.5/day at the time).

Utility functions: We set the parameters of the Stone–Geary utility function $U_{\sigma r} \propto \left(\left(s - v_\sigma s_0 \right)^+ \right)^\alpha q^\beta z^\gamma \tau^\delta$ as follows: $\alpha = 0.12$, $\beta = 0$, $\gamma = 0.5$ and $d = 0.2$. The quality of housing is not considered in the application. In addition, the minimum floor area is 10 m² per person. All these parameters were estimated using the 2002 French housing survey (Coulombel, 2012).

Parameterisation of the algorithm: The model was coded in C++. To approach the equilibrium correctly, we carried out about 10,000 iterations of the convex combination algorithm, with a coefficient of $\zeta_k = 1/\left(\dfrac{k}{2} + 1 \right)$ at the iteration number k. With a processor at 2.77 GHz, the computation time is 5′. At convergence, the attractiveness function factor is about 100 and the duality gap is reduced to about 5,000 (Figure 12.3).

12.5.2 Simulation of the Reference State

Without any other calibration, the simulation by the model reproduces the data observed quite well with regard to the population of the zones (Figure 12.4): the main downward variations concern Paris (zones 29–31), Roissy and Saclay (zones 11 and 21). The reason for this is the large number of jobs in these areas.

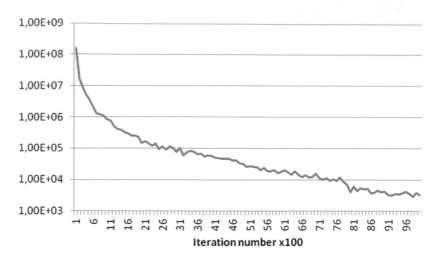

Figure 12.3 Variations of the duality gap as a function of the number of iterations.

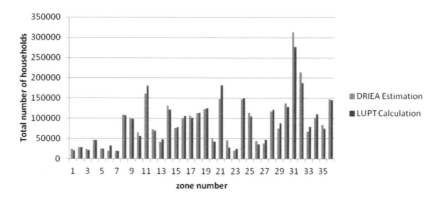

Figure 12.4 Populations of zones as simulated by the model or observed.

Figure 12.5 shows the spatial variations of prices for three-room dwellings. Prices increase with the accessibility of jobs. According to the "BIEN" notaries' database for Ile-de-France, in 2001 the ratio of home prices per square metre was 0.73 between the inner ring and the centre, and 0.6 between the outer ring and the centre. Our model predicts 0.8 and 0.58, respectively. In addition, it predicts an average monthly rent of €8.6/m² in the centre, compared to €12/m² according to INSEE.

Relative to three-room dwellings, unit prices for other types of property are as follows: 1 room +22%, 2 rooms + 1%, 4 rooms + 5% and 5 rooms + 10%. One-room dwellings are significantly more expensive as their supply is relatively rare in relation to the households of 1 or 2 people who are likely to occupy them.

In addition, the model was able to reproduce the growth of household size as a function of distance from the centre (Figure 12.6). Smaller households are more likely to

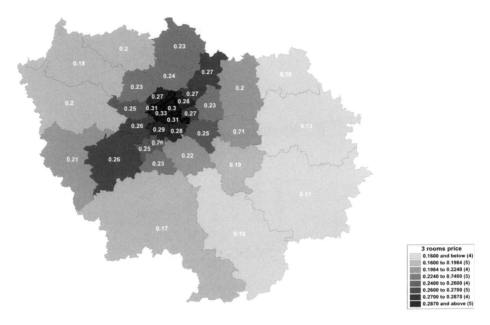

Figure 12.5 Unit prices for renting three-room dwellings.

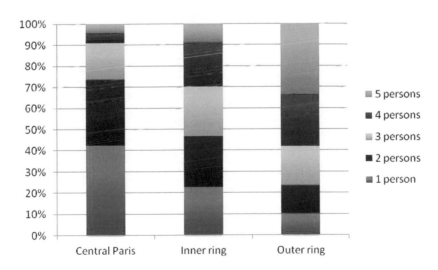

Figure 12.6 Household size distribution.

live in the centre, while larger households prefer to settle in the suburbs to benefit from more affordable prices. Figure 12.7 shows that many medium-sized households (2 or 3 people) live in the suburbs with a comfortable floor area.

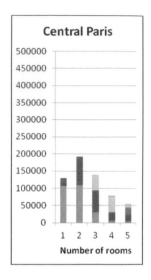
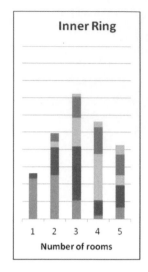
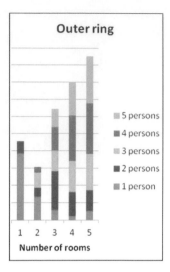

Figure 12.7 Household size distribution by type of dwelling (in number of rooms).

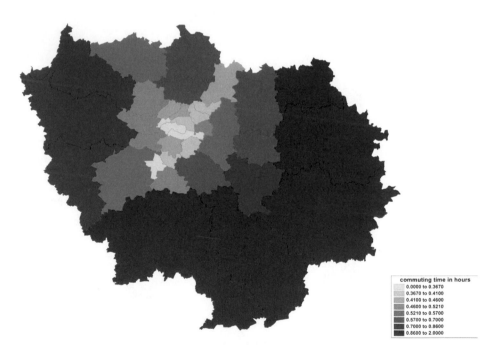

Figure 12.8 Average commute time per zone (in hours).

Finally, the model correctly reproduces the high commuting times of suburban residents, while households living in the central part enjoy lower commuting times thanks to the high job accessibility from this position (Figure 12.8).

12.5.3 A Test Scenario of an Increase in Fuel Prices

We tested the model behaviour and sensitivity by envisaging a simple scenario of an increase in fuel prices. Fuel prices are set to €2/litre, against €1 in 1999.

Table 12.3 summarises the mean relative change in commuting time by demand segment. On average, this time decreases by almost 2%, as per the intuition. In addition, the impact of the increase in fuel prices differs according to the size of the household. Large households have a larger basic need for housing than small households (the minimum is 10 m² per person), and all other things being equal, it leaves them with less disposable income than small households. Rising prices are causing large households to move closer to their place of work, including by agreeing to slightly reduce the size of their homes. On the other hand, small households are relatively less affected by price rises and can occupy the dwellings thus left vacant. This enhances the attractiveness of medium-sized dwellings around employment areas, leading to higher unit prices for this category, as opposed to the effects in suburban areas (Figure 12.9). More specifically, the rent for small dwellings would increase by more than 25% in the central

Table 12.3 Travel Time by Household Segment

Number of Household Members	1	2	3	4	5+
Variation of home-work time	+0.64%	+0.45%	−0.92%	−4.3%	−6%

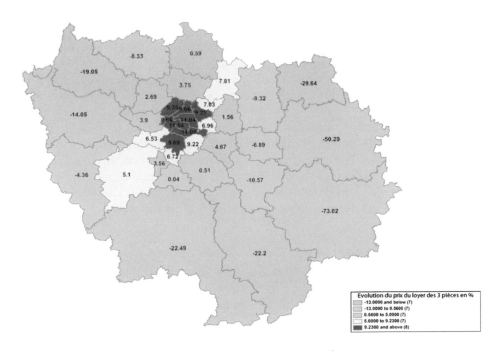

Figure 12.9 Relative change (%) in rents for three-room homes.

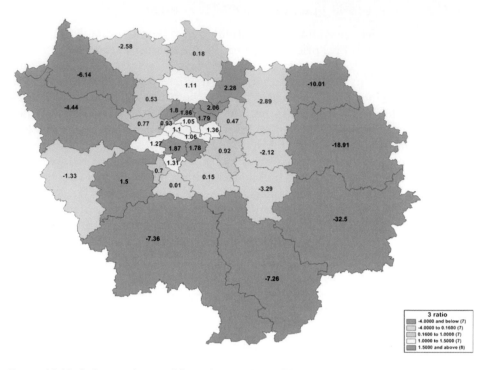

Figure 12.10 Relative change (%) in the number of homes, in case of three rooms.

zone, while that of large dwellings would only increase by 5%. On the other hand, in the suburbs, all types of housing would incur a fall in rents.

These adaptations of household locations go hand in hand with a reconfiguration of the housing stock, and this affects each zone according to the local function of the housing supply (Figure 12.10). As households tend to move closer to their places of employment, some peripheral areas would lose up to 5% of their population. However, this reconfiguration is limited by the low elasticity of the housing stock in the centre: it is the inner ring that will receive the greatest increase in the residential establishment.

To conclude, the indicators presented not only give us validation of the correct behaviour of the model, we also gain more understanding of the disaggregated behaviour of households in relation to macroscopic market conditions.

12.6 Conclusion

We modelled the structure of the housing supply according to type and zone, and the structure of the demand according to household size, job location and income. We modelled the microeconomic behaviours of demanders and macroeconomic behaviours on the supply side. The structural features that are represented are fundamental for the medium-term operation of an urban housing market: they provide a minimum core which must feature in any operational (or pre-decisional) application.

Our model respects the disaggregation of demand. The inherent complexity has been reduced mathematically and algorithmically owing to a specific aggregated treatment.

The model lends itself to various extensions, of either an economic, mathematical or algorithmic kind. From the economic standpoint, we shall simply mention the representation of households with no working member or several workers, the inclusion of social housing, the distinction between the rented sector and the "owner-occupier" sector in the case of the private sector property market, and the inclusion of complex supply behaviours, etc.

As it stands, the model provides a good compromise between explanatory power and processing simplicity. A first application has been implemented on a real case study with several household and dwelling segments. The results of the study confirm the relatively accurate predictive capacity. Local imbalances between supply and demand by category of goods were simulated, based on the limited set of parameters incorporated into the model.

Overall, the model can be used to study the real estate market in an urban area and the effects of various housing policies, as well as to simulate development and transport operations. Future developments will focus on integrating neighbourhood quality, which will lead to more complex household utility functions.

Bibliography

Anas A. and Liu Y. (2007), A regional economy, land use, and transportation model (RELU-TRAN): formulation, algorithm design, and testing, *Journal of Regional Science*, Vol. 47 (3), pp. 415–455.

Beckmann M., McGuire C.B., and Winsten C.B. (1956), *Studies in the Economics of Transportation*. Report to the Coles Commission. Yale University Press.

Delons J., Coulombel N., and Leurent F. (2009), Pirandello: an integrated transport and land-use model for the Paris area. Working document submitted to TRB Congress 2010.

DSC - David Simmonds Consultancy in collaboration with Marcial Echenique and Partners Limited (1999), Review of land - use/transport interaction models, Reports to The Standing Advisory Committee on Trunk Road Assessment (SACTRA).

King D. (2011), Developing densely: estimating the effect of subway growth on New York City land uses, *The Journal of Transport and Land Use*, Vol. 4 (2), pp. 19–32.

Leurent, F. (1993), Cost versus time equilibrium over a network. *European Journal of Operational Research*, Vol. 71, pp. 205–221.

Leurent, F.M. (1995), Un algorithme pour résoudre plusieurs modèles d'affectation du trafic: la méthode d'égalisation par transvasement. *Les Cahiers Scientifiques du Transport*, Vol. 30, pp. 31–49.

Leurent, F.M. (1996), The theory and practice of a dual criteria assignment model with a continuously distributed value-of-time, in *Transportation and Traffic Theory*, Lesort J.B. (ed.), pp. 455–477. Pergamon, Exeter, England.

Leurent, F. (1997) Analyse et mesure de l'incertitude dans un modèle de simulation : les principes, une méthode, et l'exemple de l'affectation bicritère du trafic. Thèse de l'ENPC, Paris, France.

Leurent F. (2012), Les modèles d'usage du sol et transport: où la géographie et l'économie se rejoignent. In Hégron G. (ed), *La modélisation de la ville: du modèle au projet urbain*. Special issue of the Revue du CGDD, April, pp. 154–171.

Levinson D. (2007), Density and dispersion: the co-development of land use and rail in London, Working paper.

Lowry I.S. (1964), A model of metropolis. RAND Memorandum 4025-RC.

Salvini P. and Miller E.J. (2005), ILUTE: an operational prototype of comprehensive microsimulation model of urban systems, *Network and Spatial Economics*, Vol. 5, pp. 217–234.

Simmonds D., Waddell P., and Wegener M. (2011), Equilibrium v. dynamics in urban modelling, *Paper presented at the Symposium on Applied Urban Modelling (AUM 2011)*, University of Cambridge, 23–24 May 2011.

Waddell P. et al. (2003), Microsimulation of urban development and location choices: design and implementation of Urbansim, *Networks and Spatial Economics*, Vol. 3 (1), pp. 43–67.

Wynter L. (1995), Contributions à la théorie et à l'application de l'affectation multi-classe du trafic, ENPC PhD thesis, November.

Chapter 13

Structural Design of a Hierarchical Urban Transport Network

Fabien Leurent and Sheng Li

École des Ponts ParisTech

Hugo Badia

KTH Royal Institute of Technology

13.1 Introduction

13.1.1 Context

In urban areas, passenger transport demand involves two types of travel based on spatial reach: short trips of less than 1 km, as opposed to medium and long trips above this threshold. The latter type is better served by motorised modes such as the automobile (private car, cab, car-sharing and ridesharing) or transit modes (bus, tram, metro and train).

Urban mobility planning aims to respond to the travel demand efficiently: the supply of road and rail infrastructure as well as transit services must satisfy users' needs at tolerable costs to the public purse and with acceptable environmental impacts. The optimal design of a multimodal urban mobility plan is therefore an economic programme to maximise a comprehensive well-being function. This incorporates both the social well-being of the system (users, operators) and its environmental performance, under technical constraints pertaining to quality of service in relation to the production of transport services on the one hand and the behaviour of users on the other.

13.1.2 Related Work

Previous research on this topic began at the scale of a single bus line with Mohring (1972), who considered fleet size (hence service frequency) and stop spacing as levers that could be used to maximise social welfare, defined as demand surplus plus production profit. While a large body of research has focused on the optimisation of an urban public transport network with extensive spatial detail and a very large number of discrete variables, Van Nes (2002) placed the emphasis on key structural features, including line length, the distance between neighbouring stations on a line (i.e. station spacing) and vehicle fleet size. By modelling these as continuous variables by spatial and modal subsets, he developed a complete methodology for the design of roadway networks, public transport networks and multimodal networks to optimise social welfare. He addressed multimodality as a weak interaction between modes, on the basis of a logit model of discrete choice between car and transit modes by individual users. Public transport networks at the scale of a whole urban area were addressed by Combes and Van Nes (2012), who tackled the case of Greater Paris by introducing the "component" concept, where a component is a pairing of one submode and one subarea. Comprehensive mathematical formulation and in-depth economic analysis were provided by Leurent et al. (2016). This kind of structural analysis relies on the statistical modelling of travel demand, trip-making features, network structure and traffic conditions.

A second stream of modelling was pioneered by Daganzo for two major modes of urban mobility: firstly to develop an aggregate model of the traffic performance of an urban road network by means of the macroscopic fundamental diagram (MFD) (Daganzo and Geroliminis, 2008), and secondly to model a grid network of bus lines and optimise its structural parameters (Daganzo, 2010). The two modes were combined by Estrada et al. (2012) to optimise the public transport network, taking into account the costs to both car users and transport users, as well as the costs to the public transport operators. Recently, Badia et al. (2014) adapted Daganzo's transport network design model to a city with a radial street pattern.

13.1.3 Objective

We will address the structural design of an urban transport network modelled as a set of components, in order to optimise a social well-being function that encompasses both production profit and user surplus not only for the transit network but also for the car mode. Users are either mode dependent or "flexible" in the sense that they can choose the mode that suits them best. Environmental impacts are taken into account for both modes. The model allows us to study planning trade-offs between modes based on their respective performance in terms of demand surplus, environmental benefits and production costs, and to determine the range of relevance of public transport submodes.

Outline of this chapter: The rest of this chapter consists of four parts. Section 13.2 presents the assumptions and structure of the model. Section 13.3 then defines the planning problem of optimising the structural design of the public transport network and proposes an optimisation algorithm. Section 13.4 provides a simulation demonstration for the case of Greater Paris. Finally, Section 13.5 summarises the findings and identifies directions for further research.

13.2 Model Composition

13.2.1 Urban Area and Modes of Transport

Let us consider an urban area with its population, including workers, and with its economic activity that provides job opportunities. To account for territorial heterogeneity, the study area Z is divided into subregions indexed by Z, depending on the respective densities of the population and the transportation infrastructure.

The motorised transportation system is simplified as two modes $M \in \{C, Y\}$: car mode C (for car) and public transport network T (for transit).

13.2.2 Travel Demand

Motorised transportation demand is classified into three types of users: public transport-dependent TD, car-dependent CD and mode-choosers F for flexible. Public transport-dependent users lack the capacity (i.e. have no license or vehicle) or opportunities to drive a car or carpool with others, so their trips depend totally on public transportation. Car-dependent users have access to a car but little access to public transportation, or they have logistic constraints that require the use of a car, for example the transportation of bulky items or making successive trips as part of a round. Mode-choosers have access to both modes and will choose the best option for any given trip. In the short term, both dependent types are passively exposed to variations in service quality, while mode-choosers can respond actively by adapting their choices. Demand is described in origin–destination (O-D) pairs ($i \in I$, set of O-D pairs). Trip endpoint zones are smaller than the subarea level. For each O-D pair i, the total demand Q_i^G is fixed:

$$Q_i^G = Q_i^F + Q_i^{CD} + Q_i^{TD}.$$ (13.1)

"Mode-choosing users", in number Q_i^F, choose their modes with respect to the generalised costs of public transport and car options, respectively. A logit discrete choice model is postulated, giving the respective modal volumes of public transport choosers (TF) and car choosers (CF) as follows: $\forall i \in I$,

$$Q_i^{TF} = Q_i^F \cdot \exp\left(-\theta g_i^T\right) / \left(\exp\left(-\theta g_i^T\right) + \exp\left(-\theta g_i^c\right)\right)$$ (13.2)

$$Q_i^{CF} = Q_i^F - Q_i^{TF}.$$ (13.3)

13.2.3 Public Transport System (TC)

The structure of the transit network is modelled as a set of components $r = (z, m)$ by zone z and transport submodes $m \in T_z$. The public transport submodes vary from bus to train lines, passing through tramway and metro lines.

Let R denote the set of components r. Each component is characterised by four factors: line length L_r, the number of stations σ_r, the quantity of rolling stock N_r and the tariff τ_r. Inside the component, the line spacing and stop spacing are considered homogeneous, and the rolling stock operates in ideal conditions with no specific interruption at the terminus. In a given zone z, the different r can intersect one other at transport hubs where users can make transfers – which generate penalties in terms of time spent and discomfort. Public transport operations and quality of service are modelled as a set of four technical relationships, as follows: $\forall r \in R$,

$$S_r = L_r / \sigma_r$$ (13.4a)

$$1/v_r = 1/V_r + \omega_r / S_r$$ (13.4b)

$$\varphi_r = N_r v_r / 2L_r \tag{13.4c}$$

$$d_i^{A,D} = \frac{1}{4}\left(S_r + \chi_z \frac{A_z}{L_r}\right). \tag{13.4d}$$

Relation (13.4a) derives station spacing from line length L_r and number of stations σ_r. Relation (13.4b) derives commercial speed v_r on the basis of vehicle time per unit of distance travelled, including a running part $1/V_r$ and a part of stopping at the station ω_r / S_r, depending on the stopping time per station ω_r taken as an exogenous parameter. Relation (13.4c) derives service frequency φ_r from fleet size N_r and cycle time $2L_r / v_r$. Relation (13.4d) derives the average access distance $d_i^{A,D}$ between trip endpoint and the user's access station from station spacing and line spacing, $\chi_z \cdot A_z / L_r$ which involves zone area A_z together with line length L_r and a shape parameter χ_z with a typical value between 1 and 2.

Access distance is a key factor in the quality of service for public transport users. The average generalised travel cost of a public transport trip is based on the stages of access, waiting at the station platform, journey on-board and transfer in one or more subregions r per O-D pair i: there is a sensitivity coefficient $\gamma_{r,i}^X$ that integrates both the amount of exposure (e.g. a specific distance travelled or a specific time t^X) and user sensitivity (e.g. value of time depending on comfort state) per stage X and by subregion, while for the tariff the sensitivity coefficient $\gamma_{r,i}^\tau$ (0 or 1) only measures the quantity of exposure:

$$g_i^T = \sum_{r \in R} \gamma_{r,i}^\tau \tau_r + \sum_{r \in R, X} \gamma_{r,i}^X t_{r,i}^X. \tag{13.5}$$

On the supply side, each component induces production costs that are modelled on the basis of technical factors L_r, σ_r, N_r with respective unit costs of c_r^L, c_r^σ, c_r^N and which include distance-related costs and investment costs amortised over the technical lifetime. Per year of activity, the cost per component is:

$$c_r^0 = c_r^L \cdot L_r + c_r^\sigma \cdot \sigma_r + c_r^N \cdot N_r. \tag{13.6a}$$

The transit network brings together the components and thus accumulates their respective production costs. At the overall level, the operator also provides general functions: commercial information and pricing functions, surveillance and security, and general administration. The related "general costs" can be attributed to all transit trips $Q_T = \sum_{i \in I} Q_i^{TD} + Q_i^{TF}$ on a proportional basis, at unit cost c_0^Q. With Y denoting the number of operating days in the year, the daily production cost of the TC network amounts to

$$c^0 = c_r^Q \cdot Q^T + \left(\sum_{r \in R} c_r^Q\right) \Big/ Y. \tag{13.6b}$$

13.2.4 Car Mode and Roadway System

For a trip in car mode, the generalised cost of a trip on an O-D pair for the user incorporates the monetised travel time, which depends on distances $D_{z,i}$ travelled in zone z at respective car speed v_z, and on the user's value of time, β_t, as well as parking availability and cost F_i and also the cost of possessing and using the car per unit of distance β_c, which includes purchase depreciation, energy, maintenance and insurance. In total,

$$g_i^C = F_i \sum_z D_{z,i} \cdot \left(\frac{\beta_t}{v_z} + \beta_c \right) \tag{13.7}$$

For each zone z, the average car speed is related to the quantity of traffic in two ways. On the demand side, the faster the speed, the more the demand volume, whereas on the supply side, the more the traffic, the slower the speed.

To quantify things by O-D pair i, let us denote $Q_i^C = Q_i^{CD} + Q_i^{CF}$ the volume of car trips, including dependent as well as flexible users. In zone z, the car users of the different O-D i pair require car traffic $Q_z^+ = \sum i \in I Q_i^C \cdot D_{z,i} / \eta_i$: each user requests to travel a distance $D_{z,i}$ by car in zone z, but with an influence inversely proportional to the average occupancy rate of the cars, η_i.

During the period of interest, let's say H_1, in zone z, which has length ℓ_z in traffic lanes, the total traffic flowed by the local network amounts to $Q_z^- = \ell_z q_z^- H_1$. Figure 13.1a shows the MFD retained here.

Let us now compare the traffic Q_z^- flowed during H_1, to the demand volume Q_z^+. If $Q_z^+ \leq Q_z^-$, then the duration H_1 is sufficient to accommodate the demand volume in the fluid traffic regime, without saturation. Conversely, if $Q_z^+ > Q_z^-$, then the traffic is saturated. In other words, there is a traffic bottleneck, with a duration H_z greater than H_1. The relationship between H_z and H_1 characterises the traffic regime: either $H_z = H_1$ if the traffic is in fluid regime, or $H_z > H_1$ if the traffic is at saturation.

To deal with the saturation, let us consider a bottleneck model with the entry flow rate of $q_z^+ = Q_z^+ / (\ell_z H_1)$ during $[h_0, h_0 + H_1]$ and the exit flow rate of q_z^-, for a notional road of length defined as the average distance travelled by car users in zone z:

$$u_z = \left(\sum_{i \in I} Q_i^c \cdot D_{zi} \right) \Big/ \left(\sum_{i \in I} Q_i^c \cdot 1_{\{D_{z,i} > 0\}} \right).$$

The number of vehicles, $N_z^+ = Q_z^+ \ell_z$, is flowed out from instant $h_0 + u_z^+ / v_z^*$ with $v_z^* = \arg\max_{v q \bar{z}}$ as the flow speed at capacity flow (i.e. maximum) $q^*\bar{z}$, up to instant $h_0 + u_z / v_z^* + H_z$.

The area of the rectangle drawn in Figure 13.1b, that is $N_z \cdot h_0 \left(u_z / v_z^* + H_z \right)$ is related to the average duration per user, $t = u_z / v$, by

$$N_z \cdot \left(\frac{u_z}{v_z^*} + H_z \right) = t \cdot N_z + \frac{1}{2} N_z H_1 + \frac{1}{2} N_z H_z.$$

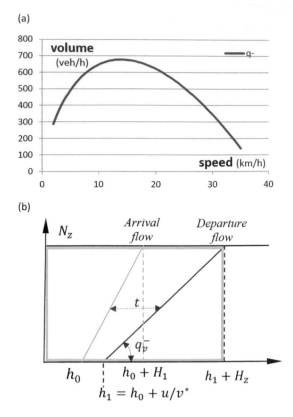

(a)

(b)

Figure 13.1 (a) Macroscopic fundamental diagram (MFD). (b) Cumulative flows in the bottleneck.

So that $= \dfrac{u_z}{v_z^*} + \dfrac{1}{2}(H_z - H_1)$, hence

$$H_z = H_1 + 2u_z\left(1/v_z - 1/v_z^*\right)^+$$

Then, from the volume conservation and the MFD, we have

$$H_1 q_z^+ = H_z q_z^- = H_z v_z R_z(v_z)s$$

which leads to the following fixed point problem:

$$v_z = \left(\dfrac{q_z^+ H_1}{R_z(v_z)} - 2u_z\right)\bigg/\left(H_1 - 2u_z/v_z\right).$$

Postulating a Daganzo and Geroliminis-type MFD, $R_z(v_z) = \omega_z \ln(v_0/v)$, which decreases from $+\infty$ to 0 as v varies from 0 to v_0, then the right-hand-side function (denoted RHS) in this fixed-point problem is a monotonic increasing function that varies from $-2u_z/\left(H_1 - 2u_z/v_z^*\right)$ to $+\infty$ sur $]0, v_0[$. As the slope on its positive part is greater than 1, it only crosses the identity function once, meaning that the fixed-point problem has one solution which is unique. Furthermore, an increase in q_z^+ increases the RHS function and makes the solution smaller: $\tilde{v}$ decreases with q_z^+, hence H_z increases with q_z^+.

As the solution is positive, necessarily $q_z^+ > 2u_z R(\tilde{v})$, hence $q_z^+ \tilde{v} > V_z\left(q_z^+ H_1/2u_z\right)$.

Figure 13.2 shows the modelled relationship between the endogenous car speed vz and the demand volume Q_z^+. This decreasing relationship is in two parts: firstly the fluid regime, for Q_z^+ varying from 0 to capacity volume $Q_z^* = H_1 q_z^*$, then the saturated regime, for values of Q_z^+ greater than the volume Q_z^*. This relation can be noted in the following mathematical form:

$$v_z = V_z^C\left(Q_z^+\right).$$ (13.8)

13.2.5 Environmental Impacts

The environmental impacts of a transportation system occur along its life cycle, and the development of mobility equipment and services must be integrated, as well as the operational transportation processes. The sources of pollution are multiple: infrastructure and rolling stock construction and maintenance, the production/distribution/consumption of energy and vehicle operation. These impacts may be local (e.g. noise and air pollution) or global (e.g. carbon emissions). The level of local impacts does not only depend on the quantities of emissions but also depend on the local populations exposed to them, and this means that population density must be taken into account in the monetisation (Quinet et al., 2013).

Here, we limit our field of analysis to two types of impacts, namely (i) air pollution by vehicles, which include cars that induce traffic as well as public transport vehicles

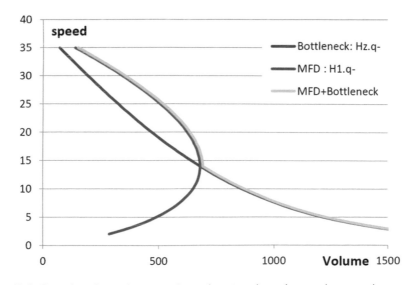

Figure 13.2 Speed–volume function for subregional roadway subnetwork.

that circulate daily for a fixed duration H_r, (ii) the greenhouse gas emissions measured in carbon equivalent, which depend locally on Q_z^+ and also on average speed v_z.

The environmental cost per day of operations is as follows, denoting as parameters $\alpha_1, \alpha_2, \alpha_3$ cost coefficients per unit of distance and $\delta_1, \delta_2, \delta_3$ sensitivities to roadway speed:

$$-P^e = \sum_{z \in Z} \alpha_{2z} \cdot Q_z^+ + \sum_{r \in R} \alpha_{1r} \cdot H_r v_r N_r + \sum_{z \in Z} \alpha_3 \cdot Q_z^+ \cdot \left(\delta_1 v_z^2 + \delta_2 v_z + \delta_3 \right).$$

(13.9)

13.3 Planning Problem and Solution Algorithm

13.3.1 Social Welfare and Overall Welfare

Mobility stakeholders in the territory include the users, the operators and also the environment stakeholders, including the residents who are exposed to local impacts. The surplus of users consists of the surplus of mode-choosers (log sum formula for a logit discrete choice model) minus the generalised costs of mode-dependent users:

$$P^u = \sum_{i \in I} \left(P_i^{uF} + P_i^{uD} \right) = \sum_{i \in I} \left(Q_i^F \cdot \frac{1}{\theta} \cdot \ln \left(\exp\left(-\theta g_i^T \right) + \exp\left(-\theta g_i^C \right) \right) - Q_i^{TD} \cdot g_i^T - Q_i^{CD} \cdot g_i^C \right).$$

(13.10)

The surplus of operators is restricted to public transport production and consists of commercial revenues minus production costs c^0. On a daily basis,

$$P^0 = \sum_{i \in I} \tau_i \left(Q_i^{TF} + Q_i^{TD} \right) - c^0.$$

(13.11)

The "benefit" for the environment is the opposite P^o of the previously mentioned cost. Two definitions of social well-being are considered: $W^0 = P^u + P^o + P^e$, or $W^1 = P^u + P^o$ only, neglecting the environmental impacts.

13.3.2 Transportation Planning and the Problem of Policy Design

In the territory under study, mobility policy is aimed to foster the overall welfare. To do so, transit modes are subsidised at level say S per year in order to make their production profitable to their operators.

Given level S, the policy-design problem is modelled as a mathematical programme for optimising the overall welfare, by acting on the levers of action, and subject to the requirement of production profitability as well as technical constraints (quality of service, generalised costs of the modes and traffic operation):

$$\max W \text{ with respect to } \tau, L, \sigma, N, v, \text{ subject to } P^o + S \geq 0 \text{ and conditions } (1) - (11).$$

13.3.3 Numerical Optimisation

Demand conditions, user surpluses, production costs and environmental impacts depend entirely on a state vector that groups together $(L_r, \sigma_r, N_r, \tau_r : r \in R)$ and car speeds $(v_z : z \in Z)$. To solve the optimisation problem numerically, we applied the generic solver of our Python library to the constrained optimisation programme. The speed variables are completely constrained, but must be included as "primal" variables in the programme.

In the application presented in the next section, there are 3 regions and 11 components, hence 11×4 variables for public transport and 3 for cars, in all 47 scalar variables.

13.4 Case Study: Planning Mobility in Greater Paris

13.4.1 Reference Situation and Its Modelling as a System State

As of 2010, Greater Paris had a population of about 12 million in an urbanised area of about $1,250\,km^2$ within the Île-de-France region, an area of $12100\,km^2$. The regional area is split into three subareas, namely the city of Paris, the inner suburbs and outer suburbs (see Figure 13.3a).

Our synthetic description of motorised transport networks is based on the MODUS transport demand model by DRIEA (the National Regional Planning Agency). It is a four-step model, from which we drew the respective lane lengths of the subregional road networks, as well as the statistical indicators of the public transport networks (Figures 13.3b and c).

The public transport system consists of four submodes (m): the RER mode is a set of train lines connecting peripheries by crossing the centre of the urban area (the green line in Figure 13.3d), the Transilien mode is a set of radial railway lines drawing on the periphery from a hub in the centre (the purple line in Figure 13.3d), the metro whose lines irrigate the central part and its immediate surroundings (the red line in Figure 13.3e) and the bus lines (Figure 13.3f).

According to the regional Household Travel Survey "EGT 2010", motorised travel demand consists of 15.5 million car trips and 8.3 million public transport trips, on average per working day. For the time distribution of car trips within the day, we

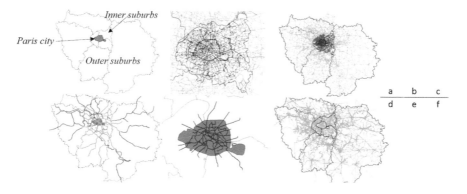

Figure 13.3 Division into subregions, road network routes and modes of public transport.

Table 13.1 Characteristics of the Road Network and Setting of Parameters

Subarea	ℓ_z (km)	ω_z	v_{0z} (km/h)	F_z (€)	Car flow M veh/day	VKT M veh*km/day
Paris city	1870	42,5	38,0	3,5	2,35	13,3
Inner suburbs	4417	54,6	51,0	1,0	6,59	39,2
Outer suburbs	19097	46,8	67,0	0,0	10,67	84,3
θ	0,27	α_{10}	4.19€/km*bus in Paris	δ_1 1,60E-05	$\gamma_{A,D}$	2,0
β_t	10€/h	α_{20}	0.55€/km*veh (inner suburbs)	δ_2 -0,00234	γ_W	1,5
β_z	0.3€/km*veh	α_3	0.1€/l	δ_3 0,1381	γ_R	1,0

Table 13.2 Components of Transit Network and Demand Indicators

Subarea	Paris city				Inner suburbs				Outer suburbs		
Transit component	1	2	3	4	5	6	7	8	9	10	11
Transport mode	Bus	Metro	RER	Train	Bus	Metro	RER	Train	Bus	RER	Train
Flow (M pax/day)	1.25	3.97	1.91	0.62	2.03	1.35	2.06	0.77	1.24	1.13	0.65
Station spacing (km)	2.06	4.01	5.02	3.15	2.27	1.85	6.94	6.61	3.46	7.72	12.7
In-vehicle time (min)	14.7	19.4	15.0	10.9	15.3	5.9	15.9	13.0	10.1	13.6	13.5
Access distance (km)	0.29	0.41	0.54	1.74	0.28	0.52	0.65	0.57	0.31	0.91	0.98
Gen. cost (€/trip)	6.84	9.24	5.64	4.23	7.36	5.84	7.72	6.19	7.70	12.6	13.4

considered a "slot" function, that is constant demand positive over a time period $H_1 = 10$ hours. Subareas are taken as trip endpoint zones for O-D pairs.

According to the Mobility Authority's 2010 activity report (Ile-de-France Mobilités (IDFM), or STIF), transit costs excluding new investment amounted to €8.1 billion, versus fare revenues of €2.4 billion actually paid by users and subsidies of €5.7 billion.

Table 13.1 sets out the major indicators and the model parameters for the roadway network (top), together with the relative parameters of user behaviours and socio-economic appraisal (bottom). Table 13.2 gives the demand and service indicators per transport component in the reference state.

13.4.2 Simulation Results and Analysis

Two planning policy scenarios were simulated: the S_0 optimisation scenario of the overall well-being function $W_0 = P^u + P^o + P^e$ and the S1 scenario to optimise social well-being function $W_1 = P^{u+}P^e$, which would not include the environment. The optimisation of zone fares and component service frequencies (via fleet size) for all public transport components, and that of line length and stop numbers for bus components, could be carried out in the short and medium term, while the optimisation of line length and station numbers for rail components is a long-term matter. In both scenarios, the short, medium and long terms were integrated. The overall level of the transport subsidy, S, is imposed, as are the O-D trip flows according to user types.

A number of salient features are common to both improvement scenarios (Table 13.3). First and foremost, network optimisation would yield significant benefits

Table 13.3 Characterization of System States by Surplus Indicators, Unit: bn €/Year

Scenario	Subarea	Operator					User			Environment		
		P^o	P^u	P^e	production cost	operator revenue	Car dependent	Transit dependent	Mode chooser	Car atmospheric pollutant	Transit all	GHG
Current situation	Paris city		-15.2	-2.4	2.7		-0.86	-7.98	-6.36	-1.96	-0.43	-0.04
	Inner suburbs	5.68	-17.7	-2.4	2.6	2.4	-2.05	-7.17	-8.51	-1.92	-0.34	-0.10
	Outer Suburbs		-23.2	-0.7	2.8		-2.05	-7.17	-8.51	-0.41	-0.09	-0.16
	Sub-total	5.68	-56.1	-5.4	8.1	2.4	-7.51	-21.48	-27.12	-4.29	-0.85	-0.30
W0	Paris	-1.58	-13.7	-1.9	3.5	1.9	-0.80	-7.63	-5.29	-1.44	-0.38	-0.03
	Inner suburbs	-3.18	-15.8	-1.8	3.4	0.2	-1.93	-6.87	-7.02	-1.50	-0.24	-0.07
	Outer Suburbs	-0.91	-20.6	-0.5	3.3	2.4	-4.51	-5.35	-10.69	-0.35	-0.07	-0.13
	Sub-total	-5.68	-50.1	-4.2	10.2	4.5	-7.25	-19.86	-23.00	-3.29	-0.70	-0.22
W1	Paris	-0.96	-14.0	-2.1	3.7	3.7	-0.80	-7.79	-5.37	-1.52	-0.51	-0.03
	Inner suburbs	-3.22	-15.9	-1.9	3.5	3.5	-1.93	-6.93	-7.03	-1.49	-0.29	-0.07
	Outer Suburbs	-1.50	-20.1	-0.5	3.5	2.0	-4.50	-5.18	-10.44	-0.33	-0.08	-0.12
	Sub-total	-5.68	-50.0	-4.4	10.8	5.1	-7.24	-19.91	-22.85	-3.34	-0.88	-0.22

o public transport users (about +10% on average) and also smaller-scale benefits for r users (about +1% on average and +5% in the central area), as better quality of public transport service would result in a modal shift from cars to public transport, which would alleviate roadway congestion and also provide benefits to "persistent" car users.

Secondly, improving public transport quality entails a significant development of these modes: the associated costs would increase by +27% for S0 and by +35% for S1. With subsidies remaining constant, this is funded by large increases in public transport fares, corresponding to +90% under S0 or +115% under S1. This amount is effectively doubled, which seems tolerable given the low levels of the reference situation.

Thirdly, public transport improvements also involve a reorganisation of these services. The changes in service plans are indicated in Table 13.4: for instance, bus line length and station numbers would be much reduced in the central area, combined with a small reduction in fleet size, which would improve service frequency, thereby modifying the trade-off between access times and waiting times.

Environmental impacts were assessed in relation to local population densities. Regarding the unit cost of air pollution in densely populated areas, the tutelary value indicated by Quinet et al., (2013) for a density of 8,500 inhabitants per km^2 has been multiplied by a factor of 3 in order to correctly represent the density of 23,000 inhabitants per km^2 in the central area – the highest urban density among European cities. In these conditions, environmental impacts generate a considerable cost of about € 5.4 billion per year, of which 16% is due to public transport: air quality represents 95%, with nearly half pertaining to the central area of Paris. This is consistent with the high level of political attention paid to this issue in the 2010s; on average, a person living in central Paris bears 3.5 times the local environmental cost of a suburbanite.

As the magnitude of environmental costs is roughly equivalent to the total transit subsidies, their inclusion in welfare function W0, but not in W1, leads to some differences between Scenarios S0 and S1, beyond their common features. Both scenarios would significantly improve environmental performance compared to the state of reference (gains of around 20%). Relieving roadway congestion would benefit not only the remaining car users but also the local environment (less air pollution) and hence the residents, so that all stakeholders would gain from this policy. Between the two policies, as a result, S0 would improve environmental performance more than the S1 policy, which is short-sighted with respect to the environment: the respective gains in environmental costs are −23% and −19%, respectively, in relation to the reference situation.

Taking the environment into account in S0 would lead to less development of public transport networks than in S1, with smaller increases in fares for these modes and also different provisions: rail-based modes would be preferred to bus lines in the central area, because of the local pollution produced by buses.

The main point to remember is that, according to the simulation of the scenario S0, there seems to be a great potential for optimising the planning of the mobility system of the Paris metropolitan area, with net benefits for all stakeholders in principle.

Table 13.4 Virtual Re-design of Public Transit Components.

Sous-région		Paris				Inner Suburb				Outer Suburb		
Service mode		Bus	Metro	RER	Train	Bus	Metro	RER	Train	Bus	RER	Train
Transit component		1	2	3	4	5	6	7	8	9	10	11
Reference situation	Line length (km)	598	171	57	13	2894	39	181	123	20032	355	761
	Station number	1795	248	29	6	7575	52	85	40	25173	128	187
	Rolling stock	1295	549	106	15	3078	104	180	108	4271	266	455
	Station spacing (km)	0.33	0.69	1.97	2.17	0.38	0.75	2.13	3.08	0.80	2.77	4.07
S0 : max Pu+Po+Pe	Line length (km)	456	350	99	16	1161	93	391	191	4005	544	532
	Station number	1271	485	43	3	3191	172	272	115	9620	386	312
	rolling stock	891	736	138	28	2029	158	250	95	4402	250	174
	Station spacing (km)	0.36	0.72	2.30	5.33	0.36	0.54	1.44	1.66	0.42	1.41	1.71
	Tariff (€/pass.)	0.73	1.12	0.88	1.28	0.18	0.27	0.00	0.00	1.78	2.83	1.98
S1 : max Pu+Po	Line length (km)	557	356	101	17	1283	93	395	193	4265	563	552
	Station number	1445	487	44	3	3390	173	274	115	10150	406	328
	Rolling stock	1334	797	154	34	2476	164	263	102	4987	268	188
	Station spacing (km)	0.39	0.73	2.30	5.67	0.38	0.54	1.44	1.68	0.42	1.39	1.68
	Tariff (€/pass.)	1.53	1.57	1.17	1.48	0.39	0.26	0.00	0.00	1.49	2.25	1.36

13.5 Conclusion

We developed a comprehensive methodology for mobility system modelling and planning design that unifies the two lines of research pioneered by Van Nes and Daganzo. Our model combines:

- A division of the urban area into subareas with specific characteristics, in particular demand density;
- Travel demand between the subareas made up of three categories of users: people dependent on cars, people dependent on public transport and those who can choose their mode of transport;
- A public transport network made up of several components with structural parameters that induce quality of service and related service production costs: these are therefore the levers of action for design;
- A roadway network with fixed capacity and production cost yet with endogenous quality of service based on an aggregate speed–flow relationship;
- The environmental impacts of vehicle traffic in terms of air pollutant emissions (local impact) and carbon emissions (overall impact).

The segmentation of travel demand, the consideration of modal choice, the hierarchical transport network and the inclusion of environmental aspects constitute improvements on the existing models. Furthermore, the economic analysis of the optimisation programme is novel and sheds light on the balancing of the diverse priorities of urban mobility planning.

The wide range of real-world issues addressed in the model, together with appropriate simplification of spatial details, make it particularly valuable for providing high-level insight into the sustainability performance of urban mobility systems. In the example discussed, it was shown that the explicit inclusion of environmental impacts leads to a noticeable shift in "policy packages".

Of course, the model is incomplete, not only because of the simplification of the spatial detail but also because of the consideration of a limited set of action levers. For instance, we did not consider parking fees or traffic tolls as levers of action. Nor did we contemplate planning options such as segregated bus lanes or bus prioritisation at roadway junctions, or electric buses which, in the application example, could yield greater environmental benefits than the transition from the baseline situation to the S0 state.

The model could be further developed along the following directions of research:

- A more detailed description of the relations between transit submodes (feeder vs. parallel use);
- Modelling the technical functioning of public transport, particularly in relation to passenger traffic (depending on the capacity of the vehicles and congestion of the lines), and drawing out the consequences for both the production costs of these modes and their quality of service for the generalised trip costs;
- Better representation of roadway networks, with an explicit hierarchy between motorways and arterial streets in each subarea, and explicit representation of roadway production costs;

- Modelling parking conditions: its technical functioning with explicit modelling of local capacity and demand, the formation of specific parking production costs, the formation of parking service quality and its inclusion in the utility functions of car modes;
- Calculation with temporal and spatial distributions for both demand and supply;
- More flexibility of demand, notably for trip timing and destination choice;
- Well-being optimisation with respect to the levers pertaining to all modes and also to demand management (orientation, transfer to alternative options according to time, place and mode).

Acknowledgements

This research has been undertaken within the framework of the ParisTech Chair in the Eco-Design of Buildings and Infrastructure, sponsored by the Vinci group. We would also like to thank IDFM for making the Household Travel Survey "EGT" available to us, and also DRIEA for providing the "MODUS" travel demand model.

Bibliography

Badia, H., Estrada, M., Robusté, F., 2014. Competitive transit network design in cities with radial street patterns. *Transportation Research Part B* 59, 161–181.

Combes, F., van Nes, R., 2012. A simple representation of a complex urban transport system based on the analysis of transport demand: the case of Region Ile-de-France. *Procedia – Social and Behavioral Sciences* 48, 3030–3039.

Commissariat général à la stratégie et à la perspective, 2013. L'évaluation socioéconomique des investissements publics, Premier Ministre de la République Française.

Daganzo, C.F., 2010. Structure of competitive transit networks. *Transportation Research Part B* 44, 434–446.

Daganzo, C.F., Geroliminis, N., 2008. An analytical approximation for the macroscopic fundamental diagram of urban traffic. *Transportation Research Part B* 42, 771–781.

Estrada, M., Robusté, F., Amat, J., Badia, H., Barceló, J., 2012. On the optimal length of the transit network with traffic performance microsimulation application to Barcelona. *Transportation Research Record: Journal of the Transportation Research Board* 2276, 9–16.

Leurent, F., Combes, F., van Nes, R., 2016. From strategic modelling of urban transit systems to golden rules for their design and management. (provisional document).

Mohring, H., 1972. Optimization and scale economies in urban bus transportation. *The American Economics Review* 62 (4), 591–604.

Quinet, E. et al. (2013). L'évaluation socioéconomique des investissements publics. Rapport du Commissariat général à la stratégie et à la perspective, République Française.

Small, K.A., Rosen, H.S., 1981. Applied welfare economics with discrete choice models. *Econometrica* 48.1, 105–130.

Van Nes, R., 2002. *Design of Multimodal Transport Networks a Hierarchical Approach*. T2002/5, TRAIL Thesis Series, Delft University Press.

Chapter 14

Application of the *ParkCap* Model to Urban Parking Planning

Houda Boujnah and Fabien Leurent

ÉCOLE DES PONTS PARISTECH, MARNE-LA-VALLÉE, FRANCE

14.1 Introduction

14.1.1 Context: Parking Supply–Demand Simulation

This chapter follows two chapters in the first collective work by the Ecodesign Chair (Peuportier et al., 2016). In the first book, Chapter 6 presented a systemic analysis of city parking and the development of a spatialised simulation model, called ParkCap, and Chapter 15 provided a detailed description of multimodal mobility in the Cité Descartes district, located in eastern Paris.

The ParkCap model addresses the interacting supply and demand of parking in an urban area. The parking supply consists of lots of spaces. Each lot is described in terms of location, number of places and management set-ups, including private/ semi-private/public access and pricing. This parking supply is connected to activity places, which are the endpoints of people's trips, and to the road network. The demand for parking is modelled by origin–destination zone pair and by segment, according to the purpose of the activity at the destination, the possibilities of access to private and semi-private lots and individual preferences – willingness to pay and the respective arduousness of the time spent travelling by car and walking times. Two forms of interaction are modelled between supply and demand. On the one hand, each of the customers wishes to access a certain place by car, and they choose a route and a lot with parking spaces for this purpose. On the other hand, the access demand in each lot is constrained by the capacity. In the event that capacity is reached, the unsuccessful seekers must turn to nearby lots. This results in trips to find a space, thus an additional car journey, more congestion for the flow in circulation and an increase in energy consumption and polluting emissions. The two interactions are jointly at work. Together, they induce a state of balance between parking supply and demand.

The supply–demand equilibrium of parking is characteristic of a configuration of demand of the locations. The activities established in the locations generate trips, some of which use the car mode and require parking. The state of equilibrium is also characteristic of the "management plan". The parking space capacity in the locations is determined, in terms of public spaces and the sizing of public car parks, or influenced, in terms of private or semi-public places associated with housing and economic establishments, through public actions: road layouts, concession of public parks, rules for the construction of new buildings and the proportioning of floor areas according to the nature of the activity and the numbers of parking spaces (e.g. a space for a two-room dwelling).

14.1.2 Problem: Quantifying and Simulating to Inform Planning

Parking planning is usually carried out at the scale of a mobility-generating hub (e.g. shopping centre) or a neighbourhood. A capacity design study is typically carried out in order to quantitatively understand the parking needs, on the one hand, that is to estimate the demand, and to conduct an inventory of the existing or planned parking spaces, on the other hand, or, in other words, to estimate the supply.

The relationship between the level of needs and the level of capacity is a pressure indicator: a higher value measures a more intensive occupation of supply, and more difficult conditions for demand. This is an important lever for managing urban mobility, by guiding the modal choices of individuals. If conditions for cars are more difficult, this will encourage an increased use of an alternative, such as public transport or an "active" mode, such as walking or using a two-wheeled vehicle (which is easier to park than a car).

However, a capacity design study only produces an observed state of the system and only provides a starting point for activating the lever. Simulation is required to understand the sensitivity of the system to the lever. This can take the form of a sample of users interviewed in a stated preference survey, or a quantitative simulation model like ParkCap. The two approaches are complementary: the stated preference survey will indicate the individual sensitivities of the demanders and be fed into the simulation model. It addresses the overall state of the system and simulates both supply and demand, as well as an interaction between them. In particular, the ParkCap simulator produces the following results:

- The filling level of each lot and, in case of saturation of capacity, the ratio between the number of candidates and this capacity, which constitutes a pressure indicator;
- The map showing how the different lots in the studied district are filled constitutes an overview of the parking situation in the territory;
- On the demand side, the average conditions of the parking obtained are calculated by segment according to the places of origin and destination, the activity purpose and the category of individuals, including the time spent looking for a space, distance travelled, the price to be paid and the walking distance to the final destination;
- The contribution of journeys in search of parking to car traffic, traffic conditions and environmental impacts.

All these characteristics, which are essential for the quality of service of a car trip, are absent from the simulation models traditionally used to inform urban mobility planning. They produce flow and congestion maps, speed and quality of service maps, as well as indications of the local passage quality of service and by demand segment. The complementarity appears evident.

14.1.3 Objective: To Study a Planning Case with ParkCap

This chapter presents a case study of parking planning addressed using the ParkCap simulator. The objective is to implement an application with demonstration value on two levels: to offer a practical demonstration of how the simulation sheds light on the concrete case and participates in the development of parking plans, and to demonstrate the "operational" capacity of the simulator methodologically.

In practical terms, we want to reveal how parking works in the study site by describing the "physiology" of the system with its "exogenous" and "endogenous" characteristics (i.e. the system state variables in the simulation), and by studying the sensitivity of the system to changes that have already been planned or which are more virtual. We consider the case of "Cité Descartes Élargie" (Cité Descartes Expanded – CDE), a district spread over $1.6\,km^2$, with an urban density that was already significant in 2010 although relatively low for the Paris urban area.

Methodologically, we will show the coding of the exogenous characteristics that describe the supply plan and the basic demand situation, and also simulate the different endogenous variables and the calculation of a state of equilibrium. We will interpret the results and look for broader implications for parking management in an urban neighbourhood. Finally, we will question the relevance of the results, and we will reflect on the "explanatory power" and the "semantic content" of the simulation model.

14.1.4 Method

Our process falls within a well-established technical tradition: study engineering to support decision-making. In practice, this engineering is conducted in a typical cycle. The cycle includes a sequence of three stages: firstly, a "variant design" stage (project, plan and policy), then a stage simulating this plan, followed by a stage to evaluate its impacts and performance.

Based on the results of the evaluation, the decision-maker decides whether to implement the project or not. If the decision is positive, the evaluation results in an implementation step, the result of which will lend itself to observations. These will certainly invite the consideration of other possibilities, leading to a return to the design stage. If the decision is negative, either the study ends or we return directly to the design stage.

In this technical process, "simulation" is an intermediate step between "variant design" and "evaluation" (see Figure 14.1).

The simulation is itself ensured by a particular technical process: the constitution and the exploitation of an applied model. In other words, the life cycle of an applied model includes a construction phase and a usage phase. The usage phase has already

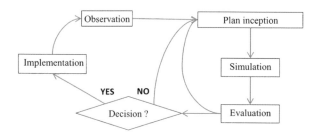

Figure 14.1 The study engineering cycle for supporting decision-making in the management of a system.

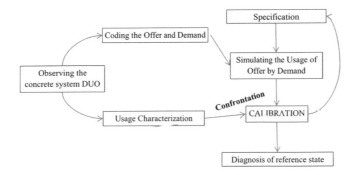

Figure 14.2 Construction of a demand–use–supply (demand–use–offer DUO) model.

been described: simulation takes place between variant design (or scripting, scenario development) and evaluation, and there is a direct or indirect feedback between evaluation and design. The construction phase is more elaborate. Its material consists of a set of observations relating to the system studied. This set is divided into two parts. One is used to describe, or code, the input variables of the model, while the other part serves as a reference against which the simulation results may be compared. To match the results to the comparison observations, the representation, the values of the parameters or even the structure of the model is adapted. This is the calibration operation. Once calibrated, the simulation can be used to establish a diagnosis of the state of the system in a reference situation. The "non-trivial" outcomes serve as the indications for this kind of diagnosis.

Figure 14.2 outlines the constructive phase of a DUO model: DUO stands for demand, usage and offer. In this demand and supply model, the interaction consists of the use of the supply by this demand – this is precisely the principle of ParkCap. In this case, the coding data is related to the supply and demand. The calibration data concern usage.

14.1.5 Outline of This Chapter

The rest of this chapter is organised into six sections. After a general presentation of the case studied (Section 14.2), we will show the construction of the model. The coding of supply and demand gives rise to both the structural specification of the model and the information concerning the numerical values that describe things (including the parameters). Application to the reference situation gives rise to the calibration (Section 14.3). This application leads to a diagnosis and provides an initial insight for decision-makers (Section 14.4).

Following this, we take full advantage of the simulation power of the model by considering variants from the reference state and evaluate the performances (Section 14.5). We then "script" virtual states to envisage the system for a future horizon (Section 14.6).

In conclusion, we will summarise the scope and limitations of the simulator and indicate areas for improvement (Section 14.7).

14.2 Presentation of the Study Area

The district of Cité Descartes has served as a case study for most of the research by the Ecodesign Chair. In the first collective book, Chapters 13–15 presented a detailed chronicle of the urbanisation and urban development projects for this territory in the state of 2010 (Leurent, 2013), its situation in relation to the urban area of Paris as a whole (Aw et al., 2013), as well as the activities, their accessibility in space and local mobility (Boujnah et al., 2013).

In this section, we first present the situation and the district's urban orientation (Section 14.2.1), then the urban functions in the state of 2010, especially in terms of population and economic activity (Section 14.2.2), then mobility demand (Section 14.2.3), and finally transport supply, quality of service and the use of modal means (Section 14.2.4).

14.2.1 Situation and Urban Orientation of Cité Descartes Élargie

Cité Descartes is an urban fragment of the "New Town" of Marne-la-Vallée. It is located at the interface of the Porte de Paris and Val Maubuée areas, specifically to the west of the municipality of Champs-sur-Marne and in the immediate vicinity of Noisy-le-Grand on the west side and Émerainville on the south side. Statistically speaking, the study area that we call "Cité Descartes Élargie" is formed by the IRIS Descartes and Nesles, south of the municipality of Champs-sur-Marne (Figure 14.3a), and covers approximately 160 hectares. This geographical position on the edge of the urban area is one of this district's major assets, especially since structural transport axes cross through it (RER (Regional Express Railway) line A and A4 motorway) ensuring its connection to the main hubs in the Parisian metropolis (Figure 14.3b).

At the regional level, the urban development of the Cité Descartes neighbourhood originated from a political desire to contribute to metropolitan dynamics, and in

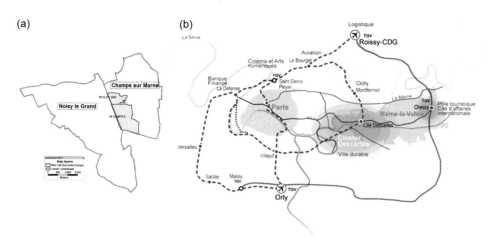

Figure 14.3 (a) Study district: IRIS Descartes and IRIS Nesles Sud in Champs-sur-Marne, (b) Cité Descartes in the Île-de-France region. (EpaMarne, (2009).)

particular to rebalance the East of Paris in terms of research and higher education, and to boost the development of the new town of Marne-la-Vallée.

Urbanisation began in the late 1980s with the construction of the Copernic building. Academic institutions then established themselves in the area, notably the University of Marne-la-Vallée in 1990, the IUT in 1993, the École nationale des ponts et chaussées in 1997 and the IFSTTAR in 2010.

This urban development was achieved through a change in land use. In the early 1980s, the site was still an area with agricultural crops and woods, close to the town of Noisy-le-Grand, which at the time was on the fringes of the Paris metropolitan area. In addition, this land without urbanisation adjoined the route of the A4 motorway, a major link between Paris and the east of France, which provides access to the heart of the urban area and its main road network.

Figure 14.4 shows the land use according to the land parcels in the state of 2008. Nearly half of the land surface is built up. The wooded areas are still present. The northern part of the district is occupied by housing, school facilities and some community amenities. In the southern part, there are large research and training institutions, and in the far south, on the edge of the motorway, there is a business zone.

The A4 motorway to the south, the RER line A in the centre and the former A199 motorway (now RD 199) further to the north constitute divisions along an east–west axis. In the north–south axis, the former RN 370 remains a major traffic way. The Ateliers Lion urban planning project (2010) intended it to become a major urban

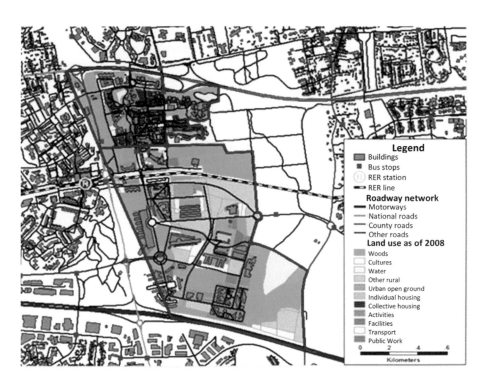

Figure 14.4 Land-use modes in 2008. (authors' map based on source from IAU-IDF.)

boulevard, lined with new housing and tertiary activities, shared between the different modes of transport and a place of calmed traffic.

14.2.2 Urban Functions; Population and Economic Activity

Four major urban functions emerge from the territorial analysis of the neighbourhood: (i) housing, (ii) professional occupation and associated employment, (iii) the research and higher education specialism and (iv) the hosting of large metropolitan transport infrastructures (for RER A and even national for A4), both crossing through and offering access (Noisy-Champs train station for RER A and the "Cité Descartes" junction for the A4 motorway).

Population: As of 2009, the neighbourhood has nearly 5,900 inhabitants, most of whom live in the northern part (60%). The average household size (2.25 people) is in line with the national average, as is the proportion of one-person households (35%). A distinctive aspect is the high proportion of the 15–29 age group, in relation to the educational function and the presence of student housing.

Workers and local employment: The dominant occupational categories among the inhabitants are those of employees and intermediate occupations. The proportion of executives is lower than the regional average, and that of industry workers is higher. Overall, the socio-professional profile of active inhabitants is medium-low for the region. By contrast, the profile of local jobs is much higher. Executive jobs are overrepresented, as opposed to workers, employees and intermediate professions (Figure 14.5b). There is therefore a qualitative imbalance between the professional skills of the inhabitants and the local employment needs. In addition, there is also a quantitative imbalance as there are 3,100 active employed inhabitants, while 5,000 jobs are established in the neighbourhood. Scientific and technical activities dominate the employer establishments (67%), followed by shops and local services (21%).

14.2.3 Mobility Demand

The presence of local amenities (including the Champy supermarket) meets certain needs of the residents. However, they carry out most of their out-of-home activities outside the neighbourhood, including work (see Figure 14.6a), shops and specialised services, and a good deal of leisure and personal relationships.

The individuals employed in the neighbourhood and the students continuing their education provide an even greater number of exchange trips between the neighbourhood and the surrounding areas (see Figure 14.6b). Figure 14.6 shows the residents' job basin and the neighbourhood's employment area: the asymmetry is clear. Resident workers mainly find their employment in neighbouring municipalities. Many of the site's employees come from distant municipalities along the east–west axis common to RER A and the A4 motorway.

In addition to internal trips within the neighbourhood and exchanges with the surroundings, the mobility demand includes through, transit trips. Car transit mainly takes the A4 motorway or the arteries that join it (including the Boulevard du Rû de Nesles). As it has a fairly dense system of junctions, the flow carried by each of them is limited. Transit travel makes greater use of public transport: line A of the RER, bus lines as feeders of the RER A (employment and school). The most noteworthy transit

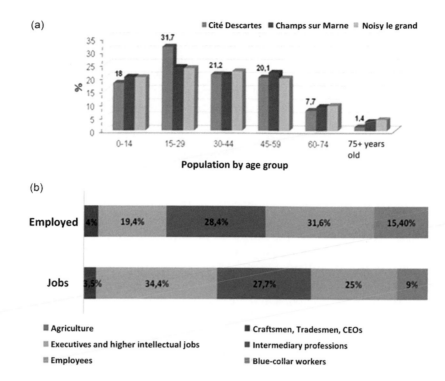

Figure 14.5 (a) Distribution of population by age group (%), (b) socio-professional group distributions of actively working inhabitants and jobs offered. (Authors based on 2009 census.)

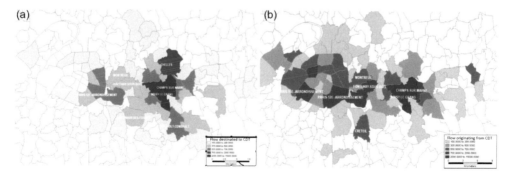

Figure 14.6 (a) CDE residents' job basin, (b) CDE neighbourhood workforce basin.

flow consists of intermodal movements: articulating a car leg and a transit leg at the Noisy-Champs station for park and ride. We will see that this particular demand plays a major role in the functioning of local parking.

14.2.4 Transport Supply, Quality of Service and Modal Split of Mobility

Transport supply: Chapter 4 of this book provides a detailed model of the transport supply in the neighbourhood. Let us remember here the opposition between the internal mobility of the neighbourhood and the longer trips, as an exchange or as a transit, which are carried out by means of mechanised modes. The two-wheeled vehicles are not very present in the district, which is a sign that the student population does not frequent the vicinity. The motorised modes, private car (VP) and public transport (TC), therefore remain to ensure the medium- or long-range mobility.

Quality of service: The respective performances of the two main motorised modes vary in space and time during the course of the day. On the public transport side, RER line A ensures fast and frequent journeys along the east–west axis, but its service is interrupted during the night. Bus connections are much slower and less frequent. In the north–south axis, they are separated on both sides of RER line A, requiring a connection to go from one half of the district to the other.

On the private car side, in the east–west axis, the A4 motorway is permanently available with a high speed, except in the direction towards the centre during the morning peak and in the reciprocal direction in the evening peak. The local road network is not congested except in the event of a major disruption on the highway.

Modal distribution of medium- and long-range mobility: For Home-Work and Home-Study purposes, public transport dominates for the employees of the site thanks to the RER A, but the car dominates among the active workers who live in the district. It also dominates for jobs located in the south of the district, which are both close to the motorway junction and relatively far from the Noisy-Champs station, meaning they are better served by cars than by the structural public transport.

By 2030, the structural public transport will be considerably strengthened. In the frame of the "Grand Paris Express" project, three new automated metro lines will connect to the Noisy-Champs RER A station and will thus serve the area: line 15 to the south and line 15 to the north, while line 11 will join in the north-east–south-west direction. This strengthened service will be of even greater benefit to the modal share of public transport, as the area will be subject to reinforced urbanisation through the addition of housing and activities in the vicinity of the station or at a short distance in the northern part of the district (in particular the subdistrict of the RN 370-RD 199 junction), which is further away from the A4 motorway. The increased use of structural public transport will justify the expansion of public transport park-and-ride services, which will strengthen the dynamic.

14.3 Constitution of the Applied Model

Chronologically, we created the study model of parking in the CDE district in 2013 and 2014, in relation to the MODUS model of the DRIEA (Direction Régionale et Interdépartementale de l'Équipement et de l'Aménagement – Inter-department and Regional Directorate for Facilities and for Land-Use Planning). Modelling the parking has led us to represent space in a more refined manner, either by refining the division of the territory into traffic analysis zones or by describing the roads in more detail. This refinement inspired the one presented in chapter 4 of this book, conducted in 2015 and 2016, for the multimodal mobility of the neighbourhood in relation to the rest of the urban area.

In this section, we retrace the coding of the supply Section 14.3.1) and that of the demand (Section 14.3.2). On each side, they concern structural specification and parameter set-ups. We will then explain the articulation of the supply and demand submodels (Section 14.3.3), the modelling of impacts on general car traffic and the environment (Section 14.3.4) and the calibration carried out (Section 14.3.5).

14.3.1 Coding of Parking Supply in the Neighbourhood

The parking study is limited to the district comprising the two IRIS areas of INSEE (the French National Institute of Statistics). Their border with the surrounding area constitutes the perimeter of our study. The reference period is an average working day in 2010.

The parking supply has been coded as lots of spaces, coded as point spatial objects in the TransCad GIS software, providing the following attributes for each of them:

- Its type, from the categories Private, Public, P + R (park and ride) and Other;
- Its physical capacity: the total number of spaces offered;
- Its management set-ups, in particular the access conditions, tariffs and permitted length of stays, the probability of checks and the amount of the fine in case of infringement.

The number of lots amounts to 390: 80 off-street lots – one per public park or private lot, grouping the neighbouring buildings that have the same profile as activity generators within a radius of 50 m – and 310 on-street lots, one per road section edge and per direction of traffic, for 40 (or 115) sections of primary (or secondary) roads.

The coding was done in the form of a geographic database, using the TransCad Geographical Information System, based on the description of the roads and buildings in BD Topo® (Ign source). The number of places was measured[1] by reading Google Earth aerial photographs for the surface spaces, completed by a field survey at the establishments for the spaces in buildings. Figure 14.7 shows the result of the parking supply coding.

Thus, the parking map is placed over that of the road. The parking supply model is superimposed on the "topological" model of the roadways in terms of nodes and arcs. This kind of graph modelling the roadways was extracted from BD Topo by slightly simplifying the topology, which was then described in terms of reference speed and flow capacity (according to the type and the width of the road) and coded under TransCad as a network of road infrastructures. Five nodes were identified as access gates between the neighbourhood and the outside, and we associated each with a fictitious parking lot reserved for transit traffic. Here is the list:

---

1 With the help of students from the ParisTech Master programme on Transport and Sustainable Development, class of 2009–2010.

(a)   (b)

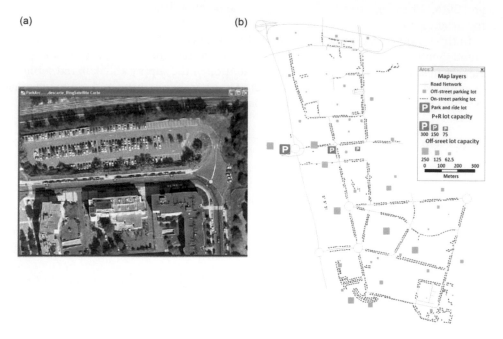

Figure 14.7 (a) Aerial photography for visual recognition. (Google.), (b) parking supply in the 2010 reference situation.

- An access point to the north, which guarantees access to Boulevard du Rû-de-Nesles (especially from RD 199);
- Access to the east, which ensures access to Avenue Blaise-Pascal (especially from the Malnoue road);
- Access to the west, via the RD 370, which borders the axis of Boulevard du Champy;
- Access to the south, via Boulevard Blaise-Pascal, which connects with the A4 motorway;
- Central access, reserved for outgoing residents.

14.3.2 Coding of the Demand

To quantify parking as demand in terms of the number of requesting trips and duration of each request, we used the generation model described in Leurent and Boujnah (2013). The model follows the five steps below:

1. *Functional generators*: By urban function, the daily number of individual activities depends on a specific headcount, called the generator. The "resident population" generator motivates the presence of residents in their homes or in a local activity. The "local employment" generator motivates the presence of the workers, plus professional visits and customers footfall for the Purchase purpose category.
2. *Local activity demand*: By activity purpose, the demand is proportional to its local generator according to a certain daily factor. For the Home purpose category, the

presence factor is the supplement to the unit of factors for the other purposes that cause a resident to leave work, for example.

3. ***Demand for parking spaces***: Again, by purpose category, the mobility conditions impose a certain modal share for the Car Driving mode. The product of the specific generator, by the daily factor and by the modal share, constitutes the daily demand for parking spaces, to be aggregated over all the activities in order to obtain the overall daily demand.

4. ***Parking occupancy load***: The demand for spaces still remains to be temporalised according to the schedule and duration of each activity. Let us focus on the daytime period from 9 AM to 5 PM, which reflects the continuous occupation of a parking space by a "commuter", that is an occupant who comes to carry out their day of work or study. Allowing one hour for a buying activity, its relative duration coefficient is 1/8 in commuting equivalents, while a 4-hour work visit is worth 1/2 commuting equivalents. The total occupancy load of the daytime period is established by adding together the masses in commuting equivalents for all the purposes.

5. ***Modal and spatial distribution of the load***: Functional generators can be disaggregated by parcel or block, and with them all their consequences on the activity demand, the parking demand and the occupation load.

Just over 100 generating sites have been coded: the railway station, 20 research and academic establishments, 17 housing plots, 60 tertiary business plots and 10 shops. By adding the five fictitious nodes, this makes a total of 103 micro-zones.

The demand was segmented by location and activity purpose. Six main purposes were distinguished:

- Employees (regular work);
- Students (higher education and training);
- Visitors (including shopping, business, leisure, support and others);
- Commuters (i.e. the demand for park-and-ride services at the Noisy-Champs RER station);
- Outgoing residents;
- Transit.

The spatial distribution model was coded in an Excel workbook by purpose category, based on specific sources and assumptions (Boujnah, 2017). Table 14.1 presents the generation coefficients retained by demand segment, as well as the resulting numbers, on average per day in 2010.

Regarding the parking behaviour of users, in the ParkCap model it is postulated that this behaviour is individual and rational. At the origin of their trip, each user chooses a target lot to start looking for parking (i.e. to make their first application). More precisely, they choose both a target lot and a route from their origin to this lot so as to minimise the generalised cost, which is the total of the generalised cost of the route and the cost associated with the target lot. By lot, the final cost includes the fare plus the generalised cost of walking to the point of destination. The expected cost is an average between the final cost of this lot and an alternative cost, weighted by their respective probability (success or failure of an application). The "alternative" cost is

Table 14.1 Parameter Estimates of the Parking Generation Rate Per Demand Segment

			Demand Segment			
	Employees	Students	Park and Ride	Visitors	Outgoing Residents	Transit
Occupants reference	Number of jobs	Number of students	Daily travellers	Incoming vehicles	Resident workers	Incoming vehicles
Parameters						
Y_{sl}	0.15(1)	0.06(1)	—	—	—	—
Y_{so}	0.85(1)	0.94(1)	—	—	—	—
α_{sl}	0.3(2[b])	0	0.05(4)	—	—	—
α_{so}	0.7(2[b])	0.13(2[a])	—	—	—	—
μs	1.1(2[b])	1.3(2[a])	1(2[b])	—	1.1(2[b])	—
β_{sj}	0.9(6)	0.8(2[a])	0.7(2[b])	—	0.9(6)	—
λ_{sj}	0.9(7)	0.7(2[a])	—	—	0.9(7)	—
Total flow (veh.)	2,694	1,593	589	136	787	815(4)

Source: Boujnah (2017).

a composite cost for all the lots attached to the destination, integrating the expected cost by lot plus the cost of access to this lot by car from the current lot *via* the road. The probabilities of an application's success and those of carrying over to another lot are calculated by the model.

The parameters of the generalised cost of trip making still remain to be specified. By segment, this cost totals tariffs and other monetary expenses, as well as the respective times spent in the car for the origin-to-first lot subpath, in the car looking for a space or walking to the final activity place, each weighted by a specific time (money) value that translates each unit of time spent in the state concerned into a monetary equivalent, according to its specific arduousness. For the travel time, the individual money value of the time was fixed at €9/hour for the Visitor purpose and at €13/hour for the other purposes. The basic level is amplified by a factor of 1.4 for search time and 2.0 for walking time (or 1.65 for a visitor or student). Finally, the hourly parking fee is multiplied by a specific average duration: 2.5 hours for employees and commuters, 1.5 for students and visitors, 1 hour for outgoing visitors (these are half durations for the activity purpose, as the other half is imputed to the associated return trip).

14.3.3 Articulation of Supply and Demand Submodels

In ParkCap, the submodels that address parking supply and demand, respectively, are linked by spatial, temporal or quantitative articulations:

- *Connection of spatial objects*: The places generating demand are nodes for the road network, in the same way as the lots dedicated to the establishments and the lots associated with the road sections. The model therefore deals with an "increased road network" that incorporates the road network, the parking supply and the locations of the mobility and parking demand.
- *A fundamental spatial relationship concerns the attachment of parking lots to the destination of the activities*: By destination, a "customer catchment area" is defined that groups all the lots within a certain radius from the centre of the place. For the application to the CDE district, we selected a radius of 900 m, which is high, because the value of 500 m did not allow for a quantitative balance between supply and demand.
- *The model considers a limited time period*: The morning peak period is between 7 and 9 AM, on average for a set of working days. This is the time when the actively employed residents go to work: a part of them use a car and free up spaces with public access, whereas workers living outside the neighbourhood come to work there by car. The same is true for students. We retained the part of the demand expressed during this period, and conversely, we considered the rest of the vehicles as a fixed pre-load that occupies part of the capacity offered by the lots.
- By individual trip, the quality of service of the different options in terms of the crossing route and the target lot is calculated according to the characteristics of the supply. Exogenous characteristics include the radius connecting the lots to the places of destination, and endogenous characteristics include the probability of success of the application for a lot. Thus, the demand submodel depends on the supply submodel.

- Conversely, by lot, the number of candidates results from the demand submodel, by the aggregation of applications for this lot issued by different customers.
- Finally, the model determines a quantitative balance between supply and demand.

14.3.4 Impacts on Mobility and the Environment

The car traffic passing through the neighbourhood during the simulated period is modelled between the access nodes located between the neighbourhood and the surrounding area: the five nodes of the cordon screenline have already been mentioned. The associated origin–destination flow matrix is derived from the regional mobility simulation model (MODUS model by DRIEA). Each transit trip is assigned to a minimum generalised cost itinerary for the motorist.

On the neighbourhood's road network, transit traffic is added to the "internal traffic", which includes intra-district trips and those to and from the surrounding areas. The resulting local traffic is increased by the journeys for looking for a parking space. For each section of road, the travel time is modelled according to the local traffic flow by an increasing function, which also depends on the flow capacity of the road and some shape parameters. We kept the time-flow functions of the MODUS model.

Finally, local car traffic generates environmental impacts, such as energy consumption, greenhouse gas emissions, air pollutants and noise. The modelling of these impacts is described in the chapter on spatial refinement by Kotelnikova-Weiler et al. (2017).

14.3.5 Calibration of the Study Model

The particular configuration of the neighbourhood limits and even excludes parking exchanges with the surrounding areas. The lots on the edge of the neighbourhood have sufficient capacity, both on the side of the neighbourhood and on the other side. There is therefore no overflow in one direction or the other. The only exception could concern park-and-ride parking at railway stations, as motorists who want to use the RER line A can choose between the Noisy-Champs station in the neighbourhood, and others in neighbouring districts. We considered this demand segment as a fixed exogenous flow.

To calibrate the ParkCap model in a particular study, the first step is to successfully calculate the joint equilibrium state between supply and demand. We achieved this by setting the radius connecting the lots to the destinations to 900 m. The calculation process is based on an algorithm of convex combinations: the convergence obtained in 1,000 iterations is satisfactory.[2]

Thereby being able to balance supply and demand accuracy, the actual calibration consists of adjusting certain input data and parameter values, and, if necessary, adapting the structural specification of the model, in order to reproduce a set of observations corresponding to the model results with sufficient accuracy. In this initial case study, we limited ourselves to broad-ranging assessments, without describing the places in too much detail. On the one hand, we found that the parking load is faithfully reproduced to a great extent (Table 14.2). On the other hand, with specific regard to

---

2 The calculation method is clearly perfectible.

Table 14.2 Average Occupancy of Spaces within 300 m Radius of Noisy-Champs Station

	Observed Situation[a]		Modelled Situation		Difference (Observed - Modelled)
	Total Demand (veh.)	Occupancy Rate	Total Demand (veh.)	Occupancy Rate	
Unregulated roads	467	98	478	100	−2%
Blue zone	46	100	46	100	0%
P + R	272	100	272	100	0%

Table 14.3 Comparison of Perceived Travel Times for the Reference Situation

	Observed Data (minutes)	Modelled Data (minutes)	Difference (Observed - Modelled) (%)
Average search time	2.14	1	+ 53
Average end journey walking time	2.76	2.17	+ 21

park-and-ride parking at the train station, we compared the travel time components as calculated with the corresponding observations from a survey dedicated to the demand. It shows a close match, of which we are satisfied (Table 14.3).

At this stage, therefore, this concerns corroboration and verification rather than calibration in the *strict sense* of the word, as with the radius connecting the lots to the places of destination, for example.

14.4 The Reference Situation: Simulation and Diagnosis

In creating the study model, we applied it to a "reference situation" for calibration purposes. The assimilation between the "calibration reference" and a "planning reference" is not a technical obligation, but simply a convenient option in carrying out the studies in practice.

In this section, we present the simulation results for this reference situation: firstly, the occupancy level of the lots of parking spaces (Section 14.4.1), then the interactions with the car traffic and the environmental impacts (Section 14.4.2), as these aspects also concern the entire neighbourhood. We will then examine the composition of generalised parking costs at the disaggregated level of users (Section 14.4.3). From all these elements, we will draw lessons for diagnosing parking in the neighbourhood (Section 14.4.4).

14.4.1 Local Parking Pressure and Occupation of Lots

By lot of spaces, the model simulates the occupancy level and therefore the usage rate of the capacity, between 0 and 1. This usage indicator is supplemented by a local pressure indicator that relates the number of candidates to the number of spaces, and can exceed the value of 1 when the use rate is close to the unit.

Figure 14.8 maps the lots and illustrates the probability of success, which is the inverse of the pressure indicator if it exceeds 100%, or which is 1 otherwise. This map is emblematic of the expressive power of ParkCap as it reveals parking congestion to be a spatial phenomenon. In this case, we can identify two areas of high pressure within the neighbourhood: one denoted as A around the Noisy-Champs train station and the other denoted as B around the research and academic institutions, which are highly frequented. The northern half of the neighbourhood is fluid, as is the extreme south.

Parking congestion around the railway station, induced by park-and-ride inter-modal trips, spreads out to the access roads, the capacity of the park-and-ride facilities is insufficient to meet the demand. In "academic" zone B, it is very difficult to find an available space on the road immediately. Some lots linked to the establishments have restricted access and residual available capacities.

Figure 14.8 Local parking pressure in the neighbourhood (simulation of 2010 reference state).

14.4.2 Traffic on the Network and Pressure on the Environment

The model simulates the search path for a parking space. Figure 14.9 shows the associated traffic flows in proportional thickness. In addition, the colour indicates the proportion of this flow in the local traffic. Flow levels reach up to 100 veh/hour (200 vehicles in 2 hours), which remains moderate compared to the theoretical flow capacity of 800 veh/hour (per lane and per direction of travel) for most of the sections concerned. There is therefore no traffic problem, only a moderate increase in local traffic. The rate of increase is significant in zone B, but the absolute levels remain moderate.

With regard to environmental impacts, it should be noted that the search paths for parking spaces contribute to up to 8% of carbon emissions (CO_2), estimated at 1.4 tonnes per day for the two-hour period.

14.4.3 User Side: The Composition of the Parking Cost

The two maps of the parking occupancy levels and the contribution to traffic present the local situation for land developers. In this case, the simulation indicates local parking difficulties in two areas, but no difficulties in terms of traffic.

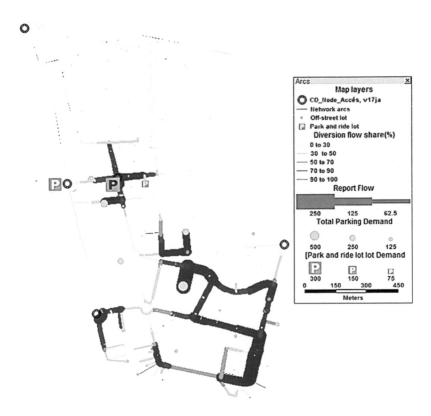

Figure 14.9 Flows of journeys to find parking in the neighbourhood (simulation of the 2010 reference state).

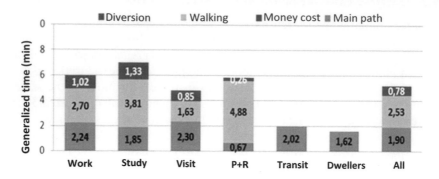

Figure 14.10 Structure of trip cost, including parking, by demand segment (simulation for 2010).

However, maps are not enough to outline the concrete conditions encountered by parking users. By demand segment according to location and activity purpose, the model calculates the times spent, distances travelled, cash expenditures and average generalised costs for parking associated with the activity. Figure 14.10 summarises the individual costs by demand segment, distinguishing between the respective contributions of the journey to find parking (designated as "Report"), the main leg in the car trip (for the part within the neighbourhood, designated "Main route") and the final trip leg on foot. Costs are estimated in equivalent time spent in the car. The overall average is approximately 5 minutes. This is a significant although generally minor amount in the overall economy of a trip, which should be made explicit in the general simulation of mobility and modal split in particular.

The levels and compositions of the individual cost are heterogeneous among the demand segments (Figure 14.10). Some characteristics are general, such as the important place of the end leg on foot (amplified by its relative arduousness) and the minor place of the research path (less than a minute, because the speed chosen for the calculation is that of the local flow, and is therefore probably overestimated).

The composition of the average cost for a given demand segment should be interpreted with finesse. Let's take the case of a train station user, who requires park-and-ride parking. The high cost of walking means that the car park is a significant distance from the station. The low cost of finding a place indicates that the user chooses their target lot effectively; they know the local conditions and have adapted to them. The cost of the "main route" is moderate because the lots close to the station are relatively close to the access nodes to the district from the surrounding area. The case of "visitors" is diametrically opposite: little walking is required because zone B, which is highly frequented and difficult to park in, is not very spread out. The journey looking for parking is longer because of the low probability of success in each lot.

14.4.4 Lessons for Parking Diagnosis

Overall, the simulation provided useful indications as to how parking works in the neighbourhood. The two situation maps are both original, easy to interpret for developers or urban managers and easy to use, including for calculating capacities that would be sufficient.

However, the limitations of a "quantitative" policy designed to meet every need are well known. In a dense urban area that is well served by public transport, planners prefer to restrict and under-size car parking to guide flexible users towards the use of other means of transport. The analysis of parking costs provides information on the individual situation of users, and provides benchmarks for estimating the level of constraint imposed on users by the state of the system. For example, the time cost of approximately 5 minutes imposed on drivers using park-and-ride facilities at the station could be compared to the quality of the public transport service available between the user's place of origin and the station.

14.5 Simulation of Variants for the Reference Situation

A fundamental advantage of the model applied lies in the simulation of virtual situations. This typically involves designing management variants (or scenarios), simulating their consequences and evaluating their performance.

In this section, we will demonstrate the simulation of virtual situations defined with respect to the reference situation itself, called "variant A". We begin by designing three alternative variants denoted B, C and D (Section 14.5.1). We compare the results with those of the reference variant, in terms of parking pressure (Section 14.5.2), traffic load (Section 14.5.3) and then user costs (Section 14.5.4). Finally, we draw general lessons from this (Section 14.5.5).

14.5.1 Design of Management Variants

The applied model makes it possible to simulate the response of users to a state of the system, as determined by an exogenous supply plan (the lots, their capacity and management) and especially by the macroscopic relationship between the capacities supplied and the volume of the demand. The supply–demand interaction dominates the determination of endogenous conditions in the system state. An individual user clearly perceives management policies such as pricing and the limitation on time a space can be occupied. They also perceive the ease or difficulty of finding a space in a place but are certainly less aware of the capacity offered, which does not concern them directly.

Three different management plans (or three variants) have been designed to influence the system state. Here are the principles, starting from the plan effective in 2010 called variant A:

- In variant A (reference variant), parking supply on the road is not regulated. The private lots are free of charge and reserved for the users of the place. The Descartes 1 and 2 park-and-ride facilities are free, and the Champy shopping centre car park offers public parking priced at €1 per hour;
- Variant B partially regulates the parking supply on public roads: within a radius of 300 m around the Noisy-Champs station, the occupancy time of a place is limited to 2 hours, under penalty of a € 11 fine with a 50% probability of being issued. It is coded "on average": a parking demand of less than 2 hours is free but another more than 2 hours is priced at $11 \times 50\% = € 5.5$;

- Variant C goes further by charging € 1/hour for the entire on-street parking supply. The provisions are identical to variant B for the others;
- Variant D charges € 1/hour for all public places, whether on the road or in the park-and-ride facilities. In addition, a new 400-space public car park will be installed at Blaise Pascal Avenue to absorb the excess demand there.

Table 14.4 summarises the characteristics of the four management variants.

14.5.2 Effects on Local Parking Pressure

Figure 14.11 shows the parking pressure maps for the three variants B, C and D, following the same principle as in Figure 14.9 for variant (A). It can be seen that the availability of public parking is improved, in a progressive way in the order of variants: relief of road congestion (B), easing of road saturation (C), freeing up of roads and public parking (D).

The effect on private lots is the opposite. Their users are led to use them as priority, from variant B onwards.

14.5.3 Effects on Search Paths and Car Traffic

For variant A, we concluded that parking search paths have a relative importance that can have a strong effect on the local traffic. However, the roads concerned have low traffic circulation, which makes the phenomenon unimportant in absolute terms.

Table 14.4 Characteristics of Simulated Parking Management Variants

Supply type	Actions	Variant A	Variant B	Variant C	Variant D
		Reference situation	Blue zone	On-street parking pricing	Global management of public parking
On-street public parking	Blue zone	× 45 places on avenue Ampère	× 350 slots within 300m from station	–	–
	Control	× Control Prob.: 0.001	Control Prob.: 0.5 × Fine fee: €11	–	–
	Pricing			× € 1/h	× € 1/h
Off-street public parking ![P]	Localisation	–	–	–	Coeur Descartes (see below)
	Capacity	–	–	–	× 500 places
	Pricing	–	–	–	× € 1/h
Park & Ride ![P+R]	Layout	–	–	–	–
	Capacity	–	–	–	–
	Pricing	× € 1/h in Noisy-le-Grand lots	× € 1/h in Noisy-le-Grand lots	× € 1/h in Noisy-le-Grand lots	× € 1/h for all P&R facilities

Free slots
Blue zone lots (2h)
Priced slots (€1/h)
RER station
Lot capacity
P   P   P
500 250 125
0   333   667   1,000
Meters

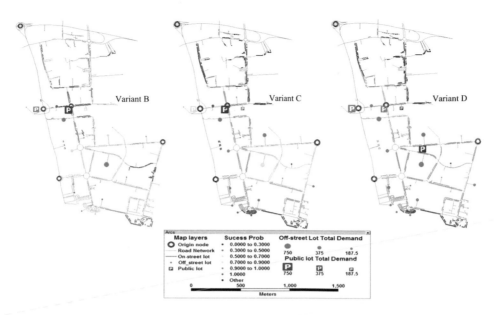

Figure 14.11 Local parking pressure according to the variants.

The results of the variants do not change this conclusion: the orders of magnitude are maintained. Figure 14.12 corresponds to Figure 14.9 for variant A, except for the flow scale. Variants B and C increase traffic looking for parking by +32% and +42%, respectively. The detail of the spatial distribution depends on the variant, but the

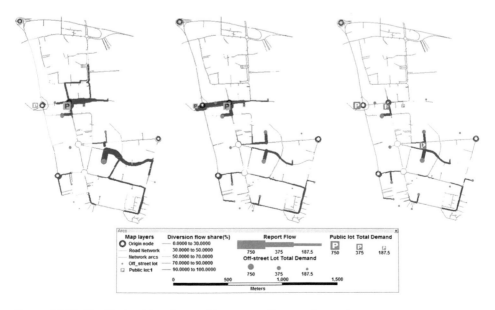

Figure 14.12 Traffic looking for parking according to the variants.

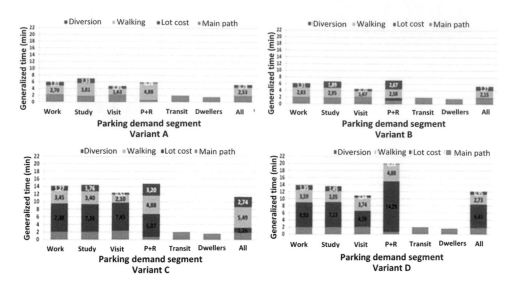

Figure 14.13 Analysis of the generalised parking costs, according to the variants.

sub-networks involved are similar. By contrast, variant D diminishes traffic looking for parking by 12% compared to variant A. It reduces parking search paths to small local systems rooted in high-capacity private or public off-street lots.

14.5.4 Effects on Generalised Costs for Users

Figure 14.13 shows the components of the generalised parking cost, by user segment, according to the four variants. Variant B is close to A. The main differences are moderate and mainly concern a reduction in walking and a corresponding increase in looking, and the emergence of payment for park-and-ride users. In other words, availability on roads is improved but at the cost of increased traffic.

Variants C and D have much stronger effects: the parking cost is doubled for the segments concerned, mainly because of the pricing. This signals the importance of pricing as a lever of action. Public parking is more fluid and more available to a marginal user, but its pricing is more expensive for all users.

Users of parking near the railway station are particularly affected in variant D, which would strongly encourage them to stop using their cars.

14.5.5 Discussion

Simulating variants makes it possible to compare them. The comparison of the results demonstrates the simulation's sensitivity to management plans. This sensitivity is consistent with common sense – after consulting the results as a whole. It is especially apparent for the increase in traffic looking for parking in scenarios B and C. The main and logical cause lies in the diversity of segments and rights of access to private lots. This provides an a posteriori, and powerful, justification of the distinction made between the segments in our specification of the applied model.

The maps of the model's original results are valuable for characterising the state of the system in each variant and for interpreting the composition of aggregates like the volume of traffic looking for parking.

The different variants each show some management trade-offs. The comparative simulation of them reveals that the fluidity of on-street parking is not an asset in itself, at least not for ordinary parking users or for the environment. Some concrete issues lie beyond the scope of the model, in particular, the facilitation of urban logistics for home deliveries and serving business establishments, or the avoidance of double parking and therefore significant obstacles to traffic.

The model simulates some possibilities for demand orientation. Variant B puts more order in the use of on-street parking, and this comes at a small cost for both the users and the environment. It seems both acceptable and useful for improving road availability. Variants C and D also put on-street parking usage in order, but under conditions that seem unacceptable for the users. At this point, we did not look for optimal pricing; rather, we only explored some possibilities.

Finally, the parking model can hardly be used in isolation if the managerial issues concern guiding the modal choice of individuals. While imposing constraints on parking encourages some users to change their mode, the parking demand will drop locally. To correctly measure the level of constraint, it is better to use a multimodal mobility model in synergy. We did this to feed our parking model, but the next step is to establish a reciprocal complementarity by importing the results of ParkCap into the urban mobility simulation.

14.6 Simulations to Explore the Long Term

A simulation model is also valuable for designing management plans for a future situation and for helping to prepare for the long term. The case study of the CDE neighbourhood is all the more favourable as it will be the subject of major urban intensification, which will double the density of human activity between 2010 and 2030 (Section 14.6.1). We have designed three parking scenarios for the future horizon (Section 14.6.2). In "trend" scenario A′, which mainly concerns the quantitative management of the supply, the parking situation would become critical, with almost total congestion of the lots on the streets (Section 14.6.3). Scenarios B′ and C′, which involve more qualitative management, would bring fluidity through pricing (Section 14.6.4). In summary, fluidity seems to be an imperative issue for the neighbourhood in its future condition of enhanced urban intensity (Section 14.6.5).

14.6.1 The Urban Project by 2030

As mentioned at the end of Section 14.2, the challenges of the metropolitan project around the Grand Paris Express particularly concern the Cité Descartes neighbourhood, which will become a "station neighbourhood" around the Noisy-Champs station, which will in turn transform into a regional hub between RER A and three new automated metro lines.

This transformation of the public transport network will greatly improve accessibility between the neighbourhood and the rest of the urbanised area. To take advantage of this quality of accessibility, the local authorities, their land operator (EPA Marne) and their urban planning consultant (Ateliers Lion) plan to intensify the urbanisation of the district, doubling it between 2010 and 2030:

*   From 5,900 inhabitants to 9,300, or +55%;
*   From 5,000 jobs to 9,800, or +96%;
*   Increasing the influence of the station on a much larger local area than the neighbourhood, reaching the level of 39,300 passengers per day arriving at Noisy-Champs station, or +134% for park-and-ride users (source RATP).

Additional housing and jobs will mostly be established near the station: on the one hand along the Boulevard du Rû de Nesles, both to the south-west and north of the station, and, on the other hand, along the RD 199, which would be transformed into an urban boulevard. An indicative general layout has been established (Figure 14.14) and makes it possible to locate new functional generators (Figure 14.15).

We modelled the generation of demand for the neighbourhood by 2030, based on urban development assumptions, by keeping the activity rates of the resident population, its distribution within demand segments according to the activity purposes and the individual mobility rates. However, we adapted the generating sites and inflected some rates. The proportion of active employees working on the site would increase to 30%, and that of students housed in the location to 10%. In terms of the modal split, private cars used for the local trips by local workers would go down to 30%, and also to 53% for the workers coming from outside. For motorists, the occupancy rate increases to 1.2 people per car for employee trips and 1.5 for students (carpooling).

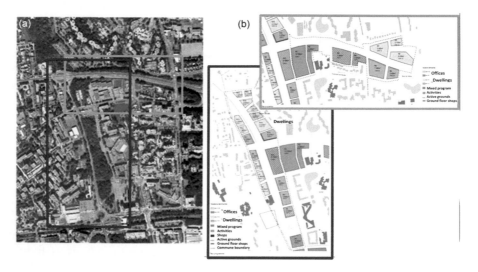

Figure 14.14 (a) Neighbourhood in 2015 situation, (b) programme plans. (Ateliers Lion, April 2013.)

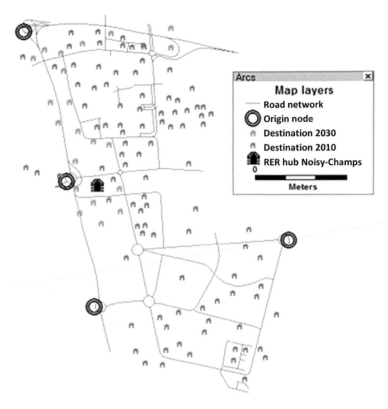

Figure 14.15 Places of destination in the CDE neighbourhood by 2030.

14.6.2 Contrasting Scenarios for Parking Management

We have designed three contrasting scenarios for parking planning and regulation in the neighbourhood by 2030 (see Table 14.5):

- Scenario A′, described as a trend scenario, only imposes the minimum standards for providing spaces to the new premises (housing and business establishments), does not charge for on-street parking or park and ride, and increases the parking capacity of the central park-and-ride facilities by 500 places;
- Scenario B′ is intermediate: it lowers the ratios for the provision of spaces in the premises (e.g. 1 place for $60\,m^2$ of net usable area), and limits the duration of free parking and charges for public car parks;
- Scenario C′ is more proactive. The ratio of the provision of spaces is further reduced (to 1 place for $100\,m^2$ of net surface area), a single rate of € 1/hour for all public parking. Hundred on-street spaces are removed, but two public cars parks with 500 spaces each are added (one in the heart of the district and the other on Boulevard du Rû de Nesles).

Table 14.5 Scenarios for the Parking Supply in Cité Descartes by 2030

Supply Type	Capacity				Available Capacity at the Beginning of the Simulation			
	Reference 2010	Scenario A'	Scenario B'	Scenario C'	Reference 2010	Scenario A'	Scenario B'	Scenario C'
Private spaces	3,908	14,246	10,326	8,268	3,664	10,955	5,863	4,292
Lots 2010	3,908	3,908	3,908	3,908	3,664	2,736	1,172	1,172
Related to housing 2030	–	3,618	3,618	2,680	–	2,171	2,171	1,608
Related to jobs 2030	–	6,720	2,800	1,680	–	6,048	2,520	1,512
On-street places	3,062	3,062	3,062	2,950	2,820	2,820	2,820	2,750
Public lots	–	–	700	1,000	–	–	630	900
Central Descartes	–	–	350	500	–	–	315	450
Rû-de-Nesles	–	–	350	500	–	–	315	450
Supply in P+R	590	1,090	1,090	1,090	531	981	981	981
P+R Descartes 1	191	691	691	691	172	622	622	622
P+R Descartes 2	81	81	81	81	73	73	73	73
P+R Champy	318	318	318	318	286	286	286	286
Total number of spaces	7,560	18,398	15,178	13,308	7,015	14,756	10,294	8,923
Evolution /2010	–	143%	100%	76%	-	95%	36%	18%

14.6.3 Results of the Trend Scenario

The simulation of trend scenario A' produces the following results:

- The saturation of on-street parking would extend widely in the neighbourhood, especially from the station to the entire northern half (see Figure 14.16);
- Traffic looking for a space would become massive (see Figure 14.17). On two main east–west axes, it would reach 300 vehicles per hour, three times the amount of the 2010 reference. Carbon emissions would increase by 20% for car traffic as a whole;
- The costs for users would increase, but in a limited proportion. Park-and-ride users would be the most affected, at a rate of +30%, with an average of nearly 4 minutes spent looking for a space.

14.6.4 Results of the Proactive Scenarios

Intermediate scenario B' and the more proactive scenario C' produce substantially equivalent simulation results, which are very similar to those of scenario C for the reference state:

- Fluidity of public on-street and off-road parking, but higher congestion of private and semi-private lots;

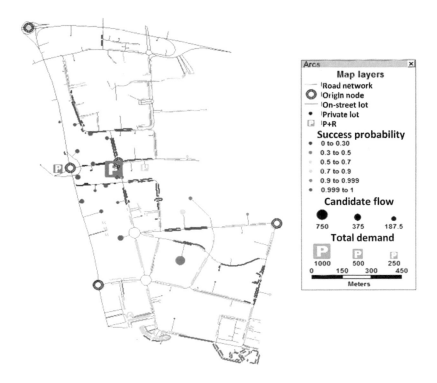

Figure 14.16 Parking pressures, variant A' by 2030.

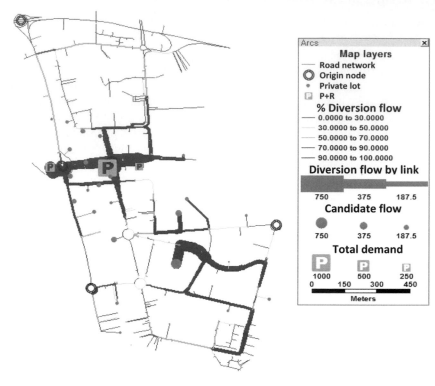

Figure 14.17 Flows of journeys looking for places, variant A' by 2030.

- Parking search paths with a limited scope. A road axis is strongly affected, namely that of Boulevard Descartes;
- User costs increase significantly compared to the reference situation (A) and the trend scenario A'. They double on average, mainly due to the pricing. For park-and-ride users at the station, the level would reach 17 or 18 minutes of equivalent time, which seems prohibitive and/or unacceptable (Figure 14.18).

14.6.5 Summary

The transformation of the district by the radical improvement of its relations to the rest of the metropolitan area *via* the structural public transport and its urban intensification will profit every occupant of the place, i.e. every inhabitant, employee, occasional visitor or user in transit through the station. Everyone will benefit from facilitated public transport conditions and a stronger local supply of urban amenities.

However, intensification will increase the tension on the rarest urban resource: usable land. This will especially affect on-street parking. We designed three contrasting parking plan scenarios and explored their potential effects using the ParkCap simulator. Trend scenario A' seems prohibitive. It saturates the on-street parking too much, and risks making urban logistics (deliveries, etc.) too detrimental for traffic. In the two

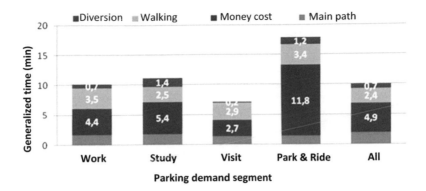

Figure 14.18 Composition of the generalised cost of parking, C' horizon 2030 variant.

more proactive scenarios, pricing would make on-street parking more fluid and easily available.

Thus, the model can identify potentialities, and is a tool for prefiguring the future and shedding light on its preparation. In practice, it will be necessary to ensure the acceptability of parking pricing, especially for the park-and-ride commuters, who will incur it every day they work. An equivalent technical solution could be to equip public car parks not only with a pricing and access control system, but also with a subscription scheme reserved for those truly needing it, which would make it possible to lower the tariffs for the commuters who are strongly dependent on cars.

14.7 Conclusion

14.7.1 ParkCap, a Tool for Planning Parking

We demonstrated the operational nature of the ParkCap simulator as a tool to inform parking planning. Using a very concrete case study, we demonstrated that the simulator is:

- *Applicable*: The input data was created at an affordable cost. For the parking supply, we used modern data (aerial photographs and BD Topo), together with a field survey of building occupants (old-fashioned crowd-sourcing). For demand, we relied on a mobility model for the entire urban area. The data was processed with a transport-oriented geographic information system, TransCad and an Excel spreadsheet.
- *Calculable*: The different submodels and the supply–demand equilibrium state could be calculated.
- *Calibratable*: The few observations of the use of parking supply by demand could be reproduced fairly accurately in reference situation.
- *Relevant*: The results obtained by simulation constitute a coherent whole. In particular, we highlight the maps of local parking pressures and local interactions with car traffic, as well as the breakdown of the generalised parking costs per

user, and the heterogeneity of these costs between the different demand segments depending on the location and the activity purpose.

• **Sensitive**: Between several scenarios graduated according to the degree of proactivity of a restrictive policy, the respective simulations produced a graduation of the results.

These operational qualities were applied to the diagnosis of a reference situation and to the study of variants, some in a reference situation and others in a virtual situation with a long-term perspective.

14.7.2 Outreach of the Model, Limitations and Areas for Improvement

Detailed modelling of parking produces very original results that are absent from the vast majority of urban mobility studies. As such, ParkCap is helping to fill an important gap. Its complementarity with an urban mobility model can and must become reciprocal. The parking costs are intended to be integrated into the travel costs by a given mode or intermodal combination, in order to simulate the modal choice on the basis of options that are better described and more representative of the concrete conditions encountered in the territory by the user.

An important lesson lies in the diversity of demand segments: the parking problem is multi-class. Even without making all the classes explicit (in particular, urban logistics and stops to pick up or drop off on its rounds are missing), we noticed the specificities of the intermodal park-and-ride trips to a railway station. These trips are distinguished by the uniqueness of the target location and the massive nature of the demand they constitute in that place (i.e. railway station as a specific generator).

The model applied here is still an initial version. It is subject to a series of limitations, which constitute an equal number of needs for future improvements:

• Only cars have been represented. The spectrum of vehicles and modes should be expanded to include small two-wheeled vehicles and large vehicles, such as trucks, buses and coaches;

• The choice of the travel mode is made by the individual on the larger scale of a round-trip tour or even of a daily programme linking several activities in one or more tours. The parking conditions should be considered in successive places, including the place of origin, depending on the vehicle parking mode;

• The filling of a parking lot is a dynamic phenomenon, as is the operation and quality of service of the different modes and the achievement of individual mobility. Dynamic modelling of parking has become a fairly active area of academic research, although operational applications are rare. As part of the Ecodesign Chair, we have developed an initial dynamic parking model to simulate the supply–demand balance of parking along an urban street during a period (Kotelnikova-Weiler and Leurent, 2016);

• The behavioural process of finding a space is an important research topic in itself. The two-step modelling in ParkCap is similar to that in specific multi-agent simulators. The representation of options and their structuring as a "universe" of modal options requires both theoretical improvements and comparisons with experience.

Acknowledgements

The C++ simulator of the ParkCap software was programmed by our colleague Alexis Poulhès (École des Ponts ParisTech, LVMT), to whom we would like to express our gratitude. Walid Chaker contributed to the investigation into parking supply and demand in Cité Descartes. We are also grateful to DRIEA for providing the MODUS model, to IAU-IDF for providing the MOS database and to INSEE for providing population census data.

Bibliography

Aw T., Coulombel N., Leurent F., Millan-Lopez S., and Poulhès A. (2013), Territoire et transport en contexte métropolitain. Chapter 14 in Peuportier B., Leurent F., and Roger-Estrade J. (eds), *Eco-conception des ensembles bâtis et des infrastructures*, Presses des Mines, Paris, 305–340.

Boujnah, H. (2017), Modélisation et simulation du système de stationnement pour la planification de la mobilité urbaine. Ph.D. thesis of Université Paris-Est, defended on 14 December.

Boujnah H. and Chaker W. (2010), Étude du stationnement à la Cité Descartes Elargie, Document de travail Enpc, 26 p.

Boujnah H., Coulombel N., Kotelnikova-Weiler N., Leurent F., Millan-Lopez S., and Poulhès A. (2013), Activités, Accessibilités et Mobilités à l'échelle du quartier. Chapter 15 in Peuportier B., Leurent F. et Roger-Estrade J. (eds), *Eco-conception des ensembles bâtis et des infrastructures*, Presses des Mines, Paris, 341–368.

EpaMarne. (2009), Cité Descartes, cœur du Cluster Descartes Ville Durable: consultation internationale de programmation urbaine, Plaquette de présentation, November 2009, 15 p.

Insee. (2009), Recensement Général de Population, exploitations principales et complémentaires.

Kotelnikova-Weiler N. and Leurent F. (2016, December), Parking equilibrium along the street. *European Transport Research Review*, 8, 24.

Kotelnikova-Weil`er N., Leurent F., and Poulhès A. (2017), Spatial refinement to better evaluate mobility and its environmental impacts inside a neighborhood. *UPLand*, 2(1), 137–151.

Lion A. (2013), Étude en vue de l'élaboration du contrat de développement territorial, April.

Leurent F. (2013), La Cité Descartes Elargie: un territoire d'éco-conception. Chapter 13 in Peuportier B., Leurent F., and Roger-Estrade J. (eds), *Eco-conception des ensembles bâtis et des infrastructures*, Presses des Mines, Paris, 283–304.

Leurent F. (2014), Modéliser le stationnement pour gérer finement la voirie et la multimodalité, Revue Techni.Cités, no. 270, 18–20.

Leurent, F. and Boujnah, H. (2014), A user equilibrium, traffic assignment model of network route and parking lot choice, with search circuits and cruising flows. *Transportation Research Part C*, 47 (1), 28–46.

Leurent F., Boujnah H., and Poulhès A. (2013), Eco-conception d'un système de stationnement. Chapter 6 in Peuportier B., Leurent F., and Roger-Estrade J. (eds), *Eco-conception des ensembles bâtis et des infrastructures*, Presses des Mines, Paris, 107–134.

Peuportier B. et al. (2013), *Eco-conception des ensembles bâtis et des infrastructures*, Presses des Mines, Paris.

Chapter 15

Modelling of Microclimates

Helge Simon and Michael Bruse

UNIVERSITY OF MAINZ, MAINZ, GERMANY

15.1 Context

Urban areas are home to millions of people of all ages, with different levels of vulnerability. These people are confronted with the microclimatic conditions of their environment every day. In order to provide a healthy and sustainable environment, architectural and urban design must minimise people's exposure to airborne pollutants and heat stress.

Thanks to the ENVI-met microclimate model, different urban design options can be evaluated and quantified with the aim of minimising health impacts and providing a thermally comfortable and attractive outdoor environment for inhabitants.

ENVI-met is a 3D digital simulation platform that aims to model the complete real cycle of micro-climatological factors in high resolution (see Figure 15.1). ENVI-met is used around the world to analyse different aspects of urban climate, architecture and landscape.

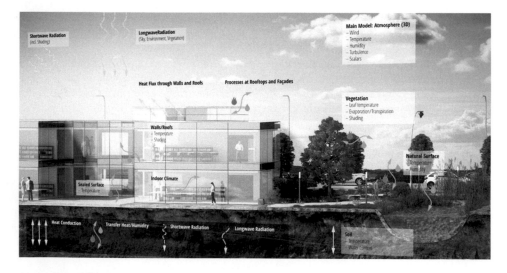

Figure 15.1 ENVI-met microclimate model design.

A specific feature of ENVI-met is its holistic approach to accurately model the complex processes of urban spaces. As a result, the ENVI-met platform includes detailed simulation models for air turbulence, soil and water assessment, building physics, and atmospheric chemistry and botany (Yang et al., 2012).

Several validation studies have been conducted (for example, Nikolova et al., 2011; Morakinyo et al., 2017; Pastore et al., 2017).

15.2 Case Study of Cité Descartes

15.2.1 Study Site

The case study presented in this chapter is located in Cité Descartes, about 20 km eastwards of the centre of Paris (Figure 15.2). The figure shows the basic vector model built in the ENVI-met environment. The planned new buildings are drawn in dark grey and the existing buildings are shown in light grey.

A special feature of the study area is the existence of a large and densely vegetated green area, the Butte Verte Park, in the western half of the modelled area. This park will eventually have a significant influence on the microclimatic conditions of the urban project studied.

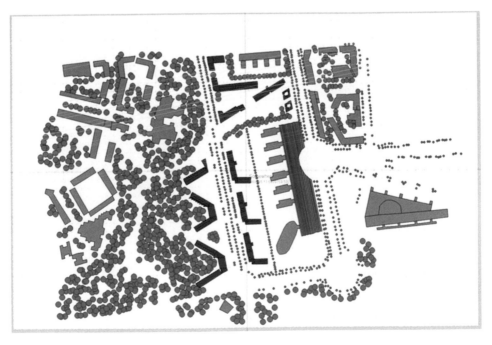

Figure 15.2 Vector model and ENVI-met representation of the study area of Cité Descartes.

15.2.2 Construction of the ENVI-met Model

In order to analyse the microclimatic conditions of a territory, an ENVI-met virtual mock-up must be generated from the geographic data provided. Three different models were generated within the framework of this case study:

- "Initial situation": this variant corresponds to the territory as it is today, without the planned new buildings.
- "Project": this time, the territory includes the planned new buildings.
- "Improved Project": this is a proposal including local changes to optimise the microclimate and thermal comfort.

The "Improved Project" scenario is generated from the simulation results of the initial project.

The ENVI-met model has $200 \times 190 \times 30$ cells. In the horizontal direction, each cell has a size of $3\,m \times 3\,m$, so the area corresponds to a surface of $600\,m \times 570\,m$. In the vertical direction, a cell size of $2\,m$ is applied, that is, a height of $60\,m$. Above the level of $60\,m$, the model is extended one-dimensionally up to a height of $2,500\,m$. Figure 15.3 shows a 2D plan of the ENVI-met model with and without the new buildings planned, for example. Figure 15.4 shows a 3D view of the model with the new buildings. Note that the model was rotated by $11°$ from north.

15.2.3 Microclimate Simulations

In order to analyse the microclimatic conditions in the neighbourhood, two scenarios are considered: (i) a microclimate corresponding to a typical summer day in 2017 (called "Scenario 2017") and (ii) a microclimate under extreme conditions comparable to the heatwave of 2003, called "Scenario 2050".

One of the likely impacts of climate change on average European climates is the occurrence of severe and long-lasting heatwaves. There have already been unusual

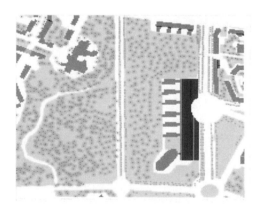

Figure 15.3 ENVI-met model 2D plan (200×190 cells) with (a) and without (b) the new buildings in the project.

Figure 15.4 3D view of the ENVI-met model, the Project scenario including new buildings.

heatwaves throughout Central Europe, not only in 2003, but also in 2007 and several years later. Urban planning should therefore take these changing climatic conditions into account by adopting climate resilience strategies.

The meteorological framework for the simulation can be summarised as follows.

Date	15 July
Vent	1.5 m/s at a height of 10 m, coming from the south-southwest
Cloud cover	no clouds
Simulation period	24 hours

The temperature and humidity profiles considered in the 2017 scenario are shown in Figure 15.5. In this scenario, the maximum air temperature is assumed to be 30°, with a relative humidity of 30% at 4 PM. The minimum temperature is set at 15°C at 6 AM, with a relative humidity of 70%.

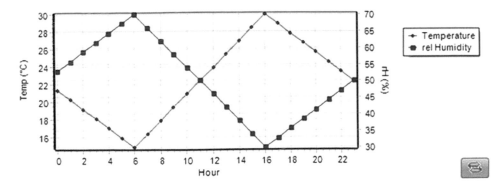

Figure 15.5 Temperature and humidity conditions imposed on the boundaries (south and west of the domain).

In the 2050 scenario representing a heatwave, a maximum air temperature of 36° is considered at 4 PM and the minimum is 18°C at 6 AM. The maximum and minimum humidities are set at 50% and 30%, respectively.

Soil moisture status is another important factor for the simulation. If sufficient extractable water is available in the soil, then the permeable surfaces may act as evaporative cooling interfaces. In addition, if the soil contains sufficient water, the transpiration of the plants will not be limited, so the leaves will stay cool and refresh the ambient air.

However, if a high-pressure situation persists with little clouds and no rain, the natural soil water reserve will start to run out. The cooling potential of the soil surface decreases, and the plants will begin to experience water stress and their transpiration rate will reduce.

To represent these conditions, the rate of accessible water in the soil was set at 30% as the maximum rate in the 2050 scenario and 50% in the 2017 scenario.

15.3 Simulation Results

Different aspects of microclimatic conditions are discussed in this section.

15.3.1 Wind Conditions

Figure 15.6 shows the wind conditions over the area studied in the case of the Project scenario (including new buildings). This figure gives two pieces of information at a time: the arrows indicate the direction of the wind, and their length indicates the wind speed. In addition, the colour of the arrows shows the relative speed of the wind in

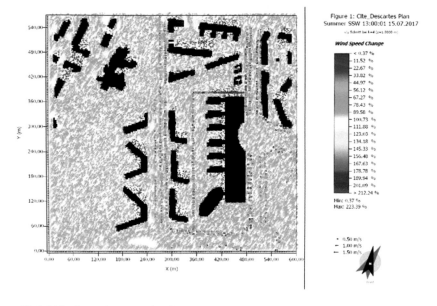

Figure 15.6 Wind conditions, the Project scenario.

different places. A value of 100% corresponds to the reference wind speed at the same height, on a flat and non-vegetated surface.

As can be seen, the wind speeds are relatively low in the modelling domain, with values less than 50% of the initial speed over a large part of the domain. To a large extent, this results from the large number of trees in the area studied, particularly in the western half corresponding to Butte Verte Park. On the other hand, the urban form includes relatively long buildings and only occasional openings, which allows limited air movements. The main passages available for the wind are the roads.

15.3.2 Temperatures, Initial Situation

The main parameter in the study of thermal comfort is the temperature. It is then useful to examine the distribution of temperatures in the territory under consideration.

Figure 15.7 shows the air temperature distribution at 1 PM for the 2017 scenario and the initial situation (without the new buildings).

The temperature range is from 23.6° to 27.2°, a difference of 3.6°. This gap is relatively large for a small area, but this can be explained by the large proportion of green space in the area and the low wind speed. At these slow speeds, the different thermal properties of the soil surfaces and the mixture of shade and sun influence the distribution of air temperatures. These differences would be much smaller in the case of higher wind speeds, due to better turbulent mixing.

To better illustrate the causes of the differences observed in air temperatures, Figure 15.8 shows the distribution of soil surface temperatures at 1 PM. The spatial variation of the temperatures reaches approximately 27 K. The main effects inducing this difference are the shading effects constituted by the numerous trees and the varying

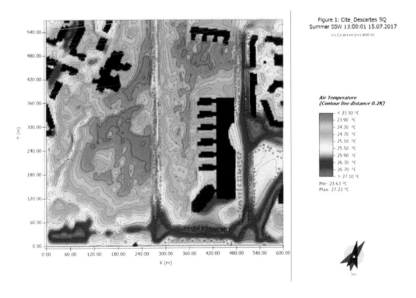

Figure 15.7 Distribution of air temperatures at 1 PM for the initial situation in the 2017 scenario.

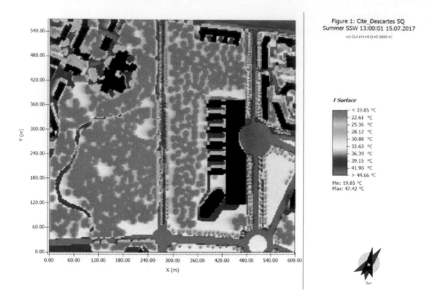

Figure 15.8 Distribution of soil surface temperatures at 1 PM for the initial situation in the 2017 scenario.

properties of the surface materials. In particular, the large roundabout to the east and the road to the south of the study area are not shaded, resulting in high surface temperatures.

15.3.3 Temperatures, Proposed Project, 2017

15.3.3.1 Air Temperature

Figure 15.9 shows the distribution of air temperatures at 1 PM on an average summer day for the 2017 scenario and the proposed project (with the new buildings).

The project clearly changes the distribution of air temperatures significantly. While in the initial situation Butte Verte Park to the west and the green space on the east side of the central street form a continuous cool space that is only slightly interrupted by the central roads, the situation has changed dramatically with the project. Because of the loss of many trees in the densified area, almost all the territory is significantly warmer and there is no connection with the park.

In order to visualise the temperature changes more clearly, Figure 15.10 shows the differences in air temperature between the initial situation and the proposed project. Positive values correspond to areas in which the air temperature increases due to the project.

The temperature increase reaches 2.8 K near the group of buildings to the north. There are small areas where the air temperature has decreased, but this is related to small variations in the air movement patterns. On the other hand, the effects at the edge of the field of study should not be taken into account, as small changes have been introduced to make the model more stable.

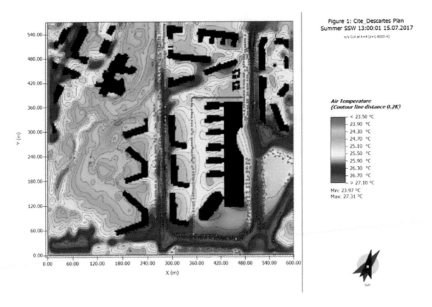

Figure 15.9 Air temperature distribution at 1 PM for the project proposed in the 2017 scenario.

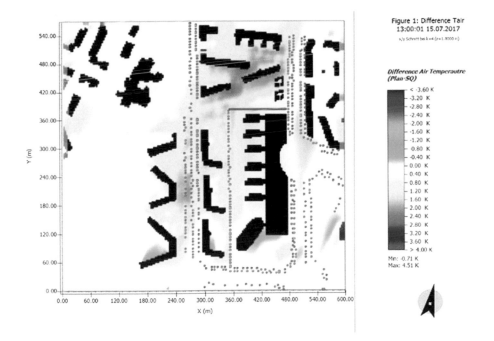

Figure 15.10 Differences in air temperature between the initial situation and the proposed project.

15.3.3.2 Thermal Comfort, Physiological Equivalent Temperature

Human thermal perception depends not only on the most commonly analysed variables, that is, air temperature and wind speed, but also to a large extent on direct (solar) and long-term energy flows (heat) exchanged between the human body and its environment. In addition, the relative humidity also influences the thermal perception by its effect on a human being's capacity to control their temperature by means of perspiration.

Finally, the occupants' perceived thermal comfort of outdoor spaces is the complex result of four interacting components: wind-related air movement, air temperature, radiative exchanges and humidity. Since these interactions are non-linear, the assessment of the comfort level is not trivial. Several indicators of thermal comfort have been developed over the years. They are all based on complex biometeorological models that make it possible to calculate the temperature of the skin as an indicator of thermal perception, among other parameters.

The most common indicator for outdoor spaces was used in this study, namely, the physiological equivalent temperature (PET) (Figure 15.11). This indicator corresponds to the (theoretical) temperature of an interior space that would produce the same thermal sensation on a human body as the assessed external condition.

This comfort scale is relevant for people exposed to constant climatic conditions for a certain time, such as people sitting on a terrace or staying in the same place for at least ten minutes. The comfort of moving pedestrians may differ significantly from the static conditions, as these people are exposed to microclimates that vary over a short period of time as they move through the urban territory.

Figure 15.12 shows the distribution of psychological equivalent temperatures calculated at 1 PM.

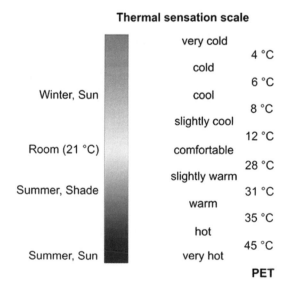

Figure 15.11 Reference scale for the psychological equivalent temperature (PET).

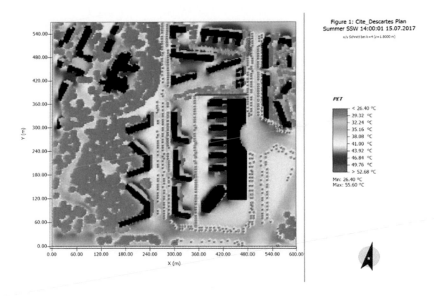

Figure 15.12 Distribution of physiological equivalent temperatures at 1 PM for the project proposed in the 2017 scenario.

This distribution of temperatures mainly results from the spatial distribution of the sunny and shaded areas. In the large green areas in the western part of the domain, the PET values are around 29°, which comes close to a feeling of warmth but is generally considered comfortable in an outdoor area in summer. However, in sunny areas, especially in the case of low wind speed, the PET rises to 45°, which indicates uncomfortable overheating.

15.3.4 Temperatures, Proposed Project, 2050

15.3.4.1 Air Temperature

The proposed project was simulated for heatwave conditions (hot and dry summer); see the 2050 scenario. Figure 15.13 shows the distribution of air temperatures at 1 PM.

Overall, we can observe a distribution similar to that obtained for the 2017 scenario but with higher temperature levels, which result from the boundary conditions set for this simulation. It can also be seen that the temperature differences between the vegetated areas, for example, the zone south of the roundabout to the west, and the roads are reduced as the vegetated soils become warmer because of a lack of water and less evaporation.

15.3.4.2 Thermal Comfort, Physiological Equivalent Temperature

As in the 2017 scenario, Figure 15.14 shows the distribution of PET calculated at 1 PM.

This figure shows the increase in PET levels, which now reach 32° in the shade of buildings and under trees with dense crowns.

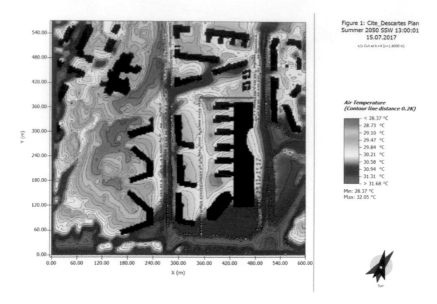

Figure 15.13 Air temperature distribution at 1 PM for the project proposed in the 2050 scenario.

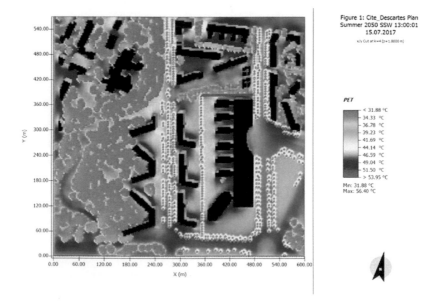

Figure 15.14 Distribution of physiological equivalent temperatures at 1 PM for the project proposed in the 2050 scenario.

15.4 Proposals for Improvement of the Project

15.4.1 Presentation of Proposals

Based on the simulation results, a proposal to modify the project was developed to improve thermal comfort conditions, especially in the most exposed areas.

Most of the proposals are as non-intrusive as possible and use the addition of trees and bodies of water as the main mitigation tool. The action with the greatest impact is the change in the design of the buildings located to the west. The solution proposed is to create air passages by raising the two portions parallel to the road by 5 m (see nos. 6 and 7, Figure 15.15), in order to allow for the circulation of the cooler air coming from Butte Verte Park. Figure 15.15 gives an overview of the improvement proposals introduced in the model.

The list of proposed measures is as follows:

1. Add vegetation to the roundabout (soil, grass and trees).
2. Add trees with dense crowns.
3. Add trees with dense crowns along the street.
4. Add bodies of water along the buildings.

Figure 15.15 Overview of the proposed improvement measures.

5. Add trees.
6. Raise the background of the buildings (parallel to the street) by 5 m.

15.4.2 Temperatures, Improved Project, 2050

15.4.2.1 Air Temperature

Figure 15.16 shows the new distribution of air temperatures at 1 PM obtained for the project with the proposed improvements. Figure 15.17 shows the differences between the initial project and the improved project. Negative values correspond to a decrease in air temperature as a result of the proposed improvements.

Thanks to the proposed improvements, the air temperature could be reduced by 1.5 K compared to the initial project. On large spaces along the central road (Boulevard de Ru de Nesles) and near the buildings, the temperature reduction is between 0.6 and 1 K. A greater reduction could be obtained by distributing the planted areas, in order to increase the latent heat flow.

The elevation of the central parts of the buildings to the west not only helped move the cooler air from the park through the row of buildings, but also increased the airflow to these buildings, reducing temperatures and benefitting the spaces upstream of these buildings.

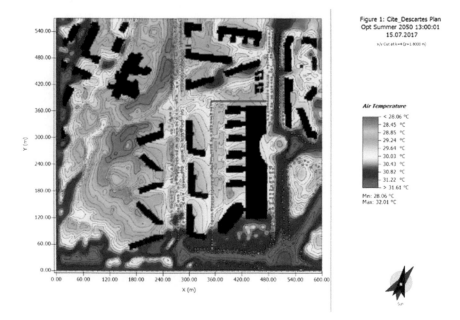

Figure 15.16 Air temperature distribution at 1 PM for the improved project in the 2050 scenario.

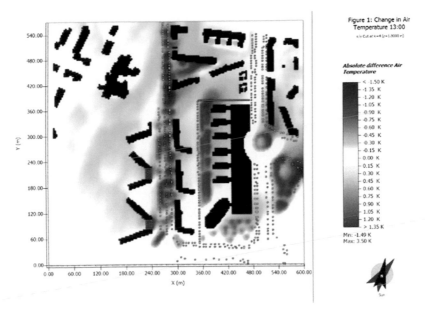

Figure 15.17 Air temperature differences between the initial project and the improved project.

15.4.2.2 Thermal Comfort, Physiological Equivalent Temperature

Figures 15.18 and 15.19 show the distribution of the PET calculated at 1 PM, and the difference in these temperatures between the initial project and the improved project.

Since PET strongly reacts to shading effects, the most significant differences correspond to the shading effects associated with the addition of trees. Since the effect on the reduction of air temperatures is much lower, it is not visible in this figure, where the PET differences are essentially linked to shading effects.

15.5 Generation of Urban Weather Data

To analyse the impact of the urban environment on building energy performance, eight test points were selected as the basis for extracting hourly temperature, humidity and wind speed data from the results of the microclimate model. These eight points (R1–R8) are shown in Figure 15.20 and their list is shown here (all the points are at a height of 2 m).

List of test points with their coordinates in the domain modelled (x, y):

R1: 74, 83	R5: 101, 83
R2: 69, 77	R6: 96, 67
R3: 86, 70	R7: 106, 67
R4: 69, 59	R8: 107, 54

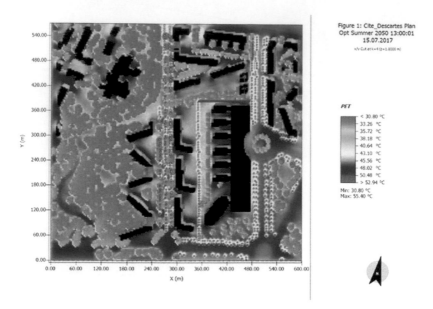

Figure 15.18 Distribution of physiological equivalent temperatures at 1 PM for the improved project in the 2050 scenario.

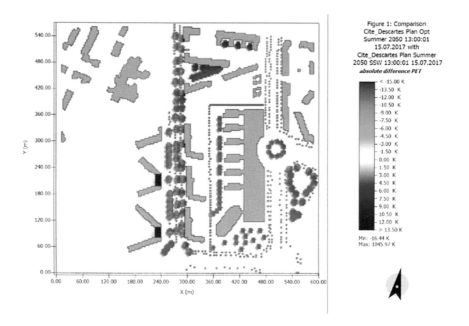

Figure 15.19 PET differences between the initial project and the improved project.

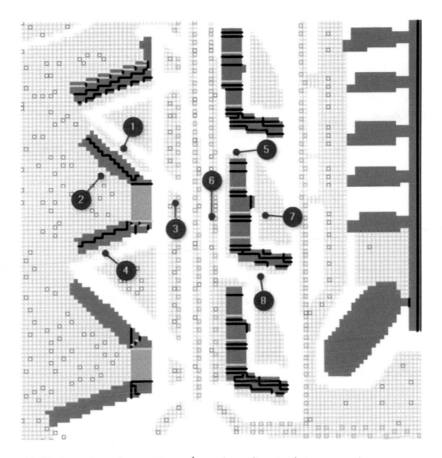

Figure 15.20 Location of test points for urban climate data generation.

A utility called GridExplorer allows users to select points of the domain studied after the simulation and extract time series of ENVI-met output variables. For example, Figure 15.21 shows the difference in air temperature between the selected points and the temperature considered as a boundary condition of the domain, which would correspond to the regional average temperature considered in a usual calculation of a dynamic building energy simulation.

During this summer day with light winds, there can be significant differences (±3 K) on the outside air temperature between certain points of the domain and the average used in the energy simulations of buildings.

15.6 Conclusion and Perspectives

This case study shows the interest and potential of a microclimate model for the eco-design of buildings. The calculation time remains high, however, because of the complexity of the models, which take the hygrothermal and aeraulic aspects into account. The 24-hour simulation presented in this chapter required 24 hours of computation on

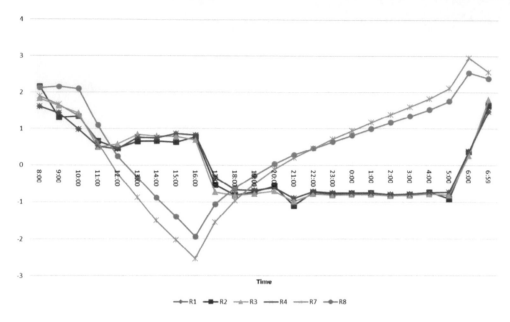

Figure 15.21 Difference between the air temperature at select points and the average temperature considered as a boundary condition.

a microcomputer. It was carried out on a summer day considering a low wind speed, which amplifies the dispersion of temperatures in the study area. It would of course be interesting to consider a representative sample of typical days (e.g. winter, spring, summer, with two wind speeds).

These data would make it possible to develop a typical climatic year, taking the microclimate into account, and to perform a dynamic energy simulation of the buildings in order to know the influence of the microclimate on the heating and cooling requirements of the premises or on their level of thermal comfort in the absence of air-conditioning.

Bibliography

Morakinyo, T. E., Dahanayake, K. C., Ng, E., and Chow, C. L. Temperature and cooling demand reduction by green-roof types in different climates and urban densities: A co-simulation parametric study. *Energy and Buildings*, 145: 226–237, 2017.

Nikolova, I., Janssen, S., Vos, P., Vranken, K., Mishra, V., and Berghmans, P. Dispersion modelling of traffic induced ultrafine particles in a street canyon in Antwerp, Belgium and comparison with observations. *Science of the Total Environment*, 412–413: 336–343, 2011.

Pastore, L., Corrao, R., and Heiselberg, P. K. The effects of vegetation on indoor thermal comfort: The application of a multi-scale simulation methodology on a residential neighborhood renovation case study. *Energy and Buildings*, 146: 1–11, 2017.

Yang, X., Zhao, L., Bruse, M., and Meng, Q. An integrated simulation method for building energy performance assessment in urban environments. *Energy and Buildings*, 54: 243–251, 2012.

Chapter 16

Development of a Methodology to Guaranteed Energy Performance

Simon Ligier, Patrick Schalbart, and Bruno Peuportier

Mines ParisTech, CES - PSL, Paris, France

16.1 Introduction

Both public and private incentives have been put in place to encourage the investments needed for the energy transition of the building sector. Professionals in the sector, energy service companies in particular, offer energy performance contracts (EPCs) that provide better control of financial returns following the work carried out and thus make it possible to safeguard the investments made. Various factors hinder the development of this type of contract, however. It is thus possible to differentiate the constraints associated with the errors coming from the modelling tools and the uncertainties in measuring the real energy consumptions and actual operating conditions.

Dynamic building energy simulation (DBES) software is widely used to predict the energy consumption of buildings, but several issues limit the reliability of the results. Substantial feedback and scientific studies show significant differences between these simulation results and the actual energy consumption measured on buildings (Macdonald, 2002). Various studies have investigated the cause of these discrepancies, identifying four main sources of error (Coakley et al., 2014): simplifications related to modelling and associated assumptions (building zoning, etc.), limited knowledge of the structure's actual characteristics of the components and systems, errors originating from the discretisation of the physical problem and its numerical resolution and the consideration of internal and external stresses, particularly those related to occupancy and use. The influence of this last point has been studied, in particular in relation to the influence of occupant behaviour and weather conditions. Wang (2012), for example, concludes that variations induce energy consumption by more than 80%.

There are other issues related to the process of measuring and verifying (M&V) energy performance. Errors related to measurement uncertainties and the choice of methods affect its reliability. These uncertainties must be incorporated into the overall methodology while risk assessment required as per the EPC process. Good practices must be established to ensure clear and transparent verification of the commitment to energy consumption. The International Performance Measurement and Verification Protocol (IPMVP), developed by EVO (2012), provides a common framework for M&V. Four options are detailed for different situations. Options C and D, corresponding to the verification of overall energy performance at the scale of the entire building, are of particular interest to us. They are differentiated by the availability of reference data in case of renovation (option C) or by the need to use detailed simulation tools to anticipate future performance (option D). In retrofitting projects, the IPMVP defines a baseline situation that allows the comparison of the performance before and after the work. In simulation, the theoretical consumption is linked to a particular context depending on the stress data used as the model input. In both cases, it is necessary to define adjustment factors (AFs) that make it possible to contextualise energy consumption and to overcome external influencing factors that are not related to the energy performance of the building. In addition to actual energy consumption, these factors will have to be measured in operation, thus raising the question of the influence of measurement uncertainties at the level of commitment of the EPC.

The processes of adjustment, risk determination and measurement planning are usually carried out independently and without coordination, while they are interdependent. The methodology developed in this work aims to associate these different processes.

This approach is made possible by a statistical analysis of multiple results of simulations describing not only a representative sample of the physical parameters of the construction, but also the variability of the conceivable operating conditions. The purpose of the methodology is to calculate a guaranteed performance linked to a quantified risk integrating upstream the impact of the uncertainties related to the measures and providing appropriate adjustment methods. Details of the methodology that meet these objectives will now be presented.

16.2 Simulation and Physico-Probabilistic Modelling

16.2.1 Physico-Probabilistic Modelling

The DBES software Pléiades+COMFIE is used to simulate the energy function of the building. The physical and geometrical parameters of the building envelope must be specified as well as the internal and external hourly stresses.

To model the uncertainty of the construction parameters, they are no longer specified with a single value but with a probability density function. The dynamic stresses that will affect the building are inherently unpredictable and variable throughout the life of the building. To take the impact of this variability into account, specific models generating coherent stresses according to the studied project are used and will be described later. The set composed of the physical model of the building, the laws of probability on the modelling parameters and models of variable stress generation. The implementation of this physico-probabilistic model relies on a large number of simulations following random draws in relation to each distribution and variability model. This makes it possible to obtain the distribution of the quantities of interest provided by the simulations, in particular the annual energy consumption.

16.2.2 Sensitivity Analysis

DBES models are composed of a very large number of input parameters, and the characterisation of the uncertainty related to each of the associated values can therefore be very heavy. Sensitivity analysis methods can be used to identify the parameters that are really influential on the output quantities of interest and thus focus the modelling effort on these parameters. Non-influencing parameters can be specified with a nominal value.

The Morris method (Morris, 1991) is an easily applicable screening sensitivity method that requires only a relatively small number of simulations. The uncertain inputs are described according to a variation range discretised into a finite (and even) number of points. Starting from an initial draw of values in the input space, simulations are performed by changing the value of only one parameter for each new simulation until each entry has been modified. This process is repeated hereafter. The elementary effects of the variations associated with a parameter are studied within each repetition. Their mean and standard deviation characterise the influence of the parameter on the output, identifying the non-linearity effects and interactions. The following Dp standard is used to characterise the influence of a parameter. The influencing parameters correspond to the high Dp values:

$$Dp = \sqrt{\sigma_p^2 + \mu_p^{*2}} \qquad (16.1)$$

Dynamic stresses can be taken into account by considering an overall variability (by translation) of the time series or can be excluded from the analysis. Various studies report the influence of meteorological stresses (Hong et al., 2013) and user behaviour (Haldi and Robinson, 2011).

16.2.3 Stochastic Modelling of Occupant Behaviour

To consider the influence of users within the physico-probabilistic model, the work of Vorger et al. (2014) is used to generate a variability of behaviours based on different statistical data (INSEE study, census, etc.). A set of probabilistic draws defines the characteristics of the users and presence and activity scenarios. The following inputs necessary for the simulation software are extracted: setpoint temperatures, dissipated power, domestic hot water consumption, window opening and blind management. Repeating this process provides a set of datasets. For example, Figure 16.1 shows the average heat loads related to electricity consumption from 200 datasets generated by the model.

16.2.4 Variability of Weather Conditions

A model generating artificial meteorological data from a typical dataset (TRY file) was developed to describe the natural variability of weather conditions for a specific geographic position. In connection with the work of Boland (1995) and Rastogi (2016), it provides hourly data of the outdoor temperature, global horizontal radiation and horizontal diffuse radiation for one year.

It is based on the decomposition of the available time series (TRY-type weather file) into a deterministic trend part and a random part. The deterministic part is identified by Fourier series analysis (discrete Fourier transform) while keeping only the characteristic frequencies corresponding to the annual average, the seasonal variations and the daily evolutions. By subtracting this trend share from the initial data, the residual share is obtained. This is characterised by a temporal logic of auto-correlation that can be modelled by a Seasonal auto-regressive moving average (SARMA) process.

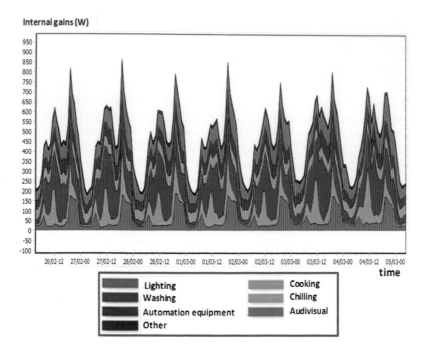

Figure 16.1 Average of 200 internal input scenarios generated by the model by post during a winter week for a flat of four inhabitants.

As the coefficients of this model are identified from the initial series, they can then be used to re-create residual time series with similar auto-correlative characteristics from random white noise. These regenerated residual shares are finally added to the trend share initially extracted, thus providing variable data.

The synthetic temperature data are first generated according to the principle described above. The time series of the typical global horizontal irradiance is also then separated into a trend part and a residual part. To respect the inter-correlation between temperature and solar irradiance, the creation of the residual irradiance series is constrained by both the self-correlation characteristics identified and the inter-correlation observed between the temperature and irradiance residues within the typical data. This process is achieved through the identification of a model called the vector autoregression model (VAR). The generated irradiance data are therefore matched to temperature data to form a coherent meteorological dataset. The diffuse horizontal irradiance is finally calculated from the global irradiance.

The meteorological data generation model thus provides semi-random data sets associated with a typical input dataset available in DES software (Figure 16.2).

16.2.5 Propagation of Uncertainties and Variabilities

The use of stochastic generation models of external (meteorological) and internal (occupant behaviour) stresses is accompanied by random draws within probability

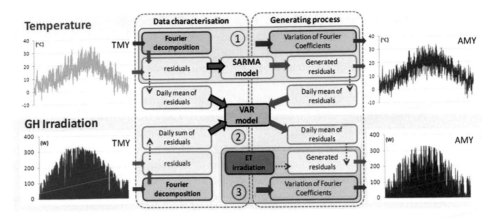

Figure 16.2 Self- and inter-correlated temperature and radiation data generation process.

density functions describing the influential uncertain parameters of the model. Building simulation with this diversity of input values allows the propagation of uncertainty and variability (PIV) within the model, describing a statistical sample of possible configurations. The statistical exploitation of the thousands of results associated with this PIV aims to provide the necessary elements for an energy performance guarantee engagement.

16.3 Methodology for the EPC

16.3.1 Objectives

The dispersion of a building's annual energy consumption distribution depends on the level of uncertainties on the input parameters. To control and reduce this dispersion, it is necessary to know the values of the parameters as accurately as possible. Some input data, such as the temperature setpoint of the housing or the outside temperature, cannot be anticipated before the operational phase. These elements, associated with meteorological or occupancy conditions, are not additionally related to the intrinsic performance of the building on which the EPG is to be carried out, and their influence should therefore be excluded from the liability of the organisation undertaking the engagement. The methodology presented here aims to characterise the link between energy consumption and these elements, to extract their influence through an adjustment process. A guaranteed energy consumption limit, for a risk of default, α (LGC$_\alpha$), is estimated according to the AFs. This maximum value, calculated from simulation results, is intended to be contractualised. The measurement of the AFs at the end of the cyclical operating period enables the calculation of the adjusted guaranteed consumption limit, which is finally compared with the actual energy consumption, which must be lower. If this value is exceeded, the contracting firm agrees to pay the difference. Penalties may also be stipulated in the contract, as well as sharing of the gains in case of under-consumption.

16.3.2 Choice of Adjustment Factors

The magnitudes associated with the AFs must be independent of the energy performance of the building and its systems, have a significant influence on the energy consumption of interest and must be measurable.

These quantities, which are related to the building stresses, are variable according to the time, and it is therefore necessary to associate them with unique characteristic indicators, the AFs, aggregating the information. For example, the degree-hours based on 18 are a potential AF in relation to the outside temperature. Different AFs can be related to the same magnitudes. The annual average is another characteristic factor of the outside temperature. Analytical methods are used to determine the best performing AFs which best explain the variations in the magnitude of interest. Comparing the mean squared error of linear models constructed with the different AF sets thus makes it possible to choose the best configurations.

16.3.3 Formulation of the Adjustment

Different types of models can be considered to simultaneously characterise a risk associated with a guaranteed limit of consumption (LGC$_\alpha$), starting from the simulation results, and the dependence of this limit on various AFs related to building stresses. The linear quantile regression model used in these works and described below responds to these objectives and is expressed as a first-degree multivariate polynomial. This simple linear formulation has the advantage of being understandable by all the stakeholders in the project. The linear dependence hypothesis will be studied in the rest of this work. The different AFs must be independent, which can be verified by a Student's t-test.

The LGC$_\alpha$ is expressed in terms of baseline conditions associated with a guaranteed reference limit (LGC$_{\alpha -\text{ref}}$) and AF reference values. These conditions are set arbitrarily and can be associated with a nominal building configuration or conditions observed before the building renovation, for example. Equation (16.2) presents the form of the adjustment model of the guaranteed energy consumption limit. This depends on the actual FA$_i$ values of the AFs. The β_i coefficients and the reference value of the LGC$_\alpha$ are determined by statistical exploitation of the simulation results of the PIV and by carrying out a multiple quantile regression, the principle of which is described as follows:

$$\text{LGC}_\alpha = \text{LGC}_\alpha - \text{ref} + \sum_{i=n}^{n} \beta_i \times \left(\frac{\text{FA}_i - \text{FA}_i - \text{ref}}{\text{FA}_i - \text{ref}} \right) \tag{16.2}$$

16.3.4 Calculation of the Adjustment Model

The regression methods aim to characterise the dependence between a variable of interest and one or more explanatory variables X. The usual linear regression provides a multidimensional linear model of the conditional expectation of Y knowing X (Equation 16.3), thanks to a least squares optimisation method (Equation 16.4). β is the vector of the coefficients in the linear model, and the series of the residuals, deviations between the initial data and the results of the model. β calculated by the least squares technique is the best estimator of β.

$$Y = X\beta + \varepsilon \tag{16.3}$$

$$\hat{\beta} = \arg\min \beta \in \mathbb{R}P \sum_{i=1}^{n}\left(y_i - \sum_{i=i}^{n}\beta_i x_{ij}\right)^2 \tag{16.4}$$

This method provides a linear model of the conditional mean of Y knowing X. The quantile regression developed by Koenker (2005) makes it possible to model the conditional quantiles of the distribution of V knowing X (Equation 16.5).

$$q\tau = (Y \mid X) = X\beta_\tau + \varepsilon \tag{16.5}$$

The linear model is calculated from the estimator presented in Equation (16.6), which is also defined based on the minimisation of the sum of the differences between the observations and the results of the model, weighted by a function depending on the quantile studied.

$$\hat{\beta}_\tau = \arg\min \beta \in \mathbb{R}P \frac{1}{n}\sum_{i=1}^{n}\rho\tau\left(Y_i - X_i\beta_{\tau,i}\right) \tag{16.6}$$

where ρ_τ is a weight function defined by

$$\rho_\tau(u) = \begin{cases} \tau \cdot u \ \{u \geq 0\} \\ (\tau - 1) \cdot u \ \{u \geq 0\} \end{cases} \tag{16.7}$$

$0 \leq \tau \leq 1$; for example $\tau = 0.95$ corresponds to a risk of 5%.

The process of carrying out the calculation according to the AFs is based on the quantile regression and is shown in Figure 16.3, in the case where only one AF is considered.

Assuming future knowledge of the values of the AFs (by measuring them during operation), the variability of energy consumption is reduced. Thanks to the quantile regression applied to the quantity of interest explained by the AFs therefore, it is possible to calculate a guaranteed consumption limit for a controlled risk of overflow.

16.3.5 Impact of Measurement Uncertainties

Data will be collected during the operation of the buildings, allowing the co-contractors to calculate the actual values of the AFs during the commitment period (typically over a year). These will make it possible to adjust the guaranteed energy consumption limit (based on the model determined during the study phase) and to compare the actual consumption, to verify the achievement of this objective. The uncertainty associated with operational measurements of the AFs, as well as the energy consumption, modifies the level of risk associated with LGC_α. The methodology therefore plans to

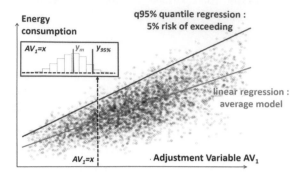

Figure 16.3 Linear models of the conditional mean (mean model) and the conditional quantile (quantile regression) for determining a limit of adjustable energy consumptions.

integrate the impact of these possible measurement errors upstream of the process, *via* a disturbance of the simulation results arising from the propagation of uncertainties and variabilities.

By specifying the measurement means associated with the selected AFs, it is possible to know the measurement uncertainty related to the actual knowledge of these values during operation. Uncertainty ranges are thus determined from the characteristics of the sensors and the measurement plan.

For each simulation, random draws are carried out on the AF and energy consumption uncertainty ranges. Virtual measurement errors are thus simulated, and the results from the simulations are disrupted by them. The methodology presented above, based on quantile regression, is applied to this noisy data set, which finally makes it possible to integrate the guaranteed default risk related to measurement errors during operation, upstream of the process.

Thus, the methodology developed in this work makes it possible to determine all the elements necessary for contracting an EPG commitment. A guaranteed energy consumption limit, associated with a default risk and dependent on AFs, is calculated from the results of a large number of simulations, taking into account all the uncertainties and variabilities of the stresses on the building studied. The selected AFs will be measured in operation and thus associated with a measurement plan. The influence of measurement uncertainties is integrated upstream of the process. This methodology makes it possible to formalise and verify an EPC in a transparent manner for all the actors, without performing new simulations at the end of each commitment period (Figure 16.4).

16.4 Example of Results

16.4.1 Case Study Description: Renovation of Collective Dwellings

This project concerns the renovation of a multifamily building consisting of 16 apartments spread over three levels. The total surface area is 1 048 m². The renovation focuses on improving the insulation of the envelope, reducing thermal bridges and

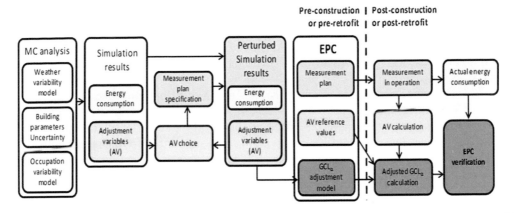

Figure 16.4 Comprehensive EPG methodology.

changing heating, ventilation and domestic hot water (DHW) systems. The building is modelled in Pleiades+COMFIE and divided into 17 thermal zones corresponding to each apartment and the common areas, as described in Figure 16.5.

The study aims to propose a guaranteed energy consumption limit related to heating and hot water with a 5% risk of being exceeded. Measurement campaigns conducted before and after renovation test the methodology as a whole, up to the verification phase. More details on the building and the corresponding model are available in Vorger (2014).

Figure 16.5 3D model of the building (Pléiades software) and thermal zoning.

16.4.2 Parameters and Stresses of the Model

A DES model of the building after renovation is made based on the project design data. The nominal values of the occupancy, internal inputs, setpoint temperatures and pumping of DHW are defined from the available data of the situation before work. To complete these reference conditions, typical meteorological data (TRY file) for the city of Mâcon are used.

A sensitivity analysis based on the Morris method is first conducted to identify the most influential uncertainties. The uncertainty affecting 39 parameters is characterised, and the range of variation is specified based on the technical data and the scientific literature. The variability of the dynamic stresses related to the uses is considered with a rate of nominal condition variation. The variability of weather conditions is not taken into account in the Morris analysis, as they are already identified as highly influential in many scientific studies. The sensitivity analysis identifies 12 parameters as highly influential, ranked according to the value of the norm defined in Equation (16.1). The results of the Morris analysis are shown in Figure 16.6. The most influential inputs are the nominal power of the heating system, internal inputs and setpoint temperatures. Indeed, these elements being very uncertain, the ranges of variation considered are wide, which explains their strong influence.

Then, more accurate probability density functions are associated with the influential parameters identified to steer the propagation of uncertainties and variabilities. With respect to occupant behaviour, the stochastic model presented above is used to generate an average representative data set of presence, internal inputs and setpoint temperatures from the generation of 200 datasets.

The stochastic model is completed according to the following specifications: the number of inhabitants in each apartment, the information known in the project and

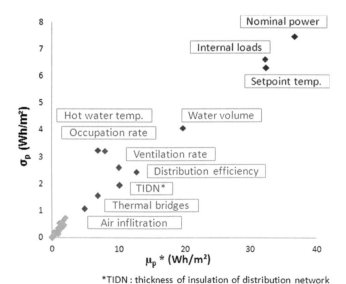

*TIDN : thickness of insulation of distribution network

Figure 16.6 Identification of influential parameters by the Morris method.

data concerning the type of building and its location. Variability is defined around the average scenarios produced by the model. The meteorological data generation model is used to generate variable weather data for each simulation propagating the uncertainties and variabilities.

16.4.3 Results of the Propagation of Uncertainties and Variabilities

Following the completion of 6,000 simulations, the distribution of energy consumption is characterised by an average of 107.2 kWh/m²/year and a standard deviation of 16.3 kWh/m²/year. The 95% quantile is 134.1 kWh/m²/year, an increase of 25% over the average value. This important difference between this extreme value and a nominal simulation result underlines the interest in forecasting an adjustment to the guaranteed energy consumption limit, to overcome the influence of elements that are unrelated to the intrinsic performance of the building, i.e. the occupant behaviour and weather stresses.

16.4.4 Definition of Adjustment Factors

To exclude the impact of parameters not related to the liability of the contracting structure, the methodology anticipates the adjustment of the $LGC_{5\%}$.

A preliminary study compared the relevance of different sets of AFs aiming to eliminate the influence of the setpoint temperatures and the indoor temperature on the heating consumption. The performances of several linear regression models associated with different AFs are studied. Indeed, different AFs can characterise the variability of the outside temperature:

- The unified heating degree-days are based on 18° (DJU_{18}). For each set of meteorological data used, the daily mean minimum and maximum temperatures are subtracted from the reference temperature of 18°. Positive differences are summed for each day of the heating season (weeks 42–18 of the year)
- The heating hours are based on 18° during the heating season (DH_{18-sc}). The positive deviations between the reference and the hourly temperature are summed at each hour of the heating season.
- The average outdoor temperature during the heating season $\left(\overline{T}_{ext-sc}\right)$.
- Other AFs can characterise the influence of indoor temperatures:
- The average indoor temperature during the heating season $\left(\overline{T}_{int-sc}\right)$. The simulations consider that the indoor temperature of a zone corresponds to the specified setpoint temperature during winter. With regard to the measurement of this magnitude, average hourly indoor temperatures are considered.
- - The heating hours are finally considered based on the indoor temperature during the heating season $\left(DH_{Tint-sc}\right)$. They characterise the influence of both indoor and outdoor temperatures. The positive deviations between the mean indoor temperature and the outdoor temperature are summed at each hour of the heating season.

Different combinations of AFs are studied to better characterise the dependence between annual energy consumption for heating and changes in indoor and outdoor temperatures. The root mean square error (RMSE) of the linear regression models is compared for each configuration. The results for this study are shown in Table 16.1.

Table 16.1 Quality Comparison of Different Adjustment Models

Adjustment Factor	RMSE (kWh/m²/year)
$\mathrm{DJU_{18}}$	5.0
$\overline{T}_{\mathrm{int-sc}}$	6.4
$\dfrac{\mathrm{DH_{18}}}{\overline{T}_{\mathrm{int-sc}}}$	4.6
$\dfrac{\mathrm{DH_{18-sc}}}{\overline{T}_{\mathrm{int-sc}}}$	4.4
$\dfrac{\overline{T}_{\mathrm{ext-sc}}}{\overline{T}_{\mathrm{int-sc}}}$	4.4
$\mathrm{DH}_{\overline{T}_{\mathrm{int-sc}}}$	4.2

The mean squared error associated with the model is smaller than that with $\mathrm{DJU_{18}}$. This parameter therefore better explains the variations in consumption due to the variability of the meteorological conditions. The best adjustment configuration is that based on the single AF $\mathrm{DH_{Tint-sc}}$. This FA is therefore selected to adjust heating consumptions according to indoor and outdoor temperatures.

Other AFs related to occupant behaviour and associated with available measurements of the building are also included in the adjustment model of the guaranteed energy consumption limit for heating and DHW being proposed. The specific electricity consumption during the heating season ($E_{\mathrm{elec-sc}}$) produces internal inputs that affect the heating needs. The volume of DHW sampled (V_{DHW}) influences the energy consumed for the preparation of DHW and the energy contributions due to losses in the networks. These two AFs are therefore also taken into account.

16.5 Measurement Procedure and Performance Verification

The methodology presented above requires a measurement plan to be chosen in relation to the AFs considered. In this case study, the building is instrumented during the first year of operation after renovation, thereby allowing access to the necessary data.

To integrate the impact of possible measurement errors during the verification phase within the methodology, attention is paid to the uncertainties associated with the sensors. For each simulation and each AF, a virtual random measurement error is obtained by drawing in a truncated normal distribution law within the measurement uncertainty ranges $\pm \Delta x$ and standard deviation $\Delta x/2$.

An uncertainty ΔT of $\pm 0.5°C$ is taken into account on the temperature measurements. This value is related to the systematic error, and the random errors are compensated for and are therefore neglected. The final error on the $\mathrm{DH_{Tint-sc}}$ value is calculated as the sum of the errors in the different hours of the heating season. With regard to V_{DHW}, a relative uncertainty $\Delta V_{\mathrm{ECS}}/V_{\mathrm{ECS}}$ of $\pm 5\%$ is considered. Again, measurement error draws are made within a normal standard deviation law $\Delta V_{\mathrm{ECS}}/2$. The same procedure is conducted on electricity consumption with a relative uncertainty of ±2%. The quantity of interest Y is composed of the annual energy consumption of heating and DHW. This is measured with an energy meter in the building substation. An uncertainty of ±5% on the measurement is taken into account.

The simulation results, AFs and energy consumptions obtained after the PIV are thus modified by integrating the measurement errors.

16.5.1 Determination of the Adjustment Model

A multiple quantile regression is carried out based on the simulation data produced by propagating the uncertainties and variabilities and is modified by integrating possible measurement errors during the verification. This makes it possible to define the polynomial adjustment of the guaranteed energy consumption limit. The reference values of the AFs are presented in Table 16.2. These correspond to the simulation results with the nominal operational conditions and the average values of the uncertainty ranges of the parameters.

Equation (16.3) below constitutes the adjustment model identified, corresponding to the conditional quantile of the energy consumption distribution as a function of the AFs. It can be associated with a guaranteed consumption limit with a 5% risk of being exceeded.

$$LGC_{5\%} = 122.6 + 77.2 \times \left(\frac{DH_{Tint-sc}}{DH_{Tint-sc-ref}} \right) - 7.7 \times \left(\frac{E_{elec-sc}}{E_{elec-sc-ref}} - 1 \right)$$
$$+ 63.6 \times \left(\frac{V_{ECS}}{V_{ECS-ref}} - 1 \right) \tag{16.8}$$

The adjustment process allows a reduction in the unexplained variability of energy consumption, as shown in Figure 16.7. Without considering any adjustment, the distribution has a standard deviation of 16.3 kWh/m². The standard deviation of the conditional distribution is 9.6 kWh/m² by considering an adjustment of the AFs set at their reference value.The adjustment forecasting methodology therefore makes it possible to reduce the gap between the value of the guaranteed consumptions and the nominal value, usually obtained by simulation, from 25% to 15%, while extracting the adjustment elements from the risk of overconsumption.

Table 16.2 Reference Values for Adjustment Factors

Adjustment Factors	Reference Values
$DH_{Tint-sc}$ (°C.h)	58,694
V_{ECS} (m³)	627.8
$E_{elec-sc}$ (kWh)	11,316

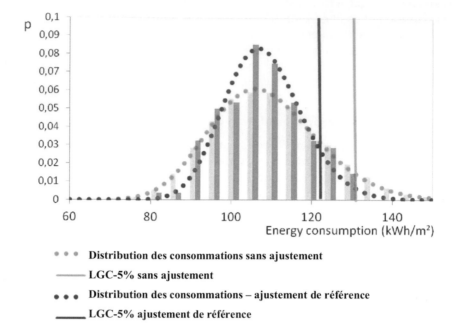

● ● ●   **Distribution des consommations sans ajustement**

────   **LGC-5% sans ajustement**

● ● ●   **Distribution des consommations – ajustement de référence**

▬▬▬   **LGC-5% ajustement de référence**

Figure 16.7 Distributions of energy consumption without adjustment (light gray) and with adjustment under reference conditions (dark gray).

16.5.2 Measurements and Verification

The actual values of the AFs during operation are obtained from the measurements made during the first year following the renovation. These are presented in Table 16.3 with the deviation from the reference values. The outside temperature was slightly above normal during the year, but the high indoor temperatures observed explain the high $DH_{Tint-sc,}$ value, the degree based on indoor temperature. The hot water consumption was lower than the reference values and the electricity consumption slightly higher. The use of the adjustment model presented in Equation (16.3) allows

Table 16.3 Adjustment Factors and Energy Consumption Measured During the First Year of Operation After Renovation

Adjustment Factors	Measured Values	Rate of Increase/Reference Conditions
$DH_{Tint\text{-}sc}\left(°C.h\right)$	64,216	+ 9%
$V_{ECS}\left(m^3\right)$	441	−30%
$E_{elec\text{-}sc}\left(kWh\right)$	11,941	+ 6%
Energy consumption $\left(\dfrac{kWh}{m^2}\right)$	109	-

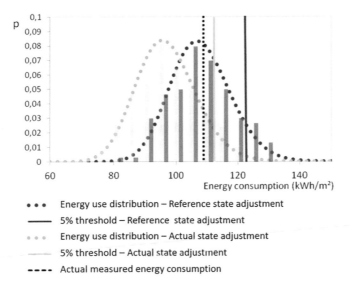

- • • •   Energy use distribution – Reference state adjustment
- ——   5% threshold – Reference state adjustment
- • • •   Energy use distribution – Actual state adjustment
- ——   5% threshold – Actual state adjustment
- ----   Actual measured energy consumption

Figure 16.8 Adjustment of LGC$_{5\%}$ and the function of associated energy consumption distributions. Comparison with the actual consumption.

the calculation of the adjusted value of the guaranteed energy consumption limit. This is 111 kWh/m²/year.

The actual consumption measured is 109 kWh/m²/year. The commitment is finally respected, as this value is lower than the 111 kWh/m²/year calculated with the adjustment model. Figure 16.8 shows the consequences of the adjustment process. The very low consumption of DHW significantly reduces the LGC value. Given that the inhabitants consume less water, it is possible to guarantee a lower value of consumption.

16.6 Conclusion

The modelling of the building studied with DES software provides a nominal energy consumption for heating and DHW of 107.2 kWh/m²/year, considering meteorological data and typical use. The methodology presented is based on a better consideration of the uncertainties concerning the parameters of the building and the systems, and the variability of the internal and external stresses. The propagation of the uncertainties and variabilities provides a probabilistic distribution of energy consumptions. There is a wide dispersion of consumption with a standard deviation of 16.3 kWh/m² and a 95% quantile of 134 kWh/m²/year. This is a significant uncertainty in terms of consumption, which is problematic given the objective of guaranteeing energy performance.

The methodology allows the calculation of a guaranteed limit value with a 5% risk, which can be adjusted according to AFs related to indoor and outdoor temperatures, specific electricity consumption and DHW pumping. The influence of measurement uncertainties during verification is integrated upstream of the process. A guaranteed reference value of 122.6 kWh/m² is associated with a set of FA reference values.

When verifying performance after the first year of operation, a consumption of 109 kWh/m^2 is recorded and can be compared to the limit adjusted by the measured AF values, which is 111 kWh/m^2/year.

The methodology is based on an exhaustive and highly detailed description of the uncertainties and variabilities related to the inputs of the building energy model, based on the knowledge available during the study phase. By using a set of results that are statistically representative of the uncertain context of the operation of the renovated building, it is possible to define a model linking a limit value of consumption associated with a risk of being exceeded to different AFs that are not part of the intrinsic energy performance of the building. A linear model that is transparent and easily understood by all parties is designed using a quantile regression method.

Thus, the methodology allows construction and operating companies to propose energy performance guarantee contracts based on scientific and statistical methods, enabling risk characterisation. This constitutes a decision support tool for defining a commitment level associated with a measurement plan that allows the contract value to be adjusted to the actual operational conditions.

Bibliography

Boland, John. 1995. "Time-Series Analysis of Climatic Variables". *Solar Energy 55* (5): 37788.

Coakley, Daniel, Paul Raftery, and Marcus Keane. 2014. "A Review of Methods to Match Energy Building Simulation Models to Measured Data". *Renewable and Sustainable Energy Reviews 37* (September): 123–41. https://doi.org/10.1016/j.rser.2014.05.007.

EVO, 2012. International Performance Measurement and Verification Protocol. Vol. 1, S.l.: s.n.

Haldi, Frederic, and Darren Robinson. 2011. "The Impact of Occupants' Behavior on Building Energy Demand". *Journal of Building Performance Simulation 4* (4): 323–38.

Hong, T., W.-K. Chang et H.-W. Lin, 2013. "A Fresh Look at Weather Impact on Peak Electricity Demandand Energy Use of Buildings Using 30-year Actual Weather Data. *Applied Energy. 111*: 333350. doi: 10.1016/j.apenergy.2013.05.019.

Koenker. 2005. *Quantile Regression.* Economic Society Monograph Series, Cambridge University Press.

Macdonald, Iain Alexander. 2002. *Quantifying the Effects of Uncertainty in Building Simulation.* University of Strathclyde. https://www.strath.ac.uk/media/departments/mechanicalengineering/esru/research/phdmphilprojects/macdonald_thesis.pdf.

Morris, Max D. 1991. "Factorial Sampling Plans for Preliminary Computational Experiments". *Technometrics 33* (April): 161–74. https://doi.org/10.2307/1269043.

Rastogi, Parag. 2016. On the Sensitivity of Buildings to Climate: The Interaction of Weather and Building Envelopes in Determining Future Building Energy Consumption. PhD thesis, EPFL. https://pdfs.semanticscholar.org/50af/ f6e04c32eb4acb83c1ebe3e155ef8885cfee.pdf.

Vorger, Eric. 2014. Étude de l'influence du comportement des occupants sur la performance énergétique des bâtiments. Ph.D. thesis, École Nationale Supérieure des Mines de Paris.

Vorger, Eric, Patrick Schalbart, and Bruno Peuportier. 2014. Integration of a Comprehensive Stochastic Model of Occupancy in Building Simulation to Study How Inhabitants Influence Energy Performance. In *Proceedings PLEA 2014*, 8. Ahmedabad (India).

Wang, Liping, Paul Mathew, and Xiufeng Pang. 2012. "Uncertainties in Energy Consumption Introduced by Building Operations and Weather for a Medium-Size Office Building". *Energy and Buildings 53* (October): 152–58. https://doi.org/10.1016/j. enbuild.2012.06.017.

Chapter 17

The Contribution of Prospective Energy Systems to the Life Cycle Assessment of Buildings

Edi Assoumou and Jérôme Gutierrez

Mines ParisTech, CMA - PSL, Paris, France

17.1 A Challenge for Building Complexes

The environmental impacts associated with energy supply and transformation are of major importance for the life cycle assessment (LCA) of building complexes. In France, the building sector mobilises 40%–45% of the final energy consumption on average, about 70% of the final consumption of electricity and is responsible for 25% of all CO_2 emissions from energy use. One of the challenges of LCA is therefore to grasp the driving forces behind the evolution of environmental impacts over the longer term, beyond the current characteristics of the energy systems studied. The contribution that forecasting can make to the various LCA approaches lies in this capacity to investigate and quantify possible futures.

Taking into account all the impacts of the construction, use and end-of-life phases of the components of a building (with potential for renovation) require analysis over a 50–100-year horizon. Over this horizon, the keys to understanding the energy systems will most likely be disrupted by the decline of the centralised state, uncertainties about the speed at which renewable energies will spread and the interconnection of grids and energy–climate policies.

The extension of LCA methods as discussed in this exploratory work is limited to the issues raised by the evolution of the CO_2 content of electricity production in the long term, in France, and at a sub-annual scale. A retrospective view of the installed capacities also reminds us that this evolution is not a characteristic of the current period, but that it is in fact a constant observable throughout the last century for the electrical sector. It allows four elements to be distinguished: the inertia of investments in the electricity sector, illustrated by the relatively old construction period of hydraulic structures; the dynamics of technological change through the adoption of innovative solutions for each era; the role played by the increase in the demand for electricity as a driver of the transformations (and the slowdown observed over the recent period); and finally, the political dimension, with the development of nuclear power and the feed-in tariffs that have enabled the recent development of renewable production sources.

Based on the current mix, the overall CO_2 impact of the electricity mix is low for France due to a production fleet consisting mainly of low-carbon sources (nuclear, hydro, wind, and solar). With an average CO_2 content of 81 g CO_2/kWh from 2005 to 2015, France is well below the average for the 28 member states of European Union, which is established at 374 g CO_2/kWh over the same period. However, as shown in Table 17.1, a 10–15-year horizon is enough to lead to significant changes in the CO_2

Table 17.1 Annual change in CO_2 content of electric kWh for several countries

g CO_2/kWh	1990	1995	2000	2005	2010	2015
Germany	694	641	577	548	516	485
France	107	76	89	101	96	52
Italy	572	541	490	522	450	385
Poland	1.043	1.002	955	853	869	803
United Kingdom	686	535	476	497	449	346
Sweden	35	53	44	50	64	35
28 EU states	507	451	413	407	370	331

Source: Key Climate Data – France, Europe and World (Commissariat général au développement durable – State Commission for Sustainable Development 2018).

impact of the electricity mix. It can be seen that an assessment of the impacts using values calibrated according to a historical approach fixes the electrical systems in the background and minimises the factors of change in the future.

This consideration of the long term is made particularly necessary by the fact that the fleet of French nuclear power plants is ageing and will require refurbishment or replacement in 2050. Rapid deployment of intermittent renewable energies is also likely to modify the evaluation of CO_2 contents and their seasonal dynamics.

Additional g CO_2/kWh emission factor analyses have been conducted for the electrical systems in different regional and national contexts. Historical data were used in the analysis proposed by Gordon and Fung (2009) to estimate the savings associated with the use of photovoltaics in a test-detached house. Data on historical CO_2 emissions and power generation have also been used in an analytical decomposition for the Philippines to identify variables (population, economic activity or structural changes) explaining the changes observed. Using half-hourly production data, the analysis conducted by Finenko and Cheah (2016) provides a typology of CO_2 emission factors for Singapore and for four typical days (weekdays, Saturdays, Sundays and public holidays). In the case of Germany, Kono, Ostermeyer and Wallbaum (2017) use historical data to study the impact of the development of renewable energies on the hourly CO_2 content of electricity. In the work of Hawkes (2014), the constraints for estimating dynamic emission factors are more particularly discussed. The method implemented by Hawkes (2014) uses a long-term energy model with a similar principle to the one we mobilised.

However, while the contribution of a long-term perspective seems necessary, it introduces an additional difficulty. Indeed, the weight of energy in the building use phase means that the determinants and unknowns specific to the energy systems become structuring for the results. In the exploratory work we conducted, the focus is on changing technological choices, as well as on the dynamic aspect of CO_2 through its variations at seasonal and hourly scales. The adopted methodology is based on an electrical system optimisation model (choice of investment and operation) in 2050 to establish and discuss some development paths for the CO_2 content per kWh. A reduced number of scenarios were then defined and simulated to illustrate the induced variability.

17.2 The Prospective Model of the French Electrical System

17.2.1 Principles of the TIMES_FR_ELC Model Mobilised

The prospective model implemented is a techno-economic model of the family TIMES models, which represents the French electricity system (metropolitan France without Corsica) with a calibration over the year 2012, allowing its evolution to be simulated up to 2050. The TIMES approach belongs to the so-called "bottom-up" models that focus on elementary technical structures, as opposed to the so-called "top-down" models that highlight macroeconomic elements with less explicit technological content. TIMES makes it possible to project the evolution of the supply of electricity production *via* an optimal paradigm. The models schematically create coherence between three dimensions:

- A network of the available options defined by all the stages of transformations and technological choices that can be mobilised over the study horizon. Each process is explicitly described by techno-economic parameters (costs are differentiated by sector, capacity, efficiency, availability factor, etc.) and its input and output flows. This is the base block of the model;
- A set of constraints that the system must respect. In particular, it is a question of meeting the demands specified for each unit of time and emission or operating constraints for the processes described;
- A criterion of choice among the available options. The optimal paradigm makes it possible to find the combination that minimises the total cost of the energy system (our objective function) over the study horizon while respecting the constraints imposed for a given scenario.

Figure 17.1 summarises the model of the transformation steps selected to represent the French electrical system. The model distinguishes the technology options for several primary resources, transformation pathways and voltage levels. The optimal solution over the long term then gives a coherent picture of the temporal dynamics involved in decommissioning existing power plants and making investment choices.

Several works have been carried out by the Centre de Mathématiques Appliquées (Centre for Applied Mathematics) of MINES ParisTech based on TIMES models for different geographical perimeters, different time steps and especially the question of future choices for electrical systems. To explore the question of the CO_2 impact per electric kWh, the model implemented in this analysis is based on a finer time resolution in 576 time steps, representing two typical days (weekdays and weekend days) for each month of the year and no hourly time. This choice notably allows a better description of the variability of wind and solar productions.

17.2.2 Ramp Constraints and Dynamics of the Means of Productions

One of the most delicate stages in modelling at this timescale concerns the representativity of the dynamic constraints on the different means of production. This representativeness is to be considered at the level of the overall annual producible as well as the finer time steps. For example, an aggregate view of production data should take

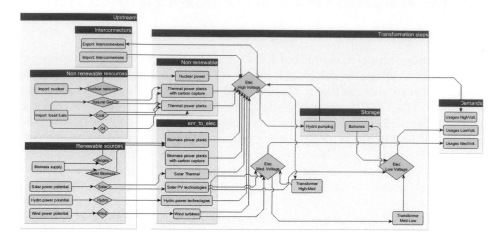

Figure 17.1 Schematic description of the electrical system represented in our model.

into account that nuclear power produces between 70% and 80% of total electricity. However, this approach does not suffice to describe the behaviour of the French electricity system. To account for this, dynamic ramp constraints are used. The following sections provide an overview of the importance of this dynamic representativeness.

17.2.2.1 Wind and Solar Energy Production Profiles

To account for their non-controllable nature, the profiles of electricity generation from renewable and intermittent resources are exogenously constrained by a maximum availability factor. Figure 17.2 illustrates the effects of calibrating model outputs against RTE (Réseau de Transport d'Electricité, French electricity transmission system operator) data.

17.2.2.2 Seasonal Availability and Daily Flexibility for Hydraulic Power

The modelling of hydraulic power plants distinguishes a seasonal scale (on which the output of the sector is determined by the water availability of dams and rivers, Figure 17.3) and a daily scale (on which this production is flexible). Seasonal availability is therefore limited by a maximum envelope similar to the intermittent resource profile, while production is controllable (Figure 17.4) with strong variations on a finer timescale.

17.2.2.3 The Case of Nuclear Power Plant Ramps

It is customary to consider that nuclear production is a basic production that does not allow for monitoring of the load curve. If this is true, the power supplied by a nuclear power plant in the first approximation can vary greatly on a daily scale, between its minimum operating power and its rated power (Figure 17.5). At the weekly scale, the

(a)

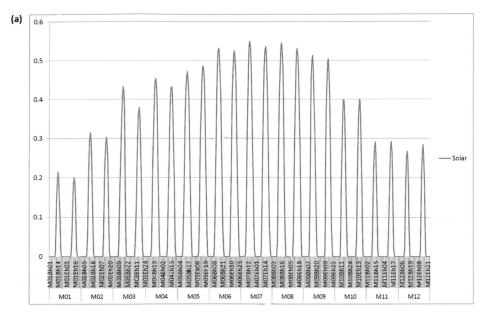

(b)

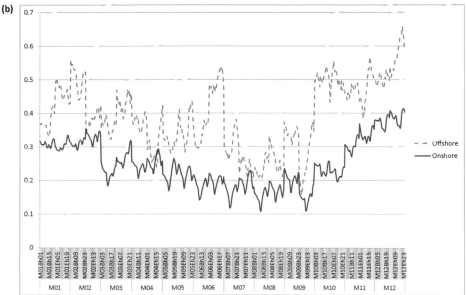

Figure 17.2 Calibration of wind and solar generation profiles by time step in 2012.

minimum lag time (1–2 days) in case of a complete stop also makes it possible to obtain additional flexibility at the scale of the total fleet.

As shown in Figure 17.6, although within the maximum technical limits of the plants over one hour, the ramps obtained do not sufficiently take those observed from the RTE data into account. This reality reflects the management habits of the

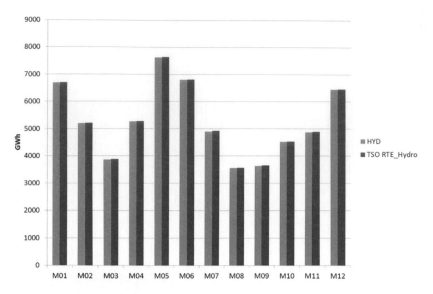

Figure 17.3 Monthly hydraulic production.

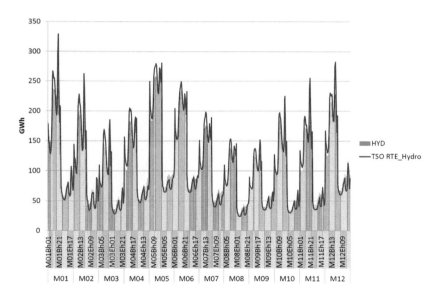

Figure 17.4 Intra-day hydraulic production.

current fleet and shows (in the simplified environment described by the model) that they could evolve in absolute terms. The addition of lower ramp values for the existing fleet then makes it possible to better describe the hourly dynamics of the existing one (Figure 17.7).

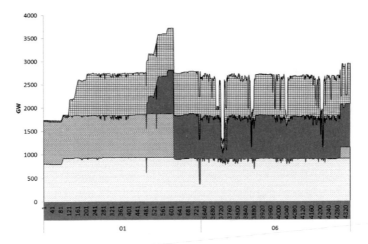

Figure 17.5 Hourly production achieved by the four reactors of the Tricastin Nuclear Power Plant: January and June 2017 (source of RTE data).

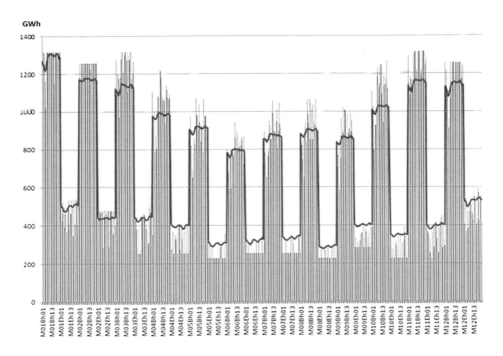

Figure 17.6 Representation of nuclear production in 2012 without ramp constraints.

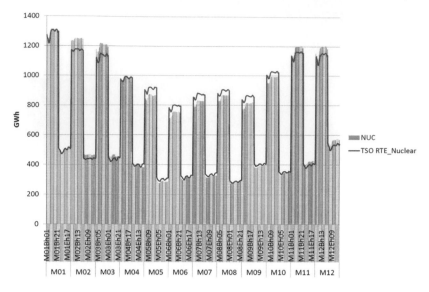

Figure 17.7 Representation of nuclear generation in 2012 with additional ramp constraints.

17.3 Key Dimensions and Choice of Prospective Scenarios

The TIMES_FR_ELC prospective model serves here as a framework for experimentation and quantitative evaluation of the French mix in the long term. To explore the question of the CO_2 content per electric kWh, about 15 scenarios are then defined to illustrate different hypotheses that we have chosen for three strategic dimensions of future choices in the electricity sector, which constitutes the background building system in a LCA. The choices made therefore only cover all the possible combinations of techno-economic parameters very partially. They are intended to illustrate the economic competitiveness of the sectors, the intensity of the CO_2 penalties and the effect of the profiles selected for intermittent production. The model minimises the discounted system costs. Here we use a discount rate of 4%. In this exercise, the projection of electricity demand corresponds to the most recent data for the time, namely the central scenario of the 2014 RTE (RTE 2014) forecast balance sheet for the 2012–2030 period. It grows from 485 TWh in 2012 to 493 TWh in 2030. A linear trend then leads to a value of 520 TWh in 2050.

17.3.1 Relative Competitiveness of Production Chains

The minimisation logic under TIMES constraints leads to the construction of the multi-year investment trajectory, and then to the selection of the technologies that are the most economically competitive for the generation at each time step. Both variants considered here highlight the cost issues for solar and nuclear power.

As recalled by the World Nuclear Association (WNA), several new generation reactors have experienced significant slides in their implementation costs around the world (World Nuclear Association n.d.). This is particularly the case for the Flamanville reactors in France and Olkiluoto reactors in Finland, both of which are the examples of EPRs (European Pressurized Reactors) built in Europe. Investment costs have been reassessed throughout the construction from $3,700 to over $7,000/kW. An important part of this drift is explained by the economic penalties induced by the drift of the construction time. The international analysis conducted in Lovering, Yip and Nordhaus (2016) provides a perspective on historical evolution, showing the wide dispersion and regional differences in investment costs. Most studies envisage a long-term cost of $4,000–5,000/kW.

This kind of debate on costs also exists in the literature about the full costs of installing solar modules. Two studies by IRENA (IRENA, 2014a and 2014b) thus highlight the average installation costs of photovoltaic panels, which range from $ 6,000 to 5,000/kW for France and a low level of about $2,000/kW for China and Germany.

Two scenarios are then constructed to reflect a more or less rapid convergence towards these long-term prices. They make it possible to test the interaction between the assumptions about the level of future costs and the need to renew the existing fleet:

- A central vision is provided with the evolving costs for the nuclear industry of $7,800/kW (to reflect the gaps due to the delays noted) for the first production units, leading to a long-term value of $5,000/kW in 2030 and $4,800/kW. In this case, the corresponding assumption for solar energy describes a slow decline from $6,000/kW in 2010 to a long-term cost of $3,000/kW in 2050 with a Passover point of $4,800/ kW in 2030.
- A more pessimistic view of the evolution of nuclear investment costs gives a value that remains high in the medium term, at $5,700/kW in 2030 and then $5,000/kW in 2050. This scenario is combined with a more optimistic view of solar energy cost development which, driven by public support, is rapidly converging towards the lowest levels in the world: $900/kW in 2030 and then $620/kW in 2050.

17.3.2 Intensity of CO_2 Emission Control Policies

Today, the Emission Trading System (ETS) market remains the practical framework for controlling CO_2 emissions from the electricity sector on a European scale, by establishing carbon penalties. However, this economic mechanism interacts with the various support mechanisms and the quantitative emission reduction targets that are decided at the national level. In fact, the future technological choices and the CO_2 content per electric kWh will be impacted as much by the form as by the intensity of the environmental policies. Three scenarios are considered in this exploratory analysis:

1. The reference case describes a CO_2 penalty that progressively increases from a low point of €10/t CO_2 in 2012 to €30/t CO_2 in 2030. Beyond this, it is kept constant. This case describes a (sectoral) market price for CO_2, which is driven by the logic of industrial competitiveness and therefore increases moderately.
2. The second assumption places a cap on the volume of emissions from the French electricity sector to reflect the status quo of the CO_2 impact of the mix. The logic of this is to proceed towards the best economic arbitration without issuing more in terms of annual volume.

3. The third environmental policy takes the liberty of introducing a CO_2 penalty for the electricity sector that is more in line with France's climate objectives. It involves the adoption of a penalty ranging from a level of €100/t CO_2 in 2030 to €240/t CO_2 in 2050.

17.3.3 Intermittent Renewable Sources

17.3.3.1 Short-Term Variability

The objective of this section is to reflect the impact of a disruption to the capacity factors retained for intermittent resources. The method involves introducing a hazard into the availability profiles used for solar and wind energy. The short-term hazard is calibrated from the forecast spreads per day published by RTE. Figure 17.8 illustrates the distribution of forecast differences for wind energy.

These observed differences then make it possible to define the so-called disrupted profiles according to a simplified method: $AF(TS_i)y = \left(1 + c_y \times err(\alpha_i) \times AF(TS_i)_{2012}\right)$, where

- AF is the coefficient of availability for the variable processes
- TS_i is timescale index of the model
- y is the year index
- c_y is a scaling factor that makes it possible to introduce a reduction in errors over the long term
- $err(\alpha_i)$ is the random distribution of errors.

For this analysis, five "disrupted" profiles are simulated for an average error of 5.5% for wind power and 2.3% for solar power (Figure 17.9). The factor is maintained at 1.

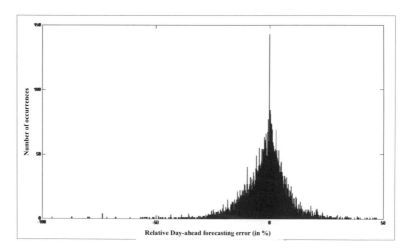

Figure 17.8 Distribution retained for the forecast deviations per day for wind energy.

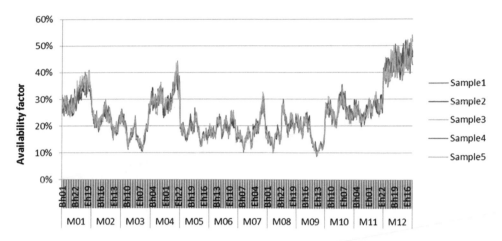

Figure 17.9 Disruptions introduced on the availability coefficients at the scale of the typical days considered: onshore wind.

17.3.3.2 Annual Variability

In contrast to the short-term variability that is reflected in the forecast differences, the inter-annual variability concerns the long-term uncertainty on the annual deterministic profile. This is shown in Figure 17.10 for three consecutive years at the temporal aggregation level of the model. It is simply simulated in this analysis by a change of reference year. A synthesis of considered scenarios is provided in Table 17.2.

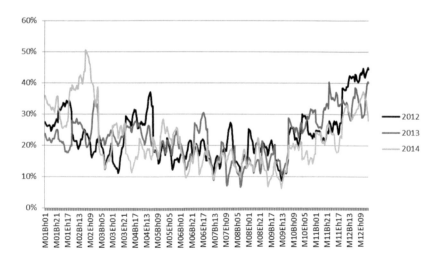

Figure 17.10 Aggregated capacity factors by model time span.

Table 17.2 Summary of the exploratory scenarios considered

Scenarios	Emission Control Measure	Economic Competitiveness	Renewable Energy Variability
BASE	Moderate tax	Nuclear central/ solar moderate decrease	2012 Profiles
NucHaut_RenBas	Moderate tax	Nuclear high/solar rapid decrease	2012 Profiles
CapCO$_2$	CO$_2$ volume	Nuclear high/solar rapid decrease	2012 Profiles
HighTax	High tax	Nuclear high/solar rapid decrease	2012 Profiles
BASE_wind2014	Moderate tax	Nuclear high/solar rapid decrease	2014 Profiles
5 VarENR scenarios	Moderate tax	Nuclear central/ solar moderate decrease	Short term: 5 "disrupted" profiles
5 Scenarios VarENR with VolCO$_2$ constraints	CO$_2$ volume	Nuclear central/ solar moderate decrease	Short term: 5 "disrupted" profiles

17.4 Results

17.4.1 Economic Dimension

Figure 17.11 compares the evolution of the French electricity mix in 2050 for our two relative cost assumptions of the nuclear and renewable sectors and for a moderate environmental policy.

Assuming an off-plan nuclear investment cost (lower than today's observed cost for the initial units of a new type of design) of €5,000/kW in 2030, the evolution of the existing fleet is first offset by the extension and then the construction of new plants. Following a purely economic approach and assuming a carbon penalty that remains low, coal is competitive and the growth of natural gas is, moreover, slowed by the faster increase in its price in the assumptions.

While the lowest cost assumption for renewables is combined with a slower decline in the nuclear sector, a strong increase in renewables can be observed. However, it is associated with the replacement of nuclear power by coal plants for base production, in the absence of any other economic or political signal.

Figure 17.12 completes these results with a vision of electricity generation at the finest time step for the year 2050. It is observed that, in the first case, coal is used as a semi-base load while the nuclear capacity is preferentially used as the base with small seasonal variations. The strong growth of renewable energy in the second case has the effect of gradually eliminating the need for a "real base". It then becomes less profitable (according to an optimising centraliser approach with perfect knowledge of the future) to invest in new power stations, including the intermediate periods if the future utilisation factor is due to a decrease. This is a situation that benefits coal. Although simplified, these results illustrate the timing issues around the announced cost reductions.

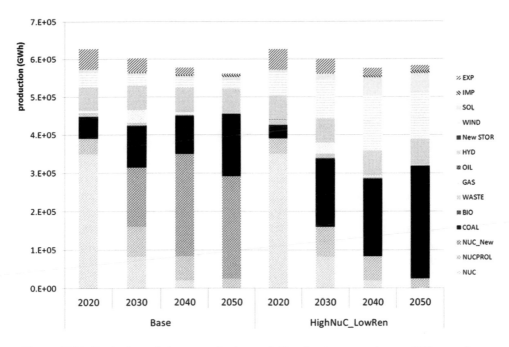

Figure 17.11 Evolution of the annual mix modelling horizon: moderate CO_2 penalty.

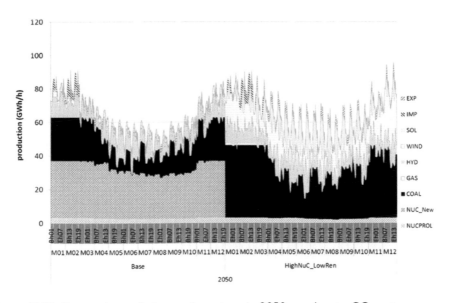

Figure 17.12 Comparison of alternative mixes in 2050: moderate CO_2 rate.

As shown in Figures 17.13 and 17.14, the average CO_2 content per electric kWh increases in both scenarios and is even greater when the renewable production is strongest. This presents a configuration that can be compared to that proposed by Germany today, with both strong renewable production and consequent fossil production. Nevertheless, the levels achieved remain 3–4 times lower than those in Germany or Italy. The seasonal variability also confirms the pertinence of a finer vision of the CO_2 content.

17.4.2 Ambitious Climate Policies

The scenarios previously mentioned illustrate what could lead to a techno-economic rationale in the context of carbon penalties that would not increase fast enough. A fall in the cost of renewable energy does not of course guarantee a reduction in emissions. Figures 17.15 and 17.16 then enable a discussion of the effect of the two more explicitly proactive environmental policies in terms of CO_2 emissions and of the worst-case investment scenario.

The economic optimum under these new constraints is then shifted. The use of coal is falling sharply as a result of higher renewable production. However, despite its high investment cost, the nuclear option is maintained. Installed capacity nevertheless drops by two-thirds. A volume limit maintains a diversified mix where fossil plants (natural gas and sequestration) play a vital role in winter. A high tax penalises fossil resources more strongly and leads to a mix that is almost carbon-free. The economic approach therefore leads to more nuclear energy replacing fossil fuels and more seasonal variability in nuclear power.

The CO_2 impact is mechanically modified (Figure 17.17). In the case of volume control, if the constraints are respected, then this results in a fixed annual volume, with

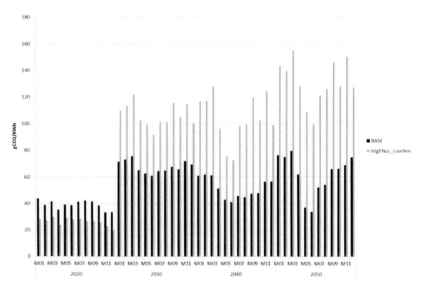

Figure 17.13 Evolution of CO_2 content: low penalty.

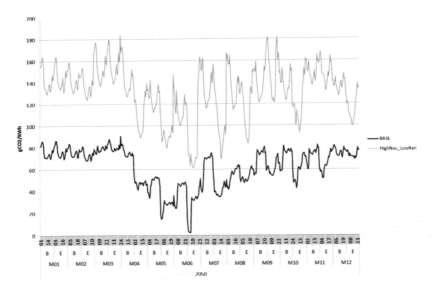

Figure 17.14 Profile at the finest timescale.

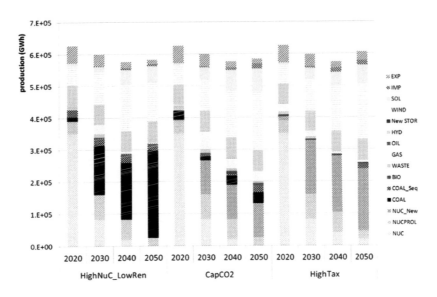

Figure 17.15 Electrical mix and environmental scenarios.

the possibility of variability at the lower timescales, however. A strong tax quickly leads to a decarbonised system with a negative impact on some time steps due to the use of even small amounts of biomass plants equipped with capture and sequestration devices.

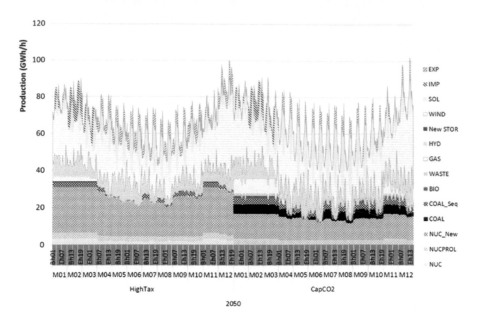

Figure 17.16 Hourly production profiles: ambitious environmental policies.

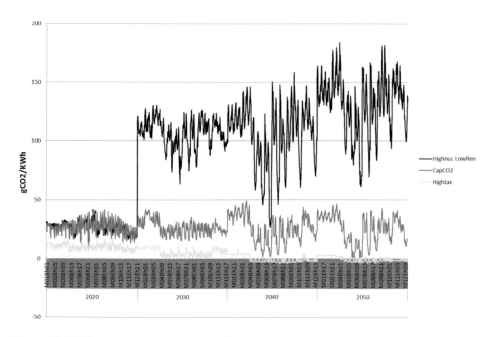

Figure 17.17 Hourly content for the different levels of environmental policies.

17.4.3 Variability of Intermittent Renewable Energies

We now discuss the results obtained by the simplified approach that we used to simulate the variability of CO_2 content induced by the intermittency of renewable resources. For the short term, Figures 17.18 and 17.19 describe the means and standard deviations of the scenarios obtained by selectively varying the capacity factors of intermittent renewable sources. The main indicator that we have sought to describe here is the standard deviation. The main result is its sensitivity to the ambition of the emission control policy. The standard deviation varies around 5.6 g CO_2/KWh with a moderate CO_2 tax and 0.95 g CO_2/KWh when the total volume emitted is constrained.

The effect of the long-term (annual) variability of intermittent renewable resource production patterns is, finally, illustrated in Figure 17.20. By moving from an annual availability of 24% to 21.4%, the difference in terms of average monthly CO_2 content is to the order of 20 g CO_2/KWh.

Here, the difference between the average monthly CO_2 content induced by a lower annual availability illustrates the issues relating to the development on less windy sites (negative effect on the overall producible power) or the improvement of the turbines (height and diameter in particular), which tends to increase the producible power.

Despite the relative simplicity of the approach used to simulate variability, these two indicators (with additional complexity introduced by intermittency) seem to be important to us for describing an order of magnitude of uncertainty around the values discussed in this exploratory exercise.

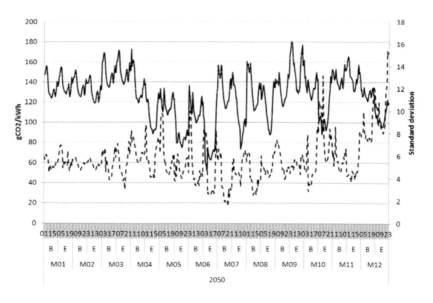

Figure 17.18 Variability induced by the disturbance of the renewable energy profiles: mean and standard deviation (moderate tax).

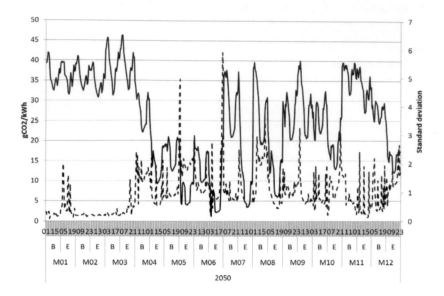

Figure 17.19 Variability induced by the disturbance of the renewable energy profiles: mean and standard deviation (capped emissions).

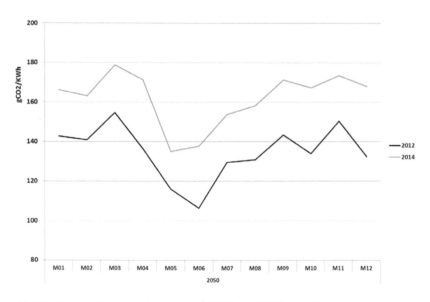

Figure 17.20 Change in monthly average: 2012 vs. 2014 profiles.

17.5 Conclusion and Perspectives

Due to the importance of the electric vector in the final energy consumption within the residential and tertiary sectors, the CO_2 content per electric kWh is a key indicator in the LCA of built complexes. The methodological difficulties involved in choosing the values to be retained for the long term are well established. In this exploratory study, we wanted to show the interest and complementarity of a prospective model of systems for the LCA of buildings and the additional complexity that emerges from them.

To do this, we tested different hypotheses in three dimensions that may seem far removed from the concerns of the building sector, but which are actually strategic for the electrical systems that *ultimately* overdetermine the impact of electrical uses during the building use phase. These are the relative competitiveness of the sectors, the intensity of the emission control policies and the potential impact of the variability of the renewable profiles on the seasonal and hourly dynamics of this indicator.

Using a future mix optimisation model and scenarios built for the analysis, we tried to quantify the mechanisms at work in the evolution of this indicator and the complexity that this extension of the perspective brings at the multi-year scale and also on a seasonal and hourly basis. The first intuitions revealed clearly show the interest of prospective association and LCA. They also show the interest of scanning a range of scenarios that are sufficiently broad and varied to allow for an informed decision. Future work will also improve the modelling for the electricity sector by considering the following, among other things:

- The potentially disruptive effect of massive renewable surpluses in neighbouring countries. This would require an increase in the geographical scope of the model, or at least the scripting of this case;
- The analysis of a broader set of demand scenarios to identify the effect of the overall volume of electricity consumed on the environmental impact per kWh;
- Characterisation of the impacts of the electrical system other than CO_2.

Finally, in the conclusion of this study, it is useful to recall once again that the simulated contexts were chosen to dissect some of the phenomena at work. They only partially cover the possible interactions between the prospective of energy systems more generally and the built systems. To name just two, we will mention the following:

- The impact of building and insulation levels on electrical demand in volume and structure. The challenge is therefore possible feedbacks;
- Beyond the development of renewable resources for the electricity sector, the LCA of buildings will also be impacted by the possible change in gas supply, particularly with biomethane or biomass gasification. The challenge is then the parallel or coupled evolution of two important systems for buildings.

Bibliography

Commissariat général au développement durable. 2018. "Chiffres Clés Du Climat - France, Europe et Monde." http://www.statistiques.developpement-durable.gouv.fr/fileadmin/documents/ Produits_editoriaux/Publications/Datalab/2017/datalab-27-CC-climat-nov2017-b. pdf.

Finenko, Anton, and Lynette Cheah. 2016. "Temporal CO_2 Emissions Associated with Electricity Generation: Case Study of Singapore." *Energy Policy* 93 (June): 70–79. https://doi.org/10.1016/j.enpol.2016.02.039.

Gordon, Christian, and Alan Fung. 2009. "Hourly Emission Factors from the Electricity Generation Sector – A Tool for Analyzing the Impact of Renewable Technologies in Ontario." *Transactions of the Canadian Society for Mechanical Engineering* 33 (1): 105–18. https://doi.org/10.1139/tcsme-2009-0010.

Harmsen, Robert, and Wina Graus. 2013. "How Much CO_2 Emissions Do We Reduce by Saving Electricity? A Focus on Methods." *Energy Policy* 60 (September): 803–12. https://doi.org/10.1016/j.enpol.2013.05.059.

Hawkes, A. D. 2014. "Long-Run Marginal CO_2 Emissions Factors in National Electricity Systems." *Applied Energy* 125 (July): 197–205. https://doi.org/10.1016/j.apenergy.2014.03.060.

IRENA. 2014a. "Renewable Power Generation Costs in 2014." http://www.irena.org/document-downloads/publications/irena_re_power_costs_2014_report.pdf.

IRENA. 2014b. "The True Costs of Solar PV: IRENA's Cost Analysis." http://www.irena.org/costing/Presentation/TAYLOR-M-EU-PVSEC-September-25–2014-Amsterdam-V2_ Oct9-WEB.pdf.

Kono, Jun, York Ostermeyer, and Holger Wallbaum. 2017. "The Trends of Hourly Carbon Emission Factors in Germany and Investigation on Relevant Consumption Patterns for Its Application." *The International Journal of Life Cycle Assessment* 22 (10): 1493–501. https://doi.org/10.1007/s11367-017-1277-z.

Lovering, Jessica R., Arthur Yip, and Ted Nordhaus. 2016. "Historical Construction Costs of Global Nuclear Power Reactors." *Energy Policy* 91 (April): 371–82. https://doi.org/10.1016/j.enpol.2016.01.011.

RTE. 2014. "Bilan Prévisionnel de l'équilibre Offre-Demande d'électricité En France." https://www.rte-france.com/sites/default/files/bilan_complet_2014.pdf.

Soimakallio, Sampo, Juha Kiviluoma, and Laura Saikku. 2011. "The Complexity and Challenges of Determining GHG (Greenhouse Gas) Emissions from Grid Electricity Consumption and Conservation in LCA (Life Cycle Assessment) – A Methodological Review." *Energy* 36 (12): 6705–13. https://doi.org/10.1016/j.energy.2011.10.028.

World Nuclear Association. n.d. "Nuclear Power Economics | Nuclear Energy Costs." Accessed May 25, 2016. http://www.world-nuclear.org/information-library/economic-aspects/economics-of-nuclear-power.aspx.

Towards a Renewal of Techniques and Systems

Chapter 18

Real-Time Energy Management Strategies for Buildings and Blocks

Maxime Robillart

KOCLIKO

Marie Frapin

MINES PARISTECH, CES – PSL

18.1 Introduction

The increase in the number of energy-efficient and "connected" buildings opens up new prospects for the development of new services related to energy management and smart grids. Of these services, those related to the optimised management of energy flows inside the habitat are particularly interesting.

As they become increasingly thermally insulated, high-performance buildings are much more sensitive to intermittent phenomena, in particular to the solar gains transmitted by windows and uses (internal inputs related to equipment and occupants). In addition, this type of building leads to sophistication of the different installed systems that can interact. Building management is no longer intuitive for the occupant and requires the use of anticipatory and reactive control devices to ensure comfort, while controlling energy consumption.

In addition, much feedback found significant differences between the theoretical consumptions calculated during the design phase and the actual consumptions measured during the operating phase (Sidler, 2011). These discrepancies can be explained by the errors of the building energy simulation model used, the uncertainties weighing on the input data (climate, occupation, physical characteristics of the building) or by poor energy management of the buildings causing a decline in performance. Optimised management of energy consumption is then one of the solutions for reducing these differences and progressing towards the guarantee of energy performance.

Finally, in France, the residential-commercial sector accounts for 68% of the final electricity consumption (CGDD, 2015), which leads to grid tensions, especially at peak electricity consumption. To meet these demands, one of the solutions is to increase the means of high-peak production. However, they are more expensive and induce high CO_2 emissions since they use fossil fuels. An alternative to this solution is to implement demand response programmes. These latter aim to smooth the temporal demand profile by shifting consumption from peak hours to off-peak hours. This "flexibility" is achieved either by encouraging consumers to shift their consumption or by taking control of electrical appliances. These power consumption peaks can be reduced by the development and implementation of advanced control systems. These energy management strategies must then be able to adapt to network constraints (and electricity prices) in real time and make it possible to withdraw high peak electricity consumption while ensuring the comfort of the occupants.

It is therefore necessary to develop methods and strategies aimed at optimised management of energy flows in real time. To ensure the comfort of the occupants and to maintain the energy performance of the building, these control strategies must be able to continuously adapt to changing environmental uses and conditions. They must also anticipate the future state of the building (we use a building thermal model) as well as the evolution of various parameters such as weather conditions, tariffs or the CO_2 content of the electricity.

The techniques commonly used to define real-time energy management strategies rely on the application of model predictive control (MPC). In this technique, the minimisation of the objective function (e.g. the minimisation of heating consumption) is carried out by taking the current and future state of the building into account, by predicting the evolution of the external and internal disturbances. This method thus consists of a repeated resolution at each regulator sampling period (set by the user) of an optimal control problem, that is, how to go from a current state to an objective state optimally while respecting the constraints. At each sampling period, an optimal control sequence (minimising the cost function) is calculated in an open loop over a fixed prediction horizon. However, only the first command is applied and the procedure is resumed at the next sampling period (receding horizon principle). Figure 18.1 illustrates the functioning of the predictive control over an NΔt prediction horizon. The use of predictive control has recently developed in the building sector, as evidenced by numerous research studies (Morosan et al., 2010; Oldewurtel et al., 2012; Robillart et al., 2018). This technique has long been penalised by the high cost of the technology required for its implementation and the difficulty of obtaining a satisfactory mathematical model of the building. However, following the rise of IoT (Internet of Things) and the availability of reliable and inexpensive sensors, as well as recent developments in computing resources (allowing fast execution of complex calculations), this technology is becoming increasingly accessible.

However, predictive control can still be far removed from the current practices of building professionals, especially regarding the use of expert rules (of the "If condition, then action" type) in conventional control systems. An alternative is to use

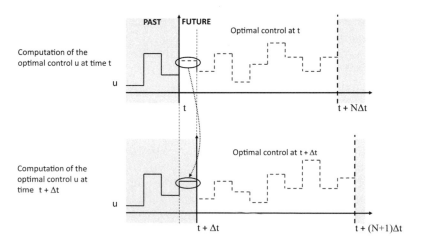

Figure 18.1 Functioning of the predictive control

offline optimisation methods to identify simplified control laws. Two offline optimisation methods exist. The first is to explicitly solve the predictive control problem, the objective being to obtain an explicit formulation of the control laws. The expression is then calculated offline and stored in a look-up table. Thus, instead of solving an online optimisation problem at each time step, it is sufficient to evaluate the explicit function stored (typically a piecewise affine function). More specifically, the conventional use of predictive control is to solve an optimisation problem parameterised by the current (measured) state of the system. On the contrary, if it is possible to solve the optimisation problem for all possible system states ("offline" optimisation), then it is no longer necessary to solve the optimisation problem at each time step ("online" optimisation). This method thus provides optimal control laws that cover the entire state space. However, this method is difficult to apply to the case of building energy regulation. An alternative has been proposed to approximate the explicit resolution of the predictive control (Coffey, 2013). To reduce the size of the problem, the idea is to solve the predictive control on a set of points determined in advance only (external disturbances, initial state of the building) and to make interpolations between the different points of the grid. Although interesting, the implementation of this method requires large computational capabilities. A second method of approximating the results of the predictive control is to minimise the objective function of the optimisation problem by simulating the building under representative outdoor and indoor disturbances of a typical year. From the results obtained, we can then identify a control law to provide to the regulation system (through heuristic control laws, for example), and use it for real-time building management. The goal here is to provide general energy management strategies to minimise energy consumption or running costs. For example, this approach has been used to identify control laws for an automated solar protection system for a building (Le et al., 2014), for controlling the opening of windows (May-Ostendorp et al., 2013) or the load shifting of high peak hour heating consumption (Robillart et al., 2017).

The use of an anticipatory regulation system becomes very important when the system has to anticipate sudden changes in the disturbances or constraints. This is especially the case for policies intended to encourage the load shifting of high peak hour electricity consumption where the electricity tariff schedule can undergo significant variations. This chapter presents the advantage of using MPC to respond to this type of problem.

18.2 Energy Management of a Monozone Building

18.2.1 Model Used for Predictive Control

18.2.1.1 Building Studied

The building studied corresponds to an I-BB (shuttered concrete) house from the "INCAS" platform of the INES (Institut National de l'Énergie Solaire – French National Institute of Solar Energy), located at Bourget-du-Lac. With simple and compact architecture, it has an interior surface of 89 m² and two floors. The house has been designed to match the performance of the "Passivhaus" label, thanks in particular to high insulation, very low thermal bridges, high-performance glazing and a high-quality airtight envelope. Erected on underfloor space, the house has an attic and is insulated from the outside by 20 cm of extruded polystyrene (vertical wall and low floor) and 20 cm of glass wool (high floor). All windows are double glazed with low emissivity, except those in the north that have triple glazing. The house is equipped with controlled mechanical ventilation of a double-flow type with a ventilation rate of 0.6 air change per hour (ACH). The plate heat exchanger allows up to 90% heat recovery. The heating is ensured by an electrical resistor located at the start of the fresh air distribution network. In order to reduce the computation time of the optimisation algorithms, a single-zone model was considered in this study.

The stochastic model developed by Vorger et al. (2014) was used to generate the occupation scenarios. In this model, the occupants' realistic characteristics and behaviours are represented by a probabilistic approach based on multiple statistical data (e.g. on time-use or socio-demographic surveys, as well as measurement campaigns). With each simulation, a different household is generated in a probabilistic manner according to the properties of the dwelling (house or apartment, number of rooms and location) and each household member is defined by a set of socio-demographic characteristics (age, sex, employment status, etc.). These influence the housing equipment in household appliances and can generate activity scenarios on the part of the inhabitants. Based on several hundred simulations of this model, it is possible to create an average occupation scenario. For this study, a household of three people with high-performance appliances and lighting was considered. An average scenario of internal gains in the building during a week in winter, obtained from a sample of 300 generated scenarios, is presented in Figure 18.2.

18.2.1.2 Dynamic Thermal Model of the Building

The dynamic thermal model of the building used in this study is based on the "thermal zone" concept, a subset of the building considered at a homogeneous temperature (Peuportier and Blanc-Sommereux, 1990). For each zone, the walls are divided into meshes (finite volume methods) to which a thermal balance is applied, assuming that the temperature of the mesh is uniform. To ensure this hypothesis, the meshing becomes finer as the mesh grows closer to the interior atmosphere. The air, the furniture

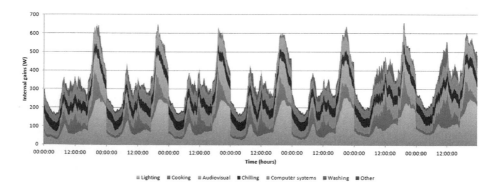

Figure 18.2 Average scenario of internal gains generated for a dwelling occupied by three people during a week in winter

and the possible light partitions contained in the zone are grouped in a single mesh. The heat balance on each mesh then takes the following form (Neveu, 1984):

$$C_m \dot{T}_m = \dot{Q}_G - \dot{Q}_L \tag{18.1}$$

where

C_m the thermal capacity of the mesh;

$\dot{T}_m$ the temperature of the mesh;

$\dot{Q}_G$ gains including solar and internal gains, as well as the heating or cooling powers of the equipment;

$\dot{Q}_L$ losses due to ventilation and heat transfer by conduction, convection and radiation.

Some non-stationary phenomena (opening of a shutter) or non-linear phenomena (air movement) are taken into account in a load vector denoted as U, which makes it possible to write the Equation (18.1) in the form of a system of linear Equation (18.2).

By repeating this balance for each mesh of the zone considered and by adding an output equation, the set of equations can thus be represented in the form of a continuous and linear time-invariant system (Bacot, 1984; Lefebvre, 1987):

$$\begin{aligned} \dot{x}(t) &= Ax(t) + Bu(t) \\ y(t) &= Cx(t) + Du(t) \end{aligned} \tag{18.2}$$

where

$x \in \mathbb{R}^q$ ($q = 28$ in this study) the state of the building (the energy stored in each mesh);

$u \in \mathbb{R}^m$ ($m = 10$ in this study) the driving forces (including climatic parameters, heating, etc.);

$y \in \mathbb{R}$ the indoor zone temperature, taking the air and the temperature of the walls into account;

A, B, C and D give the state, input, output and feedforward matrices, respectively.

The simulation of this model requires knowledge of the loads, in particular the heat emissions by the occupants and the equipment, but also the local meteorological data concerning the outside temperature and solar radiation. All these data are contained in the driving forces.

In this study, we are only interested in an electric heating system for regulation, directly integrated into the dynamic thermal model of the building. It is assumed that the heating system is 100% efficient and that it provides all the power to the air mesh. This type of system thus makes it possible to be very reactive (its dynamics will be disregarded), with no time lag between the moment when the command is sent to the system and that in which the heating power is injected into the air mesh. Not counting several validation studies (Peuportier, 2005), the reliability of the model has been studied in the particular context of high-performance houses and is comparable to international references (Brun et al., 2009; Recht et al., 2014; Munaretto et al., 2017).

18.2.1.3 Model Reduction

Dynamic thermal modelling of the building leads to the development of a large-scale system. However, it is not conceivable to solve an optimal control problem with the complete system, because its dimension is too large and can lead to problems with regard to the convergence of the optimisation algorithms. It is then preferable to reduce the order of the model while keeping its original behaviour as far as possible. Different reduction methods can be used to reduce thermal models, such as modal reduction (Lefebvre, 2007), the proper orthogonal decomposition to eigenvalues (Sempey et al., 2009) or techniques based on the singular value decomposition (Yahia and Palomo Del Barrio, 1999). Balanced realisation (Moore, 1981) is an efficient and commonly used method for model reduction in the control-command framework. This makes it possible to obtain a model adapted to control and offers good precision (Palomo Del Barrio et al., 2000). It will therefore be used in the rest of the study.

To determine the model reduction order, it is first necessary to determine the fully controllable and observable reduced models. For this, the ranking of the controllability and observability matrices has been calculated for the various reduced models. Only models reduced to less than or equal to four are fully controllable and observable (they meet Kalman's controllability and observability criteria). The time constants of the four models are reported in Table 18.1. It can be noted that the models reduced to the orders 2, 3 and 4 have well-separated time constants.

In order to determine optimal control of the heating power, it is essential to use a reduced model with performances close to those of the unreduced model. To determine the order of reduction of the appropriate model (which is a compromise between precision and model size), frequency and time-domain analyses are essential. Three driving forces were considered for the frequency analysis, namely heating power (which is the control), outside temperature and solar radiation passing through the windows (glazed flow).

Figure 18.3 shows the frequency responses (heating power → indoor temperature) of models reduced to orders 1 and 2 (the red and green curves, respectively) compared to the reference model (blue curve). It can be seen that the model reduced to order 1 exhibits a very different behaviour compared to the unreduced model, both in terms of phase and gain. For example, it has a significant time delay between one minute and several days (frequencies between 10^{-6} and 10^{-2} Hz). Since the reduced-order model has only one time constant (which can be made to physically correspond to the building's thermal capacity, divided by its loss coefficient), the latter is not able to correctly represent the variations of the indoor temperature over a few hours. The model reduced to order 2 has a phase delay in comparison with the full model, with times between one and several minutes, and a phase advance for times corresponding

Table 18.1 Time constants of the four reduced models

Order of Reduction	1	2	3	4
Time constants	16 days	17 days 13 minutes	17 days 1 hour 8 minutes	17 days 20 hours 44 minutes 6 minutes

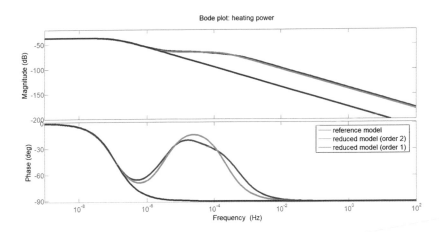

Figure 18.3 Frequency responses of the systems (not reduced, reduced to order 1 and reduced to order 2) to the heating power

to several hours. Thus, this model tends to underestimate the variation of the indoor temperature for power variations at the scale of several minutes and to overestimate the variation of the indoor temperature over several hours. Consequently, the models reduced to order 1 and order 2 are not retained in the rest of the study because their behaviours are too disparate compared to the unreduced model.

Figure 18.4 shows the frequency responses (heating power → indoor temperature) of models reduced to orders 3 and 4 (the red and green curves, respectively) compared to the unreduced model (blue curve). It can be seen that there are very few differences in terms of either phase or gain. Only slight delays or phase advances can be noted for models reduced to orders 3 and 4 according to the timescales. With regard to the frequency responses of the models reduced to orders 3 and 4 (Figure 18.5) at the outside

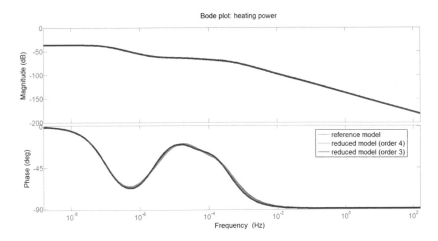

Figure 18.4 Frequency responses of the systems (not reduced, reduced to order 3 and reduced to order 4) to the heating power

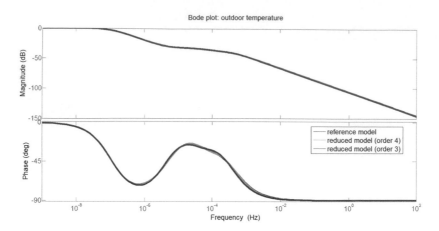

Figure 18.5 Frequency responses of systems (not reduced, reduced to order 3 and reduced to order 4) to the outdoor temperature

temperature, we can also note that their behaviour is very close to the reference model both in terms of phase and gain.

Finally, differences can be noted when considering the frequency responses of the models to the glazed flow (Figure 18.6). Thus, it can be seen that the models reduced to orders 3 and 4 show a phase advance in relation to the unreduced model for times between 1 second and a few hours. This difference can be explained from a physical point of view by the reduced models' inability to faithfully represent the impact of solar radiation passing through the windows.

By studying the frequency responses of the reduced models to the different driving forces, it was possible to eliminate the models reduced to orders 1 and 2 from the models to be used for calculating the optimal control. Indeed, the behaviour of these models is too far from the reference model to be able to ensure the reliability of the

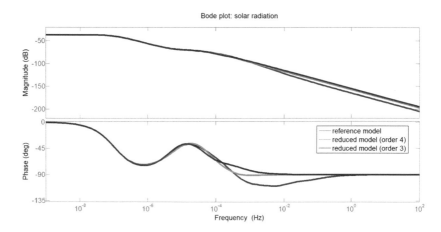

Figure 18.6 Frequency responses of systems (not reduced, reduced to order 3 and reduced to order 4) to the glazed flow

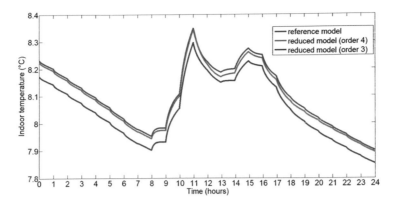

Figure 18.7 Inner temperatures simulated by models reduced to order 3 and 4 and by the reference model on the day when the mean squared difference is highest

optimal control calculated. Models reduced to orders 3 and 4 have satisfactory behaviour. However, studying frequency responses does not allow for a choice between reduction to order 3 and reduction to order 4. In fact, the frequency responses of the two reduced models are very close, including in terms of the glazed flow, where they are incapable of accurately representing the behaviour of the complete model.

A temporal analysis can be conducted in order to choose between models reduced to orders 3 and 4. The objective here is to compare the zone temperature calculated by the models reduced to orders 3 and 4, and the reference model when the building is in free evolution (unheated building). Three months of TMY-type meteorological data from the city of Chambéry (located near Le Bourget-du-Lac) were used for this analysis. Figure 18.7 presents the indoor temperatures simulated by the models reduced to orders 3 and 4 and the reference model, on the day with the highest root-mean-square error between the temperatures calculated by the model reduced to order 3 and by the reference model. It can be seen that the simulated temperatures are very close. In addition, the root-mean-square errors of the models reduced to orders 3 and 4 are low over the three months of data when compared to the reference model: 0.0368°C and 0.0082°C, respectively. Consequently, being very close to the results of the reduced models, the model's order of reduction was fixed at 3, allowing a good compromise between the precision of the results and the computation times.

18.2.2 Optimal Control Problem

In the demand response context, optimised energy management in single-family houses consists of solving a state and input constrained optimal control problem. This section presents a methodology for solving this type of problem.

We are interested in the following optimal control problem:

$$\min_{u \in U \cap X} \left[J(u) = \int_{0}^{t_f} \mathcal{L}\big(x(t), u(t)\big)\, dt \right] \tag{18.3}$$

where L is a smooth real-valued function of its arguments, under the following dynamical constraints:

$$\dot{x}^u = f\left(x^u(t), u(t)\right); \ x(0) = x_0 \tag{18.4}$$

corresponding to the balanced reduced state-space representation of the time-invariant linear system (18.2), where $x(t) \in \mathbb{R}^n$ and $u(t) \in \mathbb{R}^m$, respectively, the state of the building at time t and the heating power to be injected into the building at time t (in this study $n = 3$ and $m = 10$), and x_0 is the initial state of the building (which can be estimated by a state observer). We can thus act on state x through the control variable u on the time horizon $[0;tf]$. The set $U \cap X$ represents all the allowable commands that ensure that the control and state constraints are respected. The optimal control problem then consists of finding the command u and the state x^u associated with the differential equation solution (18.4) and minimising the integral criterion (18.3) (cost function or objective function), while respecting the state and command constraints. An interior penalty method can be used to solve this problem.

18.2.2.1 Interior Penalty Principle

The principle of penalty methods is to solve a modified optimal control problem whereby a term, called the penalty function, is added to the original cost function. This penalty function presents a divergent behaviour when the constraints are approached by a solution. Penalty methods attempt to approximate a constrained optimal control problem with a series of unconstrained optimal control problems and then apply standard techniques to obtain solutions. Constraint satisfaction is favoured by modifying the cost function and depends on the weight of the penalty functions.

The algorithm used in this study is based on an interior penalty method. This method relates to the definition of the penalty functions γ_g and γ_u as well as a generalised saturation function $\phi(v) = u$ (corresponding to a change of command variable u making it possible to remove the control constraint from the formulation of the problem) to formulate the following problem:

$$\min\nolimits_{v \in L^\infty\left([0,tf], \mathbb{R}^m\right)} \int_0^{tf} \mathcal{L}\left(x^{\phi(v)}, \phi(v)\right) dt + \varepsilon \int_0^{tf} \left(\gamma_g(x^{\phi(v)}) + \gamma_u(v)\right) dt \tag{18.5}$$

where γ_g and γ_u are the penalty functions relating to the state constraint and the control constraint, respectively. Malisani et al. (2014) showed that when $\varepsilon \to 0$, this problem generated a series of solutions converging to a solution of the constrained optimal control problem (18.3). Each solution of the series is easily characterised by the calculus of variations as the problem (18.5) is unconstrained. Specifically, the Pontryagin's minimum principle (Pontryagin et al., 1962) was used. The description of the algorithm, the convergence results and the definition of the penalty functions can be found in the article by Malisani et al. (2014). To solve the optimal control problem (18.5), we define the Hamiltonian of the penalised problem (where $p \in \mathbb{R}^n$ is the adjoint vector)

$$H_\varepsilon\left(x^{\phi(v)}, v, p\right) = \mathcal{L}\left(x^{\phi(v)}, \phi, (v)\right) + \varepsilon\left[\gamma_g\left(x^{\phi(v)}\right) + \gamma_u(v)\right] + p^t f\left(x^{\phi(v)}, \phi(v)\right) \tag{18.6}$$

18.2.2.2 Application

18.2.2.2.1 FORMULATION OF THE CONSTRAINED OPTIMAL CONTROL PROBLEM

This study proposes the consideration of the possibility of load shifting the building's electrical heating consumption building during a so-called "high peak" period ranging from 5 PM to 9 PM. The objective of the optimisation is to determine the optimal strategy that minimises the cost of energy consumption while respecting the comfort temperature constraints (state constraints) and maximum power constraints of the heating system (control constraints). Note that no constraints are added to the heating control to achieve the load shifting of high peak consumption. Load shifting is achieved exclusively by a variable cost of electricity over time.

The classical linear state-space representation was used for the model reduced to order 3, where the load vector is divided between the heating power (control variable) and vector d representing the influence of climate, occupants and appliances:

$$\dot{x}(t) = Ax(t) + B_P P(t) + B_d d(t)$$
$$T(t) = Cx(t)$$
(18.7)

where
   x is the state of the model;
   T is the indoor temperature;
   B_P and B_d are the matrices relating to $P(t)$ and $d(t)$.

To ensure an acceptable level of comfort, the indoor temperature must be between 19°C (T_{min}) and 24°C (T_{max}). The heating power can vary from 0 W (P_{min}) to 5,000 W (P_{max}). The optimisation problem to be solved is therefore as follows:

$$\min_P \int_0^{tf} C_{elec}(t) P(t) dt$$
(18.8)

where $C_{elec}(t)$ is the cost of electricity and tf the optimisation horizon. The state and control constraints are as follows:

$$T_{min} \leq T(t) \leq T_{max}$$
(18.9)

$$P_{min} \leq P(t) \leq P_{max}$$
(18.10)

In order to shift power consumption during high peak hours, the price of the kilowatt-hour of electricity considered here is double that of peak time. Since two different tariffs are already available in France according to the time of day (off-peak and peak hours), the tariff grid used in this study therefore considers three different electricity tariffs (Table 18.2).

Table 18.2 Off-peak, peak and high peak rates

	Off-peak Hours	Peak Hours	High Peak Hours
Schedule	12 PM–9 AM	9 AM–5 PM 10 PM–12 PM	5 PM–10 PM
Cost per kWh in €	0.0864	0.1275	0.255

18.2.2.2.2 NUMERICAL RESOLUTION

To solve the constrained optimal control problem, the interior-point algorithm described in the article by Malisani et al. (2014) is used. The following variable change is used for heating power (Malisani, 2012):

$$
P = \phi(v) = P_{\max}\left(\frac{e^{kv}}{1+e^{kv}}\right), k > 0 \tag{18.11}
$$

The Hamiltonian of the penalised optimal control problem is then

$$
\begin{aligned}
H_\varepsilon(x,v,p) &= C_{\text{elec}}\phi(v) + p^t\left(Ax + B_p\phi(v) + B_d d\right) \\
&\quad + \varepsilon\left[\gamma_g(Cx - T_{\min}) + \gamma_g(T_{\max} - Cx) + \gamma_u \circ \phi(v)\right]
\end{aligned} \tag{18.12}
$$

The adjoint vector p must satisfy the following differential equation:

$$
\begin{aligned}
\dot{p}(t) &= -\frac{\partial H_\varepsilon}{\partial x}(x,v,p) \\
&= -A^t p(t) - \varepsilon C^t\left[\gamma'_g(Cx(t) - T_{\min}) - \gamma'_g(T_{\max} - Cx(t))\right]
\end{aligned} \tag{18.13}
$$

where γ'_g is the derivative of the function γ'_g defined as follows:

$$
\gamma_g(s) = \left\{ s_0^{-1.1} \sinon^{\forall s > 0} \right. \tag{18.14}
$$

The algorithm used to solve the constrained optimal control problem is as follows:

Step 1: Initialise the continuous functions $x(t)$ and $p(t)$ as the initial value $C_x(t) \in [T_{\min}, T_{\max}]$ for all $t \in [0, t_f]$. On initialisation, $p(t)$ can be chosen identically equal to 0. Set $\varepsilon = \varepsilon_0$.

Step 2: Calculate $v_\varepsilon^* = \sinh^{-1}\left(-\dfrac{C_{\text{elec}}(t) + p^t(t)B_P}{\varepsilon}\right)$ the analytical solution of $\dfrac{\partial H_\varepsilon}{\partial v} = 0$[1]

where we set $\gamma'_u \circ \phi(v) = \sinh(v)$.

The optimal solution $P_\varepsilon^*(t) = \phi\left(v_\varepsilon^*(t)\right)$ is then given using Equation 18.11.

Step 3: Solve the $2n$ differential equations (two-point boundary value problem):

$$\begin{cases} \dot{x}(t) = Ax(t) + B_P P_\varepsilon^*(t) + B_d d(t) \\ P_\varepsilon^*(t) = \phi\left(\sinh^{-1}\left(-\dfrac{C_{elec}(t) + p^t(t) B_P}{\varepsilon} \right) \right) \\ \dot{p}(t) = -A^t p(t) - \varepsilon C^t \left[\gamma_g'(Cx(t) - T_{\min}) - \gamma_g'(T_{\max} - Cx(t)) \right] \end{cases}$$

With the following boundary constraints: $x(0) = x_0$ and $p(t_f) = 0$.

Step 4: Decrease ε, initialise $x(t)$ and $p(t)$ with the solutions found in step 3 and return to step 2. In this study, the sequence (ε_n) was chosen such that

$$\varepsilon = 10^{-\frac{n}{10}} \text{ with } n = -90...+70.$$

The algorithm presented was implemented in MATLAB R2012b and runs on an Intel Core i7 (2.8 GHz) with 16 GB of RAM, using a single core.

18.2.2.2.3 SAMPLING PERIOD AND OPTIMISATION HORIZON

Implementing a predictive control requires setting the period at which the optimal control is recalculated (sampling period). In the case of application to energy management of buildings, this sampling period may depend on the updating of weather forecasts (forecasts can be updated from one to several times, a day depending on the suppliers), the estimate of the state of the building or a change in the load scenario. In order to develop an algorithm that is as general as possible, the sampling period was set at 24 hours. It is also necessary to set the optimal control optimisation horizon. This choice corresponds to a compromise between the computation time and the precision of the calculated control. A sensitivity analysis conducted by Robillart et al. (2018) showed that a four-day optimisation horizon represented a good compromise for this case study.

18.2.2.3 MPC Design and Implementation

MPC design and implementation raises several difficulties, such as estimating the state of the building or setting a low-level controller tracking the trajectory to compensate for disturbances (such as meteorological or occupancy forecast errors) or building modelling errors.[1]

18.2.2.3.1 STATE OBSERVER

The use of the optimisation algorithm presented in Section 18.2.2.2.2 requires knowledge of the initial state conditions ($x(0) = x_0$), that is, the state of the building. However,

---

1 $\dfrac{\partial H_\varepsilon}{\partial v} = 0$ gives $C_{elec}(t) + p^t(t) B_P + \varepsilon \gamma_u' \circ \phi(v) = 0$

this is not directly measurable: the simple measurement of the interior temperature of the building is not sufficient to characterise it. It is essential to be able to estimate this state based on an observer. This latter aims to reconstruct the state of the building based on a dynamic model. An asymptotic observer (Luenberger observer), based on system (18.7), is used in this study. The main objective is to reconstruct an estimate t x(t) of the state x (t) solely based on the outputs and the inputs $P(t)$ and $d(t)$.

The dynamics of the estimation error of such an observer are then:

$$\frac{de_x(t)}{dt} = (A - LC)e_x(t) \tag{18.15}$$

where $e_x(t) = x(t) - \hat{x}(t)$ and L is the gain of the observer.

The objective is to ensure that the estimation error converges asymptotically to 0, that is to say, to use a judicious choice of L to make all the eigenvalues of the matrix $(A-LC)$ in the left half-plane. The construction of such an asymptotic observer is possible if and only if (A, C) is observable (which is the case in our study, where the model reduced to order 3 is completely observable and controllable). In order to ensure a faster estimation of error dynamics than the open-loop system, the eigenvalues of $A-LC$ must be chosen wisely. However, in practice, we are limited to the magnitude of this dynamic and it is not possible to include very large dynamics. Indeed, on the one hand, we can only choose achievable gains, and on the other hand, the higher the gain, the more the noise will affect the reconstruction of the state. MATLAB R2012b was used to place the poles and choose the gain of the observer. In this study, the error convergence between the estimated state and the actual state considered was of approximately four days. Thus, when estimating the state of the building, it is sufficient to have the measurements of the loads (weather conditions, occupation and heating power) and the temperature of the building over the last four days to ensure the convergence of the asymptotic observer.

18.2.2.3.2 TRAJECTORY TRACKING

The resolution of the optimal control problem provides the optimal trajectory to follow, denoted as $(T_{\text{ref}}(t), P_{\text{ref}}(t))$. In practice, when applying the optimal control to the actual building, deviations can be observed between the supposed change in temperature inside the building (reference temperature trajectory) and the actual temperature measured. These deviations can be a consequence of modelling errors (the model used is a reduced model of order 3) or of errors in the prediction loads (occupation, weather, etc.). One solution is to implement feedback that ensures the asymptotic tracking of the reference trajectory. More precisely, it is a question of correcting the heating control computed in open loop as a function of the error between the reference temperature and the measured temperature (ΔT), to compensate for the deviations between the real trajectory and the reference trajectory. This trajectory tracking can be ensured by setting up a controller.

The PI (Proportional/Integral) controller is typically used for building regulation and can be used to track the trajectory. The principle is then to add a term of anticipation, a term corresponding to the reference heating power $P_{\text{ref}}(t)$. In this context, the output power of the controller can be defined by the following:

$$\begin{cases} \Delta T(t) = T_{\text{ref}}(t) - T_{\text{mes}}(t) \\[2mm] P_{\text{reg}}(t) = \text{Sat}\left(P_{\text{ref}}(t) + K\left[\Delta T(t) + \frac{1}{T_i}\int_0^t \Delta T(t)\, dt \right] \right) \end{cases} \tag{18.16}$$

where Sat is the saturation function ensuring respect of control constraints, K is the proportional gain and T_i is the integral time. The interest of this formulation is two-fold. On the one hand, the addition of the term anticipation makes it possible to manage the variations of the reference trajectory better (e.g. during overheating of the building in off-peak hours). On the other hand, the feedback part makes it possible to compensate for the errors of the building model or the errors in the load forecasts.

18.2.2.3.3 MPC DEVELOPMENT

The model predictive controller has been implemented and tested on a case study, the objective of which is to optimise and regulate the heating power of a building in real time. In the absence of access to a real building for studying the behaviour of the predictive control, the latter will be studied through digital simulation. The building model considered corresponds to the complete model (dimension 28) modelled by COMFIE, taking non-linear phenomena into account (ventilation in particular). The MPC process developed in this study corresponded to a four-step process (Figure 18.8).

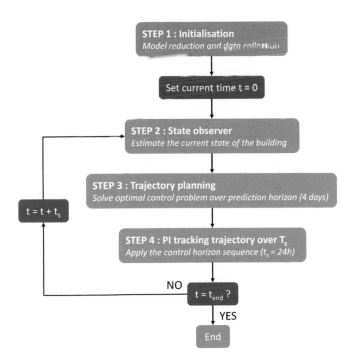

Figure 18.8 **MPC flow chart**

The first step is the initialisation. During this step, a constant heating set point (e.g. 19°C) can be applied to the building. The goal of this step is twofold. The first is to model the building to be controlled by the Pléiades+COMFIE dynamic thermal simulation software. The building model thus obtained is then reduced to order 3 (by the balanced truncation method). The second objective is to collect the data needed to estimate the state of the building (step 2). As the convergence of the asymptotic observer is approximately four days, it is necessary to have at least four days of data concerning the indoor temperature of the building, the heating power injected and the loads (weather conditions, occupancy). Following the initialisation, it is possible to estimate the initial state of the building in step 2. Thanks to the measurements collected in step 1 and the development of a reduced building model of order 3, the state of the building can be estimated from an asymptotic observer based on the dynamics of the reduced model. Being known to the state of the building, it is then possible to solve the optimal control problem (step 3) and to calculate the heating control, thanks to the algorithm presented in Section 18.2.2.2.2 for the next four days (prediction horizon). This step then makes it possible to define the heating control to be applied to the building (reference power profile) as well as the assumed evolution of the temperature inside the building (reference temperature trajectory). In step 4, the first 24 hours (control horizon) of the reference power profile (profile calculated for the next 4 days) are applied to the building. The PI regulator (defined in Section 18.2.2.3.2) is used to track the reference trajectory. Indeed, disturbances (such as meteorological or occupancy forecast errors) or building modelling errors can cause differences between the reference trajectory and the actual trajectory. During this stage, the actual temperature inside the building and the loads are measured. Finally, at the end of this 24-hour period, the current state of the building is estimated, thanks to the measurements collected: indoor temperature, heating power calculated by the PI regulator and the loads (weather conditions, occupancy). This estimated state then makes it possible to close the loop and return to step 3 to calculate the reference control for the next day.

18.2.3 Results

We propose to apply the MPC to our case study, namely the load shifting of high peak power consumption of heating. The MPC is thus tested to monitor the heating system during a cold week (measurements from 2 February to 8 February 2012, Figure 18.9), when electricity consumption is highest.

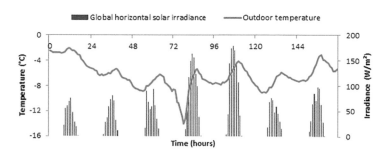

Figure 18.9 Weather conditions for the period studied

The reliability of the MPC is evaluated by a robustness analysis. The objective here is to determine the impact of forecast errors (relating to occupancy and weather) on the calculation of the heating power control and thus to evaluate the behaviour of the MPC in more realistic conditions. In this study, no measurement errors were taken into account, which is a limitation that could be considered in a future study. To apply the MPC, a four-day optimisation horizon and a one-day control horizon were considered.

Two case studies were studied that differ in the occupation scenario used during the simulation. These two scenarios are extracted from the statistical average scenario generated from the stochastic occupancy model (Section 18.2.1.1). They model two families (to recap, a household of three people with high-performance appliances and lighting) with low and high electricity consumption, respectively, in relation to appliances and lighting (identified as Family 1 and Family 2, respectively). Family 1 consumes 17% less electricity and Family 2 consumes 16% more compared to the average statistical scenario. Figure 18.10 shows the internal gains for each scenario.

Different indicators were used to evaluate the performance of the MPC:

- the cumulative cost due to heating consumption and electricity appliances (in €),
- the high peak heating consumption that has been load shifted compared to a constant set point temperature of 19°C (in %),
- the peak heating consumption that has been load shifted compared to a constant set point temperature of 19°C (in %).

Four control strategies were tested for each case study. First, a benchmark strategy was considered, corresponding to a constant set point temperature of 19°C with the actual occupancy scenario and actual weather conditions. This strategy makes it possible to evaluate the demand for electricity during a cold week. Then, the theoretical MPC was tested, defined as the optimal control calculated based on a perfect knowledge of the disturbances acting on the building. This control strategy makes it possible to determine the limit performance of the MPC in an ideal case, with a perfect prediction of the meteorological conditions and the occupancy (Family 1 or Family 2 according to the case study). Finally, two control strategies were evaluated,

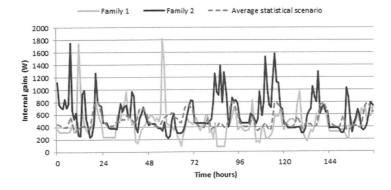

Figure 18.10 Internal gains for each occupancy scenario

each corresponding to an underestimation or an overestimate of the temperature and solar radiation forecasts, with the use of the average statistical occupancy scenario for each strategy. These last two strategies thus make it possible to evaluate the impact of the prediction errors of the occupancy and the meteorological conditions on the MPC. The different case studies are summarised in Table 18.3.

By way of example, the heating control calculated by the different strategies and the evolution of the resulting indoor temperature are presented in Figures 18.11 and 18.12, respectively. It can be seen that the indoor temperature is within the range of temperature constraints [19°C, 24°C]. As expected, the indoor temperature tends to be slightly lower when weather forecasts are overestimated (the strategy then calculates a lower heating power). The opposite is observed when weather forecasts are underestimated.

The quantified results are presented in Tables 18.4 and 18.5. For each case study, the theoretical MPC has a lower cost than the reference strategy (13% or 8% lower depending on the case study). In addition, heating is mainly used during off-peak hours.

For example, 100% of the electricity consumed during high peak hours (by the reference strategy) is load shifted, thanks to the theoretical MPC. This is also the case during peak hours, when almost 100% of the electricity consumed (by the reference strategy) is load shifted for each case study. With a constant set point temperature of 19°C (reference strategy), the electricity consumed to heat the building during peak hours represents 55% of the total energy consumed in Case Study 1 (33% in Case Study 2). Similarly, the electricity consumed for heating during high peak hours corresponds to 67% of the energy consumed in Case Study 1 (52% in Case Study 2). As a result, thanks

Table 18.3 Occupancy and weather forecasts for each strategy

Case Study 1: Family 1 Occupancy Scenario

	MPC (Theoretical)	*MPC (Underestimation)*	*MPC (Overestimation)*
Outside temperature forecast	True value	True value − 1°C	True value + 1°C
Global horizontal radiation forecast	True value	True value − 20%	True value + 20%
Occupancy forecast	Family 1 (low electricity consumption)	Average statistical scenario	Average statistical scenario

Case Study 2: Family 2 Occupancy Scenario

	MPC (Theoretical)	*MPC (Underestimation)*	*MPC (Overestimation)*
Outside temperature forecast	True value	True value − 1°C	True value + 1°C
Global horizontal radiation forecast	True value	True value − 20%	True value + 20%
Occupancy forecast	Family 2 (high electricity consumption)	Average statistical scenario	Average statistical scenario

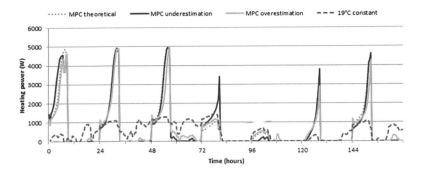

Figure 18.11 Heating power calculated for Case Study 2

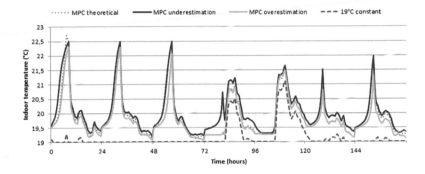

Figure 18.12 Evolution of indoor temperature for Case Study 2

Table 18.4 Results of Case Study 1 (Family 1 Occupancy Scenario)

	19°C Constant	MPC (Theoretical)	MPC (Underestimation)	MPC (Overestimation)
Cumulative cost (€)	21.9	19	19.5	20.2
Electricity consumed for heating during off-peak hours (kWh)	56.2	135.3	136.2	109.9
Electricity consumed for heating during peak hours (kWh)	29.3	0.4	0.8	13.1
Electricity consumed for heating during high peak hours (kWh)	23.6	0	1.4	6.8
High peak hour load shifted for heating (%)	-	99	97	55
Peak hour load shifted for heating (%)	-	100	94	71
Tmin/Tmax (°C)	19/20.5	19.1/23.3	19/22.6	19/22.4

Table 18.5 Results of Case Study 2 (Family 2 Occupancy Scenario)

	19°C Constant	MPC (Theoretical)	MPC (Underestimation)	MPC (Overestimation)
Cumulative cost (€)	22.8	21.1	21.5	21.2
Electricity consumed for heating during off-peak hours (kWh)	47.3	108.4	112.2	98.2
Electricity consumed for heating during peak hours (kWh)	19	0.3	0.7	3.5
Electricity consumed for heating during high peak hours (kWh)	18	0	0.3	2.6
Withdrawn high peak hours for heating (%)	-	98	96	82
Withdrawn peak hours for heating(%)	-	100	98	86
Tmin/Tmax (°C)	19/21.1	19.1/22.8	19.1/22.6	19.1/22.4

to the building's thermal inertia, energy can be stored in the building, which stops the heating system during peak and high peak times. This storage of energy is accompanied by a 24% increase in energy consumption in Case Study 1 (29% in Case Study 2). However, the cumulative cost is lower due to the variation in electricity rates.

When forecast errors are taken into account, an increase in cumulative cost can be observed in comparison with the theoretical MPC. For Case Study 1, the cumulative cost increases by 2% when weather forecasts are underestimated and by 5% when they are overestimated. However, these costs remain below the cost of the reference strategy (a reduction of 11% and 8%, respectively).

There may also be a slight decrease in the electricity consumption that has been load shifted (during peak and high peak hours) when weather forecasts are underestimated (for both case studies).

The observations are different when weather forecasts are overestimated. For example, the load-shifted electricity consumption for Case Study 2 is reduced by about 15% during peak and high peak hours compared to the theoretical MPC. As a consequence of the overestimation of weather forecasts, the predictive control tends to underestimate the heating power required. Therefore, when the strategy is applied under real-world conditions, it is necessary to heat during peak and high peak to follow the reference trajectory (due to internal gain prediction errors), resulting in an increase in cumulative cost. For Case Study 1, the same problem is reinforced because of the occupancy scenario (Family 1), which has lower internal gains. Thus, the interactions between the occupation scenarios and the weather forecasts can be observed.

In conclusion, MPC is an effective and precise method for solving the problems of load-shifting high peak power consumption in real time. Thus, considering a well-insulated building with high thermal inertia, it is possible to shift the electricity consumed by the heating during peak and high peak by storing energy in the thermal mass of the building during off-peak hours. The results of the study showed that this load shifting made it possible to reduce the heating costs by 6 to 13% compared to a strategy corresponding to a constant set point temperature of 19°C. Finally, a sensitivity analysis demonstrated the MPC's robustness to weather forecast and occupancy errors.

18.3 Extending the Methodology to a Dual-Zone Case

Extending the methodology to a dual-zone case constitutes a first step towards the development of energy management strategies at the multi-zone scale (blocks of buildings) in real time. As with the single-zone case, the application of the predictive command at a dual-zone scale involves repeatedly searching for the optimal dual-zone control. There are several possible approaches for solving the problem of optimal dual-zone control:

- The centralised approach consists of solving a single-zone optimal control problem characterised by a dynamic system and a set of state and control constraints. This single-zone optimal control problem is called the global or centralised optimal control problem. The centralised approach is not recommended at the scale of large systems because of its lack of flexibility and significant computing times. This approach is used in this part, however, as it leads to the global (or centralised) optimum, which constitutes the reference solution.
- The decentralised approach consists of dividing the global optimal control problem into two optimal control subproblems and then solving them in parallel. When the subproblems are coupled, the decentralised solution is suboptimal.
- The decomposed-coordinated approach consists of dividing the optimal control problem into two optimal control subproblems and then solving them in parallel (decomposition step). Couplings between the subproblems are disregarded during the decomposition step and then reintegrated by iterations with a coordination step.

In the presence of couplings, the decomposed-coordinated approach is preferred. To manage the thermal applications (heating or air conditioning) of a dual-zone building, optimal control subproblems can be coupled with:

- the existence of interconnection variables between the dynamic systems that can correspond to the case of two contiguous thermal zones separated by an intermediate wall where heat transfers take place. This is called thermal coupling.
- a global constraint on the control that can correspond to a limited total available power that must be shared between the two zones. This is called resource sharing.

Only the first type of coupling (thermal coupling) is studied in the rest of this chapter. This coupling has already been addressed in combinatorial optimisation (Morosan et al., 2011). The decomposition-coordination methods detailed by Carpentier et al. (2017) in the deterministic case were explored to extend the continuous algorithm developed at the single-zone scale. Of these methods, the method of decomposition-coordination by quantities was excluded because it is not recommended for continuous optimisation. The decomposition-coordination methods by price and by predictions were tested on a case study and compared to the results of the centralised approach.

Only the method of decomposition-coordination by predictions is presented in this part. The latter, inspired by the work of Mesarovic et al. (1970), was chosen because it requires less computation time and its implementation is simpler than the method of decomposition-coordination by prices.

This last part of the chapter is divided into four. The problem of dual-zone optimal control is first formulated in Section 18.3.1. The centralised algorithm and the decomposition-coordination algorithm by predictions are then presented in Sections 18.3.2 and 18.3.3, respectively. Finally, the two algorithms are compared in a case study in Section 18.3.4.

18.3.1 Formulation of the Dual-Zone Optimal Control Problem under Constraints

As with the single-zone case, the goal of the dual optimisation is to determine the optimal strategy that minimises the cost of energy consumption while respecting the state and control constraints. In the dual-zone formulation, the dynamic constraint takes the following form:

$$\dot{x}(t) = Ax(t) + \sum_{i=1}^{2} B_{P_i} P_i(t) + B_d d(t) \tag{18.17}$$

$$T(t) = Cx(t)$$

Added to the notations defined in Section 18.2.2.2.1 is P_i, the heating power associated with zone i, as well as B_{P_i} the matrix relating to P_i. In the dual-zone case, vector T contains the internal temperatures of each of the zones i denoted T_i with:

$$T_i(t) = C_i x(t) \tag{18.18}.$$

The dual-zone optimisation problem to be solved is therefore as follows:

$$\min_{P_1, P_2} \int_0^{t_f} \sum_{i=1}^{2} C_{elec}(t) P_i(t) dt \tag{18.19}$$

with a cost of electricity that is unchanged from the single-zone case. The state and control constraints are as follows:

$$T_{\min_i} \leq T_i(t) \leq T_{\max_i} \tag{18.20}$$

$$P_{\min_i} \leq P_i(t) \leq P_{\max_i} \tag{18.21}$$

18.3.2 Centralised Algorithm

The change of the variable used by Malisani (2012) and explained in Equation 18.11 was applied to each heating power P_i. The Hamiltonian of the penalised centralised optimal control problem is described as follows:

$$H_\varepsilon(x,\upsilon_1,\upsilon_2,p) = \sum_{i=1}^{2} C_{elec}\phi(\upsilon_i) + p^t\left(Ax + \sum_{i=1}^{2} B_{P_i}\phi(\upsilon_i) + B_d d\right)$$

$$+ \varepsilon\sum_{i=1}^{2}\left[\gamma_g\left(C_i x - T_{min_i}\right) + \gamma_g\left(T_{max_i} - C_i x\right) + \gamma_u \circ \phi(\upsilon_i)\right] \quad (18.22)$$

The adjoint vector must satisfy the following differential equation:

$$\dot{p}(t) = -A^t p(t) - \varepsilon C_i^t \sum_{i=1}^{2}\left[\gamma_g'\left(C_i x(t) - T_{min_i}\right) - \gamma_g'\left(T_{max_i} - C_i x(t)\right)\right] \quad (18.23)$$

where γ_g' is the function defined in Equation 18.14.

The algorithm used to solve the centralised optimal control problem is as follows:

Step 1: Initialise the continuous functions $x(t)$ and $p(t)$ as the initial values $C_i x(t) \in \left[T_{min_i}, T_{max_i}\right]$ for all $t \in \left[0, t_f\right]$ On initialisation, $p(t)$ can be chosen equal to 0. Set $\varepsilon = \varepsilon_0$

Step 2: Calculate $\upsilon_{i_\varepsilon}^* = \sinh^{-1}\left(-\dfrac{C_{elec}(t) + p^t(t) B_{P_i}}{\varepsilon}\right)$ the analytical solution of $\dfrac{\partial H_\varepsilon}{\partial \upsilon_i} = 0^2$

where we chose $\gamma_u' \circ \phi(\upsilon_i) = \sinh(\upsilon_i)$

The optimal solution $P_{i_\varepsilon}^*(t) = \phi\left(\upsilon_{i_\varepsilon}^*(t)\right)$ is then given using Equation 18.11.

Step 3: Solve the $2n$ differential equations (two-point boundary value problem):

$$\dot{x}(t) = Ax(t) + \sum_{i=1}^{2} B_{P_i} P_{i_\varepsilon}^*(t) + B_d d(t)$$

$$P_{i_\varepsilon}^*(t) = \phi\left(\sinh^{-1}\left(-\frac{C_{elec}(t) + p^t(t) B_{P_i}}{\varepsilon}\right)\right)$$

$$\dot{p}(t) = -A^t p(t) - \varepsilon C_i^t \sum_{i=1}^{2}\left[\gamma_g'\left(C_i x(t) T_{min_i}\right) - \gamma' g\left(T_{max_i} - C_i x(t)\right)\right]$$

With the following boundary conditions: $x(0) = x_0$ et $p(t_f) = 0$.

Step 4: Decrease ε, initialise $x(t)$ and $p(t)$ with the solutions found in step 3 and return to step 2. In this study, the sequence (ε_n) was chosen such that

$\varepsilon_n = 10^{-\frac{n}{10}}$ with $n = +40 ... + 70$.

The algorithm presented was implemented in MATLAB R2017b and runs on an Intel Xeon e3 PC (2.8 GHz) with 32 GB of RAM, using 8 cores.

18.3.3 Decomposition-Coordination Algorithm by Predictions

Starting from the dual-zone optimal control problem formulated in Section 18.3.1, an attempt is made to obtain two optimal control subproblems associated with each of the thermal zones. The dynamic constraint for each zone can be written in the following way:

$$
\dot{x}_i(t) = A_i x_i(t) + B_{P_i} P_i(t) + B_{d_i} d_i(t) + B_{\text{adj}i}\left(\alpha_{\bar{i}} \dot{Q}_{\text{sol}\bar{i}}(t) + T_{\bar{i}}(t)\right)
$$
$$
T_i(t) = C_i x_i(t) \tag{18.24}
$$

where $\alpha_{\bar{i}} \dot{Q}_{\text{sol}\bar{i}}$ corresponds to the solar flux reaching the surface of the wall separating the two zones on the side of the adjacent zone $\bar{i}$. Subsequently, to simplify the notations, we consider that this term, which is not acted on, to be integrated into the term corresponding to the external loads $B_{d_i} d_i(t)$. The temperature of the adjacent zone $T_{\bar{i}}$ is in turn expressed according to the state of the adjacent zone $x_{\bar{i}}^2$:

$$
x_i(t) = A_i x_i(t) + B_{p_i} P_i(t) + B_{d_i} d_i(t) + B_{\text{adj}i} C_{ri} x_{ri}(t)
$$
$$
T_i(t) = C_i x_i(t) \tag{18.25}
$$

The Hamiltonian then takes the following form:

$$
H_\varepsilon(x,v_1,v_2,p) = \sum_{i=1}^{2}\left\{ C_{\text{elec}}\phi(vi) + p_i^t\left(A_i x_i + B_{P_i}\phi(vi) + B_{d_i} d_i + B_{\text{adj}i} C_{\bar{i}} x_{\bar{i}}\right)\right.
$$
$$
\left. + \varepsilon\left[\gamma_g\left(C_i x - T_{\min_i}\right) + \gamma_g\left(T_{\max_i} - C_i x\right) + \gamma_u \circ \phi(v_i)\right]\right\} \tag{18.26}
$$

The adjoint vector associated with zone i must satisfy the following differential equation:

$$
\dot{p}_i(t) = -A_i^t p_i(t) - \left(B_{\text{adj}\bar{i}} C_i\right)^T p_{\bar{i}}(t)
$$
$$
- \varepsilon C_i^t \sum_{i=1}^{2}\left[\gamma_g'\left(C_i x(t) - T_{\min_i}\right) - \gamma_g'\left(T_{\max_i} - C_i x(t)\right)\right] \tag{18.27}
$$

The dynamics of the state $x_{\bar{i}}$ associated with zone i and defined in equation 25 depend on the state $x_{\bar{i}}$ of the neighbouring zone. The dynamics of the adjoint state $\dot{p}_i$ associated with zone i and defined in Equation 18.27 depends on the adjoint state p_i of the neighbouring zone. To break the optimal control problem down into two subproblems, it is necessary to decouple the dynamics of zone i from the state and the adjoint state of the zone $\bar{i}$. For this, it is assumed that the values of the state and the adjoint state of the zone $\bar{i}$ used in the decomposition phase of zone i at iteration (k) of the decomposition-coordination algorithm are predicted by the coordinator at the iteration $(k-1)$.

2 $\dfrac{\partial H_\varepsilon}{\partial v_i} = 0$ g Implies $C_{\text{elec}}(t) + p^t(t) B_{P_i} + \varepsilon\gamma_u' \circ \phi(v_i) = 0$

The algorithm used to solve the decomposed-coordinated optimal control problem is as follows:

Level 1: Initialise predictions $x_{\bar{i}}^{(0)}$ and $p_{\bar{i}}^{(0)}$ such as initial values $C_{\bar{i}}x_{\bar{i}}(t) \in \left[T_{\min_i}, T_{\max_i}\right]$ for all $t \in \left[0, t_f\right]$. On initialisation, $p_{\bar{i}}(t)$ can be chosen equal to 0. Set $k = 1$.

Level 2 (decomposition step): for each zone i, steps 1–4 below are carried out in parallel with the MATLAB routine parfor:

Step 1: Initialise the continuous functions $x_i(t)$ and $p_i(t)$ as the initial values $C_{\bar{i}}x_{\bar{i}}(t) \in \left[T_{\min_i}, T_{\max_i}\right]$ for all $t \in \left[0, t_f\right]$. On initialisation, $p_i(t)$ can be chosen equal to 0. Set $\varepsilon = \varepsilon_0$.

Step 2: Calculate $v_{i_\varepsilon}^* = \sinh^{-1}\left(-\dfrac{C_{\text{elec}}(t) + p_i^t(t)B_{P_i}}{\varepsilon}\right)$ the analytical solution of $\dfrac{\partial H_\varepsilon}{\partial v_i} = 0$ where we chose $\gamma_u' \circ \phi(v_i) = \sinh(v_i)$.

The optimal solution $P_{i_\varepsilon}^*(t) = \phi\left(v_{i_\varepsilon}^*(t)\right)$ is then given using Equation 11.

Step 3: Solve the 2n differential equations (two-point problem):

$$\dot{x}_i(t) = A_i x_i(t) + B_{P_i} P_{i_\varepsilon}^*(t) + B_{d_i} d_i(t) + B_{\text{adj}i} C_{\bar{i}} x_{\bar{i}}^{(k-1)}(t)$$

$$P_{i_\varepsilon}^*(t) = \phi\left(\sinh^{-1}\left(-\frac{C_{\text{elec}}(t) + p_i^t(t)B_{P_i}}{\varepsilon}\right)\right)$$

$$\dot{p}_i(t) = -A_i^t p_i(t) - \left(B_{\text{adj}\bar{i}} C_i\right)^T p_{\bar{i}}^{(k-1)}(t)$$
$$- \varepsilon C_i^t \left[\gamma_g'\left(C_i x(t) - T_{\min_i}\right) - \gamma_g'\left(T_{\max_i} - C_i x(t)\right)\right]$$

With the following as boundary conditions: $x_i(0) = x_0$ et $p_i(t_f) = 0$.

Step 4: Decrease ε, initialise $x_i(t)$ and $p_i(t)$ with the solutions found in step 3 and return to step 2. In this study, the sequence (ε_n) was chosen such that

$$\varepsilon_n = 10^{-\frac{n}{10}} \text{ with } n = +40\ldots +70$$

Level 3 (coordination stage): Updating the predictions. For each zone i, the values of the state and the adjoint state that will be communicated to the zone i are as follows:

$$x_i^{(k)} = (1-\eta)_{x_i}^{(k-1)} + \eta x_i^* \tag{18.28}$$

$$p_i^{(k)} = (1-\tau)_{p_i}^{(k-1)} + \tau p_i^* \tag{18.29}$$

where η and τ, the steps of coordination, and x^* and p^* the optimal state and the optimal adjoint state obtained at Level 3 during the decomposition phase associated with zone i. The algorithm is taken back to Level 2, as long as the following assertions are not verified:

$$\int_{t_0}^{t_f} \sqrt{\left(x_i^{(k-1)}(t) - x_i^*(t)\right)^2}\, dt < \text{tol}_x \qquad (18.30)$$

$$\int_{t_0}^{t_f} \sqrt{\left(p_i^{(k-1)}(t) - p_i^*(t)\right)^2}\, dt < \text{tol}_p \qquad (18.31)$$

18.3.4 Comparison of Algorithms on a Case Study

18.3.4.1 Case Study Presentation

The case study is a six-level building. Three levels are dedicated to the use of "office zone" offices and three levels are dedicated to a residential use "housing zone" (see Figure 18.13). The total area of each zone is 2,400 m². The south facade is glazed at 45%, the north facade at 15% and the east and west facades at 20%. The structure of the building is concrete and is insulated from the outside.

The zone models used for the decomposed-coordinated algorithm are shown in Figure 18.14. The thermal zones are superimposed at the level of the intermediate wall (Blanc-Sommereux, 1991).

The centralised model and the broken-down models have been reduced by balanced reduction. To compare the centralised and decomposed-coordinated algorithms, the models reduced to the maximum orders of controllability and observability were retained, namely order 8 for the centralised model and order 4 for the decomposed models.

The weather conditions used are those of a cold week in zone H1a. The internal inputs used are derived from the weekly scenarios of the TH-BCE method for office and residential use. The test was carried out over a 3-day horizon.

With regard to the state constraints, during the occupancy periods, it is considered that the minimum temperature to be respected is 23°C in the office zone and 21°C in the housing zone, and that the maximum temperature to be respected is 28°C in the office zone and 26°C in the housing zone. A reduction of the minimum temperature

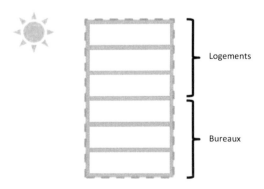

Figure 18.13 Pedagogical example for centralised model

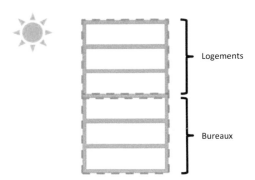

Figure 18.14 Pedagogical example for broken-down models

to 16°C is considered in both zones during the vacancy period, knowing that, over the three optimisation days considered:

- The office area is occupied from 9 AM to 7 PM;
- The housing area is occupied from 5 PM to 10 AM.

With regard to the control constraints, the heating power of each of the zones can vary between 0 kW (P_{min_i}) and 300 kW (P_{max_i}).

18.3.4.2 Results of the Comparison

Figure 18.15 (respectively Figure 18.16) compares the optimal power profiles of zone 1 (respectively zone 2) obtained with the centralised and decomposed-coordinated algorithms. Figure 18.17 (respectively Figure 18.18) compares the temperature profiles of zone 1 (respectively zone 2) resulting from the application of the optimal power

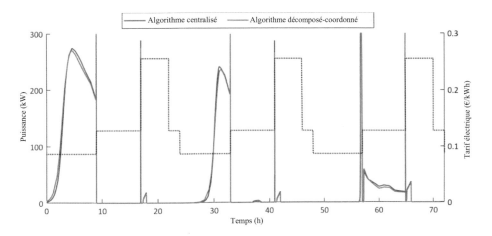

Figure 18.15 Optimal power profiles in zone 1 obtained with centralised and decomposed-coordinated algorithms

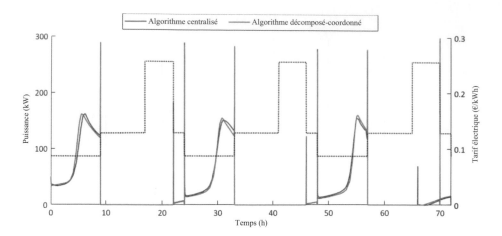

Figure 18.16 Optimal power profiles in zone 2 obtained with centralised and decomposed-coordinated algorithms

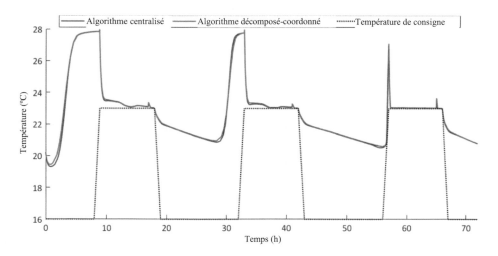

Figure 18.17 Temperature profiles in zone 1 resulting from the application of the optimal power profiles obtained with the centralised and decomposed-coordinated algorithms

profiles obtained with the centralised and decomposed-coordinated algorithms. Except for the third day of optimisation, where an edge effect is observable, each of the zones is preheated during off-peak hours so as to limit consumption during peak and high peak hours (see Figures 18.15 and 18.16) while respecting the minimum and maximum temperature set points (see Figures 18.17 and 18.18). The results in terms of energy reduction and electric load shedding are shown in Table 18.6. These results are given by the comparison of the centralised and decomposed-coordinated optimums with the so-called "no strategy" case, where simple monitoring of the set point temperatures is carried out. The differences between the centralised and

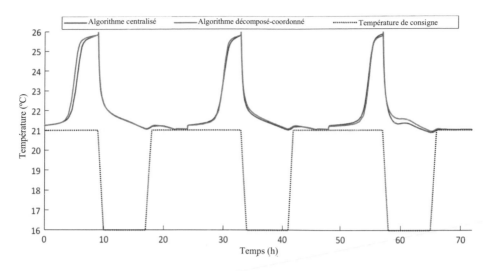

Figure 18.18 Temperature profiles in zone 2 resulting from the application of the optimal power profiles obtained with the centralised and decomposed-coordinated algorithms

Table 18.6 Comparison of the results obtained with the centralised algorithm (C) and the decomposed-coordinated algorithm (DC) in terms of the variation of the cost and withdrawal of peak and high peak consumption, compared to the case with a simple tracking of set points

	Zone 1	Zone 2	Zones 1 and 2
Cost variation (C) (%)	−14	−21	−17
Cost variations (DC) (%)	−15	−20	−17
Peak hour withdrawal (C) (%)	−86	−79	−86
Peak hour withdrawal (DC) (%)	−87	−78	−86
High peak hour withdrawal (C) (%)	−76	−98	−93
High peak withdrawal (DC) (%)	−77	−97	−92

decomposed-coordinated results of each zone do not exceed 1%. Taking the results associated with the two zones into account, the differences between the centralised approach and the decomposed-coordinated approach are more negligible, in particular with an overall reduction in the energy bill of 17% regardless of the algorithm used. With the centralised approach, the calculation time is 23 minutes, whereas for the decomposed approach, it is 21 minutes.

Thus, the decomposition-coordination algorithm by predictions makes it possible to find results close to those obtained with the centralised algorithm in a less significant time. The next step is to extend the reasoning to the multi-zone scale, in the hope of gaining more in terms of computing time through parallelisation. Finally, the case of resource sharing can be studied.

18.4 Conclusion

To conclude, an optimal control research method based on Pontryagin's minimum principle and the interior penalty method was applied to a building energy management problem. Implemented in predictive control, the method first showed good performance at the scale of a single-zone building. Then, the method for seeking the optimal control was extended to the dual-zone scale. The study carried out made it possible to show the relevance of a decomposition-coordination method in terms of optimality and calculation time through a case study of a dual zone combining tertiary and residential uses. This constitutes a first step towards the development of multi-zone energy management strategies.

Bibliography

P. Bacot, 1984. Analyse modale des systèmes thermiques, École Nationale Supérieure des Mines de Paris, PhD diss.

I. Blanc-Sommereux, 1991. Étude du couplage dynamique de composants du bâtiment par synthèse modale, École des Mines de Paris, Paris VI, PhD diss.

A. Brun, C. Spitz, E. Wurtz and L. Mora, 2009. Behavioural comparison of some predictive tools used in a low-energy building, In: *Proceedings of Elev. Int. IBPSA Conf.*, Glasgow, Scotland, July 27–30 2009.

P. Carpentier, G. Cohen, 2017. *Décomposition-coordination en optimisation déterministe et stochastique*, Springer.

CGDD. 2015. Chiffres clés de l'énergie - Edition 2014. Commissariat Général au Développement Durable, Repères, February 2015.

B. Coffey, 2013. Approximating Model Predictive Control with Existing Building Simulation Tools and Offline Optimization. *Journal of Building Performance Simulation* 6 (3): 220–235. doi:10.1080/19401493.2012.737834.

K. Le, R. Bourdais, and H. Guéguen, 2014. From Hybrid Model Predictive Control to Logical Control for Shading System: A Support Vector Machine Approach. *Energy and Buildings* 84: 352–359. https://doi.org/10.1016/j.enbuild.2014.07. 084.

G. Lefebvre, 1987. Analyse et réduction modales d'un modèle de comportement thermique du bâtiment, Université Pierre et Marie Curie, PhD diss.

G. Lefebvre, 2007. La méthode modale en thermique: Modélisation, simulation, mise en oeuvre, applications, Ellipses.

P. Malisani, 2012. Dynamic Control of Energy in Buildings Using Constrained Optimal Control by Interior Penalty, École Nationale Supérieure des Mines de Paris, PhD diss.

P. Malisani, F. Chaplais and N. Petit, 2014. An Interior Penalty Method for Optimal Control Problems with State and Input Constraints of Nonlinear Systems. *Optimal Control Applications and Methods* https://doi.org/10.1002/oca.2134.

P. T. May-Ostendorp, G. P. Henze, B. Rajagopalan, and C. D. Corbin, 2013. Extraction of Supervisory Building Control Rules from Model Predictive Control of Windows in a Mixed Mode Building. *Journal of Building Performance Simulation* 6 (3): 199–219. https://doi.org/10.1080/19401493.2012. 665481.

M.D. Mesarovic, D. Macko, and Y. Takahara, 1970. Two Coordination Principles and Their Application in Large Scale Systems Control. *Automatica* 6 (2): 261–270. https://doi.org/10.1016/0005-1098(70)90097-X.

P.-D. Morosan, R. Bourdais, D. Dumur, and J. Buisson, 2010. Building Temperature Regulation Using a Distributed Model Predictive Control. *Energy and Buildings* 42 (9): 1445–1452. https://doi.org/10.1016/j.enbuild.2010.03.014.

P.-D. Morosan, R. Bourdais, D. Dumur, and J. Buisson, 2011. A Distributed MPC Strategy Based on Benders' Decomposition Applied to Multi-source Multi-Zone Temperature Regulation. *Journal of Process Control* 21 (5): 729–737. doi.org/10.1016/j.jprocont.2010.12.002.

F. Munaretto, T. Recht, P. Schalbart and B. Peuportier, 2017. Empirical Validation of Different Internal Superficial Heat Transfer Models on a Full-Scale Passive House. *Journal of Building Performance Simulation*: 1–22. https://doi.org/10.1080/19401493.2017.1331376.

A. Neveu, 1984. Étude d'un code de calcul d'évolution thermique d'une enveloppe de bâtiment, Université Pierre et Marie Curie, Paris VI, PhD diss.

F. Oldewurtel, A. Parisio, C. N. Jones, D. Gyalistras, M. Gwerder, V. Stauch, B. Lehmann, and M. Morari, 2012. Use of Model Predictive Control and Weather Forecasts for Energy Efficient Building Climate Control. *Energy and Buildings* 45 (2): 15–27. https://doi.org/10.1016/j. enbuild.2011.09.022.

E. Palomo Del Barrio, G. Lefebvre, P. Behar and N. Bailly, 2000. Using Model Size Reduction Techniques for Thermal Control Applications in Buildings. *Energy Build* 33: 1–14, https://doi. org/10.1016/S0378-(00)00060-8.

B. Peuportier, 2005. Bancs d'essais de logiciels de simulation thermique, In: *Proceedings of Journée SFT- IBPSA Outils de simulation thermoaéraulique du bâtiment*, La Rochelle, France.

B. Peuportier and I. Blanc-Sommereux, 1990. Simulation Tool with Its Expert Interface for the Thermal Design of Multizone Buildings. *International Journal of Sustainable Energy*. 8: 109–120, https://doi.org/10.1080/01425919008909714.

L.S. Pontryagin, V.G. Boltyanski, R.V. Gamkrelidze and E.F. Mishchenko, 1962. *The Mathematical Theory of Optimal Processes, Interscience.* John Wiley & Sons, Inc: New York, London.

T. Recht, F. Munaretto, P. Schalbart and B. Peuportier, 2014. Analyse de la fiabilité de COM-FIE par comparaison à des mesures. Application à un bâtiment passif, In: *Proceedings of IBPSA*, France, Arras: 8.

M. Robillart, P. Schalbart, F. Chaplais and B. Peuportier, 2018. Model Reduction and Model Predictive Control of Energy-Efficient Buildings for Electrical Heating Load Shifting. *Journal of Process Control* https://doi.org/10.1016/j.jprocont.2018.03.007.

M. Robillart, P. Schalbart and B. Peuportier, 2017. Derivation of Simplified Control Rules from an Optimal Strategy for Electric Heating in a Residential Building. *Journal of Building Performance Simulation*: 1–15 https://doi.org/10.1080/19401493.20 17.1349835.

A. Sempey, C. Inard, C. Ghiaus and C. Allery, 2009. Fast Simulation of Temperature Distribution in Air Conditioned Rooms by Using Proper Orthogonal Decomposition. *Building and Environment* 44: 280–289 https://doi.org/10.1016/j.buildenv.2008.03.004.

O. Sidler, 2011. "De la conception à la mesure, comment expliquer les écarts?" présenté à *"Evaluer les performances des bâtiments basse consummation"*, Colloque CSTB/CETE de l'OUEST, Angers, France, January 2011.

E. Vorger, P. Schalbart and B. Peuportier, 2014. Integration of a comprehensive stochastic model of occupancy in building simulation to study how inhabitants influence energy performance, In: *Proceedings PLEA*, Ahmedabad, India, December 2014: 16–18.

A.A. Yahia and E. Palomo Del Barrio, 1999. Thermal Systems Modelling Via Singular Value Decomposition: Direct and Modular Approach. *Applied Mathematical Modelling* 23: 447–468 https://doi.org/10.1016/S0307-904X(98)10091-4.

Chapter 19

Urban Farming

From Discovery to Knowledge in the EEBI Chair

*Christine Aubry, Anne Cécile Daniel, Baptiste Grard,
Agnès Lelièvre, Nathalie Frascaria-Lacoste and Claire Chenu*

AGROPARISTECH, SIAFEE

When we created the "urban agriculture" research team within the INRA SAD-APT research unit (Science for Action and Development – Activities Products Territories) in June 2012, in France, this form of agriculture (UA) was mainly represented by (sub)urban farms, mostly market gardens, in local distribution channels that grew increasingly diversified with the city. Over the past decade, they have been boosted by the explosion of urban demand for hampers and other forms of local provisions (box schemes, farm shops etc.; Aubry and Chiffoleau 2009; Aubry and Kebir 2013). In parallel, at this time, we observed the multiplication of various forms of community, family, shared, collective, inclusive garden schemes etc., revealing pressing urban demand for more links to nature in the city, more social links between inhabitants, and pedagogy and knowledge of the origin of food. In several European countries, including France, an economic need for self-production has also became apparent (Pourias et al., 2015). At that time, and with an international outlook, the EEBI Chair supported a first review (Daniel, 2013) of the forms of AU. It was also in 2012 that we initiated the now famous "AgroParisTech experimental rooftop garden", known as Bertrand Ney,[1] with the support of AgroParisTech and assistance from the future founders of the Topager© company. Its primary objective was to test the relevance of extending the offer of Parisian spaces to roofs to create community gardens.

Six years later, the landscape has changed completely. While in 2012 we could gather the five Parisian urban agriculture project leaders in a small room of AgroParisTech (and these were putative figures, as three of them had no achievements to show), in 2015 we had to use the boardroom and in 2016 we needed an amphitheater, as we brought together more than 120 project leaders, not all of them from Paris! The creation of the AFAUP (Association Française de l'Agriculture Urbaine Professionnelle, French Association of Professional Urban Agriculture) in January 2017 was then a response to our wish to be relieved of the task of networking. Although this social enthusiasm inspired a lot of passion, it was nevertheless time-consuming and would potentially cause us to spread our resources too thin. As such, it quickly became urgent for our team to focus on its primary task: to produce knowledge on urban agriculture, formulate research questions and process them. The contribution of the EEBI

---

1 Named after Professor of Agronomy Bertrand Ney (1956–2013), who ardently supported the launch of this experimental roof and suggested we respond to the call for tenders from the AgroParisTech scientific council in 2012.

Chair from 2013 onwards has been decisive. Alongside other contributions, such as the ANR JASSUR research project (Urban Community Gardens 2013–2016[2]) or the DIM ASTREA regional research programme, the Chair's support for topics within urban agriculture led to further research into two major and related themes: on the one hand, an analysis of the sustainability of a then-emerging form of urban agriculture, namely multifunctional micro-farms; on the other hand, the ecosystem services provided by urban agriculture on roofs.

Why these choices? Within the forms of UA that can now be classified into five major types,[3] urban micro-farms in particular highlight a major expectation for city dwellers: multifunctionality. A founding hypothesis of our research – one which is widely shared at the global level (Duchemin et al., 2010, Zasada 2011) – is that, in order to exist in urban environment with a high level of competition for resources, and especially land resources, urban agriculture must be intrinsically "sustainable" (economically, socially and environmentally). However, at the same time, it must also ensure that the city dwellers recognise that the functions it fulfils for the city are important and cannot be substituted for other uses of the space (Nahmias and Le Caro 2012; Aubry et al., 2012; Aubry 2015; Poulot-Moreau 2015). Small in size and spatially and socially integrated within the urban environment, micro-farms are essentially multifunctional, but how do they function precisely, and what are the criteria for their economic, social and environmental sustainability? Though participatory research carried out with meticulous detail (and a lot of patience!) and thanks to the support of the EEBI Chair, Anne-Cécile Daniel has highlighted the mechanisms, the strengths and also the many weaknesses of these innovative, evolutionary and highly hybrid forms of urban agriculture. This work provoked a fully fledged craze, as these micro-farms echo the "miniaturisation" of vegetable farms in organic farming but in an urban context, as well as their inclusion in other paradigms (including "permaculture"), a phenomenon notably studied in the dissertation by Kevin Morel (2016). Today we can see that many communities engaged in urban renewal operations want to set up micro-farms, as the diversity of services rendered are very attractive for making the city more "sustainable".

Services provided are the central theme of the other project supported by the EEBI Chair during this period, namely, Baptiste Grard's thesis on "Technosols", forms built of urban "soils", especially on rooftops, to produce urban agriculture. As has been said, the "Bertrand Ney" experimental rooftop garden was initially designed to use substrate engineering studies in order to propose an "on the frame" alternative to the scarcity of ground space for establishing community gardens. However, while the first two years of experimentation showed the potential productivity of these container-grown crops and the limited risks related to the urban pollution (Grard et al., 2015),

---

2 Note that the idea for this project was first submitted to the ANR Agronomy programme, which considered it non-priority. This was not the opinion of the "Villes et Bâtiments Durables" (Cities and Sustainable Buildings) programme, which took it, as the people in charge of this programme felt better about the strong questions involved in this "return" to cultivation in the city than their colleagues from the world of agriculture.

3 Suburban farms in short supply chains, community gardens, multifunctional micro farms, urban greenhouses, "indoor farming". In addition, there are various forms of individual urban agriculture (balconies, small hydro or family aquaponics), interstitial agriculture (of the "Incredible edibles" type) and urban animal farming (beekeeping, hens, sheep for eco-grazing).

it soon appeared necessary to better understand and quantify the functioning and ecosystem services provided by these innovative forms of rooftop agriculture. The thesis carried out, which was co-financed by the Chair alongside the DIM ASTREA regional programme, has made it possible to go beyond the food supply service, and it expanded with the temporal evolution of urban substrates in containers and the diversification of produced crops (associations and successions of crops) to quantify other environmental-type services: recovery of urban waste (one of the central themes of the circular economy), a water retention service (which is considerable, as it often exceeds 80% of the water received), a carbon storage service and recovery of urban by-products. However, the thesis also allowed disservices to be highlighted, prompting a scientific and operational reflection on their future control, including the loss of soluble carbon in water percolation by these systems in containers on urban substrates. It is a thesis with considerable resonance, especially given how widely it was shared after publication.[4]

It therefore marks a societal "success story" for the EEBI Chair's support for work on urban agriculture. It is also a scientific success story. This is because this twofold support now fulfils the main role that we expected from the point of view of research: to ask new questions, to lead to pathways of development and thus to encourage new research. This pioneering work on micro-farms is now leading to a research programme on sustainability in urban agriculture, proposals for indicators derived from the work of Anne-Cécile Daniel, the questioning of "classical" methods for assessing the sustainability of agricultural holdings (Fargue-Lelièvre and Clarino, 2018) and the in-depth study of economic models of urban agriculture (start-up work by V. Saint-Ges). The growing development of these productive micro-farms on rooftops also encourages the knowledge of the ecosystem services provided to be expanded and comparisons to be made with "ground" micro-farms (SEMOIRS research project 2018–2020 supported by ADEME).[5] These services themselves will evolve (the quantification of thermal regulation services by green and productive roofs is the subject of a current post-doctoral thesis under the direction of Patrick Stella). Another avenue to be explored is the increasing order to recover biowaste in cities, for which it will be necessary to test new composts or digestates and introduce them into new cultivation systems, and the Bertrand Ney experimental rooftop garden is an indispensable device for doing this. Beyond this, methods such as life cycle assessment (LCA), which do not obviously lend themselves to being adapted to forms of urban agriculture, show their promising nature *through* their application on the roof (Dorr et al., 2018), especially in terms of questioning the adaptation of urban agriculture to climate change or urban agriculture's reduction of the effects of climate change. A European project on the LCA analysis of urban farms was created from the rooftop in 2017–2018 (Kic Climate), and probably a new thesis as well!

In the rest of this chapter, we will present a summary of the main results obtained through the two axes of work.

---

4 A strong recovery was reported by the press through the release of articles and the thesis (http:// presse. inra.fr/Communiques-de-presse/Toits-potagers-et-ville-durable; https://theconversation.com/ toits-potagers-en-ville-ce-nest-pas-que-pour-faire-joli-88457) and an interview with the doctor in the famous France Inter science programme "la Tête au carré" (23/02/2018).

5 The co-facilitator of which is Baptiste Grard and Anne Cécile Daniel is responsible for a role.

19.1 Functioning and Sustainability of Urban Micro-Farms

For the past ten years, urban micro-farms have been developing throughout France, Europe, Asia and North America (Morel 2016). Located in small plots in urban areas, often in neglected interstices, they generate an economic activity by placing the food-stuffs they produce on the market (Figure 19.1). However, the production and sale of foodstuffs is not sufficient for the economic balance of the farm, so they offer other activities in parallel to it, such as educational, recreational, cultural and/or garden design activities, including many that provide complementary sources of income. They are rightly called "multifunctional" urban farms, providing in *situ* (recreational, educational) and environmental services *via* flows related to urban metabolism (Simon-Rojo et al., 2016).

Given the many constraints imposed by the city, a partnership between the operator and the private or public owner of the host site is essential for their existence, as is the involvement of volunteers. This in particular distinguishes them from market garden micro-farms (Morel 2016), small farms in suburban or rural areas included in alternative food systems. Urban micro-farms are therefore inter-mediate forms that lie between community gardens (of allotment or shared gardens) and vegetable farms recognised as such by their profession and the agricultural authorities.

Despite their growth, urban micro-farms in France remain limited in number, and their heterogeneity does not allow a unique typology to be outlined. In order to differentiate them, the main activity of the structure is of interest. Thus our work has made it possible to distinguish the following types among the most widespread urban farms: "Agricultural" farms, which mainly produce fruits and vegetables; "Cultural" farms, the core business of which is event programming; "Discovery" farms that focus their strategy on experience, sharing and shared intelligence; "Education" farms, where training and awareness are the main activities; "Integration" farms, which aim to include people in the professional world *through* gardening activities; and "Restorative" farms, the economy of which is based on promoting crops and catering activities.

Figure 19.1 Cover of Anne Cécile Daniel's book (in French) on urban micro-farms. (Available for download from the Chair's website www.EEBI.fr.)

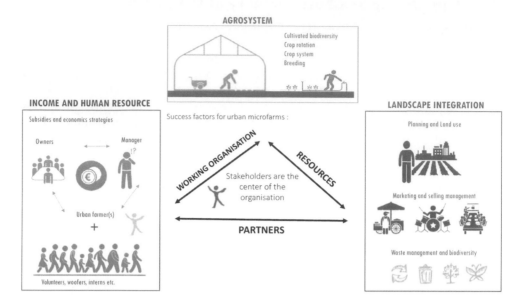

Figure 19.2 General functioning of urban micro-farms (Daniel, 2017).

There are also "Nursery" farms that produce seedlings for urban or other productive farms in or around urban areas, which are currently being developed in France but already established in the United States.

Drawing inspiration from the tools for assessing agricultural sustainability (IDEA and FADEAR), this study generated an original reading grid to understand the functioning of these urban micro-farms. It took 18 months of participatory research, including numerous surveys of producers, follow-up and the collection of field data concerning production, production costs and revenue and working time. Five urban farms were used to inform the main pillars that govern the functioning of an urban micro-farm: (i) the territorial anchorage analyses the association's interactions with the neighbourhood and the city, (ii) the agricultural technical system describes the technical choices and the practices following on from them, and finally (iii) the financial and human resources give an understanding of the structures' organisational choices. To be successful, urban micro- farms must find an overall coherence between these strategic choices (Figure 19.2).

Projects must therefore integrate work organisation adapted to the project objectives, continuous and quality access to resources[6] and a stable and solid partnership network.

---

6 The productive resources of which, such as water and fertilisers, must adapt to the urban environment. For the fertilisers, much is achieved through relationship with the urban organic waste that the supply of the micro farms and still more will be done in the future, thus differentiating them strongly from their rural counterparts. With regard to water, a strong concern arises from this research: everybody now uses city water to water crops, which can be considered as a waste, and few people store rain water at present, in particular because of a lack of space. One of the challenges for tomorrow is to better recover this rainwater and recycle wastewater from cities.

Figure 19.3 Recreational space of the cultural farm on a wasteland site (Daniel, 2017).

It is often the farmers themselves, the project leaders, who are the keystone of an urban micro-farm's longevity. This analysis of the three pillars structuring urban and contemporary micro-farms offers a unique insight into their ongoing operation in our cities (Figure 19.3).

Becoming part of the territory: urban micro-farms create original dynamics[7] within a district and more widely within the city, or even the region. They are nodes for very complex exchanges and contacts, in which a panel of activities is offered to the city dwellers. They generally develop in the spaces forgotten by town planning, thus contributing to their individual uniqueness. However, securing land is still a very delicate process, and the spaces on which they are located are not always conducive to their development (including the possible presence of contamination in the urban soil).[8] They often implement improvements, especially those favouring biodiversity, which can be especially appreciated by locals and developers (Tables 19.1 and 19.2).

Cultural practices: the cultivation system of each urban micro-farm depends very much on the objectives of the project, but also on the urban farmers themselves, the

7 In order to be convinced of this, it is sufficient to observe the "urban transhumance" offered by the "cultural" farm on Sunday morning with the small flock of sheep, in a suburban neighbourhood, or to see the students' parents visit the vegetable garden set up by the Educational farm.

8 In 2016, the study of micro farms led to the research programme INRA/AgroParisTech REFUGE: Risks in Urban Farms, management and evaluation (led by C Aubry and N Manouchehri), which is now supported by the Ile de France Region, Ademe and INERIS.

Table 19.1 Developments Fostering Biodiversity Implemented in Urban Micro-Farms

	F. Agricultural	F. Cultural	F. Discovery	F. Educational	F. Inclusive
Hives	X	X		X	X
Insect hotel	X	X	X		X
Low wall		X			X
Pond			X	In the planning stage	In the planning stage
Nesting box			X	X	X
Meadow	X	X	X	X	X
Wastelands					
Fruit trees/ orchards	X	X		X	X

Daniel (2017).

products grown, the location, the resources available and the quality of the soil. Given that the projects have not matured yet, the technical trading rules are not stabilised. The cultural practices are very close to organic farming (and/or permaculture) and there is a very strong interest in the recovery of organic waste in the city. Thus, two of the farms studied have set up composting platforms that are regularly supplied with coffee grounds, unsold goods from markets and supermarkets, wood shavings, branches and dead leaves. One of the associations produced between 6 and 8 m^3 of compost in 2015, or about 3.5T of compost; 10.5 tons of waste was recovered as fertiliser for crops.

An urban micro-farm has a variety of production objectives, however, depending on the constraints and opportunities it faces. In general, there is a great diversity of species and cultivated varieties, with "taste" and "authenticity" (including old species and varieties) as determining choices (Figure 19.4).

Human and economic resources: the surface area does not seem to be a criterion for determining the viability of an urban micro-farm, which instead depends on its strategy for combining activities, linked to the multifunctional nature of the projects. Using the total weight of sales and size of the cultivated area, we obtained approximate values of the annual yield for each farm, ranging from 1.7 to 5.6 kg/m² (17–56 tonnes per hectare per year), all vegetables combined. We also counted all the activities proposed over the course of one year in each of the urban micro-farms. These data show that they diversify their activities considerably to bring in additional revenue.

These orders of magnitude confirm the strategic choices of each farm, however. Although relatively high per unit area, these levels of production mean it is not conceivable for the sale of vegetables to cover all the wage costs, given the small surface areas concerned. To cope with this, these farms all receive varying amounts of aid, which is justified in terms of the services they provide: job creation, social connections, maintenance of parks and green spaces, enhancement of heritage, cultural activities, circular economy, etc. The association's structure makes it easy to obtain this funding from public institutions and private foundations, or even businesses. Each urban micro-farm obtains aid through various channels, depending on meetings and requests for funding assistance. There is, however, a climate of uncertainty with regard to state

Table 19.2 Nature of Trade and Source of Materials for the Cropping System According to the Urban Micro-Farms Studied

	Agricultural	Cultural	Discovery	Education	Inclusion
Self-production	Plants Manure Compost	Compost Sheep/horse manure Wool Coffee grounds Plants Potting soil	Compost Dead leaves Plants Manure	Pine needle Plants Manure	Compost Plants
Donations	Wood shavings/wood Dead leaves	Animal feed Wood shavings/wood Unsold market goods Straw/hay Seeds	Bins/containers Wood shavings/wood Sheep/horse Manure Unsold market Goods Straw/hay Plants Topsoil	Wood shavings/wood containers Boxes Beer grains Coffee grounds Plants Manure Seeds Potting soil	Manure/fertiliser Unsold market goods Coffee grounds Coffee grounds Straw/hay Manure Seeds Seeds Topsoil Potting soil
Purchase	Compost Manure/fertiliser Sheep/horse manure Straw/hay Plants Seeds Potting soil Provenance of materials	Animal feed Plants Potting soil	Manure/fertilizer Seeds Potting soil	Animal feed Compost Manure/fertiliser Plants Seeds Topsoil	
On site City Region Outside the region					

Daniel (2017).

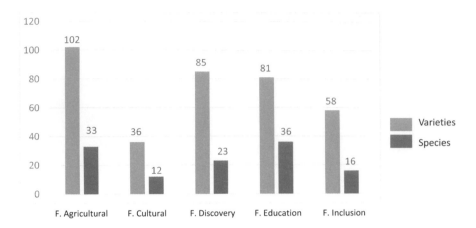

Figure 19.4 Number of species and varieties grown by the urban micro-farm types in the study (Daniel, 2017).

aid, as its instability and downward trend are a major impediment to the promotion of these initiatives.

At a time when urban policy makers and planners are increasingly taking urban agriculture into account operationally, urban micro-farms are appealing. They represent real opportunities to reclaim neglected spaces in cities and revive some obsolete spaces. As they are often lacking concepts and tools, communities and developers invent new ways of carrying out projects. We hypothesised that understanding the inner workings of urban micro-farms would provide an entry point for understanding the conditions for their sustainability and also for stimulating reflections on how to integrate these new agricultural factories into the city. Pagès (2016) inscribes them in the process of agro-urban change, bearers of an agro-cultural alternative for writing metropolises in an associative and activist manner.

Without a doubt, these micro-farms will be able to develop thanks to a better understanding of the ecosystem, social and economic services they render to the community, as well as the institutional recognition of these urban agricultures (Figures 19.5 and 19.6).

19.2 Technosols for Urban Agriculture

In recent years, a new form of green roof has emerged in cities in industrialised countries: the productive green roof (producing food biomass, for example). Although still little developed and largely unknown, the interest in these productive roofs continues to grow. Their design, their development and especially the ecosystem services they can render need to be better understood. The cornerstone of green roofs, the "soil" in place directly and indirectly influences the ecosystem services rendered by them, yet few studies have been devoted to it. In addition, non-renewable products such as pozzolan, expanded clay balls or peat soil are now largely used in their composition, bringing their sustainability into question. These artificial roofing soils are called *Technosols* (see the WRB international soil classification for a precise definition).

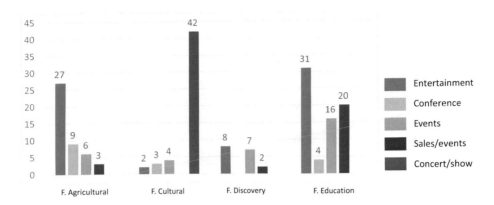

Figure 19.5 Activities proposed to the general public in 2015 (Daniel, 2017).

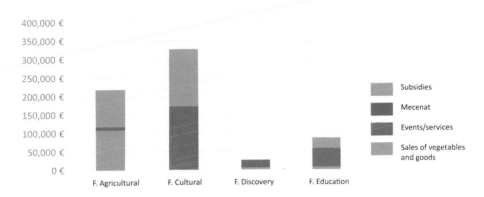

Figure 19.6 Operating revenues of the five urban micro-farms in the study (Daniel, 2017).

In the wake of AgroParistech's first work on the "Bertrand Ney" rooftop (Figure 19.7), we studied the services rendered by Technosols for a green roof producing food biomass. Setting ourselves the constraint of only using urban residues as a substrate, the aim of this work was twofold: (i) to quantitatively evaluate the ecosystem services rendered and (ii) to understand the first phases of the evolution of Technosol. Experimental devices installed on the roof were used for this. These are growing boxes in which Technosols were created, with various arrangements of five types of urban by-products: green waste compost, crushed wood, crushed brick and tiles, spent mushroom substrate based on spent coffee grounds and bio-waste compost. Using three experimental devices, we studied the effect of these different Technosols on (i) the ecosystem services delivered, (ii) food production (quantitative and qualitative) and (iii) physicochemical fertility. In addition, we also followed the evolution of their state over time, under cultivation. Figure 19.8 gives an example of the four experimental processes implemented in experimental device no. 1.

Figure 19.7 Experimental device on the rooftop of AgroParisTech in April 2017. (Photo credit: Baptiste Grard.)

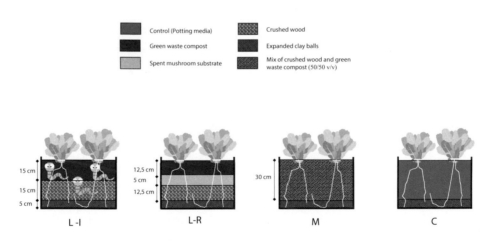

Figure 19.8 Schematic representation of the four experimental processes present in experimental device no. 1.

19.2.1 Principal Results

19.2.1.1 Regarding Food Production

The Technosols created are fertile and allow for significant food production (between 5.6 and 20.8 kg/m²/year for certain treatments) over one to five years (Figure 19.9), characterised by low levels of trace metal elements (pollutants can come from urban

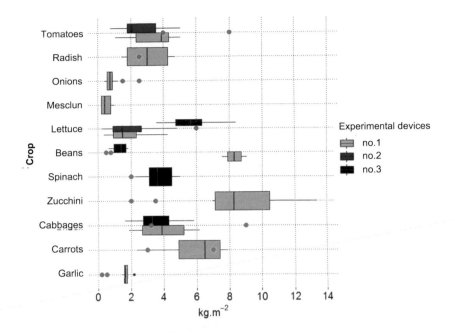

Figure 19.9 Comparison of the yields obtained on the three experimental devices compared to those of professional market gardeners (indicated by the green points issued by ITAB (2017).

traffic in particular) in vegetables, equivalent to the average for market gardening. For example, in terms of lead, the concentrations in the vegetables are 3–30 times lower than the standards defined for marketing products, the values being otherwise equivalent to or even lower than the concentrations found in French food (Figure 19.10).

19.2.1.2 Regarding Other Ecosystem Services Rendered

Four ecosystem services were studied from a quantitative point of view: food production (quantity and quality of production), retention of rainwater (quantity of water retained and quality of seepage water), the recovery of urban waste and the storage of carbon. The research resulted in the publication of the first article quantifying the ecosystem services rendered by a productive roof (Grard et al., 2018). Figure 19.11 summarises the main results of this study on one of the experimental devices. The approach used highlighted the multifunctionality of these roofs and made it possible to distinguish certain levers for optimising the ecosystem services rendered.

19.3 Balance Sheet

These results open up a new field of study, linked to a practice that is now undergoing full expansion. The quantitative study of the ecosystem services delivered made it possible to highlight the interest and also the limits of productive roofing designed with a

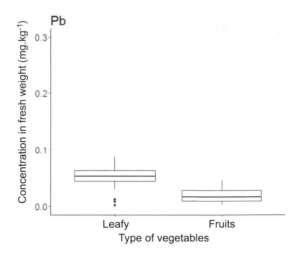

Figure 19.10 Graphs showing the lead concentrations according to the type of vegetable and for all the measurements made on the three experimental devices. Standards (EU 2011) refers to a European standard. ANSES (ANSES 2011) refers to an EAT2 (Étude de l'alimentation totale – Total Food Study) by ANSES in 2011 on the contamination of French food.

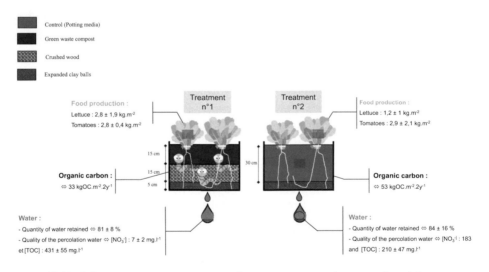

Figure 19.11 Schematic representation of ecosystem services rendered by two experimental processes in experimental device no. 2. OC: organic carbon; TOC: total organic carbon and NO₃: nitrate.

Technosol composed solely of urban by-products. Henceforth, it appears necessary to optimise the cropping systems put in place with regard to the ecosystem services rendered. Three major areas of optimisation can be distinguished:

- **Technosol**: This work has shown that Technosols can be optimised in relation to existing elements in the literature, in terms both of the nature of the parent materials (i.e. the nature of the compost, organic or mineral residue) and of their proportions and arrangements (i.e. layered or not). In a complementary way, the integration of economic (e.g. cost of materials), spatial (e.g. availability of materials) and regulatory criteria (e.g. limit to the use of certain by-products) is essential to allow for large-scale development. In parallel, the characterisation of new deposits of by-products (such as spent coffee grounds from mushroom farms) and the adaptation of composted products to the constraints of a Technosol producing food biomass seem to indicate an interesting line of research. Current composting methods are indeed designed to add exogenous organic matter to soil and not for direct use as Technosol components, as we were able to do in these studies.
- **Cultivation practices**: must be adapted to the systems in place and the services expected. Thus, in connection to the evolution of Technosols, the fertilisation of the systems must be studied at the level of the contribution of waste products, for example, considering the frequency (i.e. annual or multiyear), the type (i.e. fertiliser in solid or liquid form – type of fertiliser used) and the method of providing the fertiliser. Other elements involved in managing the system also require research efforts to be made in connection with the composition of Technosols: irrigation (type and frequency in order to minimise the water input), mulching, reuse of seepage water, etc.
- **Cultivation**: The choice and association of plants are essential elements of the cultivation system and must also be considered according to different criteria: economic, productive (quantity and quality), landscape, etc. During these studies, the crop type was not a directly tested variable, rather crop rotation was defined by taking predominantly agronomic criteria into account. Nevertheless, the choice of this rotation and the vegetables that compose it can be discussed, and it could be interesting to carry out tests on the effect that the type of vegetables grown has on the ecosystem services rendered.

In view of the operational development of the model studied here, it is necessary to take other services into account: energy balance of buildings, mitigation of urban air pollution, biodiversity reserve, the social, educational and pedagogical function, etc. The integration of the first feedback from existing micro-urban farms (Daniel 2017) seems to constitute an essential element for adapting practices and systems accordingly.

From a more general point of view, integrating tools such as LCA into the design of these systems will minimise their environmental impacts. Thus, an LCA carried out on one of the systems analysed in this thesis shows the reduction in the carbon impact per kilogram of biomass produced with the use of urban by-products, such as Technosol, compared to that of potting soil. (Dorr et al., 2017). The use of this type of analysis should guide the choices made for the different components of the system.

Beyond the topic of green roofs, on which this thesis was based, the conception of productive Technosols can and must be understood in other contexts. Thus, the use of abandoned urban areas such as wastelands with polluted soils represents an opportunity to extend these forms of soilless cultivations from urban by-products. This has also been done in the "Les fermes en villes" (Farms in Cities) demonstration project set up by the Vivant et la ville association. Such projects offer significant development potential for this type of Technosol, beyond roofs.

Even without feeding the cities of tomorrow, the relocation of urban micro-farms provides an opportunity to develop multifunctional spaces and biodiversity and to reconnect urban populations to food production. Moreover, the possibility of setting these systems up on buildings as well as on the ground offers many development alternatives in the future, which means that they are likely to become increasingly essential in the future development of the cities of tomorrow. They lead to the emergence of a new disciplinary field: urban agronomy.

Bibliography

Aubry C. 2015. Les agricultures urbaines et les questionnements de la recherche. *Review for special issue of Agricultures urbaines no.* 224: 35–49.

Aubry C., and Chiffoleau Y. 2009. Le développement des circuits courts et l'agriculture périurbaine: histoire, évolution en cours et questions actuelles. *Innovations Agronomiques* 5: 41–51.

Aubry C., Dabat M.H., Ramamonjisoa J., Rakotoarisoa J., Rakotondraibe J., and Rabeharisoa L. 2012.Urban agriculture and land use in cities: an approach with the multi- functionality and sustainability concepts in the case of Antananarivo (Madagascar). *Land Use Policy* 29: 429–439. doi: 10.1016/j.landusepol.2011.08.009.

Aubry C., and Kebir L. 2013. Shortening food supply chain: a way for maintening agriculture close to urban areas? The case of the French metropolitan areas of Paris. *Food Policy* 41 (2013): 85–93. http://dx.doi.org/10.1016/j.foodpol.2013.04.006.

Daniel A.-C. 2013. *Aperçu de l'agriculture urbaine en Europe et en Amérique du nord.* Rapport Chaire Éco-conception des ensembles bâtis et des infrastructures, AgroParisTech, Paris, 75 p.

Dorr E., Sanyé-Mengual E., Gabrielle B., and Aubry C., 2017: Proper selection of substrates and crops enhances the sustainability of Paris rooftop garden. *Agronomy for Sustainable Development* 37: 51. https://doi.org/10.1007/s13593-017-0459-1.

Duchemin E., Wegmuller F., and Legault A.M. 2010. Agriculture urbaine: Un outil multidimensionnel pour le développement des quartiers. *VertigO* 10(2). http://dx.doi.org/10.4000/vertigo.10436.

Fargue-Lelièvre A., Clerino P. Developing a tool to evaluate the sustainability of intra-urban farms. 13rd European IFSA Symposium. Farming systems: facing uncertainties and enhancing opportunities, IFSA, International Farming Systems Association, Jul 2018, Chania, Greece. HAL: 02318311.

Grard B., Bel N., Marchal N., Madre F., Castell J.F., Cambier P., Houot S., Manoucheri N., Besançon S., Michel JC, Chenu C., Frascaria Lacoste N., and Aubry C. 2015. Recycling urban waste as possible use for rooftop vegetable garden. *Future of Food: Journal on Food, Agriculture and Society* 3(1): 21–34.

Grard B.J-P., Chenu C., Manouchehri N., Houot S., Frascaria-Lacoste N., and Aubry C. 2018. Rooftop farming on urban waste provides many ecosystem services. *Agronomy for Sustainable Development*. doi: 10.1007/s13593-017-0474-2.

Joimel S., Grard B.J-P., Auclerc A., Hedde M., Le Doaré N., Salmon S., and Chenu C. 2018. Are Collembola 'flying' onto green roofs? *Ecological Engineering* 111: 117–124 doi: 10.1016/j.écoleng.2017.12.002.

Koegler M., Grard B.J-P., and Christine A. 2017. Climate Innovation Potentials of Urban Agriculture (CIPUrA) Geographic Pathfinder. Climate KIC report.

Morel K. 2016. Viabilité des microfermes maraîchères biologiques. Une étude inductive combinant méthodes qualitatives et modélisation. Thesis of Université Paris-Saclay under the direction of François Léger, defended on 15 December 2016.

Morel K., and Léger F. 2015. Aspirations, stratégies et compromis des microfermes maraichères biologiques. 14 p. |ONLINE] URL: https://hal.archives-ouvertes.fr/hal-01206302/ document.

Morel K., and Léger F. 2016. A conceptual framework for alternative farmers' strategic choices: the case of French organic market gardening microfarms. *Agroecology and Sustainable Food Systems* 40(5): 466–492. DOI: 10.1080/21683565.2016.1140695.

Nahmias P., and Le Caro Y. 2012. Pour une définition de l'agriculture urbaine: réciprocité fonctionnelle et diversité des formes spatiales. In Urban Environment [Online], Volume 6 | 2012, posted on 16 September 2012, accessed 14 April 2018. URL: http://journals.openedition.org/eue/437.

Pagès D. 2016. Le Grand Paris qui mange: un projet politique culturel en émergence? [Part 2] La fabrique symbolique, des processus aux écritures, Quaderni 2016/2.

Poulot-Moreau M. 2015, March. Agriculture et ville: des relations spatiales et fonctionnelles en réaménagement, Une approche diachronique. In Review FOR Special Issue 224 "Agricultures Urbaines": 51–66.

Pourias J., Aubry C., and Duchemin E., 2015. Is food a motivation for urban gardeners? Multifunctionality and the relative importance of the food function in urban collective gardens of Paris and Montreal. *Agriculture and Human Values* DOI 10.1007/s10460-015-9606-y.

Rahmanian M., Daniel A., Grard B.J-P., Juvin A., Besancon S., Bosch A., Aubry C., Cambier P., and Manouchehri N. 2016. Edible production on rooftop gardens in Paris? Assessment of heavy metal contamination in vegetables growing on recycled organic wastes substrates in 5 experimental roofgardens. 2–6. *ICFAE International Symposium paper.*

Simon-Rojo M., et al., 2016, From Urban food gardening to urban farming, In: Lohrberg, F., L. Licka, L. Scazzosi, and A. Timple (eds.), *Urban Agriculture Europe*, Jovis: Berlin.

Zasada I. 2011. Multifunctional peri-urban agriculture -a review of societal demands and the provision of goods and services by farming. *Land Use Policy* doi:10.1016/j.landusepol.2011.01.008. http://www2.agroparistech.fr/T4P-un-Projet-de-recherche-innovant-pour-des-Toits-Parisiens- Productifs.html.

Chapter 20

Collecting, Classifying and Visualising Big Data for Biodiversity and Mobility

Madjid Maidi, Hela Marouane, Salma Rebai, and Sébastien Herry

ESME Sudria, Ivry-sur-Seine, France

20.1 Context

The analysis and use of Big Data represents a major challenge for both industry and science today. This booming phenomenon is due to the increase in the variety and number of data sources (social networks, sensors, web application logs, etc.) generating a growing amount of data. Indeed, it is estimated that about 50 TB of data were generated every second in 2018. These data must be stored, processed and analysed to derive usable information from them. However, traditional data management methods are not designed to handle this large volume of data and must make room for better-suited Big Data tools.

This is the context surrounding the Big Data application proposed by the ESME Sudria team as part of the Eco-design of Buildings and Infrastructures Chair. The purpose of this application is to establish a solution for the collection, storage and analysis of data in the areas of biodiversity and transport. The application is broken down into three parts:

- an Android mobile application for collecting and reporting biodiversity and transport data
- a Big Data application for storing and processing collected data
- an application for the analysis and visualisation of data by topic and/or geographical position.

This chapter details the different parts of the creation of this application in terms of design, technical choices and results obtained.

20.2 Data Classification Application for Biodiversity

20.2.1 Technical Choices

The technical development of the application dedicated to biodiversity fully implements all the modules and functions detailed in the functional logic diagram described by BiodiVstrict. The user interface has been designed to respect a homogeneous and standardised graphic chart for all the stages of exploring the natural species classification tree. The interactivity and ergonomics of the interface are also considered in all stages of selection and/or validation to allow better interaction and responsiveness on the part of the user.

The application was developed on Android, a choice supported by a large number of mobile devices using this operating system. Indeed, Android is the most used mobile operating system in the world, with more than 80% of the smartphone market share. In addition, the use of Google services and APIs is natively accessible and used and managed optimally, as Android is developed by Google.

20.2.2 The BiodiVstrict App

When the application is launched, the user first selects their region (Figure 20.1), and then, in the next step, they choose the BiodiVstrict classification path (Figure 20.2).

In the BiodiVstrict tree diagram, we first select the environment and validate it (Figure 20.3). A list of habitats is then displayed with the names and images of each habitat (Figure 20.4). The user can select an image of a particular habitat for more details (Figure 20.5). If the habitat is confirmed, then a new page is displayed and prompts the user to take a picture (Figure 20.6). This operation is required, and the next step cannot be taken if the user does not take a picture of the habitat found in their environment. However, it is possible to return to the previous step in case of uncertainty, and the user can then check their selection again. Starting the camera requires GPS activation to capture a location-based image and to associate the GPS coordinates with the habitat (Figure 20.7). When the photograph of the habitat is confirmed, a data record is made in the form of a frame consisting of the name of the habitat, the image file and the geolocation coordinates, saved in the phone memory (Figure 20.8).

Figure 20.1 Selection of the region.

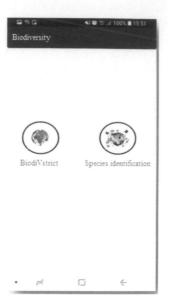

Figure 20.2 Main menu of the biodiversity application.

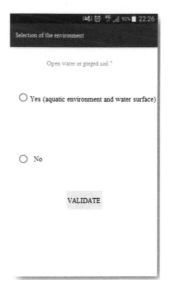

Figure 20.3 Selection of the environment.

If no habitat corresponds to the proposed list, the user can edit an additional entry corresponding to a new habitat, and the selection procedure remains identical to that of an existing habitat. If the selected habitat contains remarkable elements, then they are displayed (Figure 20.9). The user can see more details on a remarkable item by selecting their picture, and they can then validate their choice if the item corresponds to

Figure 20.4 Display of the list of habitats.

Figure 20.5 Display of a detailed description of a selected habitat.

the observation (Figure 20.10). They are then asked to take a picture and confirm their action to save their data frame in memory (Figure 20.11). However, if no remarkable element is present in the proposed list, then the user specifies that there is no remarkable element. On the other hand, if the habitat has a remarkable element that is not in

Figure 20.6 Request for a photograph of the habitat.

Figure 20.7 Enabling GPS from phone settings.

the proposed list, it is then possible to take a picture of this new element and then provide its name in a specific field. When a remarkable element is confirmed, a dialogue box invites the user to verify the existence of other remarkable elements, if they want (Figure 20.12). When validation is complete for this step, the list of invasive elements is

Figure 20.8 Habitat confirmation.

Figure 20.9 Display of the list of remarkable elements.

displayed (Figure 20.13). In this step, the user can explore all the invasive elements and read their descriptions. They confirm their choice if the invasive element corresponds to the observed species (Figure 20.14); then, they take a photograph and confirms the action to record the data frame (Figure 20.15). A dialogue box gives control back to

Figure 20.10 Selection of a remarkable element.

Figure 20.11 Confirmation of the remarkable element.

the user by allowing them to select other existing invasive elements (Figure 20.16) or to add a new invasive element if it is not in the list (Figure 20.17). Otherwise, at this step, the user can return to the beginning of the application for another experience or exit the application (Figure 20.18).

Figure 20.12 Verification of the existence of other remarkable elements.

Figure 20.13 Display of the list of invasive elements.

Figure 20.14 Selection of an invasive element.

Figure 20.15 Confirmation of the invasive element.

20.2.3 Lambert 93 Coordinates

The geolocation coordinates provided by the GPS (longitude and latitude) are transformed into Cartesian coordinates (X, Y) according to the Lambert 93 system, which represents the official projection for maps of mainland France. The transition

Figure 20.16 Verification of the existence of other invasive elements.

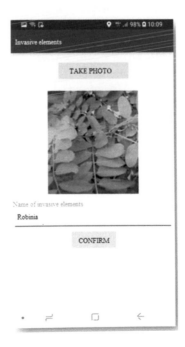

Figure 20.17 Adding a new invasive element.

Figure 20.18 Termination of the application.

from GPS coordinates to Lambert 93 coordinates is transcribed by a mathematical formalism detailed as follows.

We first define the following parameters:

- semimajor axis $a = 6,378,137$ m
- flattening $f = 1 - b/a$, where $f = 1/298.257222101$
- first scale parallel $\varphi_1 = 44°$
- second scale parallel $\varphi_2 = 49°$
- the longitude of origin given by the Greenwich central meridian $\lambda_0 = 3°$
- latitude of origin $\varphi_0 = 46°30$
- x-coordinate at origin $X_0 = 700,000$ m
- y-coordinate at origin $Y_0 = 6,600,000$ m

With these parameters, we can calculate

- the semiminor axis $b = a(1 - f)$
- the eccentricity $e = \sqrt{\dfrac{a^2 - b^2}{a^2}}$

Starting from the coordinates (latitude, longitude) ($= \lambda$) of a point on the globe supposed to be an ellipsoid of revolution of semimajor axis a and semiminor axis b, we calculate its coordinates (X, Y) on the Lambert map. For this, we will go through the coordinates (ρ, θ) of the projected point on the cone. The X axis is increasing to the east and the Y axis is increasing to the north.

$$\begin{cases} X = X_0 + \rho\sin(\theta) \\ Y = Y_0 + \rho_0 - \rho\cos(\theta) \end{cases} \text{ou} \begin{cases} \theta = \eta(\lambda - \lambda_0) \\ \rho = \rho(\varphi) \\ \rho_0 = \rho(\varphi_0) \end{cases}$$

The origin corresponds to the values $= \lambda_0$. For this point, we have $\phi = \phi_0$ and $Y = Y_0$. The parallels are defined by $\phi =$ constant, and hence $\rho =$ constant. They are represented by concentric circles.

Meridians are defined by $\lambda =$ constant, and hence $\theta =$ constant. They are represented by lines passing through the common centre of the preceding circles, with coordinates $X = X_0$ and $Y = Y_0 + \rho_0$

$$\rho(\varphi) = \rho(0)\left[\cot\left(\frac{\varphi}{2} + \frac{\pi}{4}\right)\left(\frac{1 + e\sin(\varphi)}{1 - e\sin(\varphi)}\right)^{\frac{e}{2}}\right]^{n}$$

The two constants and η and ρ_0 are calculated using the two secant parallels of reference. The scale is preserved on these parallels of latitude φ_1 and φ_2, called automecoïques:

$$\eta\,\rho_i = a\cos(t_i) = \frac{a\cos(\varphi_i)}{\sqrt{1 - e^2\sin(\varphi_i)^2}}$$

We can therefore calculate η and ρ_0:

$$\eta = \frac{\ln\left(\dfrac{\cos(\varphi_2)}{\cos(\varphi_1)}\right) + \dfrac{1}{2}\ln\left(\dfrac{1 - e^2\sin(\varphi_1)^2}{1 - e^2\sin(\varphi_2)^2}\right)}{\ln\left[\dfrac{\tan\left(\dfrac{\varphi_1}{2} + \dfrac{\pi}{4}\right)(1 - e\sin(\varphi_1))^{\frac{e}{2}}(1 + e\sin(\varphi_2))^{\frac{e}{2}}}{\tan\left(\dfrac{\varphi_2}{2} + \dfrac{\pi}{4}\right)(1 + e\sin(\varphi_1))^{\frac{e}{2}}(1 - e\sin(\varphi_2))^{\frac{e}{2}}}\right]}$$

$$\rho(0) = \frac{a\cos(\varphi_1)}{\eta\sqrt{1 - e^2\sin(\varphi_1)^2}}\left[\tan\left(\frac{\varphi_1}{2} + \frac{\pi}{4}\right)\left(\frac{1 - e\sin(\varphi_1)}{1 + e\sin(\varphi_1)}\right)^{\frac{e}{2}}\right]^{n}$$

This change in the coordinates makes it possible to transfer GPS data to the cartography used in BiodiVstrict.

20.2.4 Data Frame

The data frame consists of the various types of information provided during the application tests. This frame is composed of the date, the description, the geolocation coordinates and the image file of the natural element (Figure 20.19). Initially, the frames are saved on the phone memory after each step confirming the natural element.

```
e2018-11-16.jpg "Bretagne" "Mammal" "cat" "cat" 48.81384668033388
2.3953557976171655 677260.7433776868 6812423.730714977
h2018-12-19.jpg "Normandie" "Aquatic environment and water surface"
"Lake, pond or pond natural or semi-natural" 48.785509544464404
2.3949537657078763 677243.5806936111 6812177.997089438

r2018-12-19.jpg "Normandie" "Lake, pond or pond natural or semi-natural"
"Humid area" "Humid area" 48.785509544464404 2.3949537657078763
677243.5806936111 6812177.997089438
i2018-12-19.jpg "Normandie" "Ambrosia with sagebrush leaf"
48.785509544464404 2.3949537657078763 677243.5806936111 6812177.997089438
```

Figure 20.19 Data file.

On the other hand, a function that checks for the availability of an internet connection is added, which allows either internal storage in the absence of a connection or instantaneous delivery of the data file when the connection is established.

The data frame consists of information in a specific format:

```
ENVIRONMENT _ NAME    HABITAT _ IMAGINE    LONGITUDE    LATITUDE
X(LAMBERT   93)    Y(LAMBERT   93)    REMARKABLE _ ELEMENT _ NAME
REMARKABLE _ ELEMENT _ IMAGE LONGITUDE LATITUDE X(LAMBERT 93)
Y(LAMBERT   93)    INVASIVE _ ELEMENT _ NAME    INVASIVE _ ELEMENT _
IMAGE LONGITUDE LATITUDE X(LAMBERT 93) Y(LAMBERT 93)
```

The image file names for habitats, remarkable elements and invasive elements contain information about the date and time of the shot. Indeed, an image of an invasive element whose file name is "i20171019155208.jpg" means that the image was captured on 19/10/2017 at 15:52:08. The letter "i" at the beginning of the name of the image means that it is an invasive element, the letter "r" is for remarkable element, and the letter "h" is for habitat.

20.2.5 Identification of Species

When the application is launched, the user begins their journey into the classification tree for natural elements. First, they select their region (Figure 20.1), and they can then select either BiodiVstrict or species identification (Figure 20.2).

If the user selects species identification, a new page is displayed, which invites the user to take a photograph, select the category of the species (animal or plant) and enter the name of the species caught (Figure 20.20). Starting the camera requires GPS activation to capture a geo-localised image and associate the GPS coordinates with the species (Figure 20.21). When the species is confirmed, a data record is made in the form of a frame consisting of the region, the category of the species, its name, the image file and the geolocation coordinates and is saved in the phone memory. Finally, a dialogue box is displayed asking the user to check whether there are other species in their environment to capture (Figure 20.22).

Figure 20.20 Species identification.

Figure 20.21 Enabling GPS from phone settings.

20.3 Collection and Analysis of Mobility Data

20.3.1 Objective

The objective of this second part of the project is to propose a technical solution that makes it possible to collect mobility data in transport, then process them and analyse them to identify a person's type of mobility, to find out whether they travel on foot, by bike, by car, by bus or by train. In addition, the application aims to integrate a vision system to interpret images, detect anomalies in the environment and report specific situations.

Figure 20.22 Verification of the existence of other species.

20.3.2 User Interface

The transport application is based on the design and implementation of an Android mobile application that collects GPS location data, accelerations, images, and contextual information provided by the user (Figure 20.23).

This information constitutes a data frame sent to the server at regular intervals or specific times for analysis and processing. The application collects the movement data to determine the transport mode comprehensively according to the movement speed

Figure 20.23

variation (stop, slow, or fast). At the same time, sensor data are collected to identify the means of transport based on supervised learning or automatic classification approaches.

On the other hand, a vision-processing module has been added to collect relevant data from the environment. In this module, the camera is used to visualise the scene and observe the traffic conditions and the state of the infrastructure. This vision system automatically identifies characteristic objects on static images, such as pedestrians and vehicles. This information can also be introduced by the user or validated if already determined by the vision-processing system.

20.3.3 Features of the User Interface

The transport application is an Android mobile application that provides access to the following features:

1. The phone camera is used to take a picture of the observed scene. When the photograph is confirmed, a data record is made in the form of a frame consisting of the date, the geolocation coordinates, the description and the scene image file (Figure 20.24). Starting the camera requires GPS activation to locate the observed scene (Figure 20.25).
2. At first, the frames are recorded on the phone memory after each step confirming the observed scene. On the other hand, a function checking for the presence of an internet connection is added, which enables data to be stored in the internal storage in the absence of a connection or instantaneous delivery of the data file to the Big Data system when the connection is established.
3. Google Maps is used to locate the user and track their movements. Indeed, if the user selects geolocation, a Google Map is displayed with the current position

Figure 20.24

Figure 20.25 Enabling GPS from phone settings.

Figure 20.26 Display of the user's current position.

of the user (Figure 20.26). During a trip, a line is drawn to represent the path followed by the user (Figure 20.27). This feature also makes it possible to collect GPS, acceleration, and speed data by using different sensors.

4. The data are saved to the phone memory as a frame formed of information collected according to the following format (Figure 20.28): *LATITUDE, LONGITUDE, ACC_X (Acceleration along the x axis), ACC_Y (Acceleration along the y axis), ACC_Z (Acceleration along the z axis), SPEED, TIMESTAMP, MOVEMENT*

5. The data collected make it possible to determine the type of movement: slow, stopped or fast.

Figure 20.27 Tracking the path taken by the user.

```
"48.80247469032253", "2.3988819317844436", "4.56", "1.09", "9.03", "44.14",
"030518165319", "Fast Movement"

"48.80247460641343", "2.3988819317844777", "4.35", "1.58", "8.53", "39.55",
"030518165321", "Fast Movement"
```

Figure 20.28 Data file.

20.3.4 Determination of the Mode of Transport

The mobile app collects kinematic motion data such as GPS coordinates and linear accelerations and calculates the user's speed. These data can be used to determine the generic trip type and whether the movement is slow, fast or stopped. These data can be used for classification to predict the transport type more accurately.

The data frames are then sent to the server for analysis and used for processing. The information collected can be used for a large-scale analysis to find out the frequencies at which the paths are taken by the users, the time spent in a particular place and the associated trajectories for each route. These position data will be used to optimise the trajectories to define personalised access according to the user's objective (to arrive at a destination quickly, tourist route, etc.). In addition, the data analysis will make it possible to adapt the road infrastructure and make further modifications to the city's urban planning.

20.3.5 Description of the Urban Environment

In a second step, use of the camera is proposed to exploit the data related to the infrastructure and the external environment. Indeed, the camera makes it possible to visualise the scene, to observe the state of traffic, traffic conditions, the state of the infrastructure, etc. A vision system allows environmental information to be analysed and the visual content to be interpreted automatically, without the user's intervention (vehicle and pedestrian detection). However, it is possible to enter new descriptions, modify or validate information *via* the user interface.

This module is based on a supervised classification whereby the learning is carried out offline with the help of a database of several thousand positive and negative images using cascading classifiers [1] and convolutional neural networks [2–4]. These performance criteria are obtained by optimising the learning and training corpus to improve accuracy (classification percentage or recognition rate) and recall (ratio between the number of positive images and classified images).

The detection or test part is performed online: the image is preprocessed for restoration (colour, light, contrast, etc.). Then, the multi-resolution detection makes it possible to analyse the image on several scales, so that the detection is more robust by calculating the similarity. Finally, a post-processing phase makes it possible to present the results in a format adapted to the display.

20.3.6 Composition of the Frame

The purpose of the application is to build a datagram according to a protocol defined to aggregate the data from different sensors, namely, the camera, the GPS, the accelerometers and also the user information. These data are sent to the server for large-scale visualisation in a Big Data system. Two data frames that contain the following information are formed:

20.3.6.1 Report Frame

[### description_of_report ### report_image ### longitude ### latitude ###]

20.3.6.2 Motion Frame

[### latitude ### longitude ### acceleration_x ### acceleration_y ### acceleration_z ### speed ### timestamp ### motion ###]

20.4 Big Data System for Storage and Processing Data

Once the mobile applications dedicated to biodiversity and transport were implemented, it was necessary to design and implement a database oriented towards Big Data, for the centralisation and storage of the data collected by these applications. This database should meet several criteria, including the high availability of data with improved fault tolerance, which implies the need to replicate stored data. In addition, our Big Data system should allow reliable and secure storage of the data transmitted by mobile applications, with access to data that is as fast as possible. A distributed system therefore seemed to be the most appropriate.

We detail now outline the design of the global architecture of the Big Data system, in terms of the infrastructure and software used for data storage and processing.

20.4.1 Data Characteristics and Modelling

The data reported by the mobile application are of different types and formats: GPS traces, camera images or text data entered by users. The size of these data can vary widely between just a few KB for text data and tens of MB for images and videos sent.

These data are collected by dated frames and sent to the server instantly and continuously if data connection is established. Otherwise, the frames will be saved and accumulated in a local file on the phone, to be sent to the remote data storage server later.

Frequent use of the mobile application (biodiversity and transport segments) generates a large volume of heterogeneous and unstructured data, to the order of several terabytes a year. The storage system must be able to process this volume of data in an acceptable time. It therefore seems that the traditional database management systems (Structured Query Language (SQL) model) are not suited to the present needs. It then becomes necessary to use dedicated solutions to analyse and process these "massive data", for better optimisation of the read/write time. These systems, called Big Data, allow easy scalability and lower costs in the face of the increase in the volume of data managed.

A comparative study of the different Big Data tools and frameworks was conducted to identify the best solution for our application, in terms of performance, features and ease of use.

20.4.2 Technical Choices

The volume and heterogeneity of the data reported by the mobile application quickly led us to NoSQL (Not Only Structured Query Language) databases, which are designed for better exploitation of massive unstructured data. This model moves away from the relational model traditionally used and falls within the development of distributed databases where the data are spread across multiple servers and multiple data centres.

Architecture makes this type of database capable of scaling up quickly and cheaply in the face of increasing requests to read and/or write data. This scalability makes it possible to adapt an architecture that has already deployed to demand by simply adding (or removing) new servers to the network. In addition, one of the main interests of the NoSQL model, in addition to its high scalability, lies in the absence of an individual point of failure. This reduction in the critical failure rate allows for better availability, faster data access and increased reliability through advanced replication mechanisms.

There are different NoSQL database families, each of which responds to very specific data mining needs. As shown in Figure 20.29, NoSQL systems are generally classified into four major families, namely, key/value databases (e.g. Redis, SimpleDB), column-oriented databases (e.g. Cassandra, Hbase, BigTable), document-oriented databases (e.g. MongoDB) and graph-oriented bases (e.g. Neo4j).

A comparative study of the different NoSQL databases, according to the needs expressed by the Chair, led us to select *Apache Cassandra* [5] for storing our application data. This choice was based on the performance, flexibility, fault tolerance and high availability offered by this open source database. Thanks to its representation in dynamic columns, Cassandra makes it possible to define flexible and rapidly accessible data patterns for reading and writing thanks to advanced "commit log" mechanisms. Our choice is supported by the multitude of "drivers" that allow Cassandra to interact and interface with many programming languages, which is an essential aspect for integrating and communicating with the visualisation mobile application and web application.

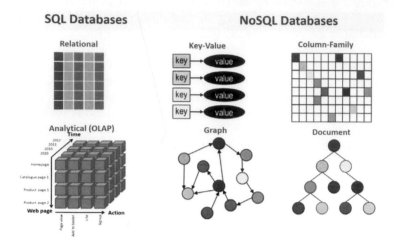

Figure 20.29 SQL and NoSQL database models.

To meet the evolving needs of the Chair, the proposed Big Data system has deployment flexibility that makes it compatible with the native system (on-premise) and with the Cloud. The overall architecture of the application is based on a *virtual machine-based solution* (VM) currently deployed on the ESME Sudria DataCenter. The latter has 64 processing servers, 150 TB of storage space and 850 GB of RAM, resources that are largely sufficient to guarantee good performance when processing big data. The choice of this VM-based solution was motivated not only by the optimisation of the infrastructure acquisition and exploitation costs in the development stage, but also by the ease of integration and implementation of the solution proposed at the end of the project. The VMs provided to the Chair will be configured with the necessary software for data processing and will be ready for use by simple deployment on local physical servers or on virtual servers in the cloud (OVH or Amazon EC2 type), according to the needs and requirements of the Chair's partners.

20.4.3 Physical Architecture of the Database

Apache Cassandra is designed to work in a *cluster* to provide Big Data storage with high availability. In general, a Cassandra database can contain one or more clusters independent of each other, which are often deployed in different geographical areas, as shown in Figure 20.30. Each cluster is composed of one or more datacenters (orange circles) that can communicate with each other and which themselves consist of a set of nodes (servers) located in a close geographical environment (local network). These nodes correspond to the physical servers on which the data will be stored.

In the design of Cassandra, the nodes composing the datacenters are completely independent. Thus, the data can be stored on all the nodes of the cluster or only on some of them. This is defined by the *replication factor* specified when creating the keyspace (data pattern), which indicates the total number of replicas of a record in the cluster. The way in which these replicas are distributed over the cluster, in terms of the physical

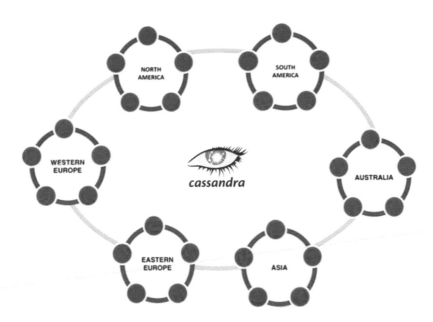

Figure 20.30 General architecture of an Apache Cassandra database [5].

location of the storage nodes and their proximity to each other, is defined according to *the replica distribution strategy,* which is also indicated when the keyspace is created.

This independence of the nodes implies that they all have the same importance and the same role. In other words, there is no master or slave node within a cluster, and all nodes can process requests received by the system. Thus, when a node receives a read or write request, it will behave as a coordinator between the client application and the nodes possessing the data, which must determine the nodes in question to make them forward the request.

This kind of decentralised architecture provides a service with high availability without a single point of failure, which is scalable according to the demand (volume of data stored). To take advantage of the intrinsic capabilities of Cassandra, we have deployed a multi-node cluster for storing and processing the first data collected. Figure 20.31 shows the physical architecture of our Cassandra cluster. Currently, the cluster is composed of a single datacenter, which itself contains four nodes spread over two different racks. A replication rate of two is chosen to allow a good robustness/efficiency compromise.

20.4.4 Logical Architecture of the Database

The functional logic chart of BiodiVstrict on which the design of the mobile application was based defines different relationships (associations) between the biodiversity elements to be inventoried, as shown in Figure 20.32. In addition to the data frame sent by the mobile application (GPS, photograph, date, element name, etc.), these associations (the belonging of a habitat to an environment, the association of one or more

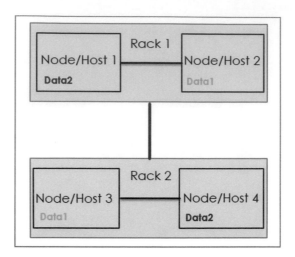

Figure 20.31 Physical architecture of the Cassandra multi-node cluster for the Chair.

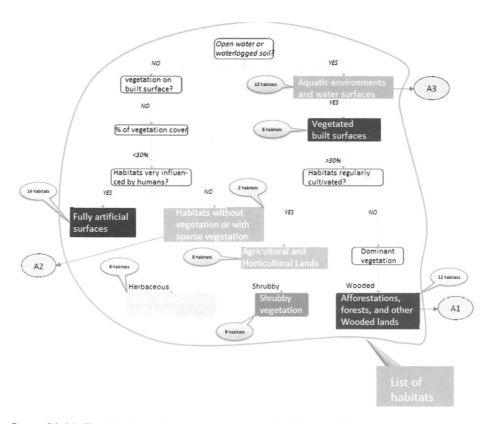

Figure 20.32 The Biodiversity element tree defined by BiodiVstrict.

remarkable elements with a habitat, etc.) were taken into account when designing the keyspace (data pattern), to avoid the loss of relevant information.

In addition, the tables in the database have been designed to respect the visualisation and data analysis features mentioned by the Chair's partners. Indeed, as the Cassandra system is distributed, it is not possible to join the tables. It is therefore preferable to group and save the "related" data in the same table so that they can be analysed and displayed. This design is based on the application's most common requests to optimise data access and led us to the "keyspaces" shown in Figures 20.33 and 20.34. Figure 20.33 describes the different tables in the "keyspace" dedicated to the "Biodiversity" component, as well as the columns of each table and their data types. Figure 20.34 describes the organisation of the keyspace tables dedicated to the "Transport" component.

Keyspace: Biodiversity

Table: Habitat

id	uuid
date	timestamp
RegionName	varchar
EnvironmentName	varchar
HabitatName	varchar
lati	double
longi	double
lambertX	double
lambertY	double
photo	blob

Table: Remarkable_Element

id	uuid
date	timestamp
RegionName	varchar
RemElemName	varchar
HabitatName	varchar
timeHabitat	timestamp
lati	double
longi	double
lambertX	double
lambertY	double
photo	blob

Table: Invasive_element

id	uuid
date	timestamp
RegionName	varchar
InvasiveElementName	varchar
lati	double
longi	double
lambertX	double
lambertY	double
photo	blob

Table: species

id	uuid
date	timestamp
RegionName	varchar
Category	varchar
SpeciesName	varchar
lati	double
longi	double
lambertX	double
lambertY	double
photo	blob

Figure 20.33 Keyspace structure for the Biodiversity component.

Keyspace: Transport				
Table: Reporting			**Table: Movement**	
id	uuid		id	uuid
date	timestamp		date	timestamp
lati	double		lati	double
longi	double		longi	double
photo	blob		Acceleration_X	double
Description	varchar		Acceleration_Y	double
			Acceleration_Z	double
			speed	double
			movement	varchar

Figure 20.34 Keyspace structure for the Transport component.

20.5 Mobile Application/Database Communication

Communication between the mobile application and the Cassandra database is illustrated in Figure 20.35. Indeed, the collected data frames are instantaneously or regularly sent to the database *via* a web service, which was developed in Java for better portability and ensures reliable and secure data management.

The storage of data collected by the Android application in the Big Data system is based on a function that checks the presence of an Internet connection. This function makes it possible to

- Instantly send data frames in JavaScript Object Notation (JSON) format to the web service hosted on a remote server using the http protocol when the connection is established. The web service in turn sends the data to Cassandra for continued storage in the previously presented tables.
- Store the data in a text file in the phone's internal memory if there is no connection.
- When the connection is re-established, a "parsing" function is launched to analyse the file and reconstruct the requests to be transmitted to the web service in order to send the data to Cassandra. Once these data are stored in the database, they will be deleted from the file.

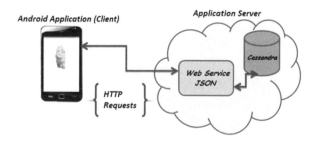

Figure 20.35 Communication between the Android application and the database.

20.6 Visualisation Application for Biodiversity

20.6.1 Technical Design and Choices

The purpose of this visualisation application is to provide the user with an ergonomic web interface for viewing and exploring the biodiversity data stored in the database. This includes an interactive web map showing the details and geographical positions of the different natural species collected by the mobile application, according to the selected filters (category of biodiversity element, date, region, etc.). The app will also provide relevant statistics on the collected data using interactive charts. These statistics come from the analysis and cross-checking of data stored according to the criteria/indicators desired by the user. Finally, the application will allow the user to export the desired data as a CSV file, to be viewed in a spreadsheet or loaded into statistical software.

The choice of a web application for visualisation is based on a logic of maximum portability, but also simplicity and lightness. Indeed, all operating systems have browsers that can use such an application, while delegating the processing part to a remote server. In addition, all modern web browsers support secure communication protocols, which facilitates design while ensuring that the set of interactions is secure.

The user interface design has been developed by keeping two key points in mind, namely simplicity and speed, to provide the user with a highly seamless experience possible. The web interface was made with *JavaScript and HTML5/CCS3*. For its part, the application part is written in JavaScript and also communicates with the Cassandra database *via* the web service described above. For the data mapping, we use *OpenStreetMap* with the *Leaflet* library. The general architecture of the application is illustrated in Figure 20.36.

20.6.2 User Interface

This section presents several interfaces of the visualisation application. To allow a flexible and intuitive choice, while remaining as exhaustive as possible, the various

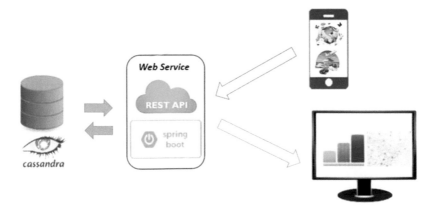

Figure 20.36 Global architecture of the Big Data application (data collection and visualisation).

biodiversity elements presented above were organised in the form of drop-down and contextual menus. This allows the user to see the inter-element hierarchy inherent to the mobile application. These drop-down menus also make it possible to just display the data desired by the user, which significantly improves the overall readability of the site. Checkboxes are presented to select the elements for display. The map then updates with the necessary markers. For the sake of ergonomics, a checkbox is available for each overall menu to filter the underlying elements. Finally, tabs are used to navigate between the different selection options, including the ability to filter the data according to the categories of biodiversity elements ("biotopes" tab), according to the location regions, dates of registration or types of animal species encountered.

Figure 20.37 shows the home page of the website with the different navigation tabs and the interactive web map.

Figure 20.38 illustrates an example of mapping a set of habitats. The figure shows the system of drop-down menus (natural environments, habitats, remarkable elements and invasive elements) as well as all the biodiversity elements selected. The map also allows the user to select the data for display related to a particular item, including their photograph, in a tooltip, by clicking on the point associated with this element on the map.

The visualisation application also allows the user to export the data recorded in Cassandra to CSV files for external use. Figure 20.39 shows an example of a CSV file, exported via the application, containing data about the invasive elements selected by the user.

20.7 Conclusion and Perspectives

As part of the Eco-design of Buildings and Infrastructures Chair, the ESME Sudria team proposed a Big Data application for collecting and classifying data in the areas of Biodiversity and Transport. The solution is based on a mobile application for collecting and reporting data to the storage and a processing system. A web application allowing the collected data to be consulted and viewed was also set up to facilitate the exploration and analysis of the stored data.

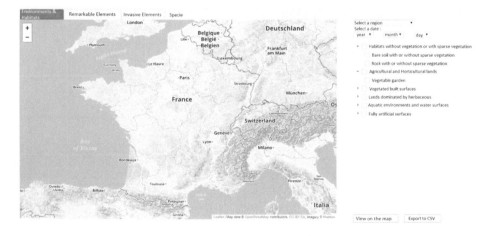

Figure 20.37 Visualisation application home page.

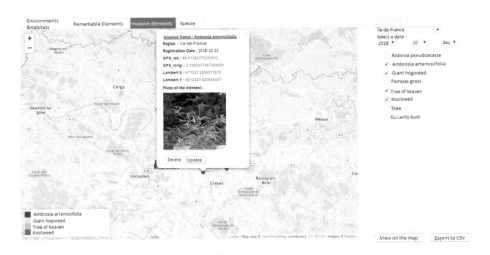

Figure 20.38 Mapping and description of selected biodiversity elements.

Figure 20.39 Exporting stored data in a CSV file.

As the mobile application is now operational and validated, the team is currently working to finalise the communication between the mobile application and the Big Data system, as well as to improve the visualisation application. Areas for improvement include the addition of statistical graphs resulting from the analysis and the identification of correlations between the reported data, using machine learning algorithms. We could also mention the export of selected data to "Shapefiles", a file format for geographical information systems (GIS).

Bibliography

Paul Viola and Michael Jones, "Rapid Object Detection Using a Boosted Cascade of Simple Features", *IEEE CVPR*, 2001, pp. 511–518.

Joseph Redmon, Santosh Divvala, Ross Girshick, and Ali Farhadi, "You Only Look Once: Unified, Real-Time Object Detection", *IEEE CVPR*, 2016, pp. 779–788.

Joseph Redmon and Ali Farhadi, "YOLO9000: Better, Faster, Stronger", *IEEE CVPR*, 2017, pp. 6517–6525.

Christian Szegedy, Wei Liu, Yangqing Jia, Pierre Sermanet, Scott Reed, Dragomir Anguelov, Dumitru Erhan, Vincent Vanhoucke, and Andrew Rabinovich, "Going Deeper with Convolutions", *IEEE CVPR*, 2015, pp. 1–9.

Documentation Apache Cassandra, http://cassandra.apache.org/, 2018.

Chapter 21

Smart Mobility

A Landscape Under Development

Fabien Leurent, Olivier Haxaire, and Gaële Lesteven

École des Ponts ParisTech, Marne-la-Vallée, France

21.1 Introduction

21.1.1 Background

Digital penetration into the mobility of people and goods has become evident over the course of a decade. Everyone inquires about travel conditions by consulting one or more mobile applications on their smartphone. In some recent cars, the dashboard has taken the form of a tablet computer. The use of taxis has been revolutionised by Uber, thanks to its mobile app. Public transport vehicles become visible on smartphone screens. More and more products can be ordered, prepaid and quickly delivered to a chosen place.

At the same time, digital technologies have continued on their great path of development. Data are now produced in abundance by diversified sensors, disseminated in vehicles, worn by people or established in the field. Large data sets are centralised and subjected to intensive calculations for various applications: operational management, customer profiling, yield management, demand simulation, etc. Functions are fulfilled by computer programmes, the "artificial intelligence" that automates them, thanks to computing devices that mimic the brain (neural network) and learning processes emblematic of data science: machine learning, deep learning, etc.

21.1.2 Objectives

In this chapter, we intend to analyse the digital transformation of transport systems and the mobility of their users, including service customers and users of physical means of travel at the same time. We describe the renewal of the "forms" of transport and mobility on three levels – technical, economic and social, respectively. On the technical level, the forms concern vehicles, infrastructures and operational processes. On the economic level, the forms relate to services, uses and commercial relations between producers and customers, as well as customer-to-supplier relationships within the transport supply. At the social level, we are interested in the individual actors that are the users in terms of use and lifestyle practices and also above all in the economic actors that constitute the transport supply. This is expanding, thanks to the development of original functions provided by new actors who take strategic positions in value chains. This results in the transformation of the actors, their relationships and their overall organisation.

In our description, we will look for explanatory factors: between the availability of technologies, the technical improvement of equipment, the qualitative improvement of the service rendered, the creation of value for the customers, the collection of this value and its distribution between actors, the profitability strategy carried out by one supply actor or another according to their position in the productive circuit and in the organisation of the mobility system. Therefore, the key concepts are value and quality: this *ultimately* coincides with the notion of performance that is essential in eco-design approaches. Our final objective will be to characterise the environmental consequences and discuss the potential of the digital transformation of transport and mobility systems for improving environmental performance.

21.1.3 Method

Our analysis is both systemic and techno-economic. Transport and mobility systems are socio-technical systems. They are technical forms designed, implemented and used for and by social actors. Their technical functioning involves and implements physical space, and collectively managed public space in particular. Public action is important at several spatial scales: from the microlocal scale of a road element to large territories, through corridors such as transport lines and urban scales. The layout in physical space is fundamental. Transport is aimed at crossing the space and its technical equipment refers to it in an essential way, as well as the infrastructures that are inscribed there and the vehicles that carry out movements and displacements according to special technical conditions. To the physical hardware triangle – infrastructure, vehicle and service – now corresponds the software triangle – Internet as infrastructure, smartphones and other computer processing units as vehicles, and computer applications as services, all in a virtual space that seems limitless but also requires collective management. The hardware and software triangles will serve as our grid of technical analysis. At the economic and social levels, we will consider the actors, their relationships and forms of organisation to highlight both the opportunities offered by the digital transformation and the challenges it brings.

We mobilised three sources of information to ground our analysis: (i) technical knowledge of transportation systems, as taught to engineering students at École des Ponts ParisTech (Leurent et al., 2015); (ii) the expertise of business strategy consultants and transportation researchers (Haxaire et al., 2018); and (iii) consultations with actors in the economic and political world (Archéry et al., 2016).

The resulting overall picture is original in terms of the width and depth of the field covered. It results in a certain complexity, which we manage by structuring the subject and also by systematic recourse to abstraction – at the risk of generalisations that might only be appropriate for some of the concrete cases covered by the same name.

21.1.4 Outline of the chapter

The body of the chapter is organised into four main parts followed by a conclusion. We first address digital technologies, the factors of their development, their distribution around the world and the concrete conditions that result from them. In fact, they act as equipment for innervating technical and social systems in the territories (Section 21.1).

We then explore their contribution to the technical subsystems that comprise mobility: vehicles become connected and "helpful", infrastructures lend themselves to expanded services and more dynamic management and technical processes are reinforced. Thus, the digital transformation reinforces the functions fulfilled and makes the subsystems more powerful, in short, empowering them (Section 21.2) (Figure 21.1).

Figure 21.1 Complexity of an urban mobility system (Jean-Vincent Sénac and DREIF, 2003)

The most radical changes concern the interactions between the transport supply and its customers. Digital technologies allow the development of multisided platforms, which are a fundamental component of on-demand transport services. More broadly, the relationship between service and customer is empowered by digital technologies in all mobility services. The "customer journey" is redesigned according to the sequence "inspire, plan, book, ticket and evaluate". The potentialities of supply management, revenue management and yield management are also greatly empowered (Section 21.3).

After having characterised the empowerment of technical and commercial forms, we examine the actors and organisations involved. Innovative functions in mobility services and services for mobility are offered by actors from the digital world. Their inclusion transforms the relationship patterns between the supply actors in relation to the demand and puts the value chains into question. These can be all the more reconfigured as pricing becomes more flexible. All that remains is to take the initiative of this kind of reconfiguration and a systemic recomposition of mobility (Section 21.4).

All in all, the digital opportunities for mobility are real, very important and even fundamental. That is why their implementation raises considerable technical, economic, societal and, ultimately, managerial challenges for the communities responsible for organising mobility systems in the territories (Section 21.5). In conclusion, we suggest a bottom-up approach to using digital technologies to serve eco-design for these systems (Section 21.6).

21.2 Enabling Technologies and Practical Arrangements

21.2.1 Global Economic Development

The human population is now very large with a high standard of living (despite disparities). It has come to impact the climate of the planet, among other environmental impacts. There is a massive demand for economic products and services. This demand is gradually being standardised, although of course there are local variations (e.g. the language chosen for the interface of a mobile application). Technologies and services are spreading rapidly, as the Internet is a great sounding box for information (in push mode or as word of mouth) and for the provision of mobile applications. An individual's equipment for accessing the Internet on their smartphone connects them to the world of economic opportunities. In other words, not only are services becoming more abundant, but also the way they are distributed to the general public has changed.

The products offered are relatively affordable and can be delivered in a few days or hours. The advent of the general public Internet was preceded by the globalisation of industrial supply chains, both in the very large-scale industrial production with huge economies of scale (see the mega-factories in China), and in the logistics of goods and products, with the facilitation of freight transport and the complexification and hierarchical constitution of logistics chains. For example, we could mention the development of container transport, the expansion of container ships or the modernisation of urban deliveries.

21.2.2 Industrial Hyperpower

Major industrial and logistical rationalisation has taken place gradually, and there has also been a massive expansion of industrial capacity across all the stages of the product life cycle. At the design stage, computer-aided design (CAD) (equipped with powerful systems, see the Catia solution by Dassault Systèmes) accrues digital avatars of physical objects – automobiles, aircraft, etc., up to packaging products – so that simulations can then be made of their usage phase as well as their manufacturing process. These avatars are immaterial assets that are used in the design phase by imagining different variants in order to then select those that increase attractiveness and lower the production cost. The manufacturing stage has also undergone an in-depth transformation, with rationalisation of production lines in factories and the increased presence of robots in the development process, workshop by workshop. The expansion of manufacturing by means of robotisation has taken place in accordance with the cumulative expansion of the design and supply-chains. The overall expansion is immense. Considered as a whole, the global industrial system probably gained a factor of 100 in 40 years, increasing tenfold in every 20-year period.

The development of the supply is therefore even more massive in terms of potential capacity than market size. Not all the potentialities have been amassed as yet, far from it. The selection is severe as a result of increased economic competition at the international level. Modern management practices have also spread worldwide in search of ever more production efficiency.

Let us mention some impressive examples of industrial achievement:

- The smartphone is a veritable mass of sophisticated telecommunication, sensor, computing and multimedia interaction technologies for users in terms of its hardware. Its software is just as powerful, with free or almost free access to a plethora of applications. This extraordinary tool for interaction is offered for a price of just a few hundred euros. As the price-performance ratio is irresistible, its use has become widespread around the globe in less than a decade.

- Electric or hybrid cars from the 2010 generation have become an established part of the commercial ranges of all major car manufacturers. Some models have been totally designed for electric use, with mass manufacturing and affordable purchase prices (Renault Zoe, Nissan Leaf, BMW iON, Chevrolet Volt and the Tesla Model 3).
- The range of playful or sporty small one- or two-wheeled vehicles has become abundant, diversified and affordable. An accessory of this kind can be acquired for just a few hundred euros, making it cheaper than a conventional bicycle but with the addition of an electric motor and a whole dynamic stability control system.

21.2.3 IT: A Major Industry and Service Sector

IT includes hardware (calculators and telecommunication means) and software methods (operating systems, programming languages, algorithms and programmes), processes (applications) and systems for processing information. Since the mid-20th century, IT has emerged as both a fundamental technology and a technical infrastructure in the lives of businesses and individuals.

Computer products include hardware and software:

- "Hardware" relates to processing equipment: computing units, user interaction terminals and transmission devices for networked systems, which have developed with the digitisation of telecommunications.
- "Software" refers to processing software, developed according to a logic of vertical accumulation. On a computer, "low-level" operations that exploit the hardware components constitute a "basement" below the user-accessible operating system, which acts like a "ground floor". Office-type applications or "web browsers" form a final floor. More specialised types of software such as database management systems (DBMS) are situated on intermediate floors.

In terms of computer subsystems, microcomputers have been available to households and businesses alike since the 1990s. Networked processing has developed in the business world since the 1960s. Its growth has been reinforced by the advent of microcomputing and the expansion of the Internet, the network of networks, and by the marketing of software for business management, which assists the accounting departments, the administrative departments (including human resources management and tax relations) and the management of customer relations and the sales force (e.g. Salesforce), not to mention multifunctional Enterprise Resource Planning software (ERP, e.g. SAP).

IT products are developed by a range of suppliers. Some have become global technology giants such as HP, Lenovo or Cisco for hardware; Microsoft, Oracle or Google for operating systems and software; and Apple on both sides. Other groups have specialised in the provision of business services, in particular for the design of information systems adapted to the specific needs of the company (e.g. IBM, computer service companies such as Atos, Stéria), possibly coupled with management consulting services (e.g. Capgemini).

Thus, IT infrastructure has progressively become highly developed equipment that is used extensively in various social and economic activities.

21.2.4 Web 2.0: A Digital and Real World Where One Has to Be Established

IT infrastructure became globalised with the great development of the Internet in the 1990s. The flagship software of this period was "web browsers", interactive communication tools for individuals equipped with their microcomputers. As such, companies and other organisations began to make themselves accessible in the virtual world, in the form of "web pages" available to everyone.

From the year 2000 onwards, the 2.0 generation of the Internet greatly increased the possibilities of interaction, thanks to specific formats for "websites" and tools for constructing them that were both simpler and more powerful.

Many individuals, both private and professional, developed "blogs", i.e., web pages gradually fed with "news". Conversely, "RSS feeds" allowed the latest information to be extracted from specified source sites. Faced with the profusion of information made available "online", "search engines" became indispensable. The Google search engine has come to dominate in the Western world, but other engines lead in other regions (Baidu in China, Yandex in Russia).

These new technical provisions radically renewed the conditions for interactions between individuals, for all forms of relationships: cooperation in the business world, private communication, B2B or B2C relations between customers and suppliers, P2P cooperation between individuals.

The economic world established itself on the web. For companies, web presence is no longer simply a matter of signalling its existence and the nature of its activities, but also of offering its products and services, and even of attracting talent by offering job opportunities. "E-commerce" has become an activity in its own right, and specialised distributors such as Amazon and Alibaba have achieved a global reach.

For individuals, web-based presentation formats became standardised under the impetus of social media. Individual "pages" on Facebook for private use or LinkedIn for professional purposes are pre-formatted to allow easier consultation by other members of the network, and the platform offers powerful tools for special requests and getting in touch with others.

In short, the web 2.0 generation took the form of a sort of urbanisation of the digital world: social and economic activities took a root with specific forms of interaction, communication and even "politeness".

21.2.5 Being Mobile and Connected: Yes We Can

Mobile telecommunication and its networks have been dynamically developed since the 1990s. The familiar question "Where are you?" is emblematic of the potential for interaction that has opened up in any place, becoming "ubiquitous" in other words. Connection to the web was proposed quickly. It found its ideal form, thanks to the appearance of smartphones, with the first iPhone dating back to 2007.

Its shape is ideal on the material level in terms of portability, manoeuvrability and usability. It is also ideal from the perspective of software, thanks to the very rapid development of a myriad of "mobile applications", most of which are intended for web interactions.

Many of the companies on the Internet have developed their own mobile application, to make themselves accessible from smartphones. They have become the preferred interactive terminals for individuals, especially for social media activities.

The ultra-fast spreading of smartphones testifies to the extraordinary advantage offered by the convergence of mobility and connectivity. This advantage is multifaceted. On the one hand, individuals can connect to any service available on the web at any place and at any time. They can therefore reuse "passive time" and also access dynamic information in any situation, especially while on the move. On the other hand, individuals can report dynamic events voluntarily or involuntarily if their digital traces are exploited by a third party (Figures 21.2 and 21.3).

On the contrary, smartphones can take full advantage of geolocation technologies. Automatic capturing of satellite signals (GPS) or by antenna networks (GSM, Wi-fi, Bluetooth) allows for instant geolocation that can be related to a geographical information system (GIS) and presented by a mobile application as a point on an interactive map of places.

The relationship between individuals and geographical space has been profoundly altered. Geographic information has become an abundant deposit, and mobile applications constitute the platform for its exploitation. The information obtained by individuals is not only particularised, but is also even customised, thanks to the possibility of making requests to identify opportunities in places – routes, public transport stations, shops, restaurants, relay points, individual addresses, etc.

We will address the potentialities opened up by the convergence between mobility and connectivity later on, as well as the renewal of the relationship to the physical space and its synergy with the virtual space of the world established on the web. At this point, let us remember that the Internet triplet–smartphone–mobile application constitutes a "software triangle" for every mobile and connected individual.

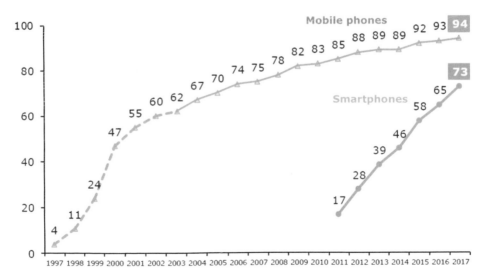

Figure 21.2 Rapid spreading of smartphones in France (Credoc, 2018)

Figure 21.3 Dataïku Big Data Platform: an overview of an application (Florian Servaux, Coyote, 2018)

21.2.6 The Era of Big Data

Big Data refers both to the data now available in abundance and to those data processing techniques that are the equivalent of extraction platforms and refineries for oil and gas fields.

Data are produced with a wide variety of content and forms, including digital traces left on telecommunication networks and websites by individuals in private or professional situations, and "observations" collected by various fixed or mobile sensors. Mass-produced equipment is inexpensive, and its production cost essentially includes field implementation, energy supply and data collection. Video or infrared cameras, mobile network tags, RFID chips and detectors, various ticketing badges, microtemperature sensors, pressure sensors, pollutant or smoke detectors, or the various status sensors in all bodies of vehicles and other machines, etc., – the list is far from exhaustive.

These deposits constitute commercial opportunities for companies. The companies have powerful motivations:

- to study their own operational functioning, retrace it in real time and use it dynamically, or to better plan it in real time;
- to study their clientele: to observe the products purchased and the circumstances in terms of time and space, to trace the intentions attested by information requests made on the website, to identify a customer by means of any medium and to analyse their traces for "profiling" him or her as part of a "demand segment";
- to simulate prospective product and product range scenarios by predicting their potential demand;
- to jointly analyse supply and demand data, as well as production data, in order to optimise combinations and improve the respective provisions.

The range of opportunities is extremely wide and companies are gladly signing up for them. In medium and large companies, the role of Chief Information Officer has appeared beside that of Chief Data Officer, as the person responsible for leading the digital transformation of the company. This is an obvious sign of the attention paid to data and the interest in comprehensive processing, unifying the functions carried out in the company.

These opportunities for companies lead to the development of a whole range of data, infrastructures and processing software:

- Data brokers have specialised in the acquisition, analysis and provision of data sets on individuals in various everyday situations. Tracking and digital profiling can characterise behaviour and also influence it, for example, by predicting health status from customer data, or by offering bank credit on terms that depend on consumer behaviour.
- Clustered servers. Large data sets require large digital storage spaces. Their systematic analysis includes particular data matching, which multiplies the need for memory space and computing capacity. Few companies have a specific hardware infrastructure for storage and calculations, so many prefer to outsource these functions to specialised suppliers, which include several web giants (Amazon, Microsoft, Google, etc.). The associated hardware infrastructure consists of "server farms", accessible via the web and therefore in the cloud, and exploited using the service's software.
- Cloud storage and computing. The cloud's Big Data services enable resilient and distributed processing of unstructured and Big Data sets. To this end, Google has created the MapReduce framework, which is implemented in particular in the Hadoop software proposed by the Apache Foundation. Since its foundation in 1999, it has been one of the major open-source institutions. In a cluster, each computer is a node with its own storage space. To process a particular request, the MapReduce model distributes the processing as tasks to be processed by different nodes (Map), then extracts the results and summarises them (Reduce) into a single coherent response.
- NOSQL DBMS. The data sets are not only large but also diverse. Relational DBMS manage and link tables of data organised in rows (by entity) and columns (by attribute). This notion is used to handle certain types of requests in Structured Query Language (SQL) mode. They are not very suitable for Big Data, which is why Not Only SQL (NOSQL) DBMS have been developed, with four main types that differ in terms of the orientation of their "documents", "graphs", "columns" and "key-value". Since 2003, Google has developed the Bigtable column-oriented DBMS, the open-source version of which is called HBase and has been adapted by Facebook in particular. Amazon Web Services has developed its own Dynamo solution (key-value oriented), as have Microsoft Azure, Salesforce, IBM, Oracle, LinkedIn, etc.
- Data lakes. Traditional data warehouses process information according to pre-defined data patterns. When it is difficult to anticipate all the applications and thus specify the associated data patterns, the "data lake" solution is to store the raw data by subsets, each with metadata describing its content but regardless of source and format. A variety of specific tools can quickly exploit data as new

requirements arise. The main advantage of these data lakes is their flexibility, which allows the "scalability" of the solution provided to the client company.

- Machine Learning (ML). Machine learning is intended to create predictive models for decision-making support. It comprises a set of artificial intelligence techniques, associating typical forms of models (including neural networks, Bayesian networks and other Markov chains) and algorithms to set specific parameters (e.g. relative weights of connections between neurons) and update them automatically, by learning from data or events. The algorithms differ depending on whether the learning mode is supervised (based on a set of application examples) or "unsupervised". This restricted mode is intended to explore the data, to group them according to proximity criteria defined by the user of the method. Reinforcement learning is common in robotics. Each of the model's predictions for a given data point is subject to a quality evaluation, the score of which determines a "reward signal" which is then fed back to the algorithm strategy to optimise the prediction quality.
- ML software suites and Big Data industrial platforms. ML tools process data in the form of large matrices (Tensors). There are a variety of existing specialised software suites: SciPy, Torch, CNTK, Theano or TensorFlow developed by Google. Most are available as open source. "Turnkey" solutions are offered by major cloud players or specialised companies (in particular the French company Dataïku). These are generally built from open-source "software stacks", supplemented by advanced configuration and administration functions, as well as specific programming and user interface tools.

In short, a whole ecosystem of hardware and software has progressively built up around Big Data. Automation of processing is a major issue, because companies that invest in this kind of computer equipment want to make their investment profitable over several years, which leads to the gradual incorporation of the data produced over time and progressive updates to the predictive models. In addition, the companies' digital needs are likely to grow, which encourages the decision to invest in a flexible solution, minimising the capital invested, and the payment-per-use of the IT operating costs, while ensuring a level of reliability, availability and security that is difficult to achieve for companies not specialised in IT.

21.2.7 Robots and Internet of Things (IoT)

A robot is an automaton equipped with abilities to interact with its environment, and with a certain autonomy. There are purely software-based automatons, such as "chatbots" or conversational agents able to "converse" in textual mode, and those with hardware support, capable of mechanical movements and actions, which are robots in the original sense of the word. A chatbot programme associated with hardware to provide an audio and voice mode constitutes an intermediate category. The same applies to video cameras that automatically adjust their orientation and shooting mode.

Mechanical robots have spread throughout the world of industry to make up mass-produced products. The transition to the digital mode from analogue modes with poor flexibility has been gradual. The miniaturisation and low prices of computing units greatly expand the possibilities of decentralising technical or management processes to particular entities. There are multiple fields of application, in particular for the activation of devices "in context" according to the ambient conditions. The empowerment to perform basic functions allows the robot to be integrated into higher level processes, where it will be controlled by instructions from a central control station.

Most industrial robots have moving parts but a fixed position. By giving the robot the ability to move, its field of intervention and field of application are broadened. Android robots have versatile vocations. Remote-controlled drones and self-driving vehicles therefore constitute robots, among many other types. The economic stakes are obvious for services such as public transport: we will come back to this later.

Between the different automated types of equipment within a technical system, the centralisation of orders ensures the coherence of the respective actions and their cooperation for a common goal. It is increasingly operated *via* the Internet. The result is an IoT made up of subsystems. The Amazon IoT solution allows "objects" to be connected, to acquire the information that they generate and to interact with one another. Several subsystems of this kind may coexist in shared places, especially as the connected entities will be able to move. This will result in a generalised IoT, in which interactions between entities will not all be centrally coordinated, far from it.

The technical potentialities are extremely broad and promising for functions that operate in the open environment. However, principles for presence, action and circulation in such an environment will have to be put into effect. One solution will be able to establish protocols for different potential interactions.

21.2.8 Summary

We have presented digital-based innovations according to the chronological order in which they spread through the economy and society. This chronology differs from that of particular inventions as it basically involves their complementarity, the mutual reinforcement that results from it and the gradual build-up.

An entire digital-based technical ecosystem has been gradually built up and continues to grow. This ecosystem equips social life and economic activity and now constitutes a basic infrastructure. In the sections that follow, we will explore the facets, forms and issues in the field of transportation and mobility (Figure 21.4).

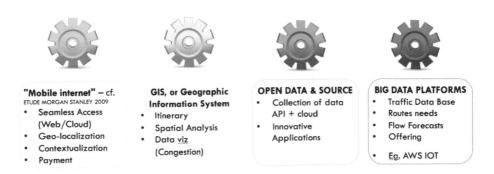

Figure 21.4 Summary of enabling technologies (Haxaire et al., 2018)

21.3 Diversified Physical Means That Are Better Innervated and More Flexible

As a technique for crossing space, transport has an irreducible physical nature. Its technical composition as three interacting subsystems – infrastructures, vehicles and processes (services and protocols), respectively – is fundamental. Improvements and innovations are added onto this basic organisation. We will sift through each subsystem to identify the places, roles and influences of the digital world. We will first consider vehicles, then infrastructure and finally protocols and services.

21.3.1 Vehicles: Diversification and Empowerment

<u>Diversification of vehicles</u>. The new industrial order expands the design, manufacture and distribution of vehicles. These opportunities are seized by the manufacturers of "transport equipment": cars that are medium-sized vehicles, large vehicles (road, rail, air, as well as river-based or maritime) and also small vehicles intended primarily for the urban environment. For the mobility of people, various means of transport are proposed to augment pedestrians, including those with one or two wheels, with two wheels arranged in tandem or in parallel, with or without seats, with or without engines and even without wheels, such as "Seven leagues boots", a board without wheels (Figure 21.5).

The third dimension of space becomes accessible. Digitally controlling several engines linked to propellers makes it possible to design utility drones or flying cabs, which broadly update the family of helicopters.

Digital technology is an essential part of hardware design, thanks to digital mock-ups. In some cases, it is also used in day-to-day operation, especially in terms of

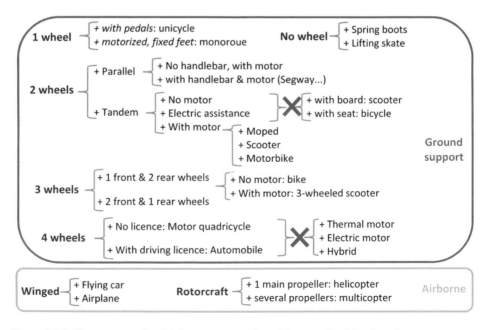

Figure 21.5 Taxonomy of vehicles, restricted to "domains" of land and air

managing the stability of vehicles without tandem wheels, and to manage the traction of vehicles with several engines and propellers.

The diversification of vehicle types goes hand in hand with the expansion of well-established types. Let us focus on the automobile, a mid-sized vehicle with a majority share in the mobility of people. In functional terms, the characteristic capabilities of an automobile are related to its capacity, traction, sturdiness, sobriety, safety and comfort.

Capacity. This means the ability to accommodate one person or several people in the passenger compartment and offer luggage space. Digital technology intervenes in a poor manner. CAD optimises the passenger compartment and small storage spaces. Autonomous driving may eventually dispense with a driving position, which will free up space, especially in taxis.

Comfort. The automatic personalisation of seats for an occupant according to previously recorded individual parameters and fine thermal regulation are improvements that are possible, thanks to the coupling of devices (mechanical or thermodynamic) and digital management based on the sensors. Other improvements specifically concern the driver: automating the activation of headlights or wiper blades, adjusting mirrors, closing doors and seat belts, centralised door opening or closing control. Many aids are offered for the current use of cars. The dashboard is completed, sometimes even completely replaced, by an interactive tablet (a touch screen with audio and voice functions) that facilitates access to vehicle functions. Digital technology also lends itself to entertainment devices, to the direct benefit of the passengers and the indirect benefit of the driver.

Security. Digital technology extends the supervision of the state of the vehicle. Some of the devices mentioned in relation to driver comfort also concern the management of security. The physical state of the vehicle as a set of elements is now controlled by a set of sensors, which monitor the tyre pressure, the liquid levels for the braking system or for the combustion engine cooling system, for example. These sensors fulfil the alert functions and help to make the vehicle more reliable, thus avoiding potentially dangerous accidents. Intelligent driver-assistance systems participate more actively in safety, through speed control, lateral trajectory control, spacing control in relation to the vehicle in front, obstacle detection, etc.

Sturdiness: digital technology contributes to this indirectly by both the optimisation of the design and the improvement of reliability and avoidance of accidents.

Sobriety. Electronic control of the engine speed has become very refined, in order to reduce energy consumption, noise and pollution emissions. The automation of the gearbox is leading in the same direction. The fullness of the tank or battery is measured and reported to the driver, as are the real-time energy consumption rates, which allows the driving style to be adjusted.

Traction. various driving aids previously mentioned with regard to safety contribute to the management of movement. Digital technology also makes it possible to manage motor capacity, especially for vehicles with multiple engines (hybrid combustion and electric engines in particular). Other devices assist the driver during parking manoeuvres by observing the immediate environment around the vehicle with cameras and providing an overview on a dedicated screen or window on the tablet.

Thus, digital technology amplifies the essential characteristics of an automobile, assisting interactions with its occupants, with traffic and parking conditions and with the environment (energy consumption, pollutant emissions). More profoundly, digital technology is radically transforming automobiles by adding two major characteristics: connectivity and serviceability (or helpfulness).

Connectivity can be defined as the ability to exchange information in real time to interact and cooperate with other entities, including other vehicles, infrastructure operators or a service system manager. On-board units (OBUs) are emblematic of this, as they allow drivers to benefit from electronic toll collection and to obtain dynamic traffic information, route recommendations and guidance services in the field. The interaction has a reciprocal way: the driver and/or the vehicle can provide information to an external system, to report a traffic incident, local congestion, signalling failures, etc. Geolocation plays a key role in these relationships.

"Serviceability" concerns the pooling of usage: pooling in real time for carpooling, or delayed time for car sharing. This faculty is comparable to the integration of a taxi within a central reservation system. It requires the vehicle to be integrated with an intelligent system that centralises customers' requests and assigns some of them to the vehicle. Serviceability refers to the set-up within the vehicle to interact with the system and foster the requests for reservation or hailing (in digital or visual mode), to allow the vehicle (in the absence of the driver) to be accessed by other users, to participate in one or more systems.

Overall, the digital transformation of the vehicles is essentially a managerial transformation: (i) internal management, (ii) interaction with immediate or remote local conditions, in a system ensuring traffic flow and/or parking and (iii) inclusion in a system providing one or more mobility services as a mobile element.

To make an analogy with biology, digitisation tends to reinforce the vehicle's nervous system as an organism, and then, on a larger scale, to include the vehicle in an intelligent, innervated system that exceeds and transcends it. Advanced driver-assistance systems do contribute to driver comfort, safety and fluidity of journeys. Fully autonomous driving will lead to the increased potential for cooperative use of the automobile and for various logistics services (Figures 21.6 and 21.7).

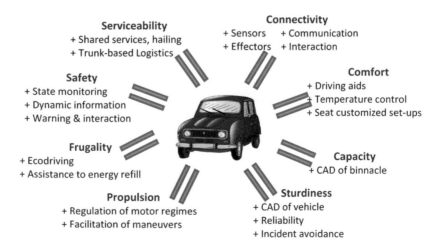

Figure 21.6 Automobile and its digital expansion

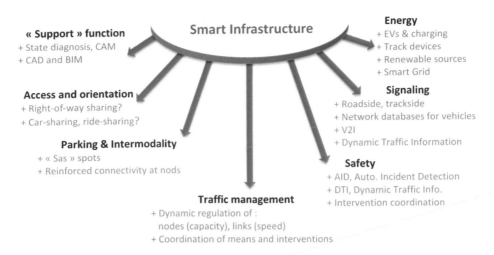

« Support » function
+ State diagnosis, CAM
+ CAD and BIM

Smart Infrastructure

Energy
+ EVs & charging
+ Track devices
+ Renewable sources
+ Smart Grid

Access and orientation
+ Right-of-way sharing?
+ Car-sharing, ride-sharing?

Signaling
+ Roadside, trackside
+ Network databases for vehicles
+ V2I
+ Dynamic Traffic Information

Parking & Intermodality
+ « Sas » spots
+ Reinforced connectivity at nods

Safety
+ AID, Auto. Incident Detection
+ DTI, Dynamic Traffic Info.
+ Intervention coordination

Traffic management
+ Dynamic regulation of :
   nodes (capacity), links (speed)
+ Coordination of means and interventions

Figure 21.7 Smart Network, a system of systems, established on the basis of the physical infrastructure

21.3.2 Infrastructures

A transport infrastructure is a physical object located and inscribed in space, to provide traffic, access and/or parking functions for the mobile entities that use it. There can be two levels of entities: vehicles and cargo units, which are either travellers or goods.

The fundamental functions of the infrastructure are fulfilled through the physical constitution in several technical subsystems, each of which ensures a particular function: (i) support, (ii) energy processing, (iii) signalling, (iv) access management and ground guidance, (v) traffic management, (vi) security management, (vii) parking, waiting to board and intermodality. We shall review these subsystems to mark the sites of software and the influence of digital technology to be found in each of them.

Support function. The loaded or unloaded vehicles have a large mass that solicits traffic support, increased by dynamic movement. For the road and rail modes, the support is built on and in the ground in a solid way, in order to offer a long-lasting geometric form and mechanical properties of rigidity (particularly for rails), strength and adhesion. Without going into the details of the physical constitution, digital technology transforms the diagnosis of the current state (measurement of wear and detection of anomalies) by multiplying the sensors and increasing the dynamism of the maintenance, which is not only preventive but also and especially predictive. Digital technology is also involved in design (CAD) and construction (site scheduling, automated progress monitoring). The design and construction functions are renewed, thanks to the BIM (building information model). The digital model of the infrastructure is shared between the actors who participate in these functions.

Functions relating to energy. Rail modes for electric motor vehicles require a power supply along the track. More generally, the various devices that equip the infrastructure, such as traffic signals or variable message signs, require electricity, but in much smaller quantities than those required for the circulation of vehicles (the so-called weak currents, as opposed to the so-called strong currents). In addition, with its large footprint, the transport infrastructure offers great opportunities for harvesting renewable, solar or wind energies. Moreover, its deep body could lend itself to heat exchanges with the surface and thus allow the storage throughout the seasons. These opportunities can be enhanced by the advent of energy "smart grids".

Signalling function. In the railway mode, signalling is used for trains in order to manage their progress on the track. In the road mode, signalling is intended to inform drivers about the destination places accessible at the next junctions. The respective systems along the infrastructure are very different in material terms. However, the development of digital technologies tends to bring the two modes together technically, transferring signalling to vehicles, thanks to embedded devices connected to information and management centres. Embedded systems are more powerful as the information is personalised and better adapted to particular needs.

Access management and ground guidance. These functions are strictly performed in rail mode, by means of rail and switch guidance, as well as strong cooperation between mobile units and centralised traffic management. Digital technology makes the interaction between the mobile unit and the management centre more flexible and more versatile, in particular for playing more finely with the different sections of track when one of them undergoes an accident (specific faults, train breakdown, intrusion on the road, etc.).

In the road mode, the corresponding devices include: (i) ground marking to identify traffic lanes and special places (pedestrian crossings, weaving sections, bus or taxi stations); (ii) junction management to divide the right of way between the different traffic flows in time; (iii) organisation of parking along the road and lateral access from the buildings bordering it; and (iv) access authorisation according to vehicle types and, where applicable, payment of a toll. Digital technology greatly increases the technical possibilities:

- Dynamic marking devices make it possible to vary the assignment of a lane to a direction of traffic, or to certain categories of traffic, over time.
- Access authorisations can also be dynamically managed, as lanes dedicated to multiple occupancy vehicles can be managed by varying the threshold of the number of occupants per eligible vehicle over time.
- These so-called HOV (high-occupancy vehicle) lanes can be shared with toll vehicles. This is the HOT (high occupancy or toll) concept.
- Similarly, parking along a road or in the car park can be managed dynamically, in terms of pricing and access conditions.

In general, digital technology can expand control and boost the detection and punishment of offences, to the benefit of collective management. Communication between the management centre and the vehicles or individuals (smartphones) makes it possible to provide upstream information, to prevent and to guide, in cooperation with the previous function.

Traffic management. While access management is aimed at each mobile entity individually, traffic management is specifically aimed at interactions between the mobile entities. Some devices participate in both functions. In particular, traffic signals at a junction distribute the access rights over time and also serve to measure the circulation of the different flow currents per cycle over time. Digital technology allows finer dosing and adaptability to needs in real time, for example, the prioritisation of trams or buses or, in times of very low traffic, changing to a green light for cars detected remotely by a camera or by a specific form of telecommunication. It also enables vehicles' runs to be better coordinated along a transit line, to ensure safety intervals as well as to deliver more regular service headways at the stations. Finally, it allows better detection of all types of incidents in order to react quickly in an adapted way that is most conducive to resilience.

Security management. The speeds and masses involved in vehicle traffic mean that any accident is potentially dangerous, including any type of collision, a vehicle exiting a lane, a pedestrian falling in the way, etc. The latent risks are managed on several temporal levels. Primary safety aims to avoid accidents and concerns the design of the road and its operating modes, as well as its equipment with safety barriers, or the equipment of the vehicles with protective devices. Secondary safety aims to reduce the severity of an accident, by activating a full solution through emergency braking and triggering defences. Tertiary safety aims to mitigate the consequences of an accident by intervening as quickly and efficiently as possible to treat the wounded (by means of emergency vehicles in road mode).

Digital technology is a privileged tool for enhancing security in all its states, which it does by means of risk detection, anticipation, immediately alerting emergency services and facilitating their arrival to the spot, alerting other users preventively and avoiding secondary accidents.

Parking, waiting for boarding and intermodality. Infrastructures also provide parking functions for vehicles, waiting spaces for incoming or connecting travellers and for changing modes of transport. Transition contexts are particularly sensitive for travellers, who are exposed to uncertainty in a place that is unfamiliar to them, and to situations of congestion and discomfort. Digital technology that allows for personalised dynamic information is certainly valuable and requires a telecommunications infrastructure that has the appropriate dimensions for rapidly serving large numbers of requests.

This already long inventory is not exhaustive. It demonstrates the very wide domain of digital opportunities for the different functions expected of infrastructures. Digital technology makes each of the functions more responsive and productive, in a word, more nervous, and also causes them to cooperate more closely in the service of greater overall efficiency. Overall, it expands functions and cooperation between infrastructure and mobile entities at both vehicle and passenger levels. It also expands synergies between transport infrastructures, energy infrastructures and telecommunications infrastructures.

21.3.3 Operational Processes

A process is a complex action composed of elementary operations that mobilise various technical means and involve agents, and potentially also other actors. Transport services rely on processes that engage infrastructure and vehicles, among other resources: agents, telecommunications, etc. We distinguish between offline design and construction processes and online actions for routine maintenance, the operational use of an infrastructure or public transport service, the operation of an on-demand transport service, emergency interventions and, finally, for the "internal" logistics of the service. We will review each of the types before summarising the influences of digital technology on the whole process.

Design process. CAD has been developed for both vehicles and infrastructure to capitalise on technical knowledge (numerical models of components, subsystems) and to simulate "compositions" with a strong emphasis on interactions, improvements and revisions. The manufacture takes place on location for a particular site. The development of BIM promotes the digital design of infrastructure elements, from design through to construction and use. Digital technology still makes it possible to conceive original transport solutions in an integrated way, whereby the vehicle, the infrastructure and their interactions would be innovative: see personal rapid transit (PRT) or Hyperloop projects.

Construction process. Digital tools are involved in the construction of infrastructure, to calculate the scheduling of the site and also to monitor the progress (on the site manager's tablet). For complex constructions such as a large bridge, the high precision assembly of large elements requires sensors, a calculator and positioning algorithms. An embryo of robotisation can be detected in this. However, robotisation is still far from being fully exploited and will require the forms that have yet to be invented. One major challenge is that of dividing manufacturing between remote sites and on-site construction in order to achieve economies of a scale comparable to those obtained in automotive or aeronautic construction.

Maintenance and servicing process. Orientation towards saving money, driven by low-cost airlines in the world of aviation, has profoundly transformed aeronautical maintenance processes. These are organised at three levels of depth, with cycles that are longer when the intervention is more profound. At each level, the sensor equipment reveals wear conditions and allows for the targeting and dosing of replacements. The "commercial state" must also be restored at the end of each flight, by cleaning the interiors, emptying and filling with water and other consumables, refuelling and replacing meal trays.

These processes are transposed to the maintenance and upkeep of other types of vehicles, and even infrastructures. For these, major maintenance is carried out in long cycles of 30 or 40 years, while medium maintenance (e.g. road resurfacing, rail polishing, renovation of a technical subsystem) has cycles lasting several years (between 5 and 15 years). Small maintenance is carried out through interventions targeted to correct specific defects, or through relatively short cycles (e.g. maintenance of green spaces). "Cleaning" is important for spaces intended for travellers, as is "maintaining viability" (e.g. the removal of objects hindering traffic, snow removal, preventive salting). In this respect, digital technology is a key factor in the weather forecasts that condition preventive actions to cope with bad weather.

Emergency intervention process. The maintenance and upkeep of the infrastructure, as well as its construction, require large or small, short or long on-site projects. The question of robotisation is not only posed to increase economic productivity, but also to allow faster recovery, thus minimising disruption to traffic, and moreover to avoid physical risks to human agents. Thus, flying or wheel-driven drones can intervene in dangerous conditions (e.g. repositioning a road sign moved by high wind). When addressing the issue of infrastructure security, we have already discussed the issue of fixed or mobile sensors spread out in the field to detect incidents and report them to the operating system, as well as the role of digital media in disseminating alert messages for service users. We also mentioned the interest in quick intervention in the field and the contribution of digital technology in this sense. Another source of opportunities concerns the preliminary positioning of the means of intervention. The analysis of accident data and intervention needs, considered in relation to the traffic conditions if possible, makes it possible to identify positions that are "advantageous on average" for possible means. These positions may vary according to the day and the hour for a concrete network.

Infrastructure operations in the current system. This involves bringing the operational conditions together (e.g. the energy supply) and ensuring access and traffic management. We have already noted the interest in centralised management based on an information and action system, combining sensors and actuators. In addition to their role as receivers of instructions and recommendations, users' participation in the acquisition of information turns this kind of system into a draft of a multisided platform, in this case for the passage *through* the infrastructure.

Public transport (PT) operations. We have already indicated the role played by digital technology in the regulation of PT vehicle traffic, which is especially used to regulate the intervals between two consecutive service runs. This helps to improve the regularity and reliability of the service, and therefore the quality for users. Full driving automation has already been implemented on metro lines, as their closed-circuit operation favours the principle of better regulation. The installation of landing doors allows better respect for the processes on the part of the users and also increases security. Other forms of automation are used to regulate the temperature in spaces hosting travellers, making it possible to improve comfort levels. Other sensors measure the filling of vehicles. The centralisation of this information and its retransmission to travellers in the summary form allows them to place themselves better inside the trains and again increases the comfort of use.

The full potential gains are still far from being reaped. On a given line, the dynamic management of traffic does not yet take into account the numbers of passengers concerned on board or in the station. Coordination also remains weak between different lines. Some stations have dynamic systems to synchronise the respective service runs as much as possible, thus minimising waiting times for connecting travellers. However, between two closely parallel lines, there is little or no dynamic reinforcement of the service of one in case of failure of the other.

Transport on-demand (ToD) operations. A ToD system connects a fleet of vehicles and drivers, with requests spread out in space and time, to transport each customer individually (taxi) or in small groups (shuttles). In addition to the requests, the routes of the vehicles are widely dispersed, which differs from PT by lines. The interaction between supply and demand is mediated by a platform that centralises information and allows each customer to be assigned to a vehicle. Transaction operations concerning information about pricing, expected waiting time and travel time, as well as vehicle reservation and payment, have been radically simplified through digitisation by companies such as Uber and Lyft, quickly followed by large taxi companies that had to adapt (e.g. G7). Companies such as Uber have also radically transformed the management of vehicles and drivers as production means. The form of employment is reduced to a sort of admission to a club and to involvement in a kind of race to serve customers, with remuneration also paid out on a "gig" basis. Each driver is responsible for their vehicle, which they must keep and maintain in good operational condition (refuelling, maintenance, parking, cleaning), which allows the company to avoid operating a depot comprised of a garage and workshop.

In these conditions, the coordination of the means is primarily played out in the virtual world, replacing a central hub where agents would traditionally be managed. This "uberisation" of the involvement of agents in production has led public authorities to redefine the forms of employment. However, the fundamental fact remains that the centralisation and ubiquitous accessibility of information allow ToD services to cover the field and the market much better. The benefit to the customers is obvious. The prospect of completely autonomous driving vehicles on the road promises a significant reduction in production costs. However, it will require the fleet to be maintained in good operational and commercial conditions (cleanliness) in a different way, presumably by the establishment of *ad hoc* workshops in the field.

Internal logistics processes. This analysis of Uber-style service illustrates the potential of an information centralisation platform in dynamic interaction with productive resources equipped with a geolocation and information processing terminal, and with connected customers. This kind of a communicative system opens up very broad perspectives for all "hire or reward" transportation services: for customer service in public transport, individual on-demand transportation in various forms, including car sharing, bike sharing, park-sharing and carpooling, as well as urban logistics (parcel collection and delivery). More broadly, it effects any logistical process distributed in space, for example the reception and care of travellers in a terminal, or tourists in a place to visit, or the processing workshops in a production line in factory, etc. The ubiquitous and instantaneous provision of information, coupled with its automated processing, reduces transaction costs to nothing or almost nothing. This then radically transforms the technical nature of production and leads to a fundamental overhauling of roles and work positions in productive processes, while also affecting the professions of the agents within the organisation. In other words, the relationship to physical space is transformed by information connectivity: the notion of logistics is deeply renewed.

Summary for the technical processes. The new information order expands the logistical potentialities of playing with time and physical space in a general way. This transformation is far from being over in some services – it is the most advanced in the world of air transport. The redesign of a productive service is an operation of an architectural order. It concerns the architecture of information systems, which are classically analysed on the three levels – functional, application and technical, respectively. The technical architecture will highlight (i) the fixed and mobile sensors and actuators; (ii) the connected and geolocated terminals, and the productive resources that are equipped with them; (iii) physical processing centres and information processing centres; and (iv) transmission channels and networks, which are again physical on the one hand and informational on the other.

The application architecture maps physical processes to computer applications performing associated processing. In the next part, we will see that computer applications increase the possibilities for commercial operations and are increasingly replacing interventions by human agents. Beyond simple automation, it is a question of robotisation for processing of an administrative nature.

Functionally, the architecture will distinguish between the things in the two physical and virtual worlds less, or not at all. It will highlight the types of clients, the services rendered and the major productive functions, as well as their links.

Finally, the development of robotisation in the physical world as well as in the virtual world will in turn motivate the architectural redesign of productive systems. Welcome to the world of the IoT!

21.4 Service-Level and Commercial Overhaul of Mobility

In the traditionally technical world of transportation, innervation through digital technologies empower vehicles, infrastructure and operational and managerial processes. This joint empowerment represents a new opportunity for business initiatives aimed at the demand for mobility, both in terms of extending and renewing the forms of services, and reinventing the relationship with customers, in particular through personalisation.

21.4.1 Diversification of Mobility Services

Let us consider the mobility of people in the city. We talked about the emergence of very small individual vehicles that augment pedestrians. These means, namely bicycles and scooters, can be used by a large part of the population and can be shared by networked services. We also mentioned the improvement of the quality of public transport service, in terms of the reliability of the service and the provision of dynamic information.

In the world of mobility services, the most significant transformation concerns the development of "intermediate" mobility services. These are hybrid forms between the two classic types, namely the individual vehicle (especially passenger cars) on the one hand and public transport operated according to lines on the other. The intermediate services fall into the category of public transport and thus the functional economy. They take very diverse forms:

- Bike sharing. Fleets of bicycles with or without an electric motor are offered for hire by platform services. Some services are based on the stations for anchoring bicycles (e.g. Vélib'), while others are free-floating (e.g. Mobike).
- Car sharing. Fleets of cars with combustions or electric engines are offered for hire by platform services, with stations (e.g. Autolib') or else free-floating (e.g. car2go). For the services with stations, a distinction is made between hiring with and without the constraint of returning the vehicle to the station from which it was borrowed (with: two way, without: one way).
- The taxi mode has been reinvented by the so-called hailing platforms: the customer calls and "hails" the service via the Internet. The debate between taxi or passenger car with driver is mainly administrative and legal. Customers have become less concerned as the taxi companies have adapted to the new transactional format.
- Likewise, the shuttle mode has been reshaped on the basis of hailing platforms. The traditional offer of point-to-point connections is now expanded by collective

taxi services, such as Uber Pool, Lyft Line, Bridj and Padam, which collect several people on the way in order to share the cost and reduce the price for each of them.
*   The principle of cost sharing also characterises BlaBlaCar-style ridesharing. It differs from a taxi service as the driving is not billed, and the economic principle is the sharing of costs. Modern ridesharing is also based on a platform application, which makes connections by offering certain guarantees: by registering the transaction between driver and passenger in the system, as well as allowing passengers to assess their driver, and likewise allowing the drivers to assess the passengers.

The development of intermediate mobility services is based on several factors that concern both supply and demand. On the supply side, the platform is a centralised information and management system that strongly automates the contact with customers. This makes it possible to largely discharge the drivers and to improve their productivity, and thus also that of their vehicle. The connectivity of the vehicles makes it possible to follow them up in an automated manner and also to dynamically reposition them closer to the demand, either through the drivers or by special repositioning means (jockeys equipped with foldable two wheels for cars, trucks for bicycles). The service literally goes to the customer. On the demand side, the consumer saves waiting time and also benefits from the facilitation of transactions: smartphone touch screens are the new standard for visualising the positions of the vacant vehicles and making a choice. Applications are designed for easy use, which becomes standardised and routine. These intermediate mobility services constitute public transport options, particularly for connections that are poorly or not served by transit lines, or outside of service hours. As opposed to individual vehicles, the user is relieved of the need to keep and maintain a vehicle. They may also be deprived of certain possibilities for personalisation (e.g. fitting seats for young children) (Figure 21.8).

Intermediate mobility services have been classified according to three types of business:

*   P2P, Peer To Peer, between individuals for ridesharing, for the sharing of cars or parking spaces.

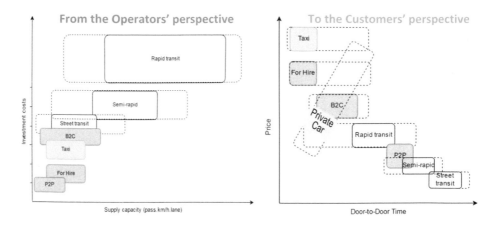

Figure 21.8 Positioning of intermediate mobility services (Berrada et al., 2017)

- <u>B2C, Business To Consumer</u>. A company provides vehicles it holds and maintains. This is an investment in the economic sense (which is broader than the accounting sense) that is strictly business-specific and requires profitability at a much higher level than the simple intermediation platform between customers and service providers.
- in <u>For Hire</u>, the company also provides the driving service, which obviously raises the production costs and therefore the price level. Taxi and passenger car with driver services fall into this category.

The prospect of autonomous driving obviously stimulates the imagination of business developers. It would cause B2C and For Hire to converge, as well as ridesharing and car sharing. The distinguishing features will *ultimately* be reduced to the type of vehicle and the mode of sharing: pooling including real-time, as for ridesharing and shuttles, or in alternation, as for individual taxis.

21.4.2 The Overhaul of the "Customer Journey"

The digitisation of transaction operations allows advanced automation and significant productivity gains for the operator, as well as a strong simplification for the customer. This is especially the case when the interaction proceeds via a mobile application, through the finger and the eye *via* a smartphone. Ergonomically, it marks a real revolution: elaborated and personalised information requests have become possible and easy, at minimal cost for the two related parties. The time cost of the transaction becomes affordable on the scale of any trip, even for movements of just a few minutes in urban areas.

In fact, the transformation is even more important, as the new ergonomics can empower the entire "<u>customer journey</u>" at the commercial level. In marketing theory, the "customer journey" is constituted by a sequence of five legs that form complementary stages: "inspire, plan, book, ticket and evaluate". Digitisation empowers the different functions and thus the service rendered:

- The <u>inspire</u> stage proposes places and activities likely to interest the traveller. This practice is common for air travel, is in development for railways and is still embryonic for urban modes, the developing form being to highlight amenities located near a given position, for example on Google Maps.
- The <u>plan</u> stage, dedicated to the actual organisation of the journey, is the most advanced for all modes of transport. Thanks to their mastery of cartographic information and the centralisation of dynamic information on the state of availability of the routes, such as public transport services, "search engines" find an ideal field of application in calculating routes between places specified by the user, as well as certain usage preferences (travel times, exclusion of certain modes, avoidance of tolls or connections).
- The <u>book</u> stage is being extended, starting from traditional car rental and long-distance public transport (planes, trains), thanks to the development of car-based intermediary services, such as ridesharing, taxis, shuttles and car sharing. Infrastructures and the mass modes represented by highways and urban and regional mass transit constitute a field of innovation that remains to be explored.

- The <u>ticket</u> stage has been made very fluid by the diversification of innovative means (PayPal, Google Pay, Apple Pay, WeChat, etc.) which are often based on the credit cards. The use of smartphones as a tariff support can make it possible to dematerialise tickets, and most services with a reservation now offer it. However, the transposition to mass modes requires the technical adaptation of all tariff validation points, which are very numerous in a large urban area.

- The <u>evaluate</u> stage, in which the service is assessed by the customer, has become commonplace in the form of a quick questionnaire in the operator's mobile application, or through personal comments posted on the service's websites or social media pages. This "customer feedback" comprises a sharing of experiences that is valuable for potential users and for viral marketing. It makes it possible to discriminate between unsatisfactory providers, particularly for intermediate mobility services. However, there is still little integration with the redesigning of services in traditional modes.

In this overhaul of commercial customer journeys associated with mobility services, the keywords are <u>information</u> and <u>personalisation</u>, but also <u>merchandising</u>. As long as the transaction operations had prohibitive costs, there could be no question of pricing the services finely. With the new technical and commercial situation, it becomes relatively easy to price the service as delivered, according to the sections of the network used and the schedules, as well as the congestion level and environmental impacts. Among the ToD companies, Uber has distinguished itself by its uninhibited pricing. The real-time rate increase ("surge pricing") exploits the scarcity of means of transport in a given place at a given time in order to capture the highest willingness to pay.

21.4.3 The Revolution of Services to Mobility

In the field of mobility, the expansion of customer journeys has been strongly driven by the development of innovative services to mobility. The Google Maps application is a vector that has become increasingly powerful, thanks to the progressive enhancement of its functions:

- The <u>cartography of places</u> has been enhanced by the identification of "amenities". A given business, service or public facility – in a word, an "amenity" – can be presented on the map, which thus becomes a directory that locates it in the field and gives the precise address (making traditional postal directories largely obsolete). Being listed allows the amenity to be identifiable in an information request triggered by the application user, putting them in contact.

- Cartography has been augmented by <u>access conditions</u>: travel times by section, on foot, by car or by bike, and even times and travel times via public transport lines. The application offers route calculations and progressive guidance in the field, from very refined complementary mapping (navigation databases, including all the traffic signs visible in the field).

- Finally, the application <u>gives access to mobility services</u> that have negotiated a partnership. Google Maps highlights several modes of travel, including driving, public transport, cycling and walking, as well as some on-demand transport services. With just one click on the icon for this mode, a special screen page will open to offer the user the various services offered by Uber (see Figure 21.9)!

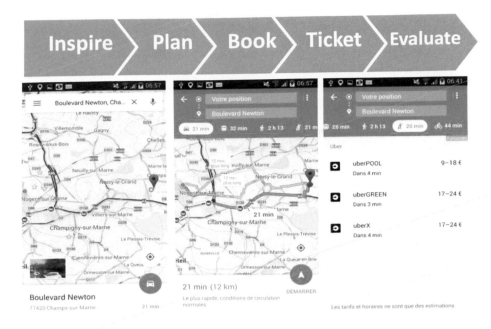

Figure 21.9 "Customer journey": the stages and an application under Google Maps

Thus, the application is not restricted to giving information, or to playing the role of mobility assistance in other words. It offers accompaniment for the trip, fulfilling the role of a travel companion. It gives access to mobility services, fulfilling the role of a travel agency.

All of these roles mobilise powerful functions. Some are developed by start-ups that Google has gradually integrated into its ecosystem, notably the Waze application for dynamic road traffic information and incident reporting.

The expansion of the customer journey at one stage or another marks an area of innovation for mobility and thus a field of potential conquest for innovators. The takeover by Google is synonymous with the objective of taking advantage of the whole planet (or almost), and the price of the takeover is proportional to this "commercial surface" and to the brand's notoriety, which goes well beyond the financial means of a transport service operator based mainly on an urban area or a country of origin.

21.4.4 Customisation, Avatars and Tracking

The user is known by the application providing the service. It records their requests as well as recurring information elements, such as the identification of the means of payment and personal characteristics (e.g. date of birth and gender, usage preferences), and even includes family (home) and professional details.

The digital "client file" comprises an avatar of the user, at a static stage. Its systematic integration facilitates transaction operations and serves to personalise processing, in other words, to treat requests with respect for the customer. Within this, automated processing is very efficient compared to human advisors, as the automaton offers a permanent quality of attention. Human intervention can then only be advantageous for complex operations, to explain special subtleties.

During successive uses, or even just successive connections with or without a request, the application gleans more and more information about the client. The avatar becomes dynamic, with a history of activities and a trajectory in time and in space that reveals their rhythms and places to the service. This is only one step away from inferring and inducing activities, uses and habits of consumption – which the service providers happily take. Web giants also know the user of their web browser and their search engines, as well as the websites they visit and their interests, which are deduced by text recognition tools. Commercial channels can analyse consumption bundles based on the digital traces of product purchases and by identifying the consumer according to their credit card or "loyalty card". For an individual, the merger of their respective avatars on Google (or Apple) and Amazon (or a competitor) is more than a potential possibility. The use of voice-activated digital assistants could well trigger such a marriage, for example.

In any case, knowledge of dynamic avatars allows the service to anticipate the needs and even the desires of its clients. Targeted advertising messages is emblematic of the Inspire link, i.e., the dynamic avatar becomes proactive in relation to the individual it is simulating (Figure 21.10). To sum up, from simulation to stimulation, the path is short.

21.4.5 Customer Relations: Expansion and Development

In terms of the form, the facilitation of transaction operations and the amplification of the customer journey services are equivalent to the privileged treatment of the user. The privilege given distinguishes the connected individual from the old situation and is available to everyone.

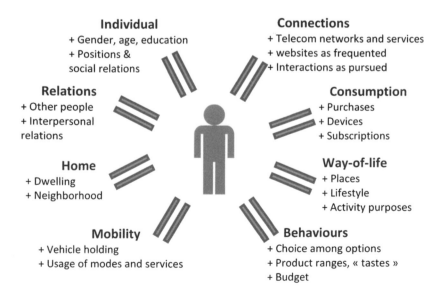

Individual
+ Gender, age, education
+ Positions &
social relations

Connections
+ Telecom networks and services
+ websites as frequented
+ Interactions as pursued

Relations
+ Other people
+ Interpersonal
relations

Consumption
+ Purchases
+ Devices
+ Subscriptions

Home
+ Dwelling
+ Neighborhood

Way-of-life
+ Places
+ Lifestyle
+ Activity purposes

Mobility
+ Vehicle holding
+ Usage of modes and services

Behaviours
+ Choice among options
+ Product ranges, « tastes »
+ Budget

Figure 21.10 Customer-user and their computer avatar (or "digital twin")

Essentially, the relationship between the customer and the service is transformed. Customers have amazingly extensive access to information. Services are expanded and their use is simplified. Their knowledge of the service now not only passes through their personal and physical experience of its use, for many it is formed by consulting the potentialities and experiences of other users.

On the service side, knowledge of the customer is considerably increased (see the notions of avatars mentioned earlier). This new form of relationship creates value. Thanks to the near-elimination of transaction costs, the feasible field is greatly expanded, especially as the physical and commercial offer is diversified by innovative solutions.

The new opportunities opened up to customers stimulate consumption. For some, the transaction times saved are worth the money and justify the use of more efficient or more comfortable services, by accepting higher rates.

The willingness to pay more attests to a surplus for the customer, and for the offer of a created value to be divided between the intermediary "software" and "hardware" producer. The latter is obviously interested in ensuring the "software" part of the service. In order to do so, it is responsible for designing and deploying customer loyalty marketing policies through various "mutually beneficial" schemes. Stimulation of the physical consumption will likely follow, with relatively stable monetary expenditure (flat rate) or one directed towards the increase (usage component in the tariff), and, on average, a decrease in the ratio between the expenditure and the quantity of traffic concerned, in other words, a lower unit price for the customer.

Many customers will find themselves engaged in deeper commercial relations with each of their suppliers, hence a new source of complexity, which will require specific solutions.

21.4.6 Revenue Management and Yield Management

Innervation, the expansion of technical subsystems and services, as well as the expansion of automated processing comprise transformations by the expansion of production activities. The management is also affected by the transformation in its two sides, which are oriented towards the market (revenue and yield) and towards production (supply), respectively.

For a producer, revenue management takes place at the level of a customer's basic demand. The maximum value must be created through the tariff revenue, within the limit of the willingness to pay. There are various techniques that can be automated in a digital application for customer interaction:

• Division of the service into a base that is as thin as possible and into a series of options billed individually, in order to offer a call price that is as low as possible but restricted to the base only. Existing applications include air travel in low-cost airlines or new car sales.

• Variation of the price according to the reservation date and time, according to the retroplanning until the service and the already constituted volume of customers.

• On a booking website, variation of the price based on the requests previously made by the same customer for the same service: the application progressively estimates the willingness to pay and the preferences of the individual for this particular trip and consequently adjusts its price proposal for one option or another. The proposed options are based on intensive digital computation in real time. This is version 2.0 or 3.0 of the "carpet salesman".

- The tariff modulation mentioned with regard to Uber adapts the tariff according to the place and the moment, depending on the objective conditions of scarcity.

These tactics are applied to individual customers with no market power. The sale of "lots of seats" between carriers and travel agents is less imbalanced.

Yield management operates at a more strategic level. Based on tactical revenue management, it is specifically aimed at stimulating overall demand (inspire and needs) and coverage by a grid of services intended to be cast wide and deep. Charter flights for a certain attractive tourist destination from a certain large city and at a certain period constitute a basic form. Air flights offered with several different comfort and benefit classes constitute a more advanced form, especially with dynamic tariff discrimination. The digital tool is valuable not only for proposing its own offers to potential customers, but also for monitoring the offers of competitors (web-scrapping) and for studying opportunities for cooperation in the form of code-sharing or by organising connections.

21.4.7 Supply Management is Amplified and Energised in Feedback

On the production side, we have noted the significant development of digitisation-based processing operations for vehicles, infrastructures and operational processes, as well as for service offers and customer journey and customer relationship management. Automation promises to increase resource productivity and timeliness. It also reduces the place of human interventions. The number of agents can be reduced in the field and in the back office. The roles to be played and the types of intervention to be employed must be redefined.

In these conditions, management must intervene at two levels. On the one hand, it must ensure the production, quality and proper functioning of the process, and good conditions for the resources. On the other hand, quality must be assured in all respects in the general management of the means and in the very design of the processes. Digital tools facilitate the supervision of production. Others can constitute equipment for redesign studies, but their role will remain purely utilitarian in the managers' reflections, in codevelopment with the agents concerned, clients and consultants.

Supply management must also support and strengthen yield management: the challenge is to make production as agile as possible to satisfy the market. For example, for a certain air transport link, yield management will explore the demand for as long as possible, thereby shortening the residual time for programming the offer, in particular for choosing the model of aircraft that will be sufficient to serve the customers gathered in this way. In accounting terms, after optimising revenues, costs must be minimised in order to maximise profit.

For this, each mode of transport has specific levers and means of action, knowing that the technical solution for travellers must ensure a nominal level of quality. For the air mode, a model of airplane can be chosen according to the fleet available at the departure airport, without compromising the speed of the flight. In rail transport, the selection of transport units (single or double trains, simplex or duplex on high-speed lines) and the composition of convoys of cars or wagons are factors. Similarly, convoys of barges are formed in river freight transport. In road freight transport, both vehicle models and connection frequency are involved. In passenger transport by coach, the link frequency is modulated. Intermediate mobility services offer increased possibilities: vehicle transfer strategies (trucks, jockeys, coproduction with customers) for car sharing and bike sharing; pre-positioning of cars and drivers for taxis and passenger cars with drivers. In short, the internal logistics of transport services becomes more dynamic and interacts with market construction in or almost real time. Digital processing of it is now unavoidable.

21.5 Actors and Organisations

So far, we have analysed the expansion of mobility through digital technology by subsystem, service or function. We will now examine the implication of the actors in mobility as transformed in this way: we will distinguish between the demand side, i.e. the users who are customers of the services (Sections 21.5.1 and 21.5.2), and the supply side, in which various producers (Section 21.5.3) take positions in the value chains and participate in relational schemas (Section 21.5.4). Value chains can be all the more easily reconfigured as pricing becomes more flexible (Section 21.5.5). The initiative still needs to be taken for this kind of reconfiguration and systemic recomposition of mobility (Section 21.5.6).

21.5.1 The Demand Side: The Logistics of Everyday Life

In their daily life, each person is immersed in the technical, service and commercial conditions of their time, and this especially takes place in the territory that constitutes their living area, including their home, places of constrained activities (work, studies), utilitarian activities (commercial, administrative) or optional activities (leisure, family or friendly relations).

In less than a decade, the smartphone has become widespread among people in developed countries. The reason for this is twofold. On the one hand, the package of services offered is extremely broad and abundant and concerns almost all the activities that people can carry out and mobility in particular. On the other hand, the user interface is very ergonomic, and both purchase prices and subscription packages to mobile telecommunication networks are suitable for all budgets.

Placement of the smartphone in their hand or pocket allows the user to be connected at any time and in any place. The "mobile" state is no longer a restriction on connectivity, except in a very specific situation (tunnels in land transport, air flight, sea crossing). The individual's relationship to activities and mobility has been transformed. In particular, the physical appointment to lead two or more people to a common activity becomes more flexible and can partly be managed at a distance by taking part telephonically or by informing the other people concerned that they will arrive later than expected.

Beyond this basic function of mobile telephony, smartphones give access to the Internet and mobile services. The services previously mentioned for mapping places and activities, itinerary advice and guidance, suggestions for activities to be carried out and the proposal of travel solutions can thus be used at the discretion of the individual. As such, dynamic information is available along the whole route, making it possible to know the current state of the system (including congestion and incidents) and to adapt if necessary, by adjusting the travel solution or the activity that motivates it. Upstream of activity and movement, the provision of information on places, on the offer of activities and on mobility solutions has enabled the individual to choose their activity (insofar as it is optional) and prepare for their trip. The investment in preparation is inversely proportional to the knowledge already acquired by the user and the routinisation of their daily life. It is significant for exceptional or rare circumstances, short for occasional circumstances and almost immediate for familiar or recurring circumstances – for which a "push" application will spontaneously provide concise and relevant information.

Smartphones and their applications also transform the individual's relationship to transport vehicles. Driving may be assisted by a guidance application. Rental services for cars or two-wheel vehicles, or pooling clubs, are accessible with a few clicks on the touch screen – subject obviously to their presence and availability in the field.

The automation of transactions can ease the constraints of the appointments with a rental service or facilitate making an appointment with the person delivering a product. In other words, smartphones open the doors to pooling: in terms of the use of the vehicle as well as logistical cooperation between people, between the user as a customer and the various providers.

It therefore marks a profound transformation of daily life, in situations of mobility, as in situations of activity. These situations are being recomposed, especially with the possibilities of teleworking for active workers, either at home or in a third location. Teleshopping and telepayment coupled with delivery in a place of their choice are also a factor. The distinction between mobility situation and activity situation is partially blurred. The connected individual can reuse the "passive time" of their travels to carry out personal activities (reading, music, videos, etc.) or private or professional activities. This potentiality makes it possible to value the transport time by reducing the inconvenience and the cost for the individual, as well as for their employer if the superimposed activity is of a professional nature.

Finally, beyond the organisation of activities and the transformation of movements, widespread connectivity transforms social relations. The individual is made virtually present through messages and posts on social media, communicating their position to their friends in real time. It is on Internet platforms that they carry out a large part of their interpersonal activities.

All these conditions and basic situations at the individual level create a strong synergy between transport and telecommunications (by means of the Internet–smartphone– mobile application triangle). At the same time, this synergy has functional complementarity and substitutability, especially for facing an unforeseen situation and finding solutions. Overall, individual mobility becomes more flexible, more resilient and doubtless more diverse and more "experimental" in relation to unfamiliar options, which tends towards the adoption of service innovations.

21.5.2 In Business: What Mobility and Logistics, for One's Own Account and for Hire and Reward

Professionally, employers can take advantage of connectivity and associated flexibility. The notion of a counter stand for serving customers is undergoing a second transformation. After having connected the counter stand to a centralised information system by a microcomputer, the smartphone or tablet positioning delivers the physical constraint from a fixed point to the agent, which makes them available to go to meet the customers in their current location. This change is already evident in air terminals and major railway and road stations.

Another case of transformation is that of curative maintenance processes, in which connectivity allows tasks to be distributed between a mobile agent in the field, who accesses the site of a failure, and a specialist who is available remotely to diagnose the problem and find a solution on the basis of both a telephone conversation and photographs of the dysfunctional equipment.

Similarly, surveillance activities are reconstructed between remote surveillance, patrolling and the response team. More generally, robotisation and remote control transform many trades: in freight port logistics equipment handling controls, in public transport for operating lines with automated driving and in air transport for piloting drones.

Thus, the connectivity of individuals and materials lends itself to a transformation of production processes. The agents are mobilisable, while the vehicles are shareable and "serviceable". The services of or to mobility, which are available to the general public, can obviously also be used by agents within the context of their organisation. A company's fundamental alternative for developing its products and services, namely buy or make, is being refined, and with the economy of functionality, the options are expanded to include share.

In addition, the internal coordination of mobile means can be questioned: should a company keep their own fleet of service vehicles? If so, can it be pooled very effectively, by centralised management that is as efficient as the general public services of on-demand transport? Is it necessary to provide certain agents with company cars, while subscriptions to intermediate mobility services would make it possible to replace them? The Mobility as a Service (MAAS) concept was primarily proposed for the general public; however, it would be particularly relevant for business customers in order to make access to vehicles, car parks, toll roads and transport more fluid, collective or on-demand!

Large logistics companies are obviously at the forefront of the transformation process, to meet their own needs as well as to serve their customers. In particular, La Poste coordinates the management of its mobile agents (the "postmen") and their vehicles at a level that is sufficiently integrated to allow a large part of the fleet to transition to fully electric vehicles, overcoming the constraints of range and recharge time. This large logistics company is also concerned with its agents' missions in the field, i.e. those of the postmen. It is expanding them by going beyond the delivery of mail and parcels to include typical post office services (sending mail, banking transactions).

The redesign and recomposition of company-specific mobility will benefit from being combined with the business travel plan (Plan de Déplacement d'Entreprise, PDE), which primarily relates to home-to-work mobility for agents. The joint scheme can also be extended to integrate the mobility needs of customers and suppliers who come to the company.

21.5.3 The Diversification of the Actors on the Supply Side

On the supply side, between transport services, mobility services and services to mobility, products and roles have diversified and all the actors have expanded. There are now more transport service operators, infrastructure operators who broaden their scope of action, car manufacturers who are moving towards the functional economy by also becoming mobility service providers or operators, suppliers not only of material parts but also of energy, telecommunication or information, etc. We will review these actors in descending order of proximity to the end customer.

Thanks to their generic mobile application, travel advisory services are in direct contact with the user, even before the mobility decisions. This position tends to make the application a privileged crossing point for accessing the end customer, thus creating distance from the other actors involved in the offer.

Travel advisory services could expand into mobility agencies, integrating service offers: this is the goal of companies such as Citymapper or Moovel, which now locate their actions under the MAAS banner. Their vocation as integrators brings their mutual relations into question. What capacities are available and what will is there to cooperate within a unifying system?

An even more fundamental question concerns their business model. What value can they grasp from payment by mobility operators, or from subsidies from territorial actors made accessible on a specific basis (e.g. by addressing the companies served) or a collective basis (territorial collectivity) or through the dissemination of advertising? Some Google services are already established in this role and in this position, undoubtedly with an expansionist ambition.

Mobility services operators (MSOs) come right after the integrators. The diversification of service forms is driven by innovation firms. Well-established companies and groups favour "external" developments to broaden their range of operations. Among the innovation companies, international deployment in major markets (Europe, North America, China, etc.) provides the best guarantee of rapid growth. The Uber company was founded in 2008 and reached a capitalisation of $ 50 billion in 2014! Indeed, the Internet is now a formidable sounding board for the rapid distribution of a platform service worldwide. Among traditional operators, those most exposed to competition with intermediate mobility services are taxi companies and car rental companies. They are adapting their services and sometimes buy out an innovative competitor to extend their offer to include new forms of mobility (e.g. the rental company Avis bought the Zipcar car sharing service), while they have modernised their offer in a forced march (e.g. the G7 taxi company). Interurban carpooling is developing on routes with poor or expensive rail service and in competition with long-distance coach services. This competition will probably have to be regulated by a local authority of sufficient size, such as a French administrative regional authority.

An MSO of this kind is an integrator by nature. It provides the interface between customers-users to be transported and the means of transport, vehicles and drivers. The vertical integration of production raises questions, especially for small traders who operate their own vehicles to either offer a taxi service to travellers or carry freight.

The platforms facilitate not only the access of a given service to various providers, but also the access of each provider to several operators who can each bring them traffic. This symmetry of access, despite the asymmetry of roles, may allow some balance, provided that a regulator oversees correct pricing, healthy competition and the avoidance of distortions. The automation of driving represents a strategic prospect for vertical integration, not only because of the reduction in driving costs (e.g. by relaxing appointment constraints), but also because robotaxis have no calling to claim market power and can be exploited at will, without the fatigue that cannot be relieved by mechanical maintenance.

In parallel with the OSM, we can locate telecommunication operators who offer access to telephony and the Internet by targeting particular needs and mobility situations (e.g. Orange) on the one hand, and on the other the energy suppliers who engage in the mobility sector in several ways, in order to administer electric vehicles, to manage their recharging in conjunction with smart grids and to take advantage of the transport infrastructure to harvest renewable energies, etc. These two categories of actors retain their status as suppliers under the strong supervision of their regulators, which maintain a competitive game in their respective markets.

Autonomous driving of vehicles is an even more crucial issue for car manufacturers, as the functional range of an automobile will become so amplified that only customers with poor credit will live without it. The demand for personal vehicles will be increasingly linked to the forms of pooling: some will prefer to benefit from platform services rather than buying their own vehicles; others will consider acquiring a vehicle to put at the disposal of other users *via* one platform or another. This will require the vehicle operating system to be compatible with that of the platform. Finally, some shared mobility platforms choose the models in their vehicle fleet, or influence the choices of their members, meaning there is a demand segment for manufacturers. All in all, they are mobilising for both pooling services and autonomous driving. Digital transformation affects their products in terms of the uses that are made of them, as well as their internal constitution and their manufacturing processes, as explained previously. The availability of digital tools for manufacture design and organisation enabled Tesla's industrial development, whereas in the early 2000s many people believed that the "club" of car manufacturers would gradually shrink because of increasing industrial complexity. Tesla and other manufacturers, especially Chinese companies, have come to refute this opinion. These characteristic elements of the automobile sector are largely applicable to transportation equipment used in other modes, i.e. road, air, rail, inland waterway and maritime heavy goods vehicle industries.

Transport infrastructure operators (TIO) provide traffic and access functions, or parking services, or they operate logistics platforms. We have talked extensively about the potential of digital technology for the various functions of infrastructure maintenance, continued viability and access and traffic regulation. To benefit from it, operators will have to negotiate an agreement with their granting or delegating agency. One imperative concerns compatibility with vehicle innovations (driving systems and associated communication needs) and those of shared services, for example the development of parking areas for carpool users at motorway access points.

The quantitative impact of pooling on traffic and services is still uncertain. Multi-occupation of vehicles should reduce traffic volume, while the supply of P2P parking spaces should reduce the need for public parking. On the contrary, the availability of shared services will increase demand.

Many other categories of economic actors are less directly involved. The closest to both vehicles and infrastructure are on-site repair and emergency services, automotive maintenance services and automobile insurers.

On-site repair and emergency response services will be alerted more effectively, thanks to digital technology, and most likely they will be required less frequently given the trend towards improving reliability and security.

Automotive maintenance companies must adapt their services to the characteristics of the vehicles treated, including their information processing subsystem, the diversification of engines and, in future, the specific needs of robotic vehicles.

Automobile insurers must take the new usage situations and new characteristics of vehicles into account, including the question of who is responsible for an accident in the case of a driver of a rental service and, in future, in the absence of a driver. They should contribute to the design of the MAAS system to establish the new mapping of responsibilities in different risk situations.

Finally, the business scope of automotive parts manufacturers is also expanded through the sophistication of parts to achieve the functional amplification enabled by digital technology, new elements such as batteries for electric vehicles and the development of various systems to assist driving, from the automatic activation of a basic function (e.g. headlights) to proprietary autonomous driving systems for larger groups on an international scale (e.g. Valéo, Bosch, Delphi).

21.5.4 Functional Relationships and Strategic Positioning of Actors

Figure 21.11 shows the main functional relationships between the categories of actors involved in the mobility offer. Any transport service is produced by a multistage sector. Let us call it vertical – which can include the following, in a top-down order,

- the mobility agent, in direct contact with the client,
- the MSO and the vehicle operator,
- the TIO,
- telecommunication, energy, rolling stock and ancillary services suppliers.

The appearance of mobility agents *via* the virtual world makes the MSO fear "disintermediation" with respect to the end customer. However, they retain control over pricing and market conditions and can develop their own loyalty policy.

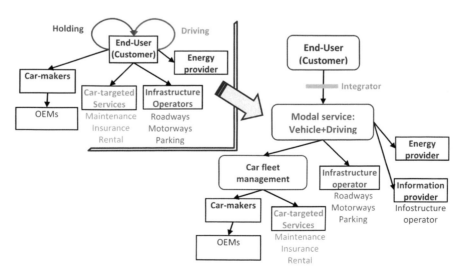

Figure 21.11 Transformation of the system of actors

At this stage, traditional actors perceive the following threats:

- A risk of disintermediation due to the multiplicity of information and evaluation platforms that are increasingly positioned as portals for transport offers, as well as the weight of distributors, of which there are many in the air sector and an emerging number in the rail sector, with the fear that a dominant distributor will capture a large part of the trade margin, like booking in the hotel industry.
- A risk of imbalance in traditional offers by the emergence of new highly effective offers (marketing, quality/price, access positioning/premium). This risk could cause traditional actors to question their nonspecialised, integrated model.
- A risk of breaking-up the integrated operator model (platform and driving), especially in urban transport, even if the transport capacity still seems protected.
- A risk of deoptimisation of the transport system by the erosion of the supervisory function. The risk is that the marginal optimisation of the effectiveness of one offer or another is carried out to the detriment of the overall optimisation capacity (e.g. the effect of Google Traffic and Waze on the supervision of motorway traffic by the infrastructure operator).

Be that as it may, in an overall context that is more dynamic and therefore more competitive in one form or another, it is important for each producer to be efficient and even excellent in their core business. The physical nature, "hardware" and technical services rendered constitute levers of action for them, which they can control live. It is necessary to constantly improve the quality of the service (in terms of comfort, regularity, security, safety, cleanliness and speed) in order to make the offers as attractive as possible and to retain customers.

The overhaul of the "customer journey" in the field of mobility amplifies the consumption by transport operators of market services not directly linked to transport on the part of operators. This is increasingly the case as potential needs and desires are better identified, on the basis of improved knowledge of the customer. It thus allows the traveller's physical journey to be made more fluid. The traveller can have a personalised route on their smartphone, just as the geolocated data of passenger flows can optimise the real-time orientation of customers within airport terminals (opening of check-in desks, police reinforcements, etc.).

In addition, it encourages multimodal integration in a door-to-door logic, both virtually, by striving to offer the most possible offers on the same platform, and physically, by developing exchange hubs.

Multimodal integration can fall within the remit of public authorities (MAAS – which originated in Finland) or come from private initiatives (SNCF, Google). It represents a real challenge:

- by the complexity of the access to partners' data and their integration in real time (see multilocal context with very little standardisation);
- by the need to demonstrate the added value of joining the platform to others;
- by the complexity of contracting and sharing value between all the actors concerned;
- by the difficulty of reaching a critical mass of supply and audience to generate demand;

- by the a priori competitive advantage of digital players over historical players (external positioning, greater agility in aggregating nontransport offers, more able to deploy a model across several countries);
- depending on the existence of a true interest in a unique and universal platform among the users.

Therefore, given limited resources, can a traditional player afford to both pursue excellence in their core business and develop an ambitious platform? Is the development of third-party multimodal integration platforms a real threat to traditional operators?

21.5.5 Pricing: A Lever for Structuring the Value Chain

In addition to optimising the customer experience, digital technology enables differentiated pricing that is closer to usage. Thus, it encourages more dynamic management of tariffs (notably, thanks to competitors' tariff tracking) and a shift to predictive yield, which reinforces the commercial challenge linked to last-minute bookings. As a corollary effect, digital technology can significantly increase tariff differentiation according to the customer's profile and usage (particularly regarding the access vs. premium dimension), by ever finer segmentation of tariffs.

Air and long-distance rail already seem to have adopted some or all of these practices. It is therefore in the urban context that the potential for creating value is the strongest. However, the diversification of tariff options and the risk of discrimination associated with differentiated pricing represent a real political issue, at a time when many cities are attracted by the principle of a single tariff for subscribers to their mass transit network.

Tariff segmentation also makes it possible to <u>take into account the environmental and social impacts</u> of the various offers through schemes such as special tariffs during off-peak hours, a reduction to encourage the use of alternative routes, the introduction of a congestion charge, the promotion of kilometric allowance for bikes, etc. However, the implementation of an increasingly fine tariff segmentation runs counter to the normative effect of comparators pushing for tariff simplification (e.g. by comparing flights according to the price and not the service offer). How can we ensure the readability and acceptability of highly differentiated pricing?

21.5.6 Which Form of Mobility System Management Should We Aim Towards?

The technological and industrial potentialities are enormous, and the proposals for innovations are varied and abundant. Towards which future should we direct mobility? Can the potentialities be mobilised, and if so how?

These major issues obviously fall within the realm of public action, both in terms of regulating supply and securing people and goods, and also the collective organisation of mobility in the territories. French territorial authorities have been pioneers in the local implementation of shared mobility systems.

<u>The role of regulations</u>. Administrative regulations are intended to regulate and stabilise the socio-technical system. They thus constitute one of the social factors of

inertia, alongside private equipment and individual routines, the internal organisation of companies and the organisation of actors' systems.

The major design challenge is a collective organisation in order to take advantage of technological potential. The social demand is indeed present: between aspirations to the "right to mobility", the demand for quality of service (in particular in terms of fluidity and security), and the quest for successful relations between quality and price.

Who will take the initiative for the integrated redevelopment of mobility systems?

Elon Musk: a strategy of conquest. American serial entrepreneur Elon Musk promotes this kind of rethinking with audacity and talent. In two master plans published 10 years apart, he proposed his vision of future mobility and announced the next actions to be taken in order to progress in the direction outlined, sizing his actions in proportion to the financial resources he manages to release and mobilise. Year after year, he has "conquered" the field of transport technically, gradually mastering the technical complexity of the subsystems that compose it. Starting from the interaction with users (PayPal service), he has invested in the manufacture of vehicles for space transport (SpaceX, since 2002) and electric cars with Tesla Motors (since 2004). In order to further promote electric mobility, he focused on the interface between the vehicle and charging infrastructure, offering batteries for domestic use (Powerwall) and Solar Roofs. He finally came to manage the transport infrastructure, which he proposes to renew as a vehicle support road ("Hyperloop", since 2013) while also reconsidering its constructive processes ("The Boring Company", since 2016) in order to tap into the greatest sources of value.

This saga of conquest is very evocative on both technical and managerial levels. It demonstrates the possibility of reinventing and rebuilding the mobility systems floor by floor (see Figure 21.12).

The demonstration is not over as there are still scale factors to be deployed.[1] The technical capital already established for transport in the territory, with road and rail

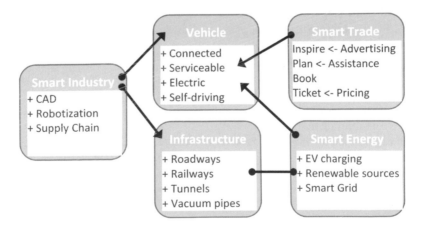

Figure 21.12 Combining systems: mobility, energy and trade

1 As of 2018, Tesla Motors was poised to increase its production of electric cars. Its share should then reach about 1% of the new car market, which will still be very low in total, but will double the market share of purely electric cars.

infrastructure stock and a fleet of vehicles, obviously imposes significant inertia. It will therefore be necessary to design renewal strategies.

A conception of this kind is a matter not only of a private initiative, but also, and more firmly, of a public action+++. Between the territorial and the national levels, it is the responsibility of the public authorities to reshape mobility systems by taking advantage of the plasticity resulting from the new industrial order. They must dare to resculpt them.

For example, the transition from a fleet of urban buses to electric engines takes a decade. The electrification of a taxi fleet can be instigated more quickly, by riding the wave of new dynamics in this mode. The renewal of the urban logistics fleets is an even bigger lever, which could be tackled immediately by acting on parking conditions and by promoting the retrofitting of vehicles towards electric motorisation. For these professional fleets as well as for private vehicles belonging to local residents, the regulation of access to the city centre is a powerful lever, just like the parking lever. Similarly, access to the network of urban expressways could be regulated, first to streamline traffic and then to impose minimal environmental performance levels.

Without further development, it seems possible to adjust let us say 20% of vehicles, providing 80% of the usages, in a decade. This would therefore affect 80% of the environmental impacts emitted in the previous configuration.

Empower the public management of mobility. The matter, the technical material of mobility, has become more plastic. Traditional functions for public management are persisting and even being extended:

- Traffic management can become smoother and more dynamic by making local supervision effective (e.g. remote monitoring of parking) and coordinating the supervision of different places and modes to achieve multimodal hypervision of mobility in the territory.
- The arrangement of places obviously continues at different scales: from the local level of road elements to be shared between the various modes of transport, to the shaping of mobility on the scale of a large territory (including its extension in space and land use), through the master network and intermodality planning, and the multimodal organisation of flows.
- The differentiation of vehicles according to energy vectors – soon according to the mode of driving and already according to their passenger occupancy – constitutes additional levers of action, to be mobilised in the development plans as well as in the traffic management schemes.
- The various sectoral policies need to be further integrated, between the multimodal mobility policy, the housing and urbanism policies, environmental policy, industrial policy and social policy (especially employment), to establish overall coherence in order to ensure greater efficiency for achieving common objectives and allowing better readability for the stakeholders.
- The stakeholders are diversifying: between individual users and major network operators, the development of services to mobility (such as dynamic and multimodal traffic information) and shared mobility services involves intermediary actors of a company or user club type. The organisation and the overall operation will involve value chains, based on the price paid by the user or the subsidy/fare

levied by the community, via the various technical stages of the supply (including services by means of a vehicle and use of infrastructure).

Value chain management is becoming absolutely necessary and essential in the MAAS era, as is easier and amplified digital ticketing and uninhibited pricing in shared services (e.g. Uber).

In all, the fields of action concerning public management are extended and their levers are amplified.

In response, it looks appropriate to provide public management with more powerful technical instruments, which will effectively enhance managerial capacity and empower control over the system. The toolkit must include sensors and actuators (effectors), a telematics system for communication and management centralisation, as well as various decision support functions that can be mobilised in real time or instantiated in deferred time. The Lyon-based project Optimod and its European extension Opticities provide an example of the partial overhaul of the instruments. It is an example to be reflected on, to be explored and transposed for Management 3.0 of Mobility 3.0.

21.6 Opportunities and Challenges: Coordination Needs

We have painted an overall picture of mobility as a socio-technical system undergoing a reinvention under the influence of digital technology: smart mobility. The picture is very broad: it is a triptych that brings together (i) technical subsystems, (ii) forms of services and relations between services and customers and (iii) actor situations. It remains for us to take a step back, to appreciate the overall perspective in a reasoned way by taking into account the virtualities, opportunities and advancement of their implementation, but also the limitations.

We will begin by summarising the scope of smart mobility as a land of opportunity. We will then highlight several types of challenges: of an economic nature for the establishment of business models and the configuration of value chains, of a technical nature, such as interoperability and cybersecurity, of a societal nature, such as the protection of privacy, and, finally, of a managerial nature for the management of physical space, virtual and media space, and the innovation and framing of service offers.

21.6.1 The Opportunities Offered by Digital Technology Are Very Real

The digital transformation is a new state of affairs that is both technological and concrete, through the arrangement of things and services. Generalised connectivity dominates the whole. The logistical foundations of social and economic life are transformed and re-established, as are the relationships between time, space, the individual, agents, vehicles and transport infrastructures. This is especially true of appointments, as meetings between several entities can first take place in the virtual world and then be orchestrated on the ground, if it still remains useful to do so.

This transformation is at the disposal of the whole mobility offer, in its various subsystems and at its different levels, which comprise (from top to bottom) the mobility agent, the transport service, the traffic or parking infrastructure and the supply of materials or services.

Digital technology amplifies the functions and services rendered by the subsystem. The development of interactivity between the components makes it possible to increase the external functionalities rendered to customers. "Internal" technical processes also profit from this transformation to gain power. Their reinforcement proceeds as an innervation: the software is incorporated into hardware piecewise and by the creation of informational circuits. The additional cost of implementation is modest because of the high overall cost of the infrastructure and the economies of scale for the vehicles.

The expansion of the functions has a profound effect on the mobile entities and on the space. Just as vehicles are redesigned and made more versatile to be used by several services, agents who become more mobile can fulfil several functions. Specific places of transportation, such as car parks, will be relatively less used by each vehicle intended to circulate more, but the stopping phases will be enhanced with other functions: electricity charging, cleaning and minor maintenance, and the car park will also be able to boost its monitoring function in relation to "unmarked" stopping places.

The innervation of the system makes the offer more reliable and more flexible and marks the first major event in the digitisation of mobility. Reliability is enhanced by better material maintenance, better process control and increased capabilities for alerting customers. Flexibility is increased in reactive mode (faster detection of incidents, faster and better adapted intervention) and could also be improved in proactive mode by better anticipating the needs of the demand (by increasing the supply plan in high-demand situations), and this anticipation is based on better knowledge acquired from massive observation.

The digitisation of transaction operations marks the second major event for mobility. It allows both the extension and the upgrading of the commercial "customer journey", reducing the transaction time and cost for both the producer and the customer. The producer benefits from performance gains and cost savings, as well as increased user numbers. For their part, customers can derive three types of benefits: first as users of existing services, then by a price reduction if the offeror concedes part of their production savings and, finally, from the provision of innovative services made possible by these new conditions, especially intermediate mobility services.

The overall magnitude of these opportunities is considerable, an obvious part of which is due to the improvement of existing services, to the benefit of both producers and demanders. This part concerns all the territories. Another part will vary in size according to the territory, depending on the size of the population concerned, its geographical distribution and its solvency. Metropolitan territories are conducive to intermediate mobility services. They are most concerned about the revolution of transport on demand. With regard to territories with widespread housing, even if an Internet platform makes it possible to attract basic requests and to develop a service schedule day after day, it will be necessary to ensure sufficient levels of clientele and revenues to make the service profitable.

21.6.2 A Big Economic Challenge: Value Sharing

We mentioned the gains in reliability and flexibility as well as transaction savings, which concern both mobility supply and demand. The logical consequence is an increase in demand and use, especially if the suppliers pass on some of their gains in efficiency to end customers through lower prices.

The value capture created at the end of the production chain and its distribution between the links that constitute the chain are strategic stakes for the suppliers and also for the community for ensuring the good constitution of the supply in terms of services, infrastructure and materials.

At the top of the chain, information integrators such as Google Maps or Citymapper do not directly charge the costs for their service to the users who consult them. They are remunerated by publishing advertising for services "outside transport" (e.g. local amenities displayed in Maps) or "internal" services, to which they bring customers. They are then "mobility agents" or "logistics agents". Platforms that offer a logistics service, such as Uber for transporting travellers on demand or Amazon for the sale and delivery of products, directly bill the service rendered. Amazon also charges the producers of goods that it sells without storing them for marketing.

The place occupied by a particular platform determines its market power. A situation of quasi-monopoly would exert a force on the market conditions in favour of the hegemonic platform, to the disadvantage of the customers and to the detriment of the other suppliers.

It is therefore necessary to regulate access to such markets. In economic terms, it is necessary to ensure the "contestability", and also the "sustainability", of an offer that would have social and economic utility but without achieving profitability from commercial revenues alone. This common situation affecting collective transport also concerns other forms of public transport, such as on-demand transport, public parking on roads or in car parks and traffic infrastructure. Public authorities can support the public transport offer not only through financial assistance but also by "in kind" aid, for example by granting rights of access and use of certain parts of the public domain, or by facilitating the establishment of clubs or networks.

The open-source provision of platform applications and open mobility data provide the first form of networking, albeit at an informal stage. The creation of a purchasing group and the sharing of a technical or accounting expertise will complement the pool-level functions needed at each node of the network in order to elevate its productive performance and to try to satisfy local demand, where feasible.

The robotisation of vehicles will avoid hiring a human agent for driving, which will significantly lower the cost of public transport. However, the absence of a driver will de facto allow less civil usage behaviours: poor cleanliness standards, material damage through misuse and damage by third parties or even other robots. To control this risk, it will be necessary to develop appropriate solutions, not only at the risk of economic losses but also the degradation of environmental performance (e.g. the environmental performance of a self-service bike whereby damage would lead to rapid replacement of the vehicle remains to be analysed over the life cycle).

21.6.3 Technical Challenges

In software, the development is fast and the updates generally follow on from one another at a faster pace than in the technical life cycle of a material. This is why a major technical challenge concerns the *upgradability* of software embedded in hardware. The manufacturer Tesla has planned to do this for its electric cars and now the traditional manufacturers are doing the same. The challenge also concerns less sophisticated objects, such as a ticketing terminal.

Interoperability is another technical challenge, one that concerns the issue of conforming the objects and subsystems to sufficiently long-lasting interaction standards in order to avoid alterations that would render hardware and its integration in the mobility system obsolete and to gradually achieve economies of scale and scope in the supply of equipment.

The systemic intelligence of mobility is also an important technical issue for the proper design of subsystems and their proper integration within a larger system. Systemic intelligence includes relevant and unambiguous terminology, as well as the "architectural" design of mobility subsystems and information systems.

We should remember that the architectural analysis of an information system is conceived on three levels: the functional architecture that shows the functions rendered, their customers, the subsystems and their relations; the application architecture for representing the processes graphically; and the technical architecture that shows the technical means and their relationships.

Cybersecurity is a crucial challenge for technical systems that implement large masses in rapid motion, with various types of potential for shocks and accidents. The solutions are restrictions on computer access, both in terms of hardware, by restricting physical stations and wireless networks, and software. The remaining "access points" must be protected by a kind of lock: cryptography is used to hide information, including the protection of digital access keys.

Indeed, the integrity of systems and the preservation of their functional state represent a broader challenge. Control of integrity includes security and cybersecurity, maintenance and upkeep processes and the avoidance of misuse (e.g. intentional vehicle damage).

21.6.4 Societal Challenges

The societal challenge most discussed in the media is respect for privacy. Each Internet or mobile network user leaves digital traces, long-term collection of which makes it possible to trace certain activities and even habits. In particular, the long-term tracing of an individual's geolocation makes it possible to characterise their profile in certain respects and perhaps even to identify the individual without the latter providing any other information.

The risks of drift are significant: intrusive advertising, distortion of access to information, commercial profiling and customisation of offers in favour of the supplier rather than consumers. In a way, the individual's avatar is alienated by the tracking service. The solution seems to be to prohibit collection, unless the user consents by a voluntary choice. To obtain this consent, some collectors have a commercial counterpart, namely the quality of their service provided free of charge.

The risk of a digital split in society is also often mentioned. This concerns the cognitive challenge for the user of the services (or also for an agent engaged in a productive process) on three levels at least: (i) the user interface of an IT application, (ii) the protocol for the use of a physical service (e.g. the use of a self-service bicycle in the take and return phases) and (iii) the tariff plans. In principle, a mobile application offers an ergonomic interface that is sufficiently intuitive for those who can master the communication codes and have the appropriate visual and motor skills. At the physical level, the richness of the smartphone's multimedia makes it possible to diversify the modes of interaction, using audio-visual mode and motion capture. The cognitive difficulty of new communication codes remains, however, especially for people who prefer the interaction with a teller. The proposed solutions range from the addition of "intelligent" agents ("chatbot" artificial intelligence programmes) in an application's user interface, up to the deployment of androids mimicking human tellers in the field.

The profusion of IT applications creates further complexity for all, which is addressed by a certain homogenisation of functions and by the creation of integrative services.

With regard to tariff plans, some plans are convoluted and the proposals for very low-cost packages with very expensive supplements are unfair. They may put people who have not paid enough attention to them in financial trouble, possibly because of cognitive deficiency.

In the near future, the presence and activity of robots in the field will pose other difficulties. In the case of self-driving cars, passengers, other road users or residents of the parking and passing places may lack control over the robot and suffer from the inconvenience caused by a computer programme that has not anticipated a particular practical difficulty. For example, if a robotaxi is stopped behind a delivery in progress and blocking a local resident's garage, will it back down and give way? Human driving behaviour is sophisticated above a generic situational and social intelligence, which is likely to be lacking in the first generations of robotaxis.

Inclusion. As long as cars have drivers, the development of on-demand transport is employment-friendly and it contributes to the social inclusion of agents in the professional sphere. On the contrary, ToD services are likely to be more expensive than public transport, which will raise the question of their affordability for low incomes. Specific financial support will be needed to compensate for possible substitution in certain parts of the territory.

Social acceptability. As with the reduction of car parking spaces on the road or increases in public transport fares, the replacement of physical counter stands or call centres with mobile application interfaces can be carried out gradually, little by little, without giving rise to insurmountable opposition, as the demand has time to adjust its behaviour under these conditions. Sudden adjustments have also been accepted seamlessly:

* the implementation of regular timing on the interurban rail network in France,
* the overall redesign of the bus network in the Lyon area,
* the establishment of the Vélib (more than 10,000 bikes) and Autolib systems (more than 1000 cars from the outset) in Paris,
* the implementation of automatic sanctions for speed checks on the French road network.

On the contrary, significant demonstrations by population, which have at times been accompanied by civil damage, have led the public authorities to abandon certain projects:

* the construction of an international airport near the urban area of Nantes (Notre-Dame des Landes),

be encouraged and stimulated, solutions proposed and experimentations facilitated; on the other hand, roll-outs on the ground must be regulated in order to seize the opportunities. By allowing the anarchic expansion of roll-outs, such as the provision of urban carpooling services, for example, the risk is not only that many services will disappear quickly, but also that excessive competition will occur over a period that is too short to allow for a consolidation of the supply, leading to the abandonment of the solution. Such consolidation is necessary to generate economies of scale and trigger network and club effects.

Managing the physical territory by integrating the levers of action. In France, economic regulation is carried out at the national level, while innovation policies are divided between the national level and the respective regional and local levels (municipalities and their groupings). The management of urban public transport is entrusted to groups of local authorities; that of parking comes from the municipal level; finally, the layout of the road network and road operations fall within a logic referring to the domaniality of networks at several spatial levels rather than metropolitan territories. It would be good to be able to combine the different levers of action by territory, at the metropolitan scale, in order to define a coherent and consistent "political package" sensitive not only to the specificities of the territory in relation to the outside world but also to the diversity of subspaces in the territory, especially the demarcation between dense urban cores and spread out zones.

Integrated management could coherently address:

- the traffic conditions of the different means of travel; in particular, defining social conventions for the safety margins to be respected between two mobile entities, according to their respective types and dynamic operating modes (thus in distance and in time).
- the sharing of roads between functions (traffic, access and parking) and modes of travel, depending on the available space and local urban density,
- parking management should integrate local specificities (e.g. local capacity in relation to urban intensity), and, in cases of complex urbanised areas, it should rather be integrated according to an overall logic that is sensitive to the general configuration and the need to serve the metropolitan flows.
- management of traffic on the network, both at the nodes and on the sections (e.g. speed regulation), by considering diversified operating solutions for "wide sections" (e.g. dedicated lanes, alternation) and also the logic of the route (mono- or multimodal corridors, delivery circuits).
- demand management: modal and temporal orientation in compliance with the logistical constraints of each and also individual budget constraints. The principle of a "personalised mobility account" is appealing in terms of operationalising economic theories of mobility management in order to centralise spending and also subsidies, and to make individuals aware of the overall needs and signals directed to them by the community. This kind of personalised account is present in the experiments with MAAS systems that have recently started in several countries.
- management of social equity: much of this issue pertains to the metropolitan level, which determines the transport and housing conditions, and also the establishment of economic activities and therefore employment. However, in France, the

majority of policy levers remain concentrated at the national level – for example, fixing the amount of minimum wage, the formats and amounts of housing benefit allowances and incentives for housing construction. The "personalised account" could be extended to the entire relationship between the individual and their metropolitan area, going far beyond the realm of transport.

Virtual space: a territory to be managed. In geographical science, the notion of territory is well defined (a triangle between space–society–public power), and a good understanding of it facilitates integrated management in the physical space – the various forms of public intervention that we have just noted, with a privileged geographical level for each of them. Both the understanding and collective management of virtual space are less advanced. The physical telecommunications infrastructure is administered in a manner comparable to the transport infrastructure. The regulation of telecommunications services is comparable to that of public transport services. However, web services and mobile applications are far less regulated than the vehicles that serve as their counterparts in the world of transportation. Transport vehicles must report their physical presence in public space. They are duly listed (through licence plates), and their possessors (through vehicle licences). They are subject to pre-commissioning checks (by the relevant technical department) and periodic technical inspections. Their use is subject to police checks (speed, parking). Their environmental impact is charged at the first entry into service (bonus/penalty for the purchase) and then in the use phase (energy taxes). Finally, the physical circulation of vehicles is restricted according to places and times, depending on the needs of local populations. The regulation and the civic education of the web are much less developed. Even disregarding the dark web, anyone can upload any kind of web page. Visceral anarchists and fanatical businessmen have a shared interest in maintaining the state of things, which has the characteristics of a pioneering era and not yet those of civil maturity.

All this serves to relativise the question of open data. Its initial claim is the unconditional and no-strings provision of public data, in particular the dynamic geolocation of public transport vehicles, or the observation of traffic conditions on road networks or other modes in real time. At the same time, web-based services form this kind of data from field observations, which are made systematically without consulting the public authorities in advance. They would benefit from the regulation of the conditions for observing public space, by reserving the right of permanent monitoring of the information produced and its industrial valorisation, together with unlimited and free right of access and use for the public management needs of the territory concerned. Major American cities impose requirements that are at least as strong as those for large web companies, which consent to them. For example, hailing services are required to post the requests made *via* its website publicly, immediately and for free. Similarly, now that all public transport vehicles are geolocated and connected, local authorities could require dynamic communication of their location on its territory. The same applies to private transport vehicles that take up public spaces (thus reducing the available traffic capacity for others) and have an environmental impact. In short, public powers must assert their prerogatives in the world of software as much as in the physical world, in compliance with the laws in force and particularly the provisions relating to respect for privacy.

21.7 Conclusion: Sustainable is Beautiful

Smart is beautiful, indeed. However, mobility systems have an inherent complexity that will remain even when they become smart.

The road promises to be long: while the proliferation of innovations incites some excitement, it above all reflects the fluidity of design associated with digital modelling and the speed of prototyping available in modern industrial conditions. The actual transformation of the entire mobility system is an entirely different matter, even if some subsystems can be subject to rapid changes (as in the case of on-demand transport services), the rhythm of which resembles that of the waves of technological innovation in the field of software.

At present, the future "smart" state of a concrete, territorialised mobility system is less at the planning stage and more on the level of prospective thinking or even science fiction. The ongoing transformation is not a big project aimed at achieving predefined objectives. It is a set of projects with a particular scope, engaged in by particular actors motivated by particular interests.

However, the challenges we have mentioned have a collective scope. Whether they may be technical, societal, economic or organisational, these challenges affect all the actors, whose respective actions are part of the same, unique reality, even if each actor may have a particular, relative and subjective representation of it.

In the absence of a grand plan for overall transformation, the actors involved can already share common issues and follow common guidelines. The common issues are known and recognised as the environmental, social and economic issues of sustainable development. In order to head towards the resolution of these issues, we must fix the direction of movement. The compass we must follow is that of the evaluation of performance indicators according to common issues. Smart in itself is a means and not an end.

In fact, as a means, smart can be used to measure performance indicators. "smart" forms of vehicles and transport subsystems lend themselves to calculating performance in a bottom-up approach. Each vehicle that is able to "trace back" a flow of detailed information to its builder can also trace certain indicators that provide a territorial coordinator with a summary of its environmental and usage performance (the paths travelled, with an estimate of the people on board). At the level of each subsystem, such as a public transport line, a logistics platform or an on-demand transport service, the indicators derived from its constituent elements can be totalled. The physical impact indicators are additive, as well as the main indicators of economic and social impacts. The aggregation process will continue at the level of the modal network and then the territory, then the "smart" reality will join the performance simulation models at the territory scale by sharing the principles of a detailed evaluation. The indicators obtained at the most aggregated level can be reciprocally transmitted to all the actors affected by their development and therefore by the common issues of the territory.

Beyond the production of quantitative results, the establishment of such a collection and aggregation process would link the actors concretely. It would contribute to their joint mobilisation, or, in other words, the innervation of the system in favour of sustainable development. *Sustainable is smart, indeed.*

Acknowledgments

This chapter, which aimed to summarise and put the topic in perspective, is based on previous knowledge that is already relatively old, on observations and recent readings and especially on two recent studies in which one or the other of the authors participated. The first study is a strategic and prospective reflection on the digital transition of transport. It was conducted by the ENPC (F. Leurent and G. Lesteven) in cooperation with the Archéry business consulting firm (B. Mouly-Aigrot and L. Fouco) and the Ethic business movement (B. Tordjman). This study itself relied on interviews with actors and a committee composed of high-level leaders from transport companies and mobility authorities. The second study was commissioned by the Eco-design Chair on the subject of Big Data and its consequences for mobility. It was conducted by a team of researchers, engineers and consultants (J. Armoogum, C. Goubet, M. Munoz and O. Richard) led by O. Haxaire. Their report contains many references that are not detailed here.

The authors would like to express their heartfelt thanks to all the people who shared their knowledge and reflection with them, as well as their observations and perceptions and their impressions and opinions. The authors declared that they are of course solely responsible for the presentation and interpretation given here.

Bibliography

Archéry, ENPC and Ethic (2016). La transformation numérique: nouvel eldorado pour les acteurs des transports? ENPC brochure published in Paris, 28 p.

Berrada J., Leurent F., Lesteven G., & Boutueil V. (2017, July) Between private cars and mass transit: the room for intermediate modes in the urban setting. *Paper Presented at the 2017 Mobi.TUM Conference*, 10 p.

Crédoc (2018) *Conditions de vie et aspirations des Français - Baromètre du numérique 2017.* Crédoc, Paris, 256 p.

DREIF (2003) *Plan de déplacements urbains d'Ile-de-France, 2003, Bilan à mi-parcours (with illustrations by Jean-Vincent Sénac).* DRIEA, Paris, 103 p.

Haxaire, O., Armoogum, J., Goubet C., Munoz M., & Richard O. (2018, April). Big Data et Mobilité; éléments sur la transformation numérique de la mobilité. Study Report for the Eco- design Chair, 100 p.

Leurent F. (2013) L'équipement territorial au prisme de l'éco-conception: quels principes et quelles méthodes pour l'aménagement et le transport? In Peuportier B. (ed.), *Eco- conception des ensembles bâtis et des infrastructures, Chapter 4.* Presse des Mines, Paris, pp. 57–78.

Leurent F. et al. (2015) Documents du cours d'Analyse et Conception des Systèmes de Transport enseigné à l'ENPC en 2014–2015.

Shaheen, S., & Cohen, A. (2016) *Innovative Mobility Carsharing Outlook - Carsharing Market Overview, Analysis and Trends.* Transportation Sustainability Research Center, University of California, Berkeley, pp. 1–6.

Conclusions and Perspectives

During the first cycle of the "Eco-design of Buildings and Infrastructures" Chair, knowledge was produced and operational tools developed and tested on a development project concerning Cité Descartes in Marne la Vallée. The second cycle has made it possible to explore a certain number of subjects and to distribute the tools thanks to the creation of new enterprises, among other things.

Four areas have been addressed. The first concerns evaluation methods implemented in eco-design approaches. Life cycle assessment (LCA) deals with global indicators concerning health, biodiversity, climate and resources. The long lifespan of buildings and infrastructure has led to the use of forward-looking modelling. It was also a question of taking into account the systemic consequences of the decisions made during the design of the projects on other sectors by applying the consequential LCA. However, the assessment would be incomplete if it did not inform decision-makers about the uncertainties of the results. Sensitivity analysis and uncertainty propagation methods have been developed and implemented in several test cases. Finally, LCA has been applied to mobility systems. The assessment of biodiversity has been refined, a tool developed, and a new company created to disseminate knowledge among the operational staff. Finally, the compensation of impacts has been studied, and from this work it emerged that practitioners lack tools to establish an equivalence between the damage caused to biodiversity and the improvement permitted in the territory where the compensation action is carried out. Various lines of enquiry have been proposed for this evaluation.

The second area corresponds to incorporating the human factor in eco-design. The environmental performance of built complexes is strongly influenced by behaviours and lifestyles. The contribution of the human sciences is then essential for taking these elements into account, and a crossover between sociology and engineering sciences has been proposed. On the other hand, the democratisation of connected sensors and data analysis techniques make it possible to progress in terms of knowledge and modelling. Field surveys remain a useful observation element in these areas, and an application centred on the theme of nature in the city has revealed some trends. Finally, a thesis discussed the residential pathway and relationships to housing.

To bolster design support through quantified assessments, the third area of work has focused on digital simulation, which offers a range of tools to predict the operation of complex systems such as buildings and infrastructures. Thus, models now exist to better understand a housing system and a mobility system, which can lead to tools for planning support. However, one innovation undertaken by the Chair has been to develop a transport model on a smaller scale than that of the classical approach, which is generally at the level of the urban area and the region. At the neighbourhood scale, we have simulated traffic related to parking to better understand the influence of the number of parking spaces on the choice of travel modes, for example. At this same scale, microclimate modelling simulates the creation of an urban project to estimate the consequences in terms of temperatures (heat island effect, for example) and air movements. It is then possible to evaluate the level of comfort in outdoor spaces, potentially including the long-term perspective based on projections that take global warming into account. Energy simulation of buildings has also been supplemented by uncertainty calculations, which has made it possible to propose a performance guarantee process that includes the determination of a level corresponding to a controlled risk as well as the verification of this performance through the measurement and adjustment of certain parameters related to climate and occupants. Models at the macro-economic scale were finally used to study prospective scenarios concerning the long-term evolution of the electricity system, the environmental impacts being strongly influenced by this parameter.

Some more technical developments have been made, representing the fourth area of the Chair in this second cycle. Optimisation has been applied to the study of management strategies, to reduce high peak demand in buildings, for example, so that they can be better integrated within a grid that is increasingly powered by renewable but intermittent sources and thus facilitate energy transition. Innovation also focused on urban agriculture techniques with the aim of optimising crop management while adopting circular economy practices, for example by using recycled products such as coffee grounds as a substrate for growing crops. A smartphone application has been developed to collect, classify and visualise big data in the spheres of biodiversity and mobility. The inhabitants will then be able to contribute to the eco-design process by sharing data about their territory. The new possibilities that this digital transformation can bring in the field of transport have been analysed by a group of experts.

The second cycle thus confirmed the usefulness of this interdisciplinary collaboration between the three schools with complementary skills. Interdisciplinarity has been further strengthened with the participation of the humanities, data science, architecture and urban planning. International scientific exchanges have also been strengthened through the organisation of several seminars.

This body of work, the results of which are materialised in the form of operational tools, leads us to consider an industrialisation stage consisting of developing the knowledge produced on a larger scale. To this end, it is necessary to respond to three major issues identified during discussions with the partner company: articulating eco-design and economic arbitrage, becoming part of the digital transition and contributing to the improvement of the services provided to users.

A number of research perspectives arise from this objective, and the following perspectives are of particular interest:

- the application of multi-criteria optimisation techniques including environmental, social and economic aspects, for the design of urban projects and for their management, including renewable energy production, shared roads or local food production

- the improvement of operational assessment tools concerning the compensation of environmental impacts (a "serious game" project on this topic is envisaged);
- the evaluation of the implementation of the circular economy in the building, transport infrastructure and agriculture sectors, through LCA;
- articulation between the tools developed, for example by means of shared digital mock-ups;
- the study of micro-climates by integrating the cross-influence of buildings, transport and vegetation;
- the enrichment of models through data collection and analysis, relying on the use of connected sensors and/or smart phone applications;
- the assessment of the improvement of services rendered in terms of comfort, health and resilience, through eco-design;
- the consideration of the evolution of lifestyles, including forward-looking aspects.

Industrialisation will no doubt also require consideration of other forms of knowledge transfer in the form of dedicated seminars, summary documents for operational staff and ongoing training activities.

T - #0267 - 111024 - C518 - 246/174/24 - PB - 9780367557713 - Gloss Lamination